Classical Electrodynamics

Problems with solutions

Classical Electrodynamics

Problems with solutions

Konstantin K Likharev

IOP Publishing, Bristol, UK

ISBN 978-0-7503-1407-7 (ebook)
ISBN 978-0-7503-1408-4 (print)
ISBN 978-0-7503-1409-1 (mobi)

DOI 10.1088/978-0-7503-1407-7

Version: 20180501

IOP Expanding Physics
ISSN 2053-2563 (online)
ISSN 2054-7315 (print)

British Library Cataloguing-in-Publication Data: A catalogue record for this book is available from the British Library.

Published by IOP Publishing, wholly owned by The Institute of Physics, London

IOP Publishing, Temple Circus, Temple Way, Bristol, BS1 6HG, UK

US Office: IOP Publishing, Inc., 190 North Independence Mall West, Suite 601, Philadelphia, PA 19106, USA

Contents

Preface to the EAP Series		**vii**
Preface to *Part EM: Problems with Solutions*		**x**
Acknowledgments		**xi**
Notation		**xii**
1	**Electric charge interaction**	**1-1**
2	**Charges and conductors**	**2-1**
	Reference	2-77
3	**Dipoles and dielectrics**	**3-1**
4	**DC current**	**4-1**
5	**Magnetism**	**5-1**
	Reference	5-41
6	**Electromagnetism**	**6-1**
	References	6-55
7	**Electromagnetic wave propagation**	**7-1**
	References	7-59
8	**Radiation, scattering, interference, and diffraction**	**8-1**
9	**Special relativity**	**9-1**
	Reference	9-52
10	**Radiation by relativistic charges**	**10-1**

Appendices

A Selected mathematical formulas A-1

B Selected physical constants B-1

Bibliography 13-1

Preface to the EAP Series

Essential Advanced Physics

Essential Advanced Physics (EAP) is a series of lecture notes and problems with solutions, consisting of the following four parts[1]:

- *Part CM*: *Classical Mechanics* (a one-semester course),
- *Part EM*: *Classical Electrodynamics* (two semesters),
- *Part QM*: *Quantum Mechanics* (two semesters), and
- *Part SM*: *Statistical Mechanics* (one semester).

Each part includes two volumes: *Lecture Notes* and *Problems with Solutions*, and an additional file *Test Problems with Solutions*.

Distinguishing features of this series—in brief

- condensed lecture notes (~250 pp per semester)—much shorter than most textbooks
- emphasis on simple explanations of the main notions and phenomena of physics
- a focus on problem solution; extensive sets of problems with detailed model solutions
- additional files with test problems, freely available to qualified university instructors
- extensive cross-referencing between all parts of the series, which share style and notation.

Level and precursors

The goal of this series is to bring the reader to a general physics knowledge level necessary for professional work in the field, regardless on whether the work is theoretical or experimental, fundamental or applied. From the formal point of view, this level (augmented by a few special topic courses in a particular field of concentration, and of course by an extensive thesis research experience) satisfies the typical PhD degree requirements. Selected parts of the series may be also valuable for graduate students and researchers of other disciplines, including astronomy, chemistry, mechanical engineering, electrical, computer and electronic engineering, and material science.

The entry level is a notch lower than that expected from a physics graduate from an average US college. In addition to physics, the series assumes the reader's familiarity with basic calculus and vector algebra, to such an extent that the meaning of the formulas listed in appendix A, 'Selected mathematical formulas' (reproduced at the end of each volume), is absolutely clear.

[1] Note that the (very ambiguous) term *mechanics* is used in these titles in its broadest sense. The acronym *EM* stems from another popular name for classical electrodynamics courses: *Electricity and Magnetism*.

Origins and motivation

The series is a by-product of the so-called 'core physics courses' I taught at Stony Brook University from 1991 to 2013. My main effort was to assist the development of students' problem-solving skills, rather than their idle memorization of formulas. (With a certain exaggeration, my lectures were not much more than introductions to problem solution.) The focus on this main objective, under the rigid time restrictions imposed by the SBU curriculum, had some negatives. First, the list of covered theoretical methods had to be limited to those necessary for the solution of the problems I had time to discuss. Second, I had no time to cover some core fields of physics—most painfully general relativity[2] and quantum field theory, beyond a few quantum electrodynamics elements at the end of *Part QM*.

The main motivation for putting my lecture notes and problems on paper, and their distribution to students, was my desperation to find textbooks and problem collections I could use, with a clear conscience, for my purposes. The available graduate textbooks, including the famous *Theoretical Physics* series by Landau and Lifshitz, did not match the minimalistic goal of my courses, mostly because they are far too long, and using them would mean hopping from one topic to another, picking up a chapter here and a section there, at a high risk of losing the necessary background material and logical connections between the course components—and the students' interest with them. In addition, many textbooks lack even brief discussions of several traditional and modern topics that I believe are necessary parts of every professional physicist's education[3].

On the problem side, most available collections are not based on particular textbooks, and the problem solutions in them either do not refer to any background material at all, or refer to the included short sets of formulas, which can hardly be used for systematic learning. Also, the solutions are frequently too short to be useful, and lack discussions of the results' physics.

Style

In an effort to comply with the Occam's Razor principle[4], and beat Malek's law[5], I have made every effort to make the discussion of each topic as clear as the time/space (and my ability :-) permitted, and as simple as the subject allowed. This effort has resulted in rather succinct lecture notes, which may be thoroughly read by a student during the semester. Despite this briefness, the introduction of every new

[2] For an introduction to this subject, I can recommend either a brief review by S Carroll, *Spacetime and Geometry* (2003, New York: Addison-Wesley) or a longer text by A Zee, *Einstein Gravity in a Nutshell* (2013, Princeton University Press).

[3] To list just a few: the statics and dynamics of elastic and fluid continua, the basics of physical kinetics, turbulence and deterministic chaos, the physics of computation, the energy relaxation and dephasing in open quantum systems, the reduced/RWA equations in classical and quantum mechanics, the physics of electrons and holes in semiconductors, optical fiber electrodynamics, macroscopic quantum effects in Bose–Einstein condensates, Bloch oscillations and Landau–Zener tunneling, cavity quantum electrodynamics, and density functional theory (DFT). All these topics are discussed, if only briefly, in my lecture notes.

[4] *Entia non sunt multiplicanda praeter necessitate*—Latin for 'Do not use more entities than necessary'.

[5] 'Any simple idea will be worded in the most complicated way'.

physical notion/effect and of every novel theoretical approach is always accompanied by an application example or two.

The additional exercises/problems listed at the end of each chapter were carefully selected[6], so that their solutions could better illustrate and enhance the lecture material. In formal classes, these problems may be used for homework, while individual learners are strongly encouraged to solve as many of them as practically possible. The few problems that require either longer calculations, or more creative approaches (or both), are marked by asterisks.

In contrast with the lecture notes, the model solutions of the problems (published in a separate volume for each part of the series) are more detailed than in most collections. In some instances they describe several alternative approaches to the problem, and frequently include discussions of the results' physics, thus augmenting the lecture notes. Additional files with sets of shorter problems (also with model solutions) more suitable for tests/exams, are available for qualified university instructors from the publisher, free of charge.

Disclaimer and encouragement

The prospective reader/instructor has to recognize the limited scope of this series (hence the qualifier *Essential* in its title), and in particular the lack of discussion of several techniques used in current theoretical physics research. On the other hand, I believe that the series gives a reasonable introduction to the *hard core* of physics—which many other sciences lack. With this hard core knowledge, today's student will always feel at home in physics, even in the often-unavoidable situations when research topics have to be changed at a career midpoint (when learning from scratch is terribly difficult—believe me :-). In addition, I have made every attempt to reveal the remarkable logic with which the basic notions and ideas of physics subfields merge into a wonderful single construct.

Most students I taught liked using my materials, so I fancy they may be useful to others as well—hence this publication, for which all texts have been carefully reviewed.

[6] Many of the problems are original, but it would be silly to avoid some old good problem ideas, with long-lost authorship, which wander from one textbook/collection to another one without references. The assignments and model solutions of all such problems have been re-worked carefully to fit my lecture material and style.

Preface to *Part EM: Problems with Solutions*

This volume of the EAP series contains model solutions of the problems formulated in volume 3, *Part EM: Lecture Notes*. For the reader's convenience, the problem assignments are reproduced in this volume as well, although the accompanying figures are frequently more detailed and extended to explain the solutions. The appendices A (Selected mathematical formulas) and B (Selected physical constants), common to all parts of the series, are also included in this volume for the reader's convenience.

Since all this series is strongly focused on the development of problem solution skills, the model solutions are rather detailed, and in some cases (especially in the harder problems marked by asterisks) extend and/or enhance the lecture material.

The solutions have numerous references to formulas in *Part EM Lecture Notes*, and occasionally to other parts of the EAP series. In these cases, the acronym of the part is included in the citation.

A file with 51 additional problems, which allow shorter solutions and hence are suitable for exams, also with model solutions, is available to university instructors from the publisher by request.

Acknowledgments

I am extremely grateful to my faculty colleagues and other readers of the preliminary (circa 2013) version of this series, who provided feedback on certain sections; here they are listed in alphabetical order[7]: A Abanov, P Allen, D Averin, S Berkovich, P-T de Boer, M Fernandez-Serra, R F Hernandez, A Korotkov, V Semenov, F Sheldon, and X Wang. (Obviously, these kind people are not responsible for any remaining deficiencies.)

A large part of my scientific background and experience, reflected in these materials, came from my education, and then research, in the Department of Physics of Moscow State University from 1960 to 1990. The Department of Physics and Astronomy of Stony Brook University provided a comfortable and friendly environment for my work during the following 25+ years.

Last but not least, I would like to thank my wife Lioudmila for all her love, care, and patience—without these, this writing project would have been impossible.

I know very well that my materials are still far from perfection. In particular, my choice of covered topics (always very subjective) may certainly be questioned. Also, it is almost certain that despite all my efforts, not all typos have been weeded out. This is why all remarks (however candid) and suggestions from readers will be greatly appreciated. All significant contributions will be gratefully acknowledged in future editions.

Konstantin K Likharev
Stony Brook, NY

[7] I am very sorry for not keeping proper records from the beginning of my lectures at Stony Brook, so I cannot list all the numerous students and TAs who have kindly attracted my attention to typos in earlier versions of these notes. Needless to say, I am very grateful to all of them as well.

Notation

Abbreviations	Fonts	Symbols
c.c. complex conjugate	F, $\mathscr{F}$ scalar variables[8]	$\dot{}$ time differentiation operator (*d*/*dt*)
h.c. Hermitian conjugate	$\mathbf{F}$, $\boldsymbol{\mathscr{F}}$ vector variables	∇ spatial differentiation vector (*del*)
	$\hat{F}$, $\hat{\mathscr{F}}$ scalar operators	$\approx$ approximately equal to
	$\hat{\mathbf{F}}$, $\hat{\boldsymbol{\mathscr{F}}}$ vector operators	$\sim$ of the same order as
	F matrix	$\propto$ proportional to
	$F_{jj'}$ matrix element	$\equiv$ equal to by definition (or evidently)
		$\cdot$ scalar ('dot-') product
		$\times$ vector ('cross-') product
		$\overline{}$ time averaging
		$\langle\ \rangle$ statistical averaging
		[,] commutator
		{ , } anticommutator

Prime signs

The prime signs (′, ″, etc) are used to distinguish similar variables or indices (such as j and j' in the matrix element above), rather than to denote derivatives.

Parts of the series

Part CM: Classical Mechanics *Part EM: Classical Electrodynamics*
Part QM: Quantum Mechanics *Part SM: Statistical Mechanics*

Appendices

Appendix A: Selected mathematical formulas
Appendix B: Selected physical constants

Formulas

The abbreviation Eq. may mean any displayed formula: either the equality, or inequality, or equation, etc.

[8] The same letter, typeset in different fonts, typically denotes different variables.

Chapter 1

Electric charge interaction

***Problem* 1.1.** Calculate the electric field created by a thin, long, straight filament, electrically charged with a constant linear density λ, using two approaches:

(i) directly from the Coulomb law, and
(ii) using the Gauss law.

Solutions:

(i) From the translational and axial symmetries of the problem, it is clear that $\mathbf{E}(\mathbf{r}) = \mathbf{n}_\rho E(\rho)$, where ρ is the shortest distance from the observation point to the filament[1]. Let us select the plane of drawing so that it contains both the filament and the observation point, and take the line of the filament for axis z (see the figure below).

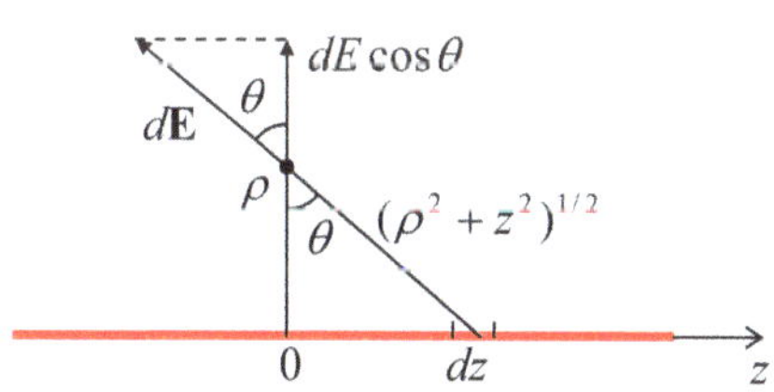

Then, according to the linear superposition principle, the field's magnitude may be calculated as

$$E(\rho) = \int_{z=-\infty}^{z=+\infty} dE_\rho = \int_{z=-\infty}^{z=+\infty} dE \cos\theta = \int_{z=-\infty}^{z=+\infty} dE \frac{\rho}{(\rho^2 + z^2)^{1/2}},$$

[1] I regret using the same letter (ρ) as for the charge density per unit volume (used in the lecture notes, but not in this problem), but both notations are traditional. The difference between these notions will be always very clear from the context.

doi:10.1088/978-0-7503-1407-7ch1

where dE is the magnitude of the elementary contribution to the field, created by a small segment dz of the filament, with electric charge λdz. According to Eq. (1.7) of the lecture notes,

$$dE = \lambda dz \frac{1}{4\pi\varepsilon_0} \frac{1}{\rho^2 + z^2},$$

so that the total field

$$E(\rho) = \frac{\lambda\rho}{4\pi\varepsilon_0} \int_{-\infty}^{+\infty} \frac{dz}{(\rho^2 + z^2)^{3/2}} \equiv \frac{\lambda}{2\pi\varepsilon_0\rho} \int_0^{+\infty} \frac{d\xi}{(1 + \xi^2)^{3/2}} = \frac{\lambda}{2\pi\varepsilon_0\rho}. \qquad (*)$$

For the last transition, I have used the well-known value (1) of this table integral—see, e.g. Eq. (A.32b)[2].

(ii) Taking a round cylinder of radius ρ and length l with its axis on the filament for the Gaussian volume, we ensure that on its round walls the electric field E is constant and perpendicular to the volume boundary, while the field flux through the cylinder's 'lids' is zero. As a result, Eq. (1.16) yields

$$2\pi\rho l E = \frac{\lambda l}{\varepsilon_0},$$

immediately giving the same result (*).

We see that for this highly symmetric problem both solution ways are straightforward, but the Gauss law makes the calculations easier.

***Problem* 1.2.** Two thin, straight parallel filaments, separated by distance ρ, carry equal and opposite uniformly distributed charges with a linear density λ—see the figure below. Calculate the electrostatic force (per unit length) of the Coulomb interaction between the wires. Compare the result with the Coulomb law for the force between the point charges.

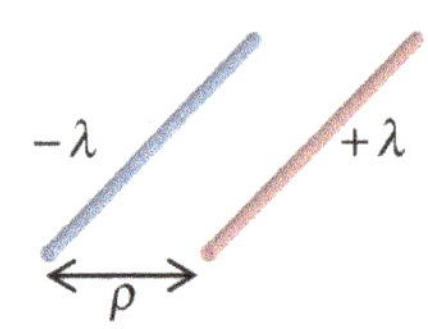

Solution: Using the result (*) of problem 1.1, we obtain

$$\frac{F}{l} = \frac{qE(\rho)}{l} = \lambda E(\rho) = \frac{\lambda^2}{2\pi\varepsilon_0\rho}.$$

Note that the force drops with distance as $1/\rho$, rather than as $1/r^2$ for point charges. Such different scaling of interaction in systems of different dimensionality is very typical for physics at large.

[2] Actually, this integral may be easily worked out by substitution $\xi \equiv \tan\varphi$, giving $d\xi = d\varphi/\cos^2\varphi = d\varphi\,(1 + \tan^2\varphi) = d\varphi\,(1 + \xi^2)$, so that $d\xi/(1 + \xi^2)^{3/2} = d\varphi/(1 + \xi^2)^{1/2} = \cos\varphi\, d\varphi = d(\sin\varphi)$.

***Problem* 1.3.** A sphere of radius R, whose volume had been charged with a constant density ρ, is split with a very narrow, planar gap passing through its center. Calculate the force of mutual repulsion of the resulting two hemispheres.

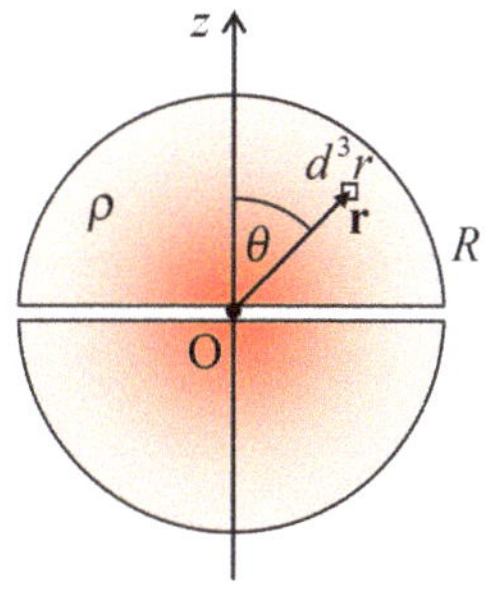

Solution: Since the gap is very narrow, we may neglect its effect on the distribution of the electric field $\mathbf{E}$ inside the sphere, and use Eq. (1.22) of the lecture notes:

$$\mathbf{E}(\mathbf{r}) = \mathbf{n}_r E(r), \qquad \text{with } E(r) = \frac{\rho r}{3\varepsilon_0},$$

where r is the distance of the observation point from the sphere's center O—see the figure above. Acting on an elementary volume $dV = d^3r$ of the sphere's material, with the elementary charge $dQ = \rho d^3r$, this electric field produces a radially directed force of magnitude

$$dF = E(r)dQ = \frac{\rho r}{3\varepsilon_0}\rho d^3r \equiv \frac{\rho^2 r}{3\varepsilon_0} d^3r.$$

Due to the axial symmetry of the problem, the net (repulsive) Coulomb force $\mathbf{F}$ acting on each hemisphere has to be normal to the gap—in the figure above, directed along the axis z. Hence only the z-component of the elementary force dF,

$$dF_z = dF \cos\theta = \frac{\rho^2 r}{3\varepsilon_0} \cos\theta \; d^3r,$$

where θ is the polar angle of the radius-vector $\mathbf{r}$ of the elementary volume d^3r (see the figure above), can contribute to the net force $\mathbf{F}$. As a result, we may use the standard spherical coordinates, with the origin in the sphere's center, to calculate the net force magnitude as

$$F = F_z = \int_{z>0} dF_z = \frac{\rho^2}{3\varepsilon_0}\int_0^{2\pi} d\varphi \int_0^{\pi/2} \cos\theta \sin\theta \, d\theta \int_0^R r^3 dr = \frac{\rho^2}{3\varepsilon_0} 2\pi \frac{1}{2}\frac{R^4}{4}$$
$$= \frac{\pi}{12}\frac{\rho^2 R^4}{\varepsilon_0}.$$

Another way to represent this result is to express it via the charge $Q = (2\pi/3)R^3\rho$ of each hemisphere:

$$F = \frac{1}{4\pi\varepsilon_0}\frac{3}{4}\frac{Q^2}{R^2} \equiv \frac{1}{4\pi\varepsilon_0}\frac{Q^2}{R_{\text{ef}}^2},$$

showing that the effective distance of their interaction (as point charges) is $R_{\text{ef}} = (2/\sqrt{3})R \approx 1.16\ R$.

One may wonder whether the above calculation properly excludes the effect of the electric field of a hemisphere on itself, since the field we have used is evidently produced by the sphere as a whole. The proper response to this concern is that due to Newton's second law, the internal forces between elementary charges of the same hemisphere compensate each other, so that these forces are automatically canceled at the integration over the hemisphere's volume:

$$\int_{z>0} \rho \mathbf{E}_{\text{full}} d^3r = \int_{z>0} \rho(\mathbf{E}_{\text{other}} + \mathbf{E}_{\text{self}}) d^3r = \int_{z>0} \rho \mathbf{E}_{\text{other}}\, d^3r + 0 = \int_{z>0} \rho \mathbf{E}_{\text{other}} d^3r = \mathbf{F}.$$

***Problem* 1.4.** A thin spherical shell of radius R, which had been charged with a constant areal density σ, is split into two equal halves by a very narrow, planar cut passing through the sphere's center. Calculate the force of electrostatic repulsion between the resulting hemispheric shells, and compare the result with that of the previous problem.

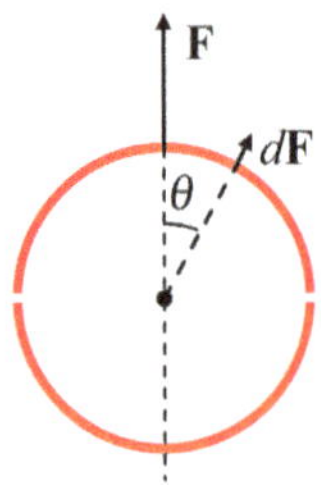

Solution: A thin cut does not alter substantially the electric field distribution, so it has the same radial direction and spherically symmetric distribution, $\mathbf{E}(\mathbf{r}) = \mathbf{n}_r E(r)$, as for the uncut shell. The function $E(r)$ may be readily found using the Gauss law applied to spheres with $r < R$ and $r > R$:

$$E(r) = \begin{cases} 0, & \text{for } r < R, \\ \sigma R^2/\varepsilon_0 r^2, & \text{for } R < r. \end{cases}$$

Hence the average field applied to the shell equals[3]

$$\overline{E} = \frac{E(R-0) + E(R+0)}{2} = \frac{\sigma}{2\varepsilon_0},$$

providing distributed normal forces **F**, with the following density (force per unit area):

$$\frac{dF}{dA} = \frac{dQ}{dA}\overline{E} = \sigma \overline{E} = \frac{\sigma^2}{2\varepsilon_0}.$$

Due to the axial symmetry of the problem, only the 'vertical' component (meaning the direction perpendicular to the cut plane—see the figure above) of

[3] A strict proof of the correctness of such averaging will be given in the model solution of problem 2.1.

these elementary forces contribute to the total force **F** acting on each hemisphere—for example the top one:

$$F = \int_{\text{hemisphere}} \left(\frac{d\mathbf{F}}{dA}\right)_z d^2r = \int_{\text{hemisphere}} \frac{dF}{dA}\cos\theta\, d^2r$$
$$= \int_0^{2\pi} d\varphi \int_0^{\pi/2} \frac{\sigma^2}{2\varepsilon_0}\cos\theta\, R^2 \sin\theta\, d\theta = \frac{\pi}{2}\frac{\sigma^2 R^2}{\varepsilon_0}. \qquad (*)$$

In order to compare this result with the solution of the previous problem, let us assume that the radius R and the total charge Q of each hemisphere are the same in both cases:

$$2\pi R^2\sigma = \frac{2\pi}{3}R^3\rho = Q.$$

In this notation, Eq. (*) may be rewritten as

$$F = \frac{1}{4\pi\varepsilon_0}\frac{1}{2}\frac{Q^2}{R^2} \equiv \frac{1}{4\pi\varepsilon_0}\frac{Q^2}{R_{\text{ef}}^2}, \qquad \text{with } R_{\text{ef}} = \sqrt{2}R \approx 1.41R,$$

while for the case of volume-distributed charge, the similarly defined effective distance equals $(2/\sqrt{3})R \approx 1.16\ R$. This difference is natural, because in the case of thin hemisphere shells, the elementary charges are, on the average, farther from each other.

***Problem* 1.5.** Calculate the distribution of the electrostatic potential created by a straight, thin filament of finite length $2\,l$, charged with a constant linear density λ, and explore the result in the limits of very small and very large distances from the filament.

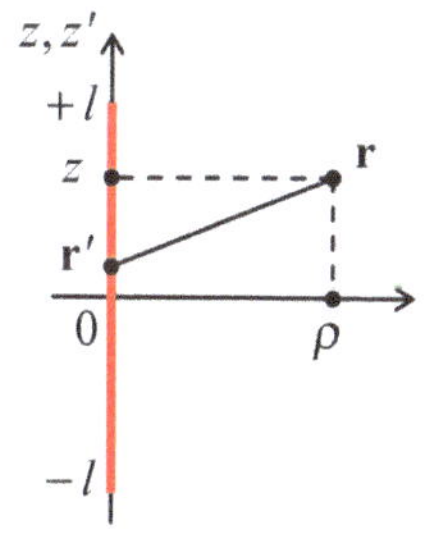

Solution: Due to the limited (axial) symmetry of the problem, applying the Gauss law to it is not very productive, so let us resort to direct summation (actually, integration) of component charge fields. Let us select the reference frame so that the filament coincides with segment $[-l, +l]$ of axis z, and the observation point $\mathbf{r}$ has Cartesian coordinates $\{\rho, 0, z\}$—see the figure above. Then the field source point $\mathbf{r}'$ has coordinates $\{0, 0, z'\}$, and Eq. (1.38) of the lecture notes, integrated across the filament cross-section, reads

$$\phi(\mathbf{r}) = \frac{\lambda}{4\pi\varepsilon_0}\int_{-l}^{+l}\frac{dz'}{|\mathbf{r}-\mathbf{r}'|} = \frac{\lambda}{4\pi\varepsilon_0}\int_{-l}^{+l}\frac{dz'}{[(z-z')^2+\rho^2]^{1/2}}.$$

Introducing the new, dimensionless variable $\xi \equiv (z' - z)/\rho$, we may reduce the integral to the table one—see, e.g. Eq. (A.29a):

$$\phi(\mathbf{r}) = \frac{\lambda}{4\pi\varepsilon_0}\int_{(-l-z)/\rho}^{(+l-z)/\rho} \frac{d\xi}{(\xi^2+1)^{1/2}} = \frac{\lambda}{4\pi\varepsilon_0}\ln\left|(\xi^2+1)^{1/2}+\xi\right|_{(-l-z)/\rho}^{(+l-z)/\rho}$$
$$= \frac{\lambda}{4\pi\varepsilon_0}\ln\frac{[(l-z)^2+\rho^2]^{1/2}+(l-z)}{[(l+z)^2+\rho^2]^{1/2}-(l+z)}.$$

For the observation points very close to the filament, and not very close to any of its ends, the denominator of the fraction becomes much smaller than its nominator. Expanding the latter in the small parameter $\rho/(l+z)$, and keeping only the leading term, we obtain the result following from the solution of problem 1.1:

$$\phi(\mathbf{r}) \approx \frac{\lambda}{4\pi\varepsilon_0}\ln\frac{2(l-z)}{\rho^2/2(l+z)} = \frac{\lambda}{4\pi\varepsilon_0}\ln\frac{4(l^2-z^2)}{\rho^2} = -\frac{\lambda}{2\pi\varepsilon_0}\ln\rho + \text{const},$$

for $|z| < l$, and $\rho \ll l$, $l-|z|$.

On the other hand, at large distances from the filament the nominator and denominator of the fraction become very close to each other. Expanding both expressions in the Taylor series in the small parameter l/r, where $r \equiv (z^2+\rho^2)^{1/2}$ is the distance from filament's middle point, in the first non-vanishing approximation we obtain

$$\phi(\mathbf{r}) \approx \frac{\lambda}{4\pi\varepsilon_0}\ln\frac{1+l/r}{1-l/r} \approx \frac{\lambda}{4\pi\varepsilon_0}\ln\left(1+\frac{2l}{r}\right) \approx \frac{\lambda}{4\pi\varepsilon_0}\frac{2l}{r}, \qquad \text{for } r \gg l,$$

i.e. just Eq. (1.35) for a point charge $q = 2l\lambda$.

***Problem* 1.6.** A thin plane sheet, perhaps of an irregular shape, carries electric charge with a constant areal density σ.

(i) Express the electric field's component normal to the plane, at a certain distance from it, via the solid angle Ω at which the sheet is visible from the observation point.
(ii) Use the result to calculate the field in the center of a cube, with one face charged with constant density σ.

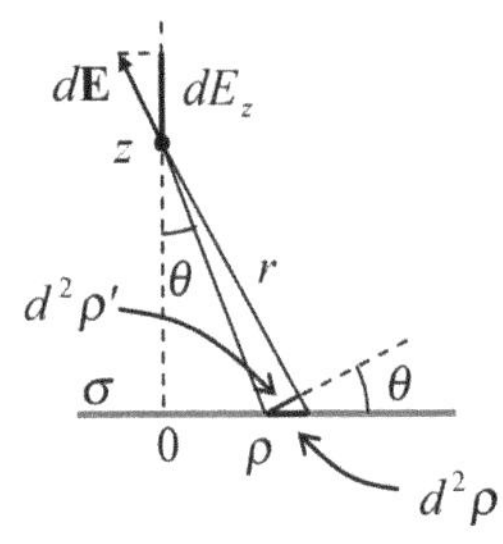

Solutions:

(i) Let us place the coordinate origin 0 at the charged plane's point that is closest to the field observation point (at distance z from the plane)—see the figure above. Then, according to Eq. (1.7) of the lecture notes, the normal (z-) component of the electric field component induced by charge element $dQ = \sigma d^2\rho$ (where $\boldsymbol{\rho}$ is the 2D radius-vector within the plane) equals

$$dE_z = dE\cos\theta = \frac{1}{4\pi\varepsilon_0}\frac{dQ}{r^2}\cos\theta = \frac{\sigma}{4\pi\varepsilon_0}\frac{d^2\rho}{r^2}\cos\theta, \qquad (*)$$

where $r = (\rho^2 + z^2)^{1/2}$ is the distance from the elementary area $d^2\rho$ and the observation point, and $\cos\theta = z/r$. But as the figure above shows, the product $d^2\rho\cos\theta$ equals $d^2\rho'$—the projection of the elementary area $d^2\rho$ on the plane normal to the vector $\mathbf{r}$. In turn, the ratio $d^2\rho'/r^2$ is just $d\Omega$—the solid angle at which the elementary area $d^2\rho$ is visible from the field measurement point[4]. As a result, Eq. (*) may be rewritten simply as

$$dE_z = \frac{\sigma}{4\pi\varepsilon_0}d\Omega,$$

and its integration over the charged sheet yields the requested result:

$$E_z = \frac{\sigma}{4\pi\varepsilon_0}\Omega. \qquad (**)$$

(ii) Since all six faces of a cube are visible from its center at equal angles Ω, and their sum constitutes the full solid angle 4π, each Ω is equal to $4\pi/6$, and Eq. (**) yields

$$E_z = \frac{\sigma}{4\pi\varepsilon_0}\frac{4\pi}{6} \equiv \frac{\sigma}{6\varepsilon_0}.$$

Due to the evident symmetry of the system, the field cannot have other Cartesian components, so that we may rewrite the result as

$$|E| - \frac{\sigma}{6\varepsilon_0}.$$

***Problem* 1.7.** Can one create, in a non-vanishing region of space, electrostatic fields with the Cartesian components proportional to the following products of Cartesian coordinates $\{x, y, z\}$,

$$\text{(i) } \{yz,\ xz,\ xy\},$$
$$\text{(ii) } \{xy,\ xy,\ yz\}?$$

[4] All this calculation is absolutely similar to the one made at the proof of the Gauss law in section 1.2 of the lecture notes—see the transition from Eq. (1.13) to Eq. (1.14).

Solution: Let us calculate the curl of both supposed fields, using the definition of that operator—see, e.g. Eq. (A.54):

$$\nabla \times \mathbf{E} = \left\{ \frac{\partial E_z}{\partial y} - \frac{\partial E_y}{\partial z}, \ \frac{\partial E_x}{\partial z} - \frac{\partial E_z}{\partial x}, \ \frac{\partial E_y}{\partial x} - \frac{\partial E_x}{\partial y} \right\}.$$

For the field (i) we obtain $\nabla \times \mathbf{E} \propto \{x - x, y - y, z - z\} \equiv 0$, while for the field (ii) $\nabla \times \mathbf{E} \propto \{z - 0, 0 - 0, y - x\}$ vanishes only at one point ($x = y = z = 0$). However, according to the homogeneous Maxwell equation (Eq. (1.28) of the lecture notes), the curl of the electrostatic field has to equal zero at any point where it exists; hence the field (i) can exist in a region of finite size, while the field (ii) cannot.

The fact that the field (i) has zero divergence as well, i.e. requires $\rho(\mathbf{r}) \equiv 0$ within the region of its existence, does not prevent it from being realistic, because the field may be created by electric charges outside that particular region. Note also that the electric field (ii) may be induced by a magnetic field changing (linearly) in time—see section 6.1 of the lecture notes, but such $\mathbf{E}$ cannot be called an *electrostatic* field.

***Problem* 1.8.** Distant sources have been used to create different electric fields on two sides of a wide and thin metallic membrane, with a round hole of radius R in it—see the figure below. Besides the local perturbation created by the hole, the fields are uniform:

$$\mathbf{E}(\mathbf{r})|_{r \gg R} = \mathbf{n}_z \times \begin{cases} E_1, & \text{at } z < 0, \\ E_2, & \text{at } z > 0. \end{cases}$$

Prove that the system may serve as an electrostatic lens for charged particles flying along axis z, at distances $\rho \ll R$ from it, and calculate the focal distance f of the lens. Spell out the conditions of validity of your result.

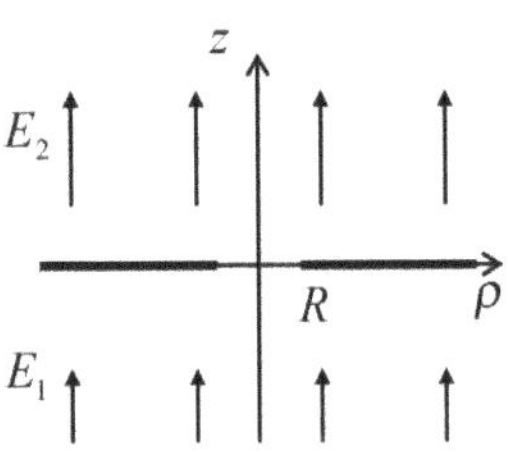

Solution: Let us select the orientation of the mutually perpendicular axes $x = \rho\cos\varphi$ and $y = \rho\sin\varphi$ so that for the particular flying particle $y = 0$. At the axis of this axially symmetric system (i.e. at $\rho = 0$), the partial derivatives $\partial E_x/\partial x$ and $\partial E_y/\partial y$ have to be equal due to the symmetry, so that the Maxwell equation (1.27), $\nabla \cdot \mathbf{E} = 0$, yields

$$2\frac{\partial E_x}{\partial x}\bigg|_{\rho=0} = -\frac{\partial E_z}{\partial z}\bigg|_{\rho=0}.$$

Since at the axis, the value of E_x has to vanish, its Taylor expansion at $x = 0$ starts with the linear term, so that at small x

$$E_x\Big|_{\substack{x<<R\\ y=0}} \approx x\frac{\partial E_x}{\partial x}\Big|_{\rho=0} = -\frac{x}{2}\frac{\partial E_z}{\partial z}\Big|_{\rho=0}.$$

According to this expression, at sufficiently small x, the field E_x is so small that particle's deviation from its initial trajectory during its flight through the perturbed field region (of the length $\Delta z \sim R$) is negligible. However, during this flight the particle picks up a transverse momentum created by the field:

$$\begin{aligned} p_x &= \int F_x\, dt = q\int E_x\, dt = q\int E_x \frac{dt}{dz}dz \\ &= -q\frac{x}{2v_z}\int_{-\infty}^{+\infty}\frac{\partial E_z}{\partial z}\Big|_{\rho=0} dz = -q\frac{x}{2v_z}(E_2 - E_1). \end{aligned}$$

(Here the change of the particle's longitudinal velocity v_z during its flight through the hole was neglected—which is legitimate if the length of the uniform field region is much larger than R, as implied by the problem's conditions.) If this momentum, and hence the acquired radial velocity $v_x = p_x/m$, are directed toward the axis, which is true at the proper sign of the combination $q(E_2 - E_1)$,[5] it would lead to the particle hitting the system's axis at the distance

$$z \equiv f = v_z t_f = v_z\frac{x}{(-v_x)} = v_z\frac{x}{(-p_x/m)} = \frac{2mv_z^2}{q(E_2 - E_1)} = \frac{4T}{q(E_2 - E_1)},$$

where $T = mv_z^2/2$ is particle's initial kinetic energy. The most important feature of this result is that the initial coordinate x of the particle has dropped out of it. This means that particles with any small x (and, by axial symmetry, all particles flying sufficiently close to axis z, and parallel to it) will be directed to the same focal point f. This is the key property of any lens—in this case the electrostatic one.

It is remarkable that this result does not depend on the exact field distribution in the hole region. (This distribution may be calculated analytically—for example, using the degenerate ellipsoidal coordinates, to be discussed in section 2.4 of the lecture notes.) It is valid if two strong conditions,

$$R \ll f \ll \frac{T}{qE_2},$$

are satisfied. The first of them has allowed us to treat x as a constant during the integration of force F_x over time, while the second one, to neglect the change of v_z

[5] If the sign is opposite, the system will disperse the parallel particle beam, but in a highly ordered way—with the distance from the axis proportional to the initial distance x. So the system will still work as a lens, though a diverging ('negative') one, with $f < 0$.

during the focusing process. Note, however, that a violation of the second condition does not ruin the focusing effect; it only makes the expression for f somewhat bulkier.

***Problem* 1.9.** Eight equal point charges q are located at the corners of a cube of side a. Calculate all Cartesian components E_j of the electric field, and their spatial derivatives $\partial E_j/\partial r_{j'}$, at cube's center, where r_j are the Cartesian coordinates oriented along the cube's sides—see the figure below. Are all your results valid for the center of a planar square, with four equal charges in its corners?

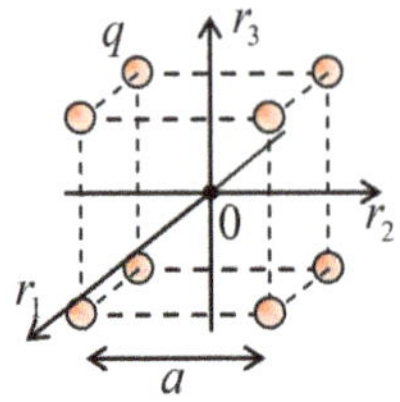

Solution: According to Eq. (1.33) of the lecture notes, the Cartesian components of the field and their derivatives may be expressed via the derivatives of the electrostatic potential:

$$E_j = -\frac{\partial \phi}{\partial r_j}, \qquad \frac{\partial E_j}{\partial r_{j'}} = -\frac{\partial^2 \phi}{\partial r_j \partial r_{j'}}. \tag{*}$$

Due to the cube's symmetry, at its center the whole vector **E** should vanish. (Indeed, if it did not, the vector would be directed toward either some face or some corner of the cube, but that would violate the equivalence of all faces and corners.) Hence, according to the first of Eqs. (*), all field components, i.e. the partial derivatives of the scalar potential, have to vanish at the center. So, if we take this point for the origin ($\mathbf{r} = 0$, i.e. $r_1 = r_2 = r_3 = 0$), the Taylor expansion[6] of function $\phi(r_1, r_2, r_3)$, apart from an inconsequential constant $\phi(0, 0, 0)$, should start with quadratic terms. Due to the cube's symmetry with respect to coordinate swaps, the coefficients at these terms have to be independent of the coordinates:

$$\phi(r_1, r_2, r_3) - \phi(0, 0, 0) = \frac{A}{2}\sum_{j=1}^{3} r_j^2 + \frac{B}{2}\sum_{\substack{j,j'=1 \\ j'\neq j}}^{3} r_j r_{j'} + O\left(r_j^3\right), \tag{**}$$

where A and B are constants:

$$A = \frac{\partial^2 \phi}{\partial r_j^2}\Big|_{\mathbf{r}=0} \equiv -\frac{\partial E_j}{\partial r_j}\Big|_{\mathbf{r}=0}, \text{ while } B = \frac{\partial^2 \phi}{\partial r_j \partial r_{j'}}\Big|_{\mathbf{r}=0} \equiv -\frac{\partial E_j}{\partial r_{j'}}\Big|_{\mathbf{r}=0}, \text{ for } j \neq j'. \tag{***}$$

[6] See, e.g. Eq. (A.14b).

But because of the system's symmetry, the potential also has to be invariant with respect to the change of sign of any single coordinate, $r_j \to -r_j$, which geometrically corresponds to its mirror reflection in the plane of two other coordinate axes[7]. As Eq. (**) shows, this is only possible if $B = 0$.

Moreover, since there is no charge in the vicinity of the cube's center, the potential has to satisfy the Laplace equation (1.42), in Cartesian coordinates reading

$$\sum_{j=1}^{3} \frac{\partial^2 \phi}{\partial r_j^2} = 0. \qquad (****)$$

Comparing this requirement with the first of Eqs. (***), we obtain $A = 0$ as well, so that all derivatives $\partial E_j / \partial r_{j'}$ at $\mathbf{r} = 0$ have to equal zero.

For the field in the center of a planar square, with four similar charges in its corners, the situation is somewhat different, due to the reduced symmetry of the system. The electric field $\mathbf{E}$ at the center (and hence all its Cartesian components) still has to equal zero, because its in-plane component (say, $\mathbf{E}_{12}$) cannot be directed toward any side of the square, and its normal component $\mathbf{E}_3$ cannot be directed to either side of the plane, due to the $r_3 \leftrightarrow -r_3$ symmetry. However, the potential may not remain the same if the coordinate r_3 directed normally to square's plane is swapped with one of in-plane coordinates (r_1 or r_2). As a result, Eq. (**), still with $B = 0$ due to the mirror symmetry, has to be generalized as

$$\phi(r_1, r_2, r_3) - \phi(0, 0, 0) = \frac{A}{2}\left(r_1^2 + r_2^2\right) + \frac{A'}{2} r_3^2 + O\left(r_j^4\right),$$

and the Laplace equation (****) imposes only the following requirement: $2A + A' = 0$, i.e.

$$\frac{\partial E_1}{\partial r_1} = \frac{\partial E_2}{\partial r_2} = -\frac{1}{2}\frac{\partial E_3}{\partial r_3}.$$

This relation is similar to the one used for the solution of problem 1.8 (where it is true due to the axial symmetry of the system).

***Problem* 1.10.** By a direct calculation, find the average electric potential of the spherical surface of radius R, created by a point charge q located at a distance $r > R$ from the sphere's center. Use the result to prove the following general *mean value theorem*: the electric potential at any point is always equal to its average value on any spherical surface with the center at that point, and containing no electric charges inside it.

[7] Actually, because of this symmetry, the higher terms of the expansion (**) should start with $O(r_j^4)$, rather than $O(r_j^3)$.

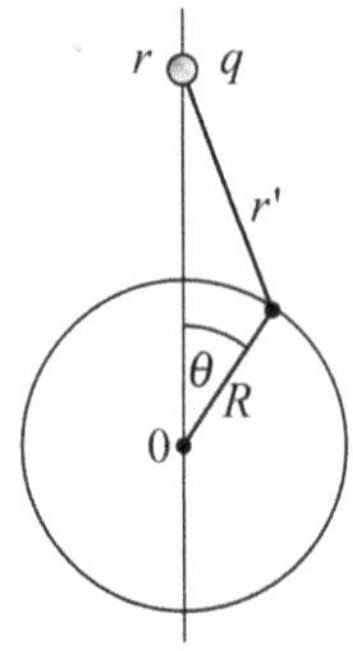

Solution: Using the evident axial symmetry of the problem (see the figure above), and Eq. (1.35) of the lecture notes, we obtain:

$$\phi_{\text{ave}} \equiv \frac{1}{4\pi}\oint \phi(\theta)d\Omega = \frac{2\pi}{4\pi}\int_0^{\pi} \phi(\theta)\sin\theta\, d\theta = \frac{1}{2}\int_0^{\pi} \frac{q}{4\pi\varepsilon_0 r'}\sin\theta\, d\theta,$$

where r' is the distance between the point charge and the observation point:

$$(r')^2 = R^2 + r^2 - 2Rr\cos\theta.$$

The integral may be readily worked out by the introduction of a new variable $\xi \equiv \cos\theta$ (so that $\sin\theta\, d\theta = d\xi$):

$$\phi_{\text{ave}} = \frac{1}{2}\frac{q}{4\pi\varepsilon_0}\int_1^{-1}\frac{d\xi}{[R^2 + r^2 - 2Rr\xi]^{1/2}} = \frac{1}{2}\frac{q}{4\pi\varepsilon_0 r'}\frac{2}{(-2Rr)}[R^2 + r^2 - 2Rr\xi]^{1/2}\Big|_{\xi=-1}^{\xi=+1}$$
$$= \frac{1}{2}\frac{q}{4\pi\varepsilon_0}\frac{2}{(-2Rr)}\left\{[R^2 + r^2 - 2Rr]^{1/2} - [R^2 + r^2 + 2Rr]^{1/2}\right\} = \frac{q}{4\pi\varepsilon_0 r}.$$

We see that the average indeed coincides with the potential's value in the middle of the sphere. (Notice that this result is only valid for the case $r > R$.) Now the proof of the mean value theorem is elementary using the linear superposition principle: since the relation in question,

$$\frac{q}{4\pi\varepsilon_0 r} = \phi_{\text{ave}},$$

holds for each point charge located outside the sphere, it is also true for any system of such charges.

***Problem* 1.11.** Two similar thin, circular, coaxial disks of radius R, separated by distance $2d$, are uniformly charged with equal and opposite areal densities $\pm\sigma$—see the figure below. Calculate and sketch the distribution of the electrostatic potential and the electric field of the disks along their common axis.

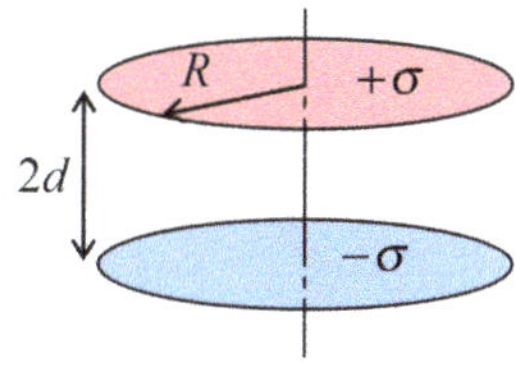

Solution: Let us start from calculating the electrostatic potential of one disk, with a constant areal charge density σ, at the axis point separated by distance z from disk's plane—see the figure below.

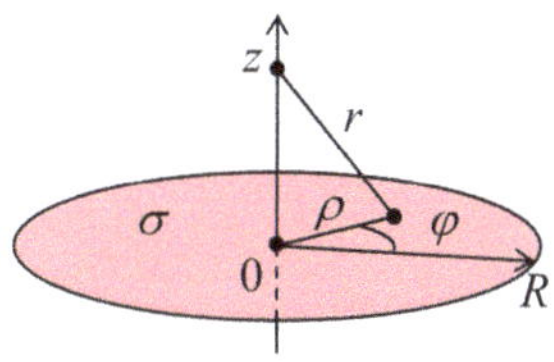

In the polar coordinates $\{\rho, \varphi\}$ within the plane of the disk (with angle φ referred to an arbitrary horizontal axis), the elementary disk area is $\rho d\varphi d\rho$, and its electric charge is $\sigma \rho d\rho d\varphi$, so that Eq. (1.38) of the lecture notes takes the form

$$\phi(z) = \frac{\sigma}{4\pi\varepsilon_0}\int_0^{2\pi} d\varphi \int_0^R \frac{\rho d\rho}{r},$$

where r is the distance between the elementary charge's location and the observation point on disks' common axis. As the figure above shows, $r = (\rho^2 + z^2)^{1/2}$, so that the function under the integral is independent of φ, and the integral may be easily worked out:

$$\phi(z) = \frac{\sigma}{4\pi\varepsilon_0}2\pi\int_0^R \frac{\rho d\rho}{(\rho^2 + z^2)^{1/2}} = \frac{\sigma}{4\varepsilon_0}\int_{\rho=0}^{\rho=R} \frac{d(\rho^2 + z^2)}{(\rho^2 + z^2)^{1/2}} = \frac{\sigma}{2\varepsilon_0}[(R^2 + z^2)^{1/2} - |z|].$$

Now using this formula and the linear superposition principle, we may readily write down an expression describing the potential created by both disks, at distance z from the center of the system (in this new reference frame, the disk center positions are $\pm d$):

$$\phi = \frac{\sigma}{2\varepsilon_0}\left\{\begin{matrix}[R^2 + (z - d)^2]^{1/2} - |z - d| \\ - [R^2 + (z + d)^2]^{1/2} + |z + d|\end{matrix}\right\}.$$

This function is plotted in the figure below for three values of the ratio R/d. For small values of this ratio, the potential is clearly separated into two peaks, of opposite polarity, created by each disk. On the other hand, at $R \gg d$ the result tends to the one for two infinite planes, with ϕ between the disks being a linear function of z, with the slope corresponding to the electric field—see the model solution of problem 1.14 below.

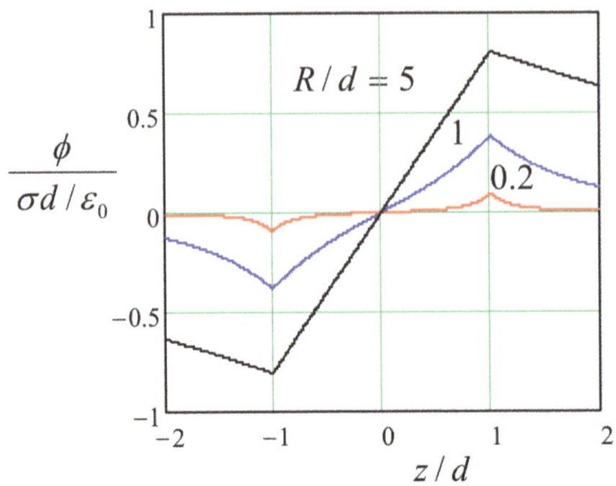

Now the electric field at the axis (which has only one, vertical component due to the axial symmetry of the problem) may be calculated by differentiation of the electrostatic potential—see Eq. (1.33) of the lecture notes:

$$E_z = -\frac{\partial \phi}{\partial z}$$
$$= \frac{\sigma}{2\varepsilon_0}\left\{-\frac{z-d}{[R^2+(z-d)^2]^{1/2}} + \text{sgn}(z-d) + \frac{z+d}{[R^2+(z+d)^2]^{1/2}} - \text{sgn}(z-d)\right\}.$$

In the figure below, this function is plotted for the same three values of the R/d ratio as the potential in the figure above. The plots show that at $R/d = 5$, the field between the disks is already pretty uniform, although its magnitude is still noticeably smaller than that (σ/ε_0) between two infinite planes. Note also that the ratio does not affect the electric field's jump by $\pm\sigma/\varepsilon_0$ as the observation point crosses a disk—just as it should be, according to Eq. (1.24) of the lecture notes.

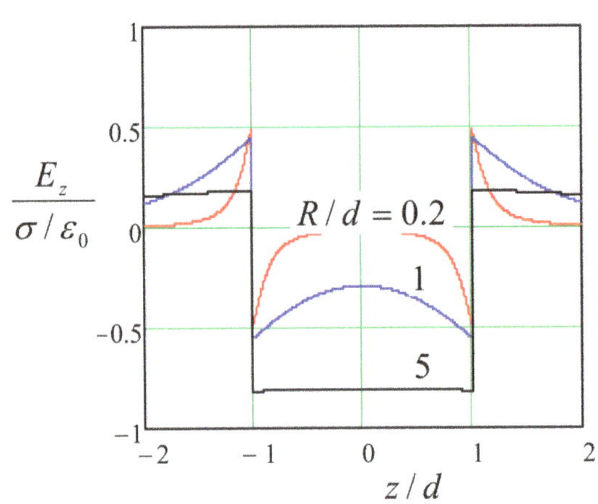

On the technical side, this solution illustrates again the advantage of calculating the electrostatic potential (a scalar function) first, and only then the electric field from it.

***Problem* 1.12.** In a certain reference frame, the electrostatic potential created by an electric charge distribution is

$$\phi(\mathbf{r}) = C\left(\frac{1}{r} + \frac{1}{2r_0}\right)\exp\left\{-\frac{r}{r_0}\right\},$$

where C and r_0 are constants, and $r \equiv |\mathbf{r}|$ is the distance from the origin. Calculate the charge distribution in space.

Solution: According to the Poisson equation (see Eq. (1.41) of the lecture notes), the charge density may be calculated as

$$\rho(\mathbf{r}) = -\varepsilon_0\nabla^2\phi(\mathbf{r}).$$

Since the potential distribution is spherically symmetric, $\phi(\mathbf{r}) = \phi(r)$, the general expression for the Laplace operator of such scalar function in spherical coordinates[8] is reduced to

$$\nabla^2\phi(r) = \frac{1}{r^2}\frac{d}{dr}\left(r^2\frac{d\phi}{dr}\right). \qquad (*)$$

so that by performing straightforward differentiation we obtain a very simple, exponential charge density distribution:

$$\rho(\mathbf{r}) = -\frac{\varepsilon_0 C}{2r_0^3}\exp\left\{-\frac{r}{r_0}\right\}. \qquad \text{(INCOMPLETE!)}$$

Note, however, that since Eq. (*) is invalid at $r = 0$, this point has to be explored separately. At $r \to 0$ the potential distribution tends to C/r and, as Eq. (1.35) of the lecture notes shows, such a potential is created by a point charge $q = 4\pi\varepsilon_0 C$, located at the origin. As a result, the complete solution of our problem is

$$\rho(\mathbf{r}) = \varepsilon_0 C\left[4\pi\delta(\mathbf{r}) - \frac{1}{2r_0^3}\exp\left\{-\frac{r}{r_0}\right\}\right] = q\left[\delta(\mathbf{r}) - \frac{1}{8\pi r_0^3}\exp\left\{-\frac{r}{r_0}\right\}\right],$$

with the total charge[9]

$$Q = \int\rho(\mathbf{r})d^3r = q\left(1 - \frac{1}{8\pi r_0^3}\int\exp\left\{-\frac{r}{r_0}\right\}d^3r\right) = q\left(1 - \frac{1}{8\pi r_0^3}4\pi\int_0^\infty\exp\left\{-\frac{r}{r_0}\right\}r^2dr\right)$$
$$= q\left(1 - \frac{1}{2}\int_0^\infty\exp\{-\xi\}\xi^2d\xi\right) = 0.$$

For the reader's reference, such a $\rho(\mathbf{r})$ is a reasonable approximation of the charge distribution in an atom, describing the screening of the positive nuclear charge

---

[8] See, e.g. Eq. (A.67).

[9] The last step of this calculation uses the table integral (A.34d) with $n = 2$.

$q = Ze$ by Z negatively charged electrons. Such an approximation, as well as a more accurate *Thomas–Fermi model* of the screening[10], essentially ignore the quantum mechanics of the electrons, and as the result work reasonably well only for very heavy atoms ($Z \gg 1$).

***Problem* 1.13.** A thin flat sheet, cut in the form of a rectangle of size $a \times b$, is electrically charged with a constant areal density σ. Without an explicit calculation of the spatial distribution $\phi(\mathbf{r})$ of the electrostatic potential induced by this charge, find the ratio of its values at the center and at the corners of the rectangle.

Hint: Consider partitioning the rectangle into several similar parts and using the linear superposition principle.

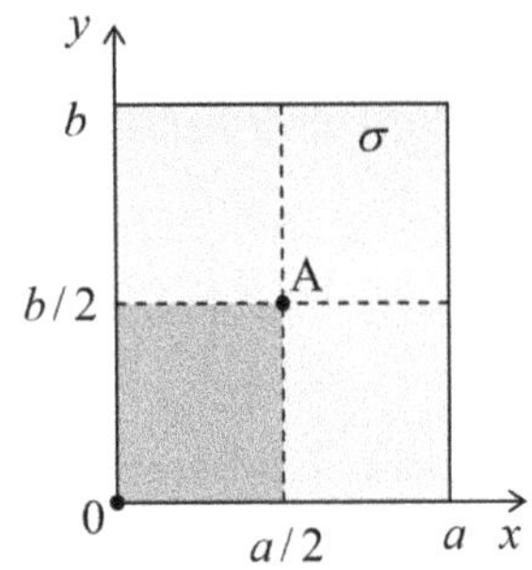

Solution: Selecting the Cartesian coordinates as shown in the figure above, we may use Eq. (1.38) of the lecture notes to calculate the potential at the origin (i.e. in one of the rectangle's corners) as

$$\phi_0 = \frac{\sigma}{4\pi\varepsilon_0}\int_0^a dx \int_0^b dy \frac{1}{(x^2 + y^2)^{1/2}} \equiv \frac{\sigma a}{4\pi\varepsilon_0} I,$$

where I is the following dimensionless integral:

$$I \equiv \int_0^1 d\xi \int_0^{b/a} d\zeta \frac{1}{(\xi^2 + \zeta^2)^{1/2}},$$

with $\xi \equiv x/a$, $\zeta \equiv y/a$.

Now let us calculate what potential would be induced, at the same point, only by the adjacent quarter of the rectangle, of size $(a/2) \times (b/2)$, marked in the figure above with darker shading:

$$\phi_0' = \frac{\sigma}{4\pi\varepsilon_0}\int_0^{a/2} dx \int_0^{b/2} dy \frac{1}{(x^2 + y^2)^{1/2}}.$$

[10] See, e.g. *Part QM*, problems 8.22 and 8.23.

Introducing dimensionless variables in a similar but modified way: $\xi \equiv x/(a/2)$, $\zeta \equiv y/(a/2)$, we may reduce this result to the same dimensionless integral I:

$$\phi_0' = \frac{\sigma(a/2)}{4\pi\varepsilon_0} I,$$

so that even without the calculation of the integral[11] we see that $\phi_0' = \phi_0/2$.

Next, we may note that due to the linear superposition principle, the potential ϕ_A at the center of the rectangle may be represented as a sum of four potentials induced by such quarters in their corners. Due to the symmetry, all these partial potentials are equal to ϕ_0', so that finally

$$\phi_A = 4\phi_0' = 2\phi_0.$$

This solution illustrates the power of *scaling* arguments, broadly used, for symmetric geometries, in all fields of physics.

***Problem* 1.14.** Calculate the electrostatic energy per unit area of a system of two thin, parallel planes with equal and opposite charges of a constant areal density σ, separated by distance d.

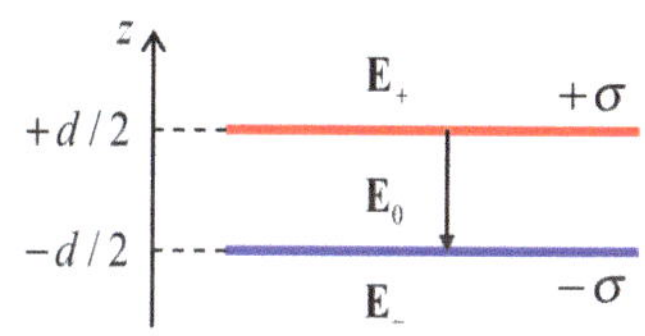

Solution: From the similar problem solved in the lecture notes (see figure 1.4 and its discussion), it is clear that the electric field everywhere should be normal to the planes and constant within each of three ranges:

$$\mathbf{E} = \mathbf{n}_z \begin{cases} E_-, & \text{at } z < -d/2, \\ E_0, & \text{at } -d/2 < z < +d/2, \\ E_+, & \text{at } +d/2 < z, \end{cases}$$

where the axis z is selected as shown in the figure below, and E_-, E_0, and E_+ are some scalar constants. These constants may be calculated by applying the Gauss law to three pillboxes with all three possible combinations of lid positions, giving three equations:

$$E_+ - E_0 = +\sigma/\varepsilon_0, \qquad E_0 - E_- = -\sigma/\varepsilon_0, \qquad E_+ - E_- = 0,$$

whose solution yields

$$E_+ = E_- = 0, \qquad E_0 = -\frac{\sigma}{\varepsilon_0}. \tag{*}$$

[11] Just for reference, this integral may be worked out analytically: $I = \frac{b}{a} \ln \frac{(a^2 + b^2)^{1/2} + a}{b} + \sinh^{-1} \frac{b}{a}$.

(An even simpler way to obtain Eq. (*) is to employ the linear superposition principle by summing up the fields of the same magnitude (1.23), $|E| = |\sigma|/2\varepsilon_0$, induced by each of the planes. Due the difference of plane charge signs, the fields add up between the plates, but cancel each other outside the system.)

Thus, the field exits only between the planes, producing the electrostatic potential

$$\phi = \phi_{|z=0} - \int_0^z E(z')dz' = \phi_{|z=0} + \frac{\sigma}{\varepsilon_0} \times \begin{cases} z, & \text{for } |z| < d/2, \\ (d/2)\text{sgn}(z), & \text{for } |z| > d/2, \end{cases}$$

so that the potential difference (called *voltage*) between the planes is

$$V \equiv \phi\left(+\frac{d}{2}\right) - \phi\left(-\frac{d}{2}\right) = \frac{\sigma}{\varepsilon_0}d \equiv \frac{\sigma}{C_0},$$

where $C_0 = \varepsilon_0/d$ is the *specific capacitance* of this simple system (which is equivalent to a *plane capacitor*—for more, see chapter 2).

Now using Eq. (*) in Eq. (1.65) for the potential energy of the field, we obtain

$$\frac{U}{A} = \frac{\varepsilon_0}{2}\int_{-d/2}^{+d/2} E^2 dz = \frac{\varepsilon_0 E_0^2}{2}d = \frac{\sigma^2}{2\varepsilon_0}d = \frac{C_0 V^2}{2}.$$

(We will repeatedly run into this expression in chapter 2.) Note that at fixed σ (of any sign) the energy grows with d. This is natural, because the oppositely charged planes attract each other, so that following the force (reducing d) corresponds to the way toward the potential energy's minimum.

***Problem* 1.15.** The system analyzed in the previous problem (two thin, parallel, oppositely charged planes) is now placed into an external, uniform, normal electric field $E_{ext} = \sigma/\varepsilon_0$—see the figure below. Find the force (per unit area) acting on each plane, by two methods:

(i) directly from the electric field distribution, and
(ii) from the potential energy of the system.

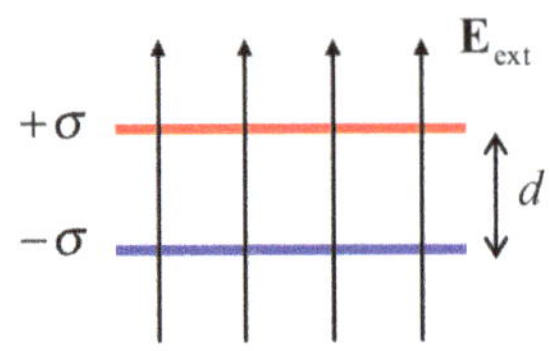

Solutions:

(i) In order to calculate the force F acting on the top plane, we have to neglect the field of this plane itself[12] and add the external field directed up and equal to σ/ε_0, and

[12] Different elementary charges of the same plane do Coulomb-interact, but the elementary forces between them are directed along the plane, and cancel at summation, due to the 2nd Newton's law.

the field of the bottom plane, directed down and equal to $\sigma/2\varepsilon_0$—see figure 1.4 and (1.24) of the lecture notes. The net field equals $\sigma/2\varepsilon_0$ and is directed up, so that the force per unit area is

$$\frac{F}{A} = \frac{q}{A}\frac{\sigma}{2\varepsilon_0} = \frac{\sigma^2}{2\varepsilon_0}, \tag{*}$$

and is directed up. A similar calculation for the bottom plane yields force of the same magnitude, but directed down.

(ii) Using the linear superposition principle, we may calculate the total field in each spatial region by adding the external field to that of the planes, which was calculated in the previous problem:

$$\mathbf{E} = \mathbf{n}_z\frac{\sigma}{\varepsilon_0} + \mathbf{n}_z\frac{\sigma}{\varepsilon_0} \times \left\{\begin{array}{ll} 0, & \text{at}\, z < -d/2, \\ -1, & \text{at}\; -d/2 < z < +d/2, \\ 0, & \text{at}\; +d/2 < z, \end{array}\right\}$$

$$= \mathbf{n}_z\frac{\sigma}{\varepsilon_0} \times \begin{cases} 1, & \text{at}\, z < -d/2, \\ 0, & \text{at}\; -d/2 < z < +d/2, \\ 1, & \text{at}\; +d/2 < z. \end{cases}$$

Using these values, let us calculate the potential energy U from Eq. (1.72) of the lecture notes, artificially limiting the integration volume to a pillbox of an arbitrary area A, with some fixed thickness $d_0 > d$. The second integral does not depend on d, and may be dropped (included into an arbitrary constant that may be always added to the energy), and the first integral yields

$$\frac{U}{A} = \frac{\varepsilon_0}{2}\left(\frac{\sigma}{\varepsilon_0}\right)^2 (d_0 - d) \equiv \text{const} - \frac{\sigma^2}{2\varepsilon_0}d.$$

Now the vertical force (per unit area) exerted at the top plane may be calculated as $F/A = -\partial(U/A)/\partial d$, giving the same result (*) as the first method. The same is true for the lower plane whose vertical coordinate is (const $-d$), so that the ratio U/A should be differentiated over $(-d)$ rather than d.

***Problem* 1.16.** Explore the relation between the Laplace equation (1.42) and the condition of minimum of the electrostatic field energy (1.65).

Solution: Let us consider a small variation $\delta\phi$ of the electrostatic potential inside a charge-free volume V limited by a closed surface S, such that the variation vanishes at the surface, but is otherwise arbitrary. Let us calculate the corresponding variation of the electric field energy (1.65):

$$\begin{aligned} \delta U &= \delta\left(\frac{\varepsilon_0}{2}\int_V (\nabla\phi)^2 d^3r\right) = \frac{\varepsilon_0}{2}\int_V \delta(\nabla\phi \cdot \nabla\phi) d^3r \\ &= \varepsilon_0\int_V \nabla\phi \cdot \delta(\nabla\phi)\, d^3r = \varepsilon_0\int_V \nabla\phi \cdot \nabla(\delta\phi) d^3r \end{aligned} \tag{*}$$

The last step above has exploited the main rule of the variational calculus that the operations of small variation and the usual differentiation are interchangeable[13]. Now let us work out the resulting 3D integration by parts—just as was done at the derivation of Eq. (1.65) of the lecture notes, i.e. by applying the divergence theorem (A.79) to the vector function $\mathbf{f} = \delta\phi\nabla\phi$:

$$\oint_S (\delta\phi\nabla\phi)_n d^2r = \int_V \nabla \cdot (\delta\phi\,\nabla\phi) d^3r \equiv \int_V \nabla(\delta\phi) \cdot \nabla\phi d^3r + \int_V \delta\phi\; \nabla^2\phi d^3r.$$

The surface integral on the left-had side of this relation has to equal zero, due to our condition $\delta\phi|_S = 0$, and the first volume integral on the right-hand side equals that in the last form of Eq. (*). Hence, that equality may be rewritten as

$$\delta U = -\varepsilon_0 \int_V \delta\phi\; \nabla^2\phi d^3r.$$

Since the variation $\delta\phi$ is arbitrary, this expression shows that the only way for the potential energy of the field to have the lowest possible value (just as it does in any stable equilibrium in classical mechanics) and hence for the variation δU to equal zero, is to have the Laplace operator of the field equal zero, i.e. to satisfy the Laplace equation (1.42) in all charge-free spatial regions.

***Problem* 1.17.** Prove the following *reciprocity theorem of electrostatics*[14]: if two spatially confined charge distributions $\rho_1(\mathbf{r})$ and $\rho_2(\mathbf{r})$ create respective distributions $\phi_1(\mathbf{r})$ and $\phi_2(\mathbf{r})$ of the electrostatic potential, then

$$\int \rho_1(\mathbf{r})\; \phi_2(\mathbf{r}) d^3r = \int \rho_2(\mathbf{r})\; \phi_1(\mathbf{r}) d^3r.$$

Hint: Consider integral $\int \mathbf{E}_1 \cdot \mathbf{E}_2\, d^3r$.

Solution: Applying Eq. (1.33) of the lecture notes to $\mathbf{E}_1$(r), let us transform the integral mentioned in the hint as

$$\int \mathbf{E}_1 \cdot \mathbf{E}_2 d^3r = -\int \nabla\phi_1 \cdot \mathbf{E}_2\, d^3r.$$

Now we may use the rule of spatial differentiation of a vector-by-scalar function product[15] to continue as follows:

$$-\int \nabla\phi_1 \cdot \mathbf{E}_2 d^3r = \int \phi_1(\nabla \cdot \mathbf{E}_2) d^3r - \int \nabla(\phi_1 \mathbf{E}_2) d^3r.$$

Next, we may use the inhomogeneous Maxwell equation (1.27) in the first integral, and the well-known divergence theorem[16] to transform the second integral to that of

[13] This rule has a simple geometric meaning—see, e.g. *Part CM* section 2.1, in particular figure 2.2.

[14] This is only the simplest one of the whole family of reciprocity theorems in electromagnetism. (Sometimes it is called the 'Green's reciprocity theorem', but historically it is fairer to reserve the last name for the generalization to surface charges, discussed in section 2.10 of the lecture notes.)

[15] See, e.g. Eq. (A.74a), with $f = \phi_1$ and $\mathbf{g} = \mathbf{E}_2$.

[16] See, e.g. Eq. (A.79).

$(\phi_1\mathbf{E}_2)_n$ over some distant, closed surface S that limits the volume of our spatial integration. As a result, our expression becomes

$$\frac{1}{\varepsilon_0}\int \phi_1\rho_2 d^3r - \oint_S (\phi_1\mathbf{E}_2)_n d^2r.$$

Since the charge (and hence the field) distributions are space-confined, we may always select the surface S so distant that the surface integral is negligible[17], and our chain of transformations may be summarized as

$$\int \mathbf{E}_1 \cdot \mathbf{E}_2 d^3r = \frac{1}{\varepsilon_0}\int \phi_1\rho_2 d^3r.$$

Now repeating the same calculation with swapped indices, we arrive at the reciprocity theorem.

Note that if some parts of these two charge distributions reside on some surface(s) S, and may be well described by surface charge densities $\sigma_1(\mathbf{r})$ and $\sigma_2(\mathbf{r})$ (as is very instrumental, for example, in systems with good conductors, to be discussed in chapter 2 of the lecture notes), the reciprocity theorem may be rewritten as

$$\int_V \rho_1(\mathbf{r})\phi_2(\mathbf{r})d^3r + \int_S \sigma_1(\mathbf{r})\phi_2(\mathbf{r})d^2r = \int_V \rho_2(\mathbf{r})\phi_1(\mathbf{r})d^3r + \int_S \sigma_2(\mathbf{r})\phi_1(\mathbf{r})d^2r,$$

where $\rho_1(\mathbf{r})$ and $\rho_2(\mathbf{r})$ are the remaining, 'genuinely volume' parts of the distributions.

***Problem* 1.18.** Calculate the energy of electrostatic interaction of two spheres, of radii R_1 and R_2, each with a spherically symmetric charge distribution, separated by distance $d > R_1 + R_2$.

Solution: According to Eq. (1.55) of the lecture notes, applied sequentially to each of the spheres, the energy U_{int} of their interaction may be calculated in either of two ways[18]:

$$U_{\text{int}} = \int_{\text{sphere 1}} \rho_1(\mathbf{r})\phi_2(\mathbf{r})d^3r, \qquad (*)$$

$$U_{\text{int}} = \int_{\text{sphere 2}} \rho_2(\mathbf{r})\phi_1(\mathbf{r})d^3r. \qquad (**)$$

But as was discussed in section 1.2 of the lecture notes (see Eqs. (1.19) and (1.20)), the electric field, and hence the electrostatic potential ϕ, created by a spherically symmetric charge distribution $\rho(\mathbf{r})$ outside of its boundary coincides with that of the

[17] Indeed, we may take S in the shape of a spherical shell, with radius R much larger than the spatial scales of distributions $\rho_1(\mathbf{r})$ and $\rho_2(\mathbf{r})$. At such large distances, their fields are close to those of point charges, so that according to Eqs. (1.6) and (1.35), $\phi_1 \propto Q_1/R$ and $E_2 \propto Q_2/R^2$, so that the integral over surface S, of area $A \propto R^2$, is decreasing with R as $1/R$ (or faster if any of the net charges $Q_{1,2}$ vanishes).

[18] Their equivalence is strictly confirmed by the reciprocity theorem (see the previous problem) even if the charge distributions $\rho_{1,2}$(r) partly or fully overlap—although in our current case, they do not.

full charge Q of the sphere concentrated in its center. Applying this fact to, for example, the first sphere, we may rewrite Eq. (**) as

$$U_{\text{int}} = \int_{\text{sphere 2}} \rho_2(\mathbf{r})\phi_1^{(\text{point})}(\mathbf{r})d^3r, \quad \text{where } \phi_1^{(\text{point})}(\mathbf{r}) = \frac{Q_1}{4\pi\varepsilon_0|\mathbf{r} - \mathbf{r}_1|},$$

$$\text{with } Q_1 \equiv \int_{\text{sphere 1}} \rho_1(\mathbf{r})d^3r,$$

where $\mathbf{r}_1$ is the position of the center of sphere 1. This means that the interaction energy, however calculated, cannot depend on whether the charge of sphere 1 is distributed as given, or compressed to point $\mathbf{r}_1$, i.e. should be invariant with respect to the replacement

$$\rho_1(\mathbf{r}) \to Q_1\delta(\mathbf{r} - \mathbf{r}_1).$$

Making this replacement in Eq. (*), we obtain

$$U_{\text{int}} = Q_1\phi_2(\mathbf{r}_1).$$

Now applying the same argument to the potential created by the spherically symmetric distribution $\rho_2(\mathbf{r})$,

$$\phi_2(\mathbf{r}) = \phi_2^{(\text{point})}(\mathbf{r}) = \frac{Q_2}{4\pi\varepsilon_0|\mathbf{r} - \mathbf{r}_2|}, \quad \text{with } Q_2 \equiv \int_{\text{sphere 2}} \rho_2(\mathbf{r})d^3r,$$

where $\mathbf{r}_2$ is the position of the center of sphere 2, we finally obtain:

$$U_{\text{int}} = \frac{Q_1Q_2}{4\pi\varepsilon_0 d}, \quad \text{with } d \equiv |\mathbf{r}_1 - \mathbf{r}_2|.$$

In plain English, the spheres interact as if their electric charges were concentrated at their centers. Remember, however, that this result is only applicable at $d > R_1 + R_2$, so that the charge distributions $\rho_1(\mathbf{r})$ and $\rho_2(\mathbf{r})$ do not overlap—even partly.

***Problem* 1.19.** Calculate the electrostatic energy U of a (generally, thick) spherical shell, with a charge Q uniformly distributed through its volume—see the figure below. Analyze and interpret the dependence of U on the inner cavity's radius R_1, at fixed Q and R_2.

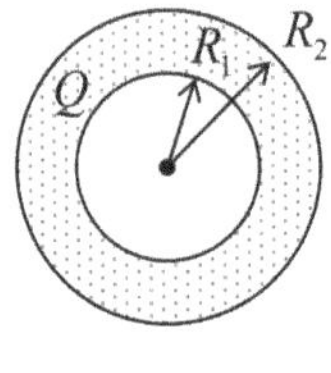

Solution: Calculating the only (radial) component E of the electric field $\mathbf{E} = \mathbf{n}_r E$ (say, using the Gauss law), we readily obtain

$$E = \frac{Q}{4\pi\varepsilon_0 r^2} \times \begin{cases} 0, & \text{for } 0 \leqslant r < R_1, \\ (r^3 - R_1^3)/(R_2^3 - R_1^3), & \text{for } R_1 < r < R_2, \\ 1, & \text{for } R_2 < r, \end{cases}$$

so that Eq. (1.65) for the electrostatic energy yields

$$U = \frac{\varepsilon_0}{2}\int E^2 d^3r = \frac{\varepsilon_0}{2}4\pi\int_0^\infty E^2 r^2 dr = \frac{\varepsilon_0}{2}4\pi\left(\frac{Q}{4\pi\varepsilon_0}\right)^2\left[\int_{R_1}^{R_2}\left(\frac{r^3 - R_1^3}{R_2^3 - R_1^3}\right)^2\frac{dr}{r^2} + \int_{R_2}^\infty \frac{dr}{r^2}\right]$$

$$= \frac{Q^2}{8\pi\varepsilon_0 R_2}\left[\int_\alpha^1\left(\frac{\xi^3 - \alpha^3}{1 - \alpha^3}\right)^2\frac{d\xi}{\xi^2} + \int_1^\infty \frac{d\xi}{\xi}\right] = \frac{Q^2}{8\pi\varepsilon_0 R_2}f(\alpha),$$

$$\text{with } f(\alpha) \equiv \frac{1/5 - \alpha^3 + (9/5)\alpha^5 - \alpha^6}{(1 - \alpha^3)^2} + 1,$$

where $\alpha \equiv R_1/R_2 \leqslant 1$.

A plot of the function $f(\alpha)$, i.e. of the normalized electrostatic energy as a function of R_1 at fixed Q and R_2, is shown in the figure below. The function reaches its maximum $f(0) = 6/5$ (so that U is given by Eq. (1.66) of the lecture notes, with $R = R_2$) at $\alpha = 0$, i.e. for a solid sphere, and tends to $f(1) = 1$, giving

$$U_{\min} = \frac{Q^2}{8\pi\varepsilon_0 R_2}$$

at $\alpha \to 1$, i.e. for the ultimately thin spherical shell. This is very natural, because the elementary charges of the sphere repulse each other and try to go apart as far as possible, in particular increasing its inner radius.

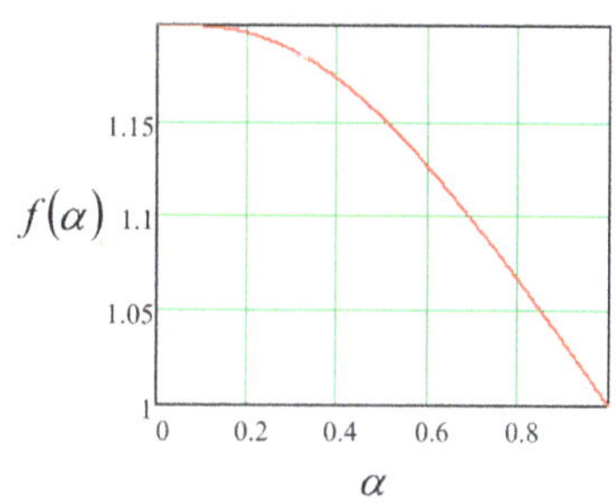

Chapter 2

Charges and conductors

***Problem* 2.1.** Calculate the force (per unit area) exerted on a conducting surface by an electric field, normal to it. Compare the result with the definition of the electric field, given by Eq. (1.6) of the lecture notes, and comment.

Solution: Let us make a natural assumption that the density $u(\mathbf{r})$ of the electrostatic energy, given by Eq. (1.65) of the lecture notes, at the conductor's surface is a smooth function of the surface's position. Then, at a small displacement δx of the surface, in the direction along vector $\mathbf{E}$, the density cannot change faster than proportional to δx.[1] As a result, the change of the field's energy Eq. (1.65), per unit area, may be calculated as

$$\frac{\delta U}{A} = \frac{1}{A}\delta\left(\frac{\varepsilon_0}{2}\int E^2 d^3r\right) = \frac{\varepsilon_0}{2}E^2\frac{1}{A}\delta V = \frac{\varepsilon_0}{2}E^2(-\delta x),$$

where V is the volume outside the conductor (filled with the field), and $\delta x \equiv \mathbf{n}\cdot\delta\mathbf{r}$ is the conductor surface's displacement toward the field-occupied space. Hence the force exerted by the field is directed out of the conductor (which is therefore *attracted* by the field—of any polarity), and equal to

$$\frac{F}{A} = -\frac{1}{A}\frac{\delta U}{\delta x} = \frac{\varepsilon_0}{2}E^2 \equiv \frac{1}{2}\sigma E, \qquad (*)$$

where σ is the surface charge density—see Eq. (*).

The apparent contradiction (the extra factor ½) between this result and the electric field's definition (1.6), reduced to a unit surface area ($F/A = \sigma E$), may be interpreted as follows. The field equals E on one side of the surface charge layer, but vanishes on its another side (inside the conductor), so that the average field inside the

[1] A small shift normal to the vector $\mathbf{E}$ evidently cannot change $u(\mathbf{r})$ at all.

doi:10.1088/978-0-7503-1407-7ch2

layer is $E_{\text{ave}} = E/2$. Hence the force per unit area may be calculated as $F/A = \sigma E_{\text{ave}} = (\varepsilon_0 E)(E/2)$, in agreement with the above result.

A possible conceptual concern about Eq. (*) is whether it means that the surface charge layer exerts a normal force on itself, if there are no explicit external sources of the field **E**, i.e. if the field may be associated with the surface charge itself. To dispel this concern, it is sufficient to recall that according to the Maxwell equation (1.27), in the free space outside the conductor $\nabla \cdot \mathbf{E} = 0$. In informal language, this equation means that the field's force lines, originating from the surface charges, cannot terminate until they have run into the corresponding opposite charges—for example on the opposite plate of a capacitor—see the figure below. The force (*) may always be attributed, in electrostatics, to the attraction of the surface charges in question to those opposite charges.

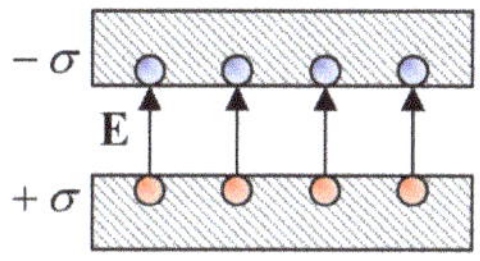

Note, however, that in the relativistic electrodynamics, where the electromagnetic field becomes an independent entity, it is more adequate to attribute the force to the field as such. (As will be shown in section 9.8 of the lecture notes, Eq. (*) remains valid in that case as well, though the magnetic field may exert an additional force on the *moving* surface charge.)

***Problem* 2.2.** Certain electric charges Q_A and Q_B have been placed on two metallic, concentric spherical shells—see the figure below. What is the full charge of each of the surfaces S_1–S_4?

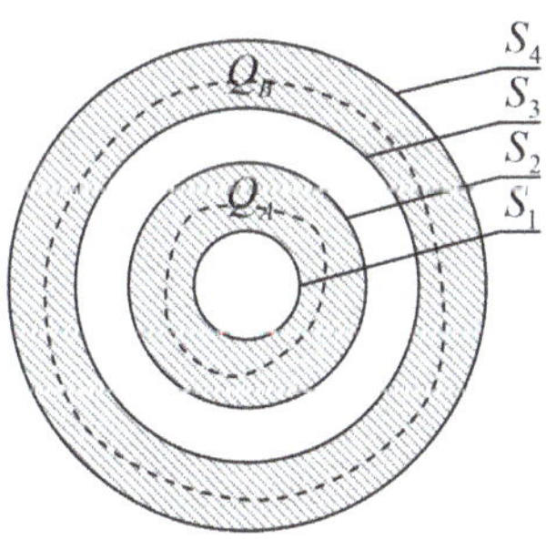

Solution: Let us start with applying the Gauss law (1.16),

$$\int_S E_n d^2r = \frac{Q}{\varepsilon_0} \qquad (*)$$

(where Q is the full charge within a closed surface S), to any volume located inside the inner sphere and containing the surface S_1 inside it—see the inner dashed line in the figure above. As was discussed in section 2.1 of the lecture notes, in statics the electric field inside any metal should be zero, so that the left-hand side of Eq. (*) for the surface vanishes, indicating that in the full charge of surface S_1 equals zero: $Q_1 = 0$. Hence the full charge Q_A of the first sphere should sit on its external surface: $Q_2 = Q_A$.

Now let us consider a similar Gaussian volume with the boundary inside the second metallic sphere, containing the surfaces S_1, S_2, and S_3. Similarly, the left-hand side of Eq. (*) equals zero as well, and the Gauss law gives

$$0 = \frac{Q_1 + Q_2 + Q_3}{\varepsilon_0}.$$

With our prior results, this means that $Q_3 = -Q_A$. Hence the charge of the last, external surface S_4 of the same sphere is

$$Q_4 = Q_B - Q_3 = Q_B - (-Q_A) = Q_A + Q_B.$$

***Problem* 2.3.** Calculate the mutual capacitance between the terminals of the lumped capacitor circuit shown in the figure below. Analyze and interpret the result for major particular cases.

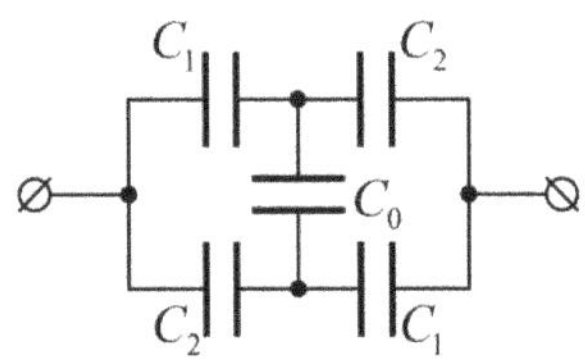

Solution: Solving the lumped-capacitor circuit problems by prescribing certain charges to the capacitor plates, it is necessary to remember that the net charge of both plates of each capacitor, and of each isolated conducting 'island' of the circuit (such as the two islands adjacent to the capacitor C_0 in our problem) have to equal zero. Indeed, the lumped capacitor model is only valid if the electric field outside the capacitors is negligible, and hence the net charge of each isolated conductor has to be negligibly small—see figure 2.5 of the lecture notes.

For example, for the elementary example of two capacitors connected in series (see the figure below), with one insulated island, these requirements immediately imply that all plate charges should have one magnitude. From this charge pattern, and Eq. (2.26) applied to each capacitor, we obtain $V_1 = Q/C_1$ and $V_2 = Q/C_2$. Since these voltages (frequently called 'voltage drops') across each capacitor are, by definition, just the differences of the electrostatic potentials of their plates, at the connection in series they just add up (see the figure below again), giving $V = V_1 + V_2 = Q(1/C_1 + 1/C_2)$ for the voltage between the external terminals. Now considering the whole circuit as the 'black box' between the terminals, we have to define its (mutual) capacitance C as the ratio of the charge Q passed through the terminals at the circuit charging, to the resulting voltage V between them. In this easy case, this definition yields a very well known answer:

$$C \equiv \frac{Q}{V} = \left(\frac{1}{C_1} + \frac{1}{C_2}\right)^{-1}, \quad \text{i.e.} \quad \frac{1}{C} = \frac{1}{C_1} + \frac{1}{C_2},$$

with an evident generalization to the case of an arbitrary number of capacitances connected in series.

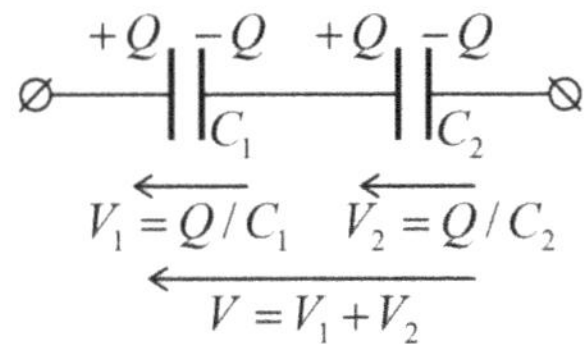

The reader has to excuse me for the discussion of this apparently trivial example, because without a clear understanding of this general approach the solution of more complex problems, such as our current problem, may be difficult, because it cannot be reduced to a chain of capacitances connected either in parallel or in series. For this problem, taking into account the capacitor and island neutrality and the circuit symmetry, we arrive at the general charge distribution shown on the top panel of the figure below, with the total inserted charge

$$Q = Q_1 + Q_2, \tag{*}$$

but Q_1 and Q_2 still unknown.

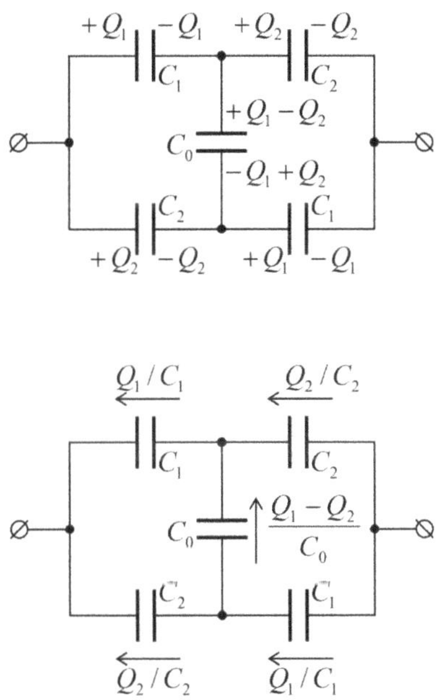

Now using Eq. (2.26) to calculate the voltage drop across each capacitor, we arrive at the voltage pattern shown in the bottom panel of the figure above, where the arrow directions show the *assumed* voltage polarities, correct if all specified charges (Q_1, Q_2, and $Q_1 - Q_2$) are positive. For these voltage drops we may write two relations: the evident equality for the voltage between the terminals:

$$V = \frac{Q_1}{C_1} + \frac{Q_2}{C_2}, \qquad (**)$$

and the condition that the net voltage drop along any of the closed loops of the circuit is zero[2]:

$$\frac{Q_1}{C_1} - \frac{Q_2}{C_2} + \frac{Q_1 - Q_2}{C_0} = 0. \qquad (***)$$

Solving the system of two linear equations (*) and (***) for $Q_{1,2}$, and plugging the result into Eq. (**), we finally obtain $V = Q/C$, where

$$C = \frac{2C_1C_2 + (C_1 + C_2)C_0}{C_1 + C_2 + 2C_0}. \qquad (****)$$

Let us start the analysis of this result from the case $C_0 = 0$. In this particular case, Eq. (****) is reduced to

$$C = \frac{2C_1C_2}{C_1 + C_2}, \quad \text{i.e. } C = 2C', \quad \text{where } \frac{1}{C'} \equiv \frac{1}{C_1} + \frac{1}{C_2}.$$

This is natural, because in this case the capacitor C_0 cannot accommodate (at a finite voltage) any charge, and may be replaced with an open circuit. As a result, our full system becomes just a parallel connection of two branches, each being just a connection in series of the capacitors C_1 and C_2, with the same branch capacitance C'.

Note that the last result (or rather its particular form) is also valid in the case of arbitrary C_0, but equal C_1 and C_2. Indeed, in this case the circuit is symmetric with respect to the horizontal line in the figure above, so that there is no voltage drop between its internal notes.

In the opposite limit, $C_0 \to \infty$ (practically meaning that this capacitance is much larger than both C_1 and C_2), Eq. (****) yields

$$C \to \frac{C''}{2}, \qquad \text{where} \quad C'' \equiv C_1 + C_2.$$

This is also natural, because (as the last figure above shows) the infinite capacitance C_0 makes the potential difference between two internal nodes of the circuit negligible, so that the circuit becomes just a connection in series of two similar groups of capacitors C_1 and C_2, connected in parallel within each group.

Finally, if one of the branch capacitances (say, C_2) equals 0, Eq. (****) is reduced to

---

[2] In the formal theory of lumped-element circuits, this condition is called the *second* (or 'loop') *Kirchhoff rule*, the *first* (or 'node') *Kirchhoff* rule saying that the algebraic sum of the charges/currents flowing into each circuit *node* (wire connection) should equal zero. (For our circuit, the latter rule yields Eq. (*), obtained above from reasoning similar to that used for the proof of the rule in the general case.) The Kirchhoff rules will be discussed in chapters 4 and 6 of the lecture notes.

$$C = \frac{C_1 C_0}{C_1 + 2C_0}, \qquad \text{so that} \quad \frac{1}{C} = \frac{1}{C_1} + \frac{1}{C_0} + \frac{1}{C_1}.$$

This result becomes clear if we redraw our circuit for this particular case, replacing both zero capacitors C_2 with open circuits. The resulting system, shown in the figure below, just a connection in series of three capacitors: C_1, C_0, and C_1 again, correctly described by the last formula for C. (The result for $C_1 = 0$ is similar, with the obvious index replacement.)

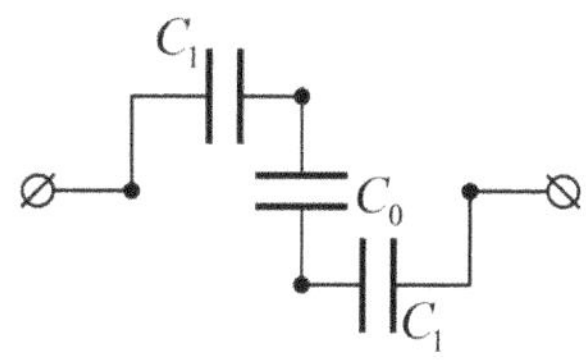

So, the circuit under analysis, despite its relatively simple topology[3], becomes either a capacitor connection in series or their connection in parallel in different particular cases, but generally it is neither.

***Problem* 2.4.** Calculate the mutual capacitance between the terminals of the semi-infinite lumped-capacitor chain shown in the figure below, and find the law of decay of the applied voltage along the system. Analyze and interpret the result.

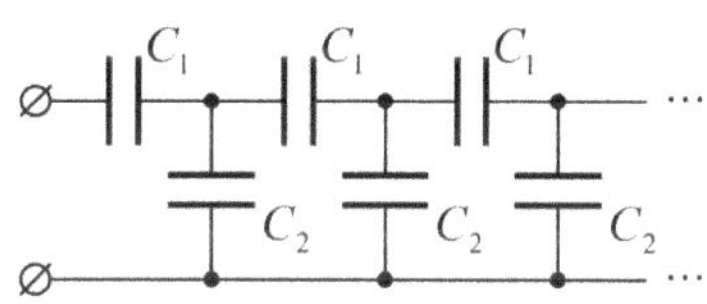

Solution: Since all links of the chain are similar, it is sufficient to analyze just a couple of them, say with numbers j and $(j + 1)$, where j is the 'distance' from the input terminals. Assigning the electric charges to the capacitor plates, with the island neutrality rule discussed in detail in the previous problem's solution, we arrive at the charge and voltage distribution pattern shown in the figure below. As it shows, we may describe the circuit by the following two relations (2.26), spelled out for two different capacitors of the system:

$$V_j - V_{j+1} = \frac{Q_{j+1}}{C_1}, \qquad Q_j - Q_{j+1} = C_2 V_j. \tag{*}$$

[3] It has a special name, the *Wheatstone bridge*, because with a galvanometer instead of the capacitor C_0, such circuits are broadly used for sensitive differential measurements.

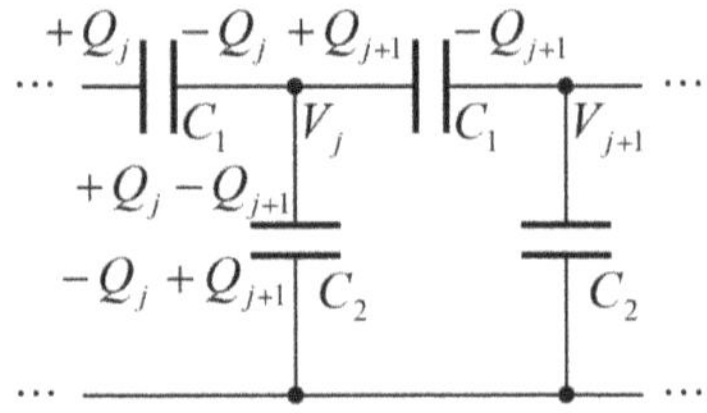

Due to the linearity of these relations, we may expect both variables V and Q to decay exponentially with the 'distance' j from the system's terminals:

$$V_j \propto Q_j \propto e^{-\lambda j}, \qquad \text{so that} \quad \frac{V_{j+1}}{V_j} = \frac{Q_{j+1}}{Q_j} = \alpha \equiv e^{-\lambda} < 1.$$

Indeed, using the last relations to eliminate V_{j+1} and Q_{j+1} from Eqs. (*), we obtain a system of two linear homogenous equations for the same two variables V_j and Q_j:

$$(1-\alpha)V_j = \frac{\alpha Q_j}{C_1}, \qquad (1-\alpha)Q_j = C_2 V_j. \qquad (**)$$

These equations are compatible (and hence the assumed solution in valid for all j) if the determinant of this system equals zero:

$$\begin{vmatrix} 1-\alpha & -\alpha/C_1 \\ -C_2 & 1-\alpha \end{vmatrix} = 0, \quad \text{i.e. if} \quad \alpha^2 - \left(2 + \frac{C_2}{C_1}\right)\alpha + 1 = 0.$$

The needed root $\alpha < 1$ of this quadratic equation[4] is

$$\alpha = 1 + \frac{C_2}{2C_1} - \left(\frac{C_2}{C_1} + \frac{C_2^2}{4C_1^2}\right)^{1/2}. \qquad (***)$$

Plugging it back to any of Eqs. (*), for example, the first one, we obtain the sought-for capacitance between the input terminals:

$$C \equiv \frac{Q_1}{V_0} = \frac{Q_{j+1}}{V_j} = \frac{\alpha Q_j}{V_j} = C_1(1-\alpha) = \left(C_1C_2 + \frac{C_2^2}{4}\right)^{1/2} - \frac{C_2}{2}. \qquad (****)$$

Let us consider what our results (***) and (****) yield in two limiting cases. If $C_1/C_2 \to 0$, then

$$\alpha \to \frac{C_2}{2C_1} \gg 1, \qquad C \to C_1.$$

---

[4] The second root $\alpha' = 1/\alpha > 1$ of the equation describes an exponential growth of V_j and Q_j with j, and would be relevant if our chain had a finite length.

This result means that the voltage between the input terminals does not propagate noticeably beyond the first capacitor C_1 of the chain, because the next, relatively large capacitance C_2 effectively 'grounds' the voltage.

In the opposite limit, $C_1/C_2 \to \infty$, Eq. (***) yields

$$\alpha \to 1 - \left(\frac{C_2}{C_1}\right)^{1/2} \to 1, \qquad \text{so that} \quad \lambda \equiv \ln\frac{1}{\alpha} \to \left(\frac{C_2}{C_1}\right)^{1/2} \ll 1.$$

According to the definition of λ, this means that the voltage applied to the chain's terminals penetrates deep into the chain, decaying at the 'distance' $N \equiv 1/\lambda \approx (C_1/C_2)^{1/2} \gg 1$. For the capacitance between the terminals, in this limit Eq. (****) gives

$$C \to (C_1C_2)^{1/2} \equiv NC_2.$$

The second form of this result may be interpreted by saying that the input voltage, penetrating by 'distance' N into the chain, 'sees' N capacitors C_2 connected in parallel.

***Problem* 2.5.** A system of two thin conducting plates is placed over a ground plane as shown in the figure below, where A_1 and A_2 are the areas of the indicated plate parts, while d' and d'' are the distances between them. Neglecting the fringe effects, calculate:

(i) the effective capacitance of each plate, and
(ii) their mutual capacitance.

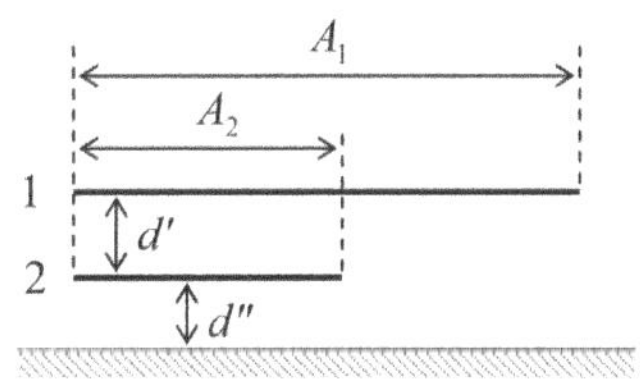

Solutions: Let us fix potentials ϕ_1 and ϕ_2 of the plates (relative to the ground). They create uniform, vertical electric fields in each of three regions of the system. Integrating Eq. (1.33) of the lecture notes, we may readily find that the field in the gap (of thickness d' and area A_2) between the plates is $E' = (\phi_1 - \phi_2)/d'$, that in the gap (of the same area A_2, but thickness d'') between plate 2 and the ground is $E'' = \phi_2/d''$, and finally in the clear gap (of thickness $d' + d''$ and area $A_1 - A_2$) between plate 1 and the ground the field is $E_0 = \phi_1/(d_1 + d_2)$. Now applying the Gauss law (1.16) to a flat pillbox drawn around each plate[5], we may calculate their total electric charges:

[5] Applying the Gauss law to plate 1, we should remember that the field above it has to be vanishingly small, because any appreciable field in that region (of volume $\sim A_1^{3/2}$) would be of prohibitive energy cost. Another explanation of the same fact is that all charges of plate 1 are strongly attracted to their opposite-sign images in the ground plane (see section 2.6 of the lecture notes), so that may reside only on the lower surface of the plate.

$$\begin{aligned}
Q_1 &= \varepsilon_0[(A_1 - A_2)E_0 + A_2E'] \\
&= \varepsilon_0\left[\frac{A_1 - A_2}{d' + d''}\phi_1 + \frac{A_2}{d'}(\phi_1 - \phi_2)\right] = \varepsilon_0\left(\frac{A_1 - A_2}{d' + d''} + \frac{A_2}{d'}\right)\phi_1 - \varepsilon_0\frac{A_2}{d'}\phi_2, \\
Q_2 &= \varepsilon_0 A_2(E'' - E') \\
&= \varepsilon_0 A_2\left(\frac{\phi_2}{d''} - \frac{\phi_1 - \phi_2}{d'}\right) = -\varepsilon_0\frac{A_2}{d'}\phi_1 + \varepsilon_0 A_2\left(\frac{1}{d'} + \frac{1}{d''}\right)\phi_2.
\end{aligned} \qquad (*)$$

These relations enable us to answer both questions.

(i) To calculate the effective capacitance of each plate, we should find its charge when the counterpart plate is grounded[6]. From Eqs. (*) we readily obtain

$$Q_1|_{\phi_2=0} = \varepsilon_0\left(\frac{A_1 - A_2}{d' + d''} + \frac{A_2}{d'}\right)\phi_1, \quad \text{i.e. } C_1^{\text{ef}} = \varepsilon_0\left(\frac{A_1 - A_2}{d' + d''} + \frac{A_2}{d'}\right), \qquad (**)$$

$$Q_2|_{\phi_1=0} = \varepsilon_0 A_2\left(\frac{1}{d'} + \frac{1}{d''}\right)\phi_2, \quad \text{i.e. } C_2^{\text{ef}} = \varepsilon_0 A_2\left(\frac{1}{d'} + \frac{1}{d''}\right). \qquad (***)$$

The first of these results, Eq. (**), may be readily understood, in the language of the equivalent circuit shown in the figure below, as the parallel connection of the 'direct' capacitance,

$$C_1 \equiv \varepsilon_0\frac{A_1 - A_2}{d' + d''},$$

of the clear gap between the plate 1 and the ground plane, with the direct capacitance between the plates:

$$C_{12} \equiv \varepsilon_0\frac{A_2}{d'},$$

while Eq. (***) describes the parallel connection of the same C_{12} with the direct capacitance,

$$C_2 \equiv \varepsilon_0\frac{A_2}{d''},$$

between plate 2 and the ground.

[6] The effective capacitances may be also calculated from the reciprocal capacitance matrix $p_{jj'}$ using Eq. (2.33) and its counterpart for the second plate, but in our case this approach is less straightforward.

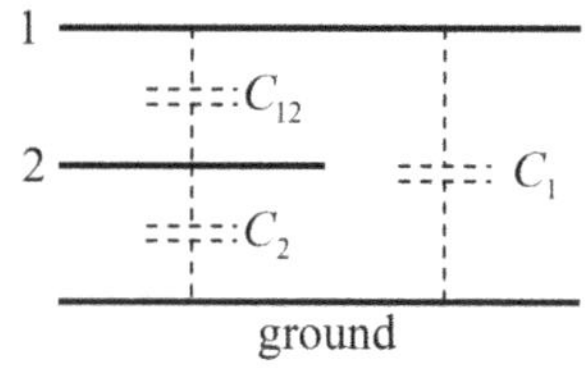

(ii) To calculate the mutual capacitance between the plates, we have to take $Q_2 = -Q_1 \equiv Q$, and calculate $V \equiv \phi_2 - \phi_1$ from Eq. (*). The result of this calculation may be represented in the form,

$$C \equiv \frac{Q}{V} = \varepsilon_0\left[\frac{A_2}{d'} + \left(\frac{d' + d''}{A_2 - A_1} + \frac{d'}{A_2}\right)^{-1}\right] \equiv C_{12} + \left(\frac{1}{C_1} + \frac{1}{C_2}\right)^{-1},$$

and understood from the same equivalent circuit (see the figure above) as the *parallel* connection of C_{12} with the *serial* connection of C_1 and C_2. This fact illustrates how convenient the language of equivalent circuits is. However, note again that this approach is valid only if the system under analysis may be partitioned into a set of independent field regions represented as lumped circuit elements—see figure 2.5 of the lecture notes and its discussion.

***Problem* 2.6.** A wide, thin plane film, carrying a uniform electric charge density σ, is placed inside a plane capacitor, whose plates are connected with a wire (see the figure below) and were initially electroneutral. Neglecting the fringe effects, calculate the surface charges of the plates, and the net force exerted on the film (per unit area).

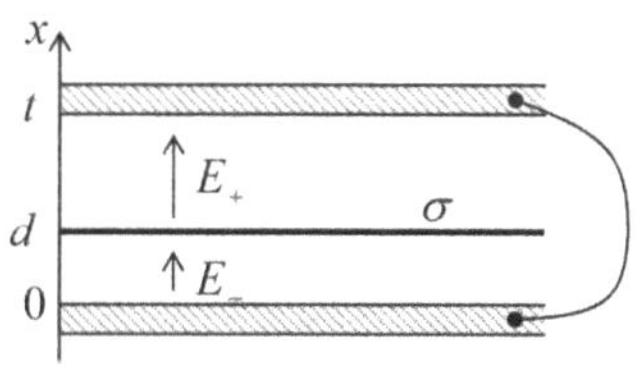

Solution: Let us pursue the most methodical (but not the shortest) way to solve the problem: solving the Poisson equation (1.41) for the electrostatic potential ϕ. Since, due to the problem's symmetry, ϕ may depend on just one Cartesian coordinate (say, x) perpendicular to the surfaces' and the film's planes, the equation is reduced to

$$\frac{d^2\phi}{dx^2} = -\frac{\sigma}{\varepsilon_0}\delta(x - d). \qquad (*)$$

Here the origin of x is taken on the bottom plate's surface, so that at the top plate's surface, $x = d$—see the figure above. Since the plates are connected by a conducting wire, in statics their potentials should be equal. Taking this value for zero, we may write the boundary conditions for the function $\phi(x)$ as

$$\phi(0) = \phi(d) = 0. \qquad (**)$$

The solution of this boundary problem is straightforward. Indeed, according to Eq. (*), in each of the two gaps between the film and the plates, $d^2\phi/d^2x = 0$, so that ϕ has to be a linear function of x:

$$\phi(x) = \begin{cases} c_- - E_- x, & \text{at } 0 < x < d, \\ c_+ - E_+ x, & \text{at } d < x < t, \end{cases}$$

where $E_\pm$ are the corresponding values of the electric field, $E = -d\phi/dx$ (see the figure above). Integrating Eq. (*) over an infinitesimal interval $[d - 0, d + 0]$, we obtain

$$\left.\frac{d\phi}{dx}\right|_{x=d+0} - \left.\frac{d\phi}{dx}\right|_{x=d-0} \equiv -E_+ + E_- = -\frac{\sigma}{\varepsilon_0}.$$

The boundary conditions (**) give two more equations for the constants $c_\pm$ and $E_\pm$:

$$c_- = 0, \quad c_+ - E_+ t = 0.$$

Finally, the potential has to be continuous, so that both expressions for $\phi(x)$ have to give the same value at $x = d$:

$$\phi(d) = c_- - E_- d = c_+ - E_+ d. \qquad (***)$$

Solving this simple system of four equations, we obtain[7]

$$E_- = -\frac{\sigma}{\varepsilon_0}\frac{t-d}{t}, \qquad E_+ = \frac{\sigma}{\varepsilon_0}\frac{d}{t}, \qquad c_- = 0, \qquad c_+ = \frac{\sigma}{\varepsilon_0}d. \qquad (****)$$

Due to the fundamental relation (2.3), the first two of these formulas immediately yield the surface charges of the capacitor's plates:

$$\sigma_{\text{bottom plate}} = \varepsilon_0 E_- = -\sigma\frac{t-d}{t}, \qquad \sigma_{\text{top plate}} = -\varepsilon_0 E_+ = -\sigma\frac{d}{t}.$$

(As a sanity check, $\sigma_{\text{bottom plate}} + \sigma_{\text{top plate}} + \sigma = 0$, as it should be for the initially electroneutral system.)

Next, plugging Eqs. (****) for the fields $E_\pm$ into Eq. (1.65) of the lecture notes, we may calculate the potential energy of the electric field in the system (per unit area):

$$\frac{U}{A} = \frac{\varepsilon_0}{2}\left[E_+^2(t-d) + E_-^2 d\right] = \frac{\sigma^2}{2\varepsilon_0}\left[\left(\frac{d}{t}\right)^2(t-d) + \left(\frac{t-d}{t}\right)^2 d\right] \equiv \frac{1}{2}\frac{\sigma^2}{\varepsilon_0}\frac{(t-d)\,d}{t}.$$

Now the force exerted on the film (also per unit area) may be found just as the (minus) gradient of its potential energy—see, e.g. Eq. (1.32). In our geometry, the only non-vanishing component of the gradient is vertical, so that $\mathbf{F} = F_x \mathbf{n}_x$, with[8]

$$\frac{F_x}{A} = -\frac{\partial(U/A)}{\partial d} = \frac{1}{2}\frac{\sigma^2}{\varepsilon_0}\frac{t-2d}{t}.$$

[7] Alternatively (and more easily), these relations for the fields $E_\pm$ may be obtained by applying the Gauss theorem to three pillboxes with various positions of their lids—see the derivation of Eq. (1.23) of the lecture notes. This exercise is highly recommended to the reader.

[8] A shorter, although less obvious, way to obtain the same result is to say that the effective electric field E_{ef} acting on the film (so that $F_x/A = E_{\text{ef}}\sigma$) is equal to the average of the fields on its two sides: $E_{\text{ef}} = (E_+ + E_-)/2$.

This result shows that the film is attracted to the nearest plate, so that at the middle between the plates, at $d = t/2$, the net force vanishes, i.e. the film is in equilibrium. Since at this point (and actually at any position d)

$$\frac{\partial^2 U}{\partial d^2} = -\frac{\sigma^2}{\varepsilon_0 t} < 0,$$

this equilibrium is unstable[9].

***Problem* 2.7.** Following up the discussion of two weakly coupled conducting spheres in section 2.2 of the lecture notes, find an approximate expression for the mutual capacitance (per unit length) between two very thin, parallel wires, both with a round cross-section, but each with its own radius. Compare the result with that for two small spheres, and interpret the difference.

Solution: In our comparison to the case of two small spheres, we should be more accurate with the selection of the additive constant in the electrostatic potential, because, as follows from Eq. (2.43) of the lecture notes (see also the model solution of problem 1.5), the potential of a single, long, uniformly charged wire is proportional to $\ln r$ and hence diverges at $r \to \infty$. As a result, the *self-capacitance* of a single wire of an infinite length is not well defined—see, e.g. the discussion following Eq. (2.49) of the lecture notes. This is why it is better to circumvent the calculation of the reciprocal capacitance matrix (2.23), and immediately proceed to the direct calculation of the *mutual* capacitance between two wires by considering them uniformly charged with an equal but opposite linear density λ—see the figure below.

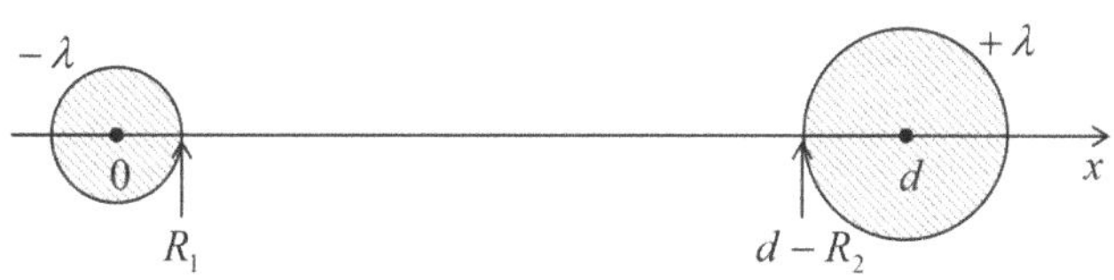

Just as in the two-spheres problem, calculating the electric field E_1 created by the left wire (charge $-\lambda$) alone, in the limit $d \gg R_{1,2}$ we may neglect the effects of the second wire. Applying the Gauss law to a round cylinder of radius r, coaxial with the first (left) wire, we obtain

$$E_{1r} = -\frac{\lambda}{2\pi\varepsilon_0 r},$$

so that the electric field along the line connecting the wire centers is

$$E_{1x} = -\frac{\lambda}{2\pi\varepsilon_0 x}.$$

[9] If the last point is not absolutely clear, please revisit the discussion of the fixed-point stability in *Part CM* section 3.2.

Similarly, with our choice of the origin of axis x (see the figure above), the field created on that line by the second wire is

$$E_{2x} = -\frac{\lambda}{2\pi\varepsilon_0(d-x)}.$$

(Here we took into account the opposite sign of the distributed charge.)

Now we can calculate voltage V between the wires by integration of the total field $E_x = E_{1x} + E_{2x}$ along axis x:

$$V \approx \phi|_{x=d-R_2} - \phi|_{x=R_1} = -\int_{R_1}^{d-R_2}(E_{1x}+E_{2x})dx = \frac{\lambda}{2\pi\varepsilon_0}\int_{R_1}^{d-R_2}\left(\frac{1}{x}+\frac{1}{d-x}\right)dx$$
$$= \frac{\lambda}{2\pi\varepsilon_0}\left(\ln\frac{d-R_2}{R_1} - \ln\frac{R_2}{d-R_1}\right) \approx \frac{\lambda}{2\pi\varepsilon_0}\ln\frac{d^2}{R_1R_2},$$

where the last step is also based on the strong inequality $d \gg R_{1,2}$. Hence the mutual capacitance per unit length

$$\frac{C}{l} \equiv \frac{\lambda}{V} \approx \frac{2\pi\varepsilon_0}{\ln(d^2/R_1R_2)} \equiv \frac{\pi\varepsilon_0}{\ln(d/R_{\text{mean}})}, \quad \text{where } R_{\text{mean}} \equiv (R_1R_2)^{1/2}. \qquad (*)$$

This result shows, that, in contrast with the similar problem for spheres, the mutual capacitance does depend on the distance between the conductors (decreasing with d), albeit weakly (logarithmically)—a result of the logarithmic divergence of the single wire's potential, mentioned above. Also note the close similarity between Eq. (*) and Eq. (2.49) of the lecture notes, for the coaxial cable geometry.

***Problem* 2.8.** Use the Gauss law to calculate the mutual capacitance of the following two-electrode systems, with the cross-section shown in figure 2.7 of the lecture notes (reproduced below):

(i) a conducting sphere inside a concentric spherical cavity in another conductor, and
(ii) a conducting cylinder inside a coaxial cavity in another conductor. (In this case, we speak about the capacitance per unit length).

Compare the results with those obtained in section 2.2 using the Laplace equation solution.

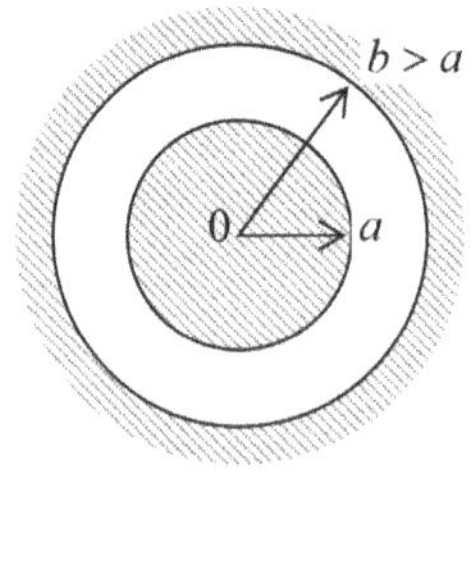

Solutions:

(i) Applying the Gauss law to a sphere of radius r in the range $a < r < b$, for the electric field $\mathbf{E} = E(r)\mathbf{n}_r$ we obtain

$$E(r) = \frac{Q}{4\pi\varepsilon_0 r^2},$$

where Q is the charge of the inner conductor. Integrating this result along the radius, we find that voltage $V \equiv \phi(b) - \phi(a)$ between the electrodes is

$$V \equiv \int_a^b E(r)dr = \frac{Q}{4\pi\varepsilon_0}\int_a^b \frac{dr}{r^2} = \frac{Q}{4\pi\varepsilon_0}\left(\frac{1}{a} - \frac{1}{b}\right),$$

so that the mutual capacitance

$$C \equiv \frac{Q}{V} = 4\pi\varepsilon_0\left(\frac{1}{a} - \frac{1}{b}\right)^{-1} = 4\pi\varepsilon_0\frac{ab}{b - a},$$

in agreement with Eq. (2.56) of the lecture notes.

(ii) An absolutely similar calculation, but applied to a cylinder, with charge $\lambda \equiv Q/l$ per unit length, yields

$$E(r) = \frac{\lambda}{2\pi\varepsilon_0 r}, \qquad \frac{C}{l} = \frac{2\pi\varepsilon_0}{\ln(b/a)},$$

in agreement with Eq. (2.49).

***Problem* 2.9.** Calculate the electrostatic potential distribution around two barely separated conductors in the form of coaxial, round cones (see the figure below), with voltage V between them. Compare the result with that of a similar 2D problem, with the cones replaced by plane-face wedges. Can you calculate the mutual capacitances between the conductors in these systems? If not, can you estimate them?

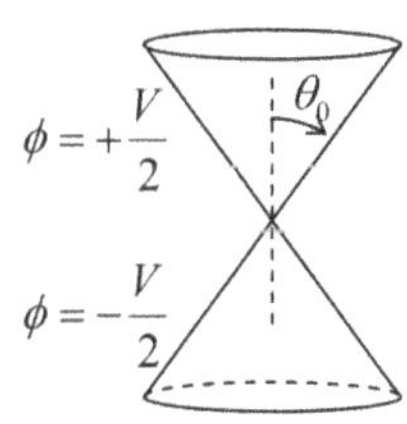

Solution: In the spherical coordinates $\{r, \theta, \varphi\}$, with axis z directed along the common axis of the cones, the conductor surfaces are described by relations $\theta = \theta_0$ and $\theta = \pi - \theta_0$, so that the values $\pm V/2$ of the electrostatic potential at these surfaces do not depend on coordinates r and φ. So it is natural to look for the solution of this boundary problem in the form $\phi = \phi(\theta)$. In this case, the Laplace equation at $\theta \neq 0, \pi$, is reduced to[10]

[10] See, e.g. Eq. (A.67) with $\partial/\partial r = \partial/\partial\varphi = 0$.

$$\frac{d}{d\theta}\left(\sin\theta\frac{d\phi}{d\theta}\right) = 0.$$

Integrating this equation sequentially twice, we obtain

$$\sin\theta\frac{d\phi}{d\theta} = c_1, \quad \text{i.e. } d\phi = c_1\frac{d\theta}{\sin\theta}, \quad \text{so that}$$

$$\phi(\theta) = c_1\int\frac{d\theta}{\sin\theta} = c_1\int\frac{\sin\theta d\theta}{\sin^2\theta} = -c_1\int\frac{d\xi}{1-\xi^2},$$

where $\xi \equiv \cos\theta$. The last integral is elementary[11], and we obtain

$$\phi(\theta) = \frac{c_1}{2}\ln\frac{1-\xi}{1+\xi} + c_2 \equiv \frac{c_1}{2}\ln\frac{1-\cos\theta}{1+\cos\theta} + c_2 \equiv c_1\ln\tan\frac{\theta}{2} + c_2.$$

With the proper choice of two integration constants $c_{1,2}$, we may indeed satisfy the boundary conditions on the surfaces of both conductors:

$$\phi(\theta_0) \equiv c_1\ln\tan\frac{\theta_0}{2} + c_2 = +\frac{V}{2},$$

$$\phi(\pi-\theta_0) \equiv c_1\ln\tan\frac{\pi-\theta_0}{2} + c_2$$

$$\equiv -c_1\ln\tan\frac{\theta_0}{2} + c_2 = -\frac{V}{2}.$$

Solving this system of two equations for the constants $c_{1,2}$, we finally obtain the potential distribution:

$$\phi(\theta) \equiv \frac{V}{2}\frac{\ln\tan(\theta/2)}{\ln\tan(\theta_0/2)}, \quad \text{for } \theta_0 \leqslant \theta \leqslant \pi-\theta_0, \qquad (*)$$

This is *a* solution, and hence *the* unique solution of this boundary problem.

The similar 2D problem, on the potential distribution between two conducting cylindrical wedges, with the cross-section shown in the figure below, may be solved using the polar coordinates. Using the same argumentation as in the above cone problem, we may expect the electrostatic potential to be the function of the polar angle alone. For the purposes of comparison with the just solved cone problem, I will denote this angle, referred to as the polar axis, by the same letter θ. Then the 2D Laplace equation is reduced to a very simple form[12],

$$\frac{d^2\phi}{d\theta^2} = 0,$$

[11] As a math reminder, it may be worked out by representing the function under the integral as the sum of two simple singularities ('poles'):

$$\frac{1}{1-\xi^2} \equiv \frac{1}{2(1-\xi)} + \frac{1}{2(1+\xi)}.$$

[12] See, e.g. Eq. (A.61) with $\partial/\partial\rho = \partial/\partial z = 0$, and $\varphi \to \theta$.

with the linear solution

$$\phi(\theta) = c_1\theta + c_2.$$

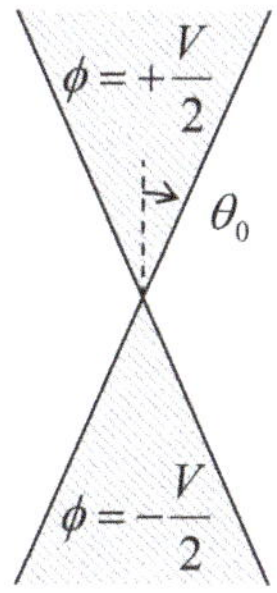

Selecting the constants $c_{1,2}$ to satisfy the same boundary conditions, $\phi(\theta_0) = +V/2$, $\phi(\pi - \theta_0) = -V/2$, we obtain

$$\phi(\theta) \equiv \frac{V}{2}\frac{\pi - 2\theta}{\pi - 2\theta_0}, \quad \text{for } \theta_0 \leqslant \theta \leqslant \pi - \theta_0. \tag{*}$$

The plots of the 3D function (2.19) and the 2D function (2.20), are shown in the figure below for two representative values of angle θ_0. They show that the 3D function's gradient (and hence the electric field) is more localized near the conducting electrodes, although for larger values of θ_0, approaching $\pi/2$, the difference is insignificant, i.e. the 3D function $\phi(\theta)$ is also virtually linear.

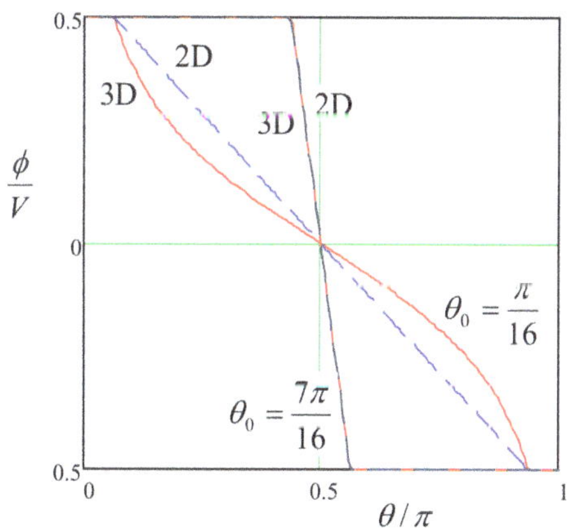

Finally, in order to calculate the mutual capacitance between the electrodes (for any geometry) we need to calculate the surface charge density on, say, the top electrode:

$$\sigma = \varepsilon_0 E_n\Big|_{\theta=\theta_0} = -\varepsilon_0\frac{\partial\phi}{\partial n}\Big|_{\theta=\theta_0},$$

where n is the linear coordinate normal to the surface. In both the 3D case and the 2D case, this coordinate is proportional to the distance from the system's center, so that σ is inversely proportional to this distance. For example, in the 3D case of two conducting cones[13],

$$\sigma = \frac{c}{r}, \quad \text{with} \quad c \equiv -\varepsilon_0 \frac{d\phi(\theta)}{d\theta}\bigg|_{\theta=\theta_0}.$$

Even without calculating the coefficient c, we see that the integral determining the total charge Q of the upper cone,

$$Q = \int_S \sigma d^2r = 2\pi \int_0^\infty \sigma r dr = 2\pi c \int_0^\infty dr,$$

strongly diverges at the upper limit. Physically, this means that in order to calculate a finite value of the system's mutual capacitance $C = Q/V$, we need to specify how exactly their spatial extent is limited.

The 2D system (of two conducting wedges) features a similar divergence not only at large, but also at small distances, although in both cases, rather mildly. Indeed, for it[14]

$$\sigma = \frac{c}{\rho}, \quad \text{with} \quad c \equiv -\varepsilon_0 \frac{d\varphi(\theta)}{d\theta}\bigg|_{\theta=\theta_0} = \varepsilon_0 \frac{1}{\pi - 2\theta_0},$$

where ρ is the distance in the cross-section's plane, so that the charge per unit length of the system,

$$\frac{Q}{l} = \frac{1}{l}\int_S \sigma d^2r = 2\int_0^\infty \sigma d\rho = 2c\int_0^\infty \frac{d\rho}{\rho} = \frac{2\varepsilon_0}{\pi - 2\theta_0}\int_0^\infty \frac{d\rho}{\rho} \to \infty.$$

So, the calculation a finite mutual capacitance between the wedges requires a specification of not only their exact shape at large distances, but also of the exact geometry of the gap between then. However, due to the weak (logarithmic) character of this singularity, a fair estimate of the capacitance may be given as

$$\frac{Q}{l} \sim \frac{2\varepsilon_0}{\pi - 2\theta_0} \ln \frac{\rho_{\max}}{\rho_{\min}},$$

where the two values of ρ are, respectively, the scales of the wedge size and the gap width.

***Problem* 2.10.** Calculate the mutual capacitance between two rectangular, plane electrodes of area $A = a \times l$, with a small angle $\varphi_0 \ll a/\rho_0$ between them—see the figure below.

---

[13] See, e.g. Eq. (A.66) with $\partial/\partial r = \partial/\partial\varphi = 0$.

[14] See, e.g. Eq. (A.60) with $\partial/\partial\rho = \partial/\partial z = 0$ and $\varphi \to \theta$.

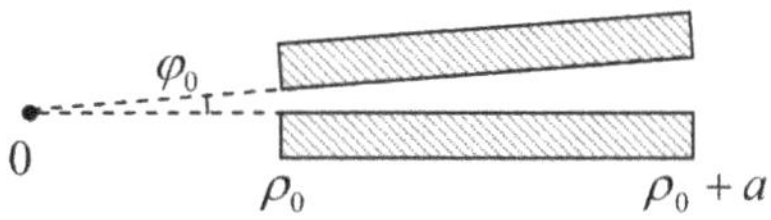

Solution: As was discussed in section 2.6 of the lecture notes, the variable separation in polar coordinates shows that the Laplace equation has a partial solution expressed by the second of Eqs. (2.110):

$$\phi = c_0 + s_0\varphi.$$

If the polar coordinate center coincides with the point 0 at which inner surface extensions cross (see the figure above), i.e. if the surfaces correspond to fixed values of angle φ (which differ from each other by φ_0), the boundary conditions on these surfaces are also satisfied by this solution, provided that $s_0\varphi_0 = V$, where V is the voltage between the electrodes, so that the following potential distribution,

$$\phi = V\frac{\varphi}{\varphi_0} + \text{const}, \qquad (*)$$

is the solution of our boundary problem. (Due to the condition $\varphi_0 \ll a/\rho_0$, the plates are very close, so that the fringe effects are minor and boundary conditions on outer surfaces are not important.)

Now the inner surface charge densities may be calculated from the basic equation (2.3):

$$\sigma = -\varepsilon_0\frac{\partial\phi}{\partial n}.$$

At the surfaces of constant φ, $\partial/\partial n = \mp(1/\rho)\partial/\partial\varphi$, where ρ is the distance from the point 0, and the upper/lower sign corresponds to the upper/lower plate, so that

$$\sigma_\pm = \pm\varepsilon_0\frac{1}{\rho}\frac{\partial\phi}{\partial\varphi} = \pm\varepsilon_0\frac{1}{\rho}\frac{V}{\varphi_0}.$$

Integrating these densities over the plate surface, we obtain their full charges.

$$Q_\pm = \int\sigma_\pm d^2r = l\int_{\rho_0}^{\rho_0+a}\sigma_\pm d\rho = \pm\varepsilon_0\frac{V}{\varphi_0}l\int_{\rho_0}^{\rho_0+a}\frac{d\rho}{\rho} = \pm\varepsilon_0\frac{V}{\varphi_0}l\ln\frac{\rho_0+a}{\rho_0},$$

so that the mutual capacitance

$$C \equiv \frac{Q_+}{V} = \varepsilon_0\frac{l}{\varphi_0}\ln\frac{\rho_0+a}{\rho_0} \equiv \varepsilon_0\frac{l}{\varphi_0}\ln\left(1+\frac{a}{\rho_0}\right). \qquad (**)$$

In the limit of a short capacitor, with $a \ll \rho_0$ (but still $\varphi_0 \ll a/\rho_0$), whose plates are essentially parallel, this expression may be approximated as

$$C \approx \varepsilon_0\frac{l}{\varphi_0}\frac{a}{\rho_0} = \varepsilon_0\frac{A}{d},$$

where $d \equiv \rho_0\varphi_0$ is the distance between the plates, thus reducing our result to the familiar Eq. (2.28).

On the other hand, at $\rho_0 \ll a$, the potential distribution (*), and hence Eq. (**), in the form

$$C \approx \varepsilon_0 \frac{l}{\varphi_0} \ln\left(\frac{a}{\rho_0}\right),$$

is valid for an arbitrary opening angle φ_0—cf. the solution of the previous problem for the 2D system.

***Problem* 2.11.** Using the results for a single thin round disk, obtained in section 2.4 of the lecture notes, consider a system of two such disks at a small distance $d \ll R$ from each other—see the figure below. In particular, calculate:

(i) the reciprocal capacitance matrix of the system,
(ii) the mutual capacitance between the disks,
(iii) the partial capacitance, and
(iv) the effective capacitance of one disk,

(all in the first non-vanishing approximation in $d/R \ll 1$). Compare the results (ii)–(iv) and interpret their similarities and differences.

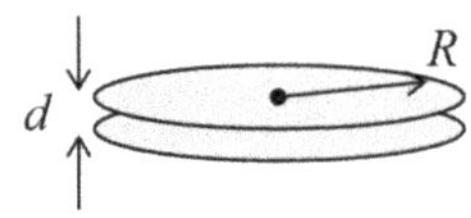

Solutions:

(i) Due to the symmetry of this system with respect to disk swap, and the general property expressed by Eq. (2.20) of the lecture notes, the general linear relations (2.19) are reduced to

$$\begin{aligned} \phi_1 &= p'Q_1 + pQ_2, \\ \phi_2 &= pQ_1 + p'Q_2. \end{aligned} \qquad (*)$$

At $d \ll R$, approximate expressions for the remaining two coefficients p' and p may be found from the two particular problems already solved in the lecture notes. Indeed, if the both disks carry equal charges, $Q_1 = Q_2 = Q/2$ (of the same sign), there is virtually no field inside the gap separating them, while the field outside is the almost same as that of a single disk charged by charge Q. According to Eq. (2.68), in this case

$$\phi_1 = \phi_2 = \frac{Q}{8\varepsilon_0 R}.$$

On the other hand, for this case Eqs. (*) give

$$\phi_1 = \phi_2 = \frac{p + p'}{2}Q.$$

Comparing these two expressions, we obtain

$$p + p' = \frac{1}{4\varepsilon_0 R}. \qquad (**)$$

In contrast, if the disk charges are equal but opposite, say $Q_1 = -Q_2 = Q$, then the system is nothing more than a plane capacitor (with a uniform field inside the gap and a negligible field outside it), so that according to Eq. (2.28),

$$\phi_1 = -\phi_2 = \frac{V}{2} = \frac{Q}{2C} = \frac{Qd}{2\pi\varepsilon_0 R^2}.$$

Comparing this result with the prediction of Eqs. (*) for this case,

$$\phi_1 = -\phi_2 = (p' - p)Q,$$

we obtain one more equation for p and p':

$$p' - p = \frac{d}{2\pi\varepsilon_0 R^2} \ll p + p'. \qquad (***)$$

Now solving the system of two linear equations (**) and (***), we obtain

$$p' = \frac{1}{8\varepsilon_0 R} + \frac{d}{4\pi\varepsilon_0 R^2} = \frac{1}{8\varepsilon_0 R}\left(1 + \frac{2d}{\pi R}\right),$$

$$p = \frac{1}{8\varepsilon_0 R} - \frac{d}{4\pi\varepsilon_0 R^2} = \frac{1}{8\varepsilon_0 R}\left(1 - \frac{2d}{\pi R}\right).$$

(ii) With these expressions, the mutual capacitance, defined by Eq. (2.26) with $p_1 = p_2 = p'$,

$$C \equiv \frac{1}{2(p' - p)},$$

becomes

$$C = \frac{\pi\varepsilon_0 R^2}{d}.$$

Note that this expression has essentially been found above—see Eq. (***).

(iii) The *partial* capacitances of the disks are defined as $C_1 = C_2 = 1/p_1 = 1/p_2 = 1/p'$, and in the first approximation are independent of $d \ll R$:

$$C_1 = C_2 \approx 8\varepsilon_0 R.$$

(iv) In contrast, the *effective* capacitances of the disks, which may be calculated from Eq. (2.34), are much larger and inversely proportional to the gap width:

$$C_{1,2}^{\text{ef}} \approx \frac{\pi\varepsilon_0 R^2}{d} = C_m \gg C_{1,2}.$$

This big difference between the partial and effective capacitance (typical for all strongly coupled systems) is very natural. If one disk is charged and the second one is not, the latter disk acquires almost the whole potential of the charge counterpart[15] and virtually does not affect the field distribution in space. However, if the second disk is grounded (i.e. set to $\phi = 0$), it kills the electric field everywhere apart from the narrow gap between the charged disk and the effective ground.

Problem* 2.12. Calculate the mutual capacitance (per unit length) between two cylindrical conductors forming a system with the cross-section shown in the figure below, in the limit $t \ll w \ll R$.

Hint: You may like to use the *elliptical* (not 'ellipsoidal'!) coordinates $\{\alpha, \beta\}$ defined by the following equation:

$$x + iy = c \cosh(\alpha + i\beta), \qquad (*)$$

with the appropriate choice of constant c. In these orthogonal 2D coordinates, the Laplace operator is very simple[16]:

$$\nabla_2^2 = \frac{1}{c^2(\cosh^2\alpha - \cos^2\beta)}\left(\frac{\partial^2}{\partial\alpha^2} + \frac{\partial^2}{\partial\beta^2}\right). \qquad (**)$$

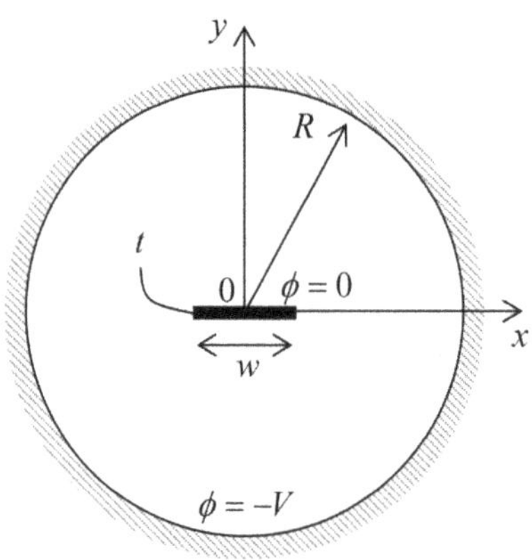

Solution: According to Eq. (*), the lines of constant α on the $[x, y]$ plane are ellipses with horizontal and vertical semi-axes $c\cosh\alpha$ and $c\sinh\alpha$, respectively. For $\alpha = 0$, the ellipse degenerates into a straight horizontal segment $-c < x < +c$, while for $\alpha \gg 1$, the ellipse is virtually a circle of radius $\rho = (c/2)\exp\alpha$. As a result, if we select the axes x and y as shown in the figure above, and take $c = w/2$, the boundary

[15] Electronic engineers call such electrodes 'floating'.

[16] This is not quite surprising, because Eq. (*) may also be considered as a conformal map $z = c \cosh w$, where $z = x + iy$, and $w = \alpha + i\beta$—see section 2.4 of the lecture notes.

conditions on the conductor surfaces may be satisfied, at $t \ll w \ll R$, by a potential distribution $\phi(\alpha)$, independent of β:

$$\phi(0) = 0, \quad \phi\left(\ln\frac{4R}{w}\right) = -V, \qquad (***)$$

where V is the voltage between the conductors. Hence the boundary problem may be satisfied by a function $\phi(\alpha)$, provided that the function satisfies the 1D Laplace equation following from Eq. (**):

$$\frac{d^2\phi}{d\alpha^2} = 0.$$

This equation shows that $\phi(\alpha)$ is just a linear function $c_1\alpha + c_2$. Selecting two constants $c_{1,2}$ to satisfy the boundary conditions (***), we obtain

$$\phi = -V\frac{\alpha}{\ln(4R/w)}.$$

To calculate the surface charge of the central conductor, using the general equation (2.3), we need to calculate the (normal) electric field on its surface:

$$E_n = -\left.\frac{\partial\phi}{\partial y}\right|_{\substack{y=+0\\ -w/2<x<w/2}} = \frac{V}{\ln(4R/w)(w/2)\sin\beta}\left.\frac{\partial\alpha}{\partial(\sinh\alpha)}\right|_{\alpha=0} = -\frac{V}{(w/2)\ln(4R/w)\sin\beta},$$

so that the two-surface charge density

$$\sigma = 2\varepsilon_0 E_n = 2\varepsilon_0\frac{V}{(w/2)\ln(4R/w)\sin\beta} = 2\varepsilon_0\frac{V}{(w/2)\ln(4R/w)[1-(2x/w)^2]^{1/2}}.$$

Integrating this distribution, we obtain the total charge (per unit length)

$$\frac{Q}{l} = \int_{-w/2}^{+w/2}\sigma dx = \frac{2\varepsilon_0 V}{(w/2)\ln(4R/w)}\int_{-w/2}^{+w/2}\frac{dx}{[1-(2x/w)^2]^{1/2}} = \frac{2\pi\varepsilon_0}{\ln(4R/w)}V,$$

so that the mutual capacitance per unit length is

$$\frac{C}{l} \equiv \frac{Q/l}{V} = \frac{2\pi\varepsilon_0}{\ln(4R/w)}.$$

Comparing this result with Eq. (2.49), we see that the only difference between the capacitance of this system and that of the round coaxial cable is a different factor under the logarithm.

***Problem* 2.13.** Formulate 2D electrostatic problems that can be solved using each of the following analytic functions of the complex variable $z \equiv x + iy$:

(i) $w = \ln z$,
(ii) $w = z^{1/2}$,

and solve these problems.

Solutions: For a 2D boundary problem of electrostatics to be readily solvable with a conformal mapping, the borders of the cross-section of conductors should coincide with lines of either the constant real or constant imaginary part of the complex function $\mathscr{w}$.

(i) Plugging the relations $\mathscr{w} = u + iv$ and $\mathscr{z} = x + iy$ into the given equality $\mathscr{w} = \ln z$, and equating the real and imaginary parts, we obtain

$$u = \ln(x^2 + y^2)^{1/2}, \quad v = \tan^{-1}\frac{y}{x}.$$

This means that on the $[x, y]$ plane the lines of constant u are concentric circles, while those of equal v are the straight lines extending from the origin to infinity. As a result, for this conformal map two examples of easily solvable boundary problems are:

Problem A: Calculate the field distribution between two concentric cylinders of radii a and b, held at different potentials (which is exactly the coaxial cable problem solved in section 2.3 of the lecture notes).

Problem B: The similar problem for two cylindrical wedges, barely separated at the origin—see the figure below.

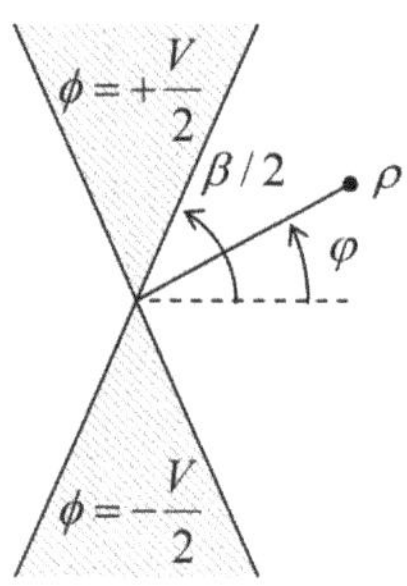

The solution of problem A is very easy: since the analytic function $u(x, y)$ satisfies the Laplace equation and the conductors' surfaces are equipotential, any line of constant u should be equipotential. Hence the Laplace equation is reduced to $d^2\phi/d^2u = 0$; its solution is a linear function of u, i.e.

$$\phi = c_1u + c_2 = c_1 \ln(x^2 + y^2)^{1/2} + c_2 = c_1 \ln \rho + c_2,$$

where constants $c_{1,2}$ may be readily calculated from the potentials fixed on the concentric conducting cylinders. This is of course just the result given by Eq. (2.43), which was obtained in section 2.3 of the lecture notes using a more general method.

For problem B, the similar argumentation yields

$$\phi = c_1v + c_2 = c_1 \tan^{-1}\frac{y}{x} + c_2 \equiv c_1\varphi + c_2,$$

which is valid in free space between the conducting wedges. For the electrode potentials shown in the figure above this gives

$$\phi = V\frac{\varphi}{\beta}.$$

Note that according to this formula, the electric field lines are just concentric circles and the electric field on the conductor surface,

$$|E_n| = \left| \frac{\partial\phi}{\partial(\rho\varphi)} \right|_{\rho=\text{const}} = \frac{V}{\rho\beta},$$

has a strong (non-integrable) divergence at $\rho \to 0$, different from the one discussed in section 2.6 for a single conducting corner/wedge—see Eq. (2.123). The difference is due to the fact that in the latter case the field was due to some remote sources, rather than the oppositely biased counterpart wedge.

(ii) In the same fashion, separating the real and imaginary parts of the function $w = z^{1/2}$, we obtain

$$u = \pm\left\{\frac{1}{2}\left[(x^2 + y^2)^{1/2} + x\right]\right\}^{1/2}, \quad v = \pm\left\{\frac{1}{2}\left[(x^2 + y^2)^{1/2} - x\right]\right\}^{1/2}, \qquad (*)$$

where the signs in both formulas should be changed simultaneously[17]. As the second of these relations shows, the lines of constant v on the $[x, y]$ plane are parabolas:

$$x = \frac{y^2}{4v^2} - v^2,$$

in the limit $v \to 0$ giving a straight segment $0 < x < +\infty$—see the figure below. (As the first of Eqs. (*) shows, the lines of constant u are similar, just with the opposite bending of the parabolas.)

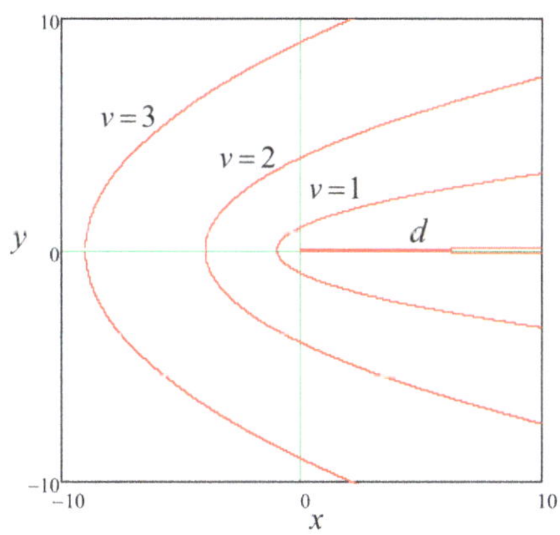

This mapping may be used, for example, to find the field distribution around the edge of a thin conducting sheet in the form of half-plane $x > 0$. For this problem, $d^2\phi/dv^2 = 0$, with the evident solution

[17] As is evident from figure 2.9b of the lecture notes, which shows the reciprocal map $z = w^{1/2}$, in our current problem each point $\{x, y\}$ is mapped on two points $\{u, v\}$.

$$\phi = c_1 v + c_2 \equiv c_1 \left\{ \frac{1}{2}\left[(x^2 + y^2)^{1/2} - x \right] \right\}^{1/2} + c_2.$$

The condition of a fixed potential of the conducting half-plane (say, $\phi = 0$), corresponding to $v = 0$, gives $c_2 = 0$, so that finally

$$\phi = c_1 \left\{ \frac{1}{2}\left[(x^2 + y^2)^{1/2} - x \right] \right\}^{1/2}, \qquad (**)$$

where the constant c_1 is determined by the potential at the (distant) external electrodes creating the electric field. This means that the figure above shows the equipotential lines, corresponding to $\phi = c_1 v$.

Near the surface of the half-plane conductor ($x > 0$, $|y| \ll x$), Eq. (**) yields

$$\phi = c_1 \frac{|y|}{2x},$$

so that both the electric field at the sheet surfaces,

$$E_n = -\left.\frac{\partial \phi}{\partial y}\right|_{y=0} = \pm \frac{c_1}{2x},$$

and the surface charge density $\sigma = \varepsilon_0 E_n$ have the same strong, non-integrable divergences at the edge, as in problem B of task (i), discussed above.

***Problem* 2.14.** On each side of a cylindrical volume with a rectangular cross-section $a \times b$, with no electric charges inside it, the electric field is uniform, normal to the side's plane, and opposite to that on the opposite side—see the figure below. Calculate the distribution of the electric potential inside the volume, provided that the field magnitude E on the vertical sides is known. Suggest a practicable method to implement such potential distribution.

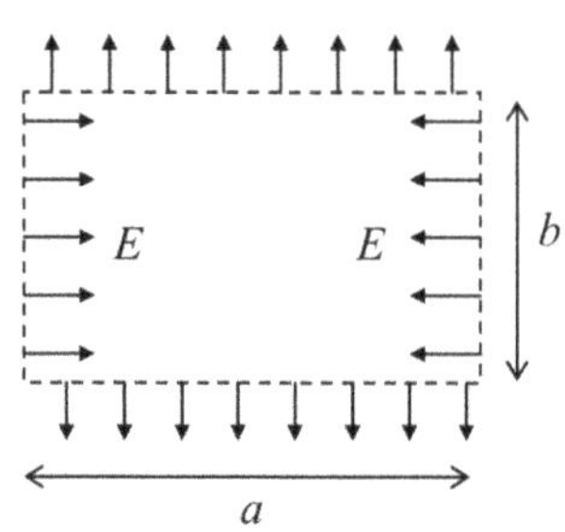

Solution: First of all, let us calculate the magnitude E' of the uniform field on the opposite walls of the volume. Since the volume is free of charge, the Gauss law (1.16), applied to its unit length, yields

$$-2Eb + 2E'a = 0, \quad \text{i.e. } E' = E\frac{b}{a}.$$

Hence, introducing suitable Cartesian coordinates (see the figure below), we obtain the following 2D boundary problem for the distribution of the electrostatic potential $\phi(x, y)$:

$$\left(\frac{\partial^2}{\partial x^2} + \frac{\partial^2}{\partial y^2}\right)\phi = 0, \quad \text{for } |x| < \frac{a}{2}, \ |y| < \frac{b}{2}; \tag{*}$$

$$\left.\frac{\partial \phi}{\partial x}\right|_{x=\pm a/2,\ |y|<b/2} = \mp E, \quad \left.\frac{\partial \phi}{\partial y}\right|_{y=\pm b/2,\ |x|<a/2} = \pm E\frac{b}{a}. \tag{**}$$

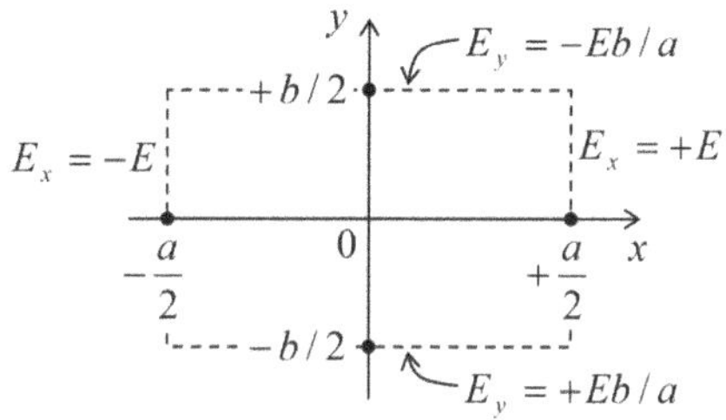

There are two ways to solve this problem. The regular (I would say 'pedestrian') way is to use the standard variable separation method[18], with due respect to the evident symmetry of our system (in the above coordinates, taking $\phi(0, 0) = 0$, we have $\phi(-x, y) = \phi(x, -y) = \phi(x, y)$, so that $\partial\phi/\partial x = 0$ at $y = 0$, and $\partial\phi/\partial y = 0$ at $x = 0$), and then use the summation formula[19]

$$\sum_{n=1}^{\infty} \frac{(-1)^n}{n^2} \cos \pi n\xi = \frac{\pi^2}{4}\left(\xi^2 - \frac{1}{3}\right), \quad \text{for } |\xi| \leqslant 1,$$

to bring the result from an infinite sum to the following very simple form:

$$\phi = E\frac{x^2 - y^2}{a}. \tag{***}$$

Although this exercise is highly recommended to the reader, he or she has probably noticed that Eq. (***) was already obtained in section 2.4 of the lecture notes—see Eq. (2.76) with $V = 2Ea$. This means, in particular, that such a potential distribution may be created in practice, for example, using the quadrupole electrostatic lens shown in figure 2.9a.

***Problem* 2.15.** Complete the solution of the problem shown in figure 2.12 of the lecture notes (reproduced below), by calculating the distribution of the surface

[18] See section 2.5 of the lecture notes.

[19] It may be readily proved by the Fourier expansion of its right-hand side over the interval $-1 \leqslant \xi \leqslant +1$, repeated periodically over all axis ξ.

charge of the semi-planes. Can you calculate the mutual capacitance between the plates (per unit length)? If not, can you estimate it?

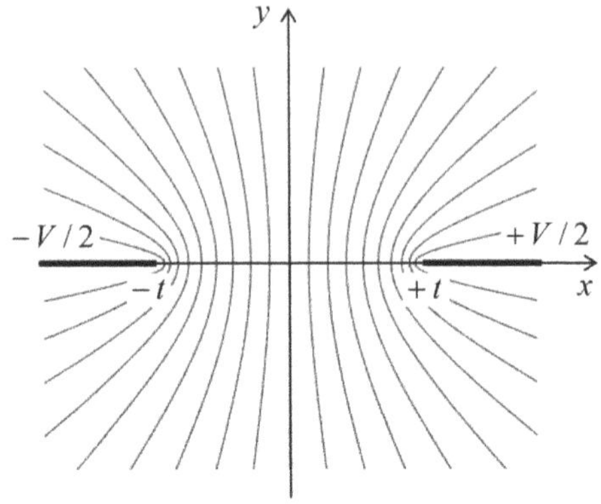

Solution: Near the electrode surface ($y \to 0$, $x^2 > t^2$), Eq. (2.83) of the lecture notes yields

$$\phi \to \frac{V}{\pi}\sin^{-1}\left\{1 - \frac{y^2}{4x}\left[\frac{1}{x+t} + \frac{1}{x-t}\right]\right\} = \frac{V}{\pi}\sin^{-1}\left\{1 - \frac{y^2}{2(x^2 - t^2)}\right\}.$$

From here, the only non-vanishing (vertical) component of the field is

$$E_n|_{y=\pm 0} = \mp\frac{\partial\phi}{\partial y}\bigg|_{y=\pm 0} = \frac{V}{\pi(x^2 - t^2)^{1/2}},$$

so that, according to Eq. (1.24) of the lecture notes, the full (double-side) surface charge density is

$$\sigma = 2\varepsilon_0 E_{n|y=\pm 0}\frac{2\varepsilon_0 V}{\pi(x^2 - t^2)^{1/2}}.$$

The divergence at $x \to t$ is not too strong (it is integrable), just as the thin-disk problem solved in section 2.4, but in contrast to that problem, the total charge of each semi-plane (per unit length) is still infinite, because the corresponding integral,

$$\frac{Q}{l} = \int_t^\infty \sigma dx = \frac{2\varepsilon_0 V}{\pi}\int_t^\infty \frac{dx}{(x^2 - t^2)^{1/2}} = \frac{2\varepsilon_0 V}{\pi}\ln\left[x + (x^2 - t^2)^{1/2}\right]_{x=t}^{x\to\infty}, \qquad (*)$$

diverges at the *upper* limit, albeit very slowly (logarithmically). Hence in order to calculate the mutual capacitance per unit length, $C/l \equiv Q/lV$, one needs to make one more step from this idealized model to reality—for example, take into account a large but finite width w of each semi-plate. However, even without solving the resulting problem exactly, we can use Eq. (*) to make a semi-quantitative prediction of the result by truncating the integral at $x = w \gg t$:

$$\frac{C}{l} = \frac{Q}{lV} \approx \frac{2\varepsilon_0}{\pi}\ln\left[x + (x^2 - t^2)^{1/2}\right]_{x=t}^{x=w\gg t} \approx \frac{2\varepsilon_0}{\pi}\ln\frac{2w}{t}.$$

Problem* 2.16. A straight, long, thin, round-cylindrical metallic pipe has been cut, along its axis, into two equal parts—see the figure below.

(i) Use the conformal mapping method to calculate the distribution of the electrostatic potential, created by voltage V applied between the two parts, both outside and inside the pipe, and of the surface charge.
(ii) Calculate the mutual capacitance between the pipe's halves (per unit length), taking into account a small width $2t \ll R$ of the cut.

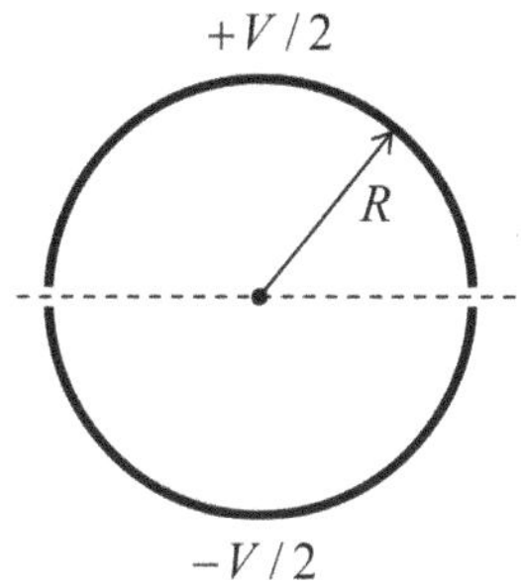

Hints: In task (i), you may like to use the following complex function:

$$w = \ln\left(\frac{R + z}{R - z}\right),$$

while in task (ii) it is advisable to use the solution of the previous problem.

Solutions:

(i) Let us consider the plane of pipe's cross-section as the plane of complex variable

$$z = x + iy = \rho e^{i\varphi},$$

where $\{x, y\}$ are the Cartesian coordinates, and $\{\rho, \varphi\}$ are the polar coordinates, introduced as shown in the figure below (so that, as usual, $x = \rho\cos\varphi$ and $y = \rho\sin\varphi$). Let us analyze the conformal mapping performed by the function given in the hints. For the points on the pipe's walls ($\rho = R$, i.e. $z = Re^{i\varphi}$),

$$\begin{aligned} w &= \ln\frac{R + Re^{i\varphi}}{R - Re^{i\varphi}} = \ln\frac{1 + e^{i\varphi} + 1}{1 - e^{i\varphi}} \\ &= \ln\frac{e^{-i\varphi/2} + e^{i\varphi/2}}{e^{-i\varphi/2} - e^{-i\varphi/2}} = \ln\frac{2\cos(\varphi/2)}{-2i\sin(\varphi/2)} = -\ln\left(-i\tan\frac{\varphi}{2}\right) \end{aligned}$$

so that, using the standard notation $w = u + iv$ (where u and v are real), we obtain

$$e^{-w} \equiv e^{-u}e^{-iv} \equiv e^{-u}(\cos v - i\sin v) = -i\tan\frac{\varphi}{2}.$$

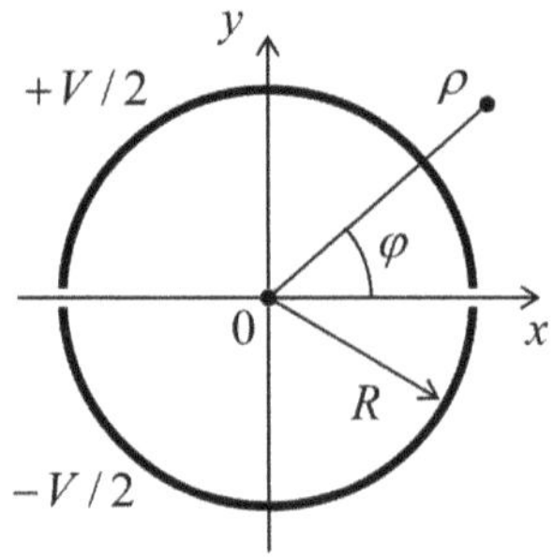

Since the last form of this equality is purely imaginary, so should be the adjacent expression, which is only possible if $\cos v = 0$, i.e. at the horizontal lines $\sin v = \pm 1$, e.g. $v = \pm\pi/2$.[20] On these two lines, the above relation yields, respectively,

$$\pm e^{-u} = \tan\frac{\varphi}{2}.$$

Since u is by definition real, so that e^{-u} may be only positive, as φ is increased from 0 to $+\pi$ (i.e. as we move, on plane z, along the pipe's top wall), the corresponding point on the plane w moves from the right to left along the horizontal line $v = +\pi/2$. Similarly, if φ is decreased from 0 to $-\pi$, the point in plane w moves in the same horizontal direction along the line $v = -\pi/2$—see the figure below.

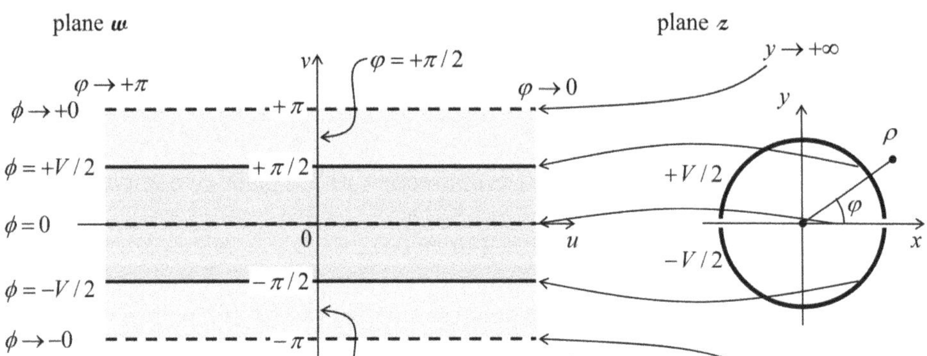

The absolutely similar analyses show that function w also maps (see the figure above):

- the horizontal axis of the z-plane ($y = 0$) onto that ($v = 0$) of the plane w,
- the vertical half-axes of the axis y (with $\varphi = \pm\pi/2$) onto the respective segments of the axis u, $0 \leqslant v \leqslant +\pi$ and $-\pi \leqslant v \leqslant 0$, and

[20] Due to the 2π-periodicity of z as a function of φ, and function $w(z)$ being logarithmic, the whole mapping has 2π-periodicity along axis v, so that we may limit our analysis to segment $-\pi \leqslant v \leqslant +\pi$, shown in the figure below.

- any horizontal segment of plane z, with $|y| \gg |x \pm R|$ onto a horizontal segment on plane w, with v approaching $\pi\text{sgn}(y)$. (Indeed, if $y \to \pm\infty$, then

$$e^{w} \equiv e^{u}(\cos v + i \sin v) = \frac{R + x + iy}{R - x - iy} \equiv -\frac{1 + (x + R)/iy}{1 + (x - R)/iy} \to -1 + 2i\frac{R}{y},$$

so that $\cos v \to -1$, while $\text{sgn}v = \text{sgn}y$.)

To summarize, the function w maps the pipe's interior on the horizontal strip $-\pi/2 \leqslant v \leqslant +\pi/2$ (shaded darker in the figure above), and two parts of its exterior, separated by the axis x, onto two strips: $+\pi/2 \leqslant v \leqslant +\pi$ and $-\pi \leqslant v \leqslant -\pi/2$ (shaded lighter). Since all the boundary conditions imposed on the electrostatic potential ϕ in our problem (including the requirement $\phi \to 0$ at $\rho \to 0$) are independent of u (see the labels on the left side of the figure above), and the function w is analytical, a linear function $\phi(v) = c_1 v + c_2$ within each strip satisfies the Laplace equation and, at the appropriate choice of coefficients $c_{1,2}$, the boundary conditions as well, so this is the (unique) solution of our boundary problem. The proper values of the coefficients are different for each of the three shown strips, and with proper selection we obtain

$$\phi = \frac{V}{\pi} \times \begin{cases} (+\pi - v), & \text{for } +\pi/2 < v < +\pi, \\ v, & \text{for } -\pi/2 < v < +\pi/2, \\ (-\pi - v), & \text{for } -\pi < v < -\pi/2, \end{cases} \qquad (*)$$

(It is straightforward to use the left panel of the figure above to verify that this solution indeed satisfies all boundary conditions.)

What remains is to express v via the coordinates on the plane z . For that we may use the fact that if $w = \ln \mathcal{F} \equiv \ln(|\mathcal{F}|\exp\{i\arg\mathcal{F}\}) \equiv \ln(|\mathcal{F}|) + \ln(\exp\{i\arg\mathcal{F}\}) = \ln(|\mathcal{F}|) + i \arg \mathcal{F}$, then $v \equiv \text{Im}w = \arg\mathcal{F}$. In our case, we can use this identity with $\mathcal{F} = (R + z)/(R - z)$, so that

$$\begin{aligned} v &= \arg\frac{R + x + iy}{R - x - iy} = \arg\frac{(R + x + iy)(R - x + iy)}{(R - x - iy)(R - x + iy)} \\ &= \arg\frac{R^2 - x^2 - y^2 + 2iRy}{(R - x)^2 + y^2}. \end{aligned}$$

The denominator of the last fraction is, by construction, real, and does not affect its argument, so that

$$\tan v = \frac{2Ry}{R^2 - x^2 - y^2} \equiv \frac{2Rr \sin \varphi}{R^2 - \rho^2},$$

and taking into account the π-periodicity of the tangent function, all Eqs. (*) may be rewritten in a simple single form

$$\phi(\rho, \varphi) = \frac{V}{\pi} \tan^{-1} \frac{2Rr \sin \varphi}{|R^2 - \rho^2|}. \qquad (**)$$

This result is plotted in the figure below as a function of φ for several values of ρ, both larger than R (red lines) and smaller than R (blue lines). These plots show, in particular[21], that small deviations from the pipe's surface to either side lead to similar deviations of the potential from its piece-constant form at the surface. This fact implies that the magnitude of the electric field on the pipe's surface, $|E_n| = \partial\phi/\partial\rho$ at $\rho = R$, and hence the surface charge density $\sigma = \varepsilon_0 E_n$ is the same, for the same φ, on both (the external and the internal) surfaces of the pipe. Indeed, a direct differentiation of Eqs. (**) gives the same result:

$$\sigma(\varphi) = \frac{\varepsilon_0 V}{\pi R \sin\varphi}. \qquad (***)$$

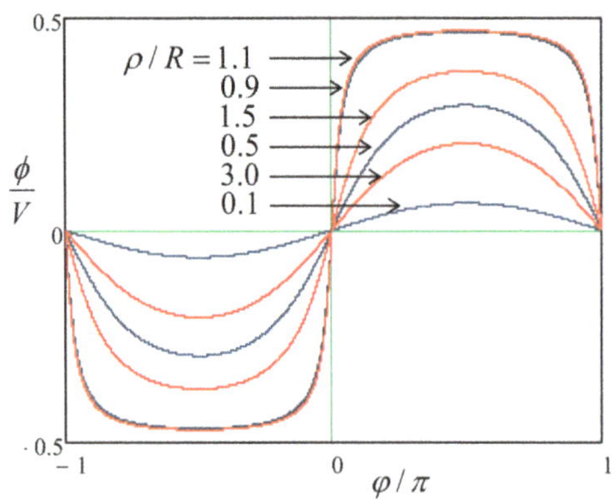

(ii) The charge density calculated above has strong (non-integrable) divergences at $\varphi \to 0$ and $\varphi \to \pm\pi$, i.e. near the gap between the two halves of the pipe; for example, at $\varphi \to 0$,

$$\sigma(\varphi) \to \frac{\varepsilon_0 V}{\pi R\varphi} = \frac{\varepsilon_0 V}{\pi\, y}, \qquad \text{for } |y| \ll R,$$

so that the total charge Q of each half of the pipe, and hence the capacitance $C = Q/V$ of the system (even per unit length along axis z) are formally infinite. This is an artifact of neglecting the cut's width $2t$ in the above calculation. In the limit $2t \ll R$, finite Q and C may be found by comparing the above solution with that of the previous problem (on the field distribution between two planar, thin electrodes separated by a cut of width $2t$), which overlap in a vicinity of the gap.

Indeed, with the notation replacement $x \to y$, the solution of the previous problem gave the following result for the single-side charge density:

$$\sigma(\varphi) = \operatorname{sgn} y \times \frac{\varepsilon_0 V}{\pi (y^2 - t^2)^{1/2}}, \qquad \text{for } |y| > t, \qquad (****)$$

which gives the same result as Eq. (***) within a relatively broad range of angles:

[21] For an analysis of the potential's behavior far beyond pipe's walls, see the model solution of the next problem.

$$\frac{t}{R} \ll \varphi \ll 1.$$

Hence Eq. (****) may be used instead of Eq. (***) at $|y| = |R\varphi| \ll R$. As a result, the total charge of the upper half of the pipe (per unit length), may be calculated by the separation of the corresponding integral into two parts[22]:

$$\frac{Q}{l} = 2\int_{\sin\varphi>0} \sigma(\varphi)Rd\varphi = 4R\int_{t/R}^{\pi/2} \sigma(\varphi)\, d\varphi = 4R\left(\int_{t/R}^{\varphi_0} \sigma(\varphi)\, d\varphi + \int_{\varphi_0}^{\pi/2} \sigma(\varphi)\, d\varphi\right),$$

where the separation boundary φ_0 is arbitrary within that solution overlap region. The inequality $t/R \ll \varphi_0$ allows us to use Eq. (****) in the first integral, while the inequality $\varphi_0 \ll 1$ makes the approximation $y = R\varphi$ valid in it. On the other hand, due to the former inequality we may use the uncorrected expression (***) for the charge density in the second integral:

$$\frac{Q}{l} = 4R\frac{\varepsilon_0 V}{\pi}\left(\int_{t/R}^{\varphi_0} \frac{d\varphi}{(R^2\varphi^2 - t^2)^{1/2}} + \frac{1}{R}\int_{\varphi_0}^{\pi/2} \frac{d\varphi}{\sin\varphi}\right).$$

The remaining calculations are simple. Using variable replacements $\varphi \equiv (t/R)\cosh\xi$ (so that $d\varphi = (t/R)\sinh\xi d\xi$) in the first integral, and $\zeta \equiv \cos\varphi$ (so that $d\varphi/\sin\varphi = -d\zeta/\sin^2\varphi = -d\zeta/(1 - \zeta^2)$) in the second one, we obtain

$$\frac{Q}{l} = \frac{4\varepsilon_0 V}{\pi}\left(\int_{\varphi=t/R}^{\varphi=\varphi_0} \frac{\sinh\xi d\xi}{(\cosh^2\xi - 1)^{1/2}} - \int_{\varphi=\varphi_0}^{\varphi=\pi/2} \frac{d\zeta}{1-\zeta^2}\right) = \frac{4\varepsilon_0 V}{\pi R}\left(\xi\Big|_{\varphi=t/R}^{\varphi=\varphi_0} - \frac{1}{2}\ln\frac{1+\zeta}{1-\zeta}\Big|_{\varphi=\varphi_0}^{\pi/2}\right)$$

$$= \frac{4\varepsilon_0 V}{\pi}\left(\cosh^{-1}\frac{R\varphi_0}{t} - \frac{1}{2}\ln\frac{1+\cos\varphi_0}{1-\cos\varphi_0}\right).$$

Due to the condition $\varphi_0 \gg t/R$, $\cosh^{-1}(R\varphi_0/t)$ may be approximated as $\ln(2R\varphi_0/t) \equiv \ln(4R/t) + \ln(\varphi_0/2)$, while due to the condition $\varphi_0 \ll 1$, the logarithm in the last term may be reduced to $\ln(4/\varphi_0^2) \equiv -2\ln(\varphi_0/2)$. As a result, the terms with φ_0 cancel (as they should, due to the relatively arbitrary choice of this separation boundary), and we finally obtain

$$\frac{Q}{l} = \frac{4\varepsilon_0 V}{\pi}\ln\frac{4R}{t}, \quad \text{i.e.} \quad \frac{C}{l} = \frac{4\varepsilon_0}{\pi}\ln\frac{4R}{t}.$$

My earnest advice to the reader is to thoroughly review the used trick of matching ('stitching') the solutions of two different problems that share a broad overlap region, because it is used in many real-life problems of not only physics, but also other qualitative sciences and engineering.

***Problem* 2.17.** Solve task (i) of the previous problem using the variable separation method, and compare the results.

---

[22] The front factor of 2 takes into account the equal densities $\sigma(\varphi)$ of the charge at the inner and outer surfaces.

Solution: Introducing the polar coordinates as shown in the figure below (i.e. just as in the model solution of the previous problem), and guided by Eq. (2.112) of the lecture notes and the symmetry of the problem, we may look for the solution of the outer problem (for $\rho \geqslant R$) in the form

$$\phi(\rho, \varphi) = \sum_{n=1}^{\infty} \frac{b_n}{\rho^n} \sin n\varphi.$$

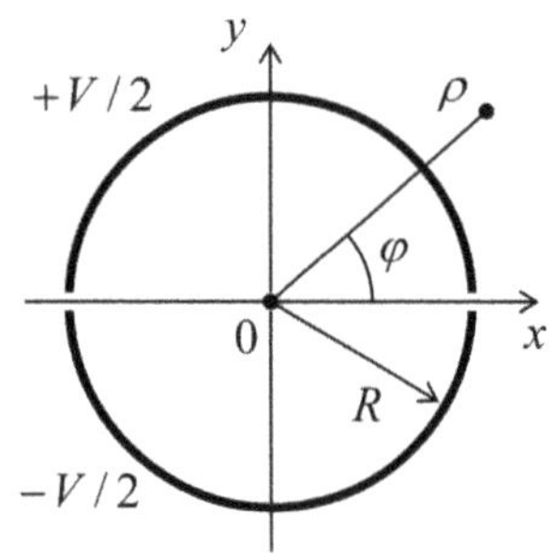

(Indeed, the inconsequential coefficient a_0 in Eq. (2.112) may be taken equal to zero for convenience. The logarithmic term in that expression, proportional to b_0, would describe the effect of the net charge of the pipe, which we evidently do not have. The terms proportional to a_n with $n > 0$ would diverge at $\rho \to \infty$ and thus should be excluded, and terms proportional to $\cos n\varphi$ would give even-function contributions into the potential, which in our problem should be an odd function of φ.)

The boundary conditions at $\rho = R$ yield the following system of equations for coefficients b_n:

$$\sum_{n=1}^{\infty} \frac{b_n}{R^n} \sin n\varphi = \phi(R, \varphi) = \frac{V}{2} \times \begin{cases} +1, & \text{for } 0 < \varphi < \pi, \\ -1, & \text{for } \pi < \varphi < 2\pi. \end{cases}$$

Solving this system in the way usual for the Fourier series, i.e. by multiplying both parts of the equation by $\sin n'\varphi$ with an arbitrary $n' > 0$, integrating them over any 2π-long interval (for example, $0 \leqslant \varphi \leqslant 2\pi$), and using the orthogonality of functions $\sin n\varphi$ with different $n > 0$:

$$\int_0^{2\pi} \sin n'\varphi \sin n\varphi \, d\varphi = \pi\delta_{nn'}, \quad \text{for } n \neq 0,$$

we obtain

$$b_n = \frac{V}{\pi} R^n \int_0^{\pi} \sin n\varphi \, d\varphi = \frac{V}{\pi n} R^n (1 - \cos n\pi) = \frac{V}{\pi n} R^n \times \begin{cases} 0, & \text{for } n \text{ even}, \\ 2, & \text{for } n \text{ odd}, \end{cases}$$

so that denoting the odd values $n = 1, 3, \ldots$ as $2m - 1$ (with $m = 1, 2, \ldots$), we finally obtain

$$\phi(\rho, \varphi) = \frac{2V}{\pi}\sum_{m=1}^{\infty}\frac{1}{2m-1}\left(\frac{R}{\rho}\right)^{2m-1}\sin(2m-1)\varphi, \quad \text{for } \rho \geqslant R. \tag{*}$$

The solution of the internal problem (for $\rho \leqslant R$) is absolutely similar, except that instead of terms proportional to b_n/ρ^n in Eq. (2.112), we need to keep terms $a_n\rho^n$, which do not diverge at $\rho \to 0$, so that

$$\phi(\rho, \varphi) = \sum_{n=1}^{\infty} a_n\rho^n \sin n\varphi.$$

Since the boundary conditions at $\rho = R$ are similar for both problems, the above result for b_n is valid for normalized coefficients a_n as well (with the replacement $b_n/R^n \to a_nR^n$), so that the final result is very similar to Eq. (*):

$$\phi(\rho, \varphi) = \frac{2V}{\pi}\sum_{m=1}^{\infty}\frac{1}{2m-1}\left(\frac{\rho}{R}\right)^{2m-1}\sin(2m-1)\varphi, \quad \text{for } \rho \leqslant R. \tag{**}$$

Numerical plots of Eqs. (*) and (**) show that they yield results identical to the explicit expressions obtained in the solution of the previous problem[23]. Moreover, these series are equally (if not more) convenient for the analysis of potential's behavior far from the pipe's walls. Indeed, as Eqs. (*) and (**) show, at both $\rho \gg R$ and $\rho \ll R$, the series terms decay fast with n, so that the potential is determined by their first terms, with $m = 1$, i.e. is proportional to $\sin\varphi$. In the former case,

$$\phi(\rho, \varphi) = \frac{2VR}{\pi\rho}\sin\varphi, \quad \text{for } \rho \gg R,$$

the potential corresponds to the electric field of magnitude $E \propto 1/\rho^2$. As will be discussed in chapter 3, such field may be interpreted as the one created by a straight line of similar electric dipoles. In contrast, near the cylinder's axis the sinusoidal dependence of the potential on angle φ,

$$\phi(\rho, \varphi) = \frac{2V}{\pi R}\rho\sin\varphi, \quad \text{for } \rho \ll R, \tag{***}$$

means that the electric field is uniform. Indeed, since $\rho\sin\varphi$ is just y (see the figure above), Eq. (***) gives

$$\phi = \frac{2V}{\pi R}y, \qquad \text{so that} \quad \mathbf{E} = -\nabla\phi = -\frac{2V}{\pi R}\mathbf{n}_y = \text{const}.$$

[23] Their equivalence may be also proven analytically, via the Fourier expansion of Eqs. (**) of the model solution of the previous problem—an optional exercise recommended to the reader.

***Problem* 2.18.** Use the variable separation method to calculate the potential distribution above the plane surface of a conductor, with a strip of width w separated by very thin cuts, and biased with voltage V—see the figure below.

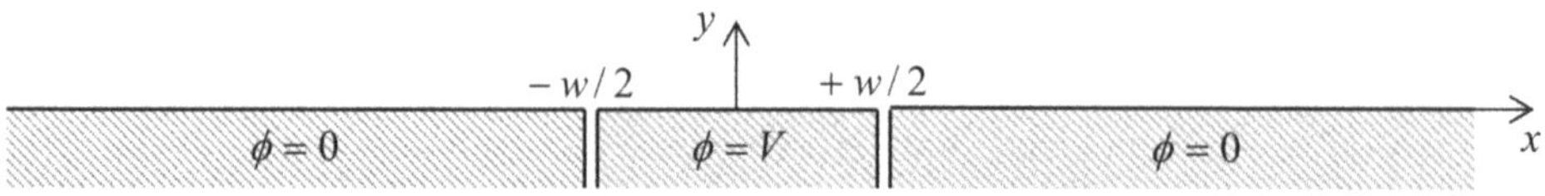

Solution: Let us introduce the Cartesian coordinates as shown in the figure above, and separate the variables similarly to Eq. (2.85) of the lecture notes, except that in this 2D problem the partial functions $Z(z)$ are evidently constant and may be taken for 1:

$$\phi_k = X_k(x)\,Y_k(y),$$

so that Eq. (2.87) may be rewritten as

$$\frac{1}{X_k}\frac{d^2X_k}{dx^2} = -\frac{1}{Y_k}\frac{d^2Y_k}{dy^2} = -k^2 = \text{const}.$$

Such notation for the separation constant is convenient in our current case, because since the electrostatic potential has to tend to zero at $y \to \infty$, only exponentially decaying solutions of the above equation for $Y_k(y)$,

$$Y_k(y) = e^{-ky},$$

with $k > 0$, are acceptable[24]. With that, the possible solutions of the above equation for functions $X_k(x)$ are $\cos kx$ and $\sin kx$. Taking into account the mirror symmetry of the system about plane $x = 0$, only the first of them are acceptable:

$$X_k(x) = \cos kx.$$

Now comes the largest difference between this problem and the one discussed in section 2.5 of the lecture notes: due to the absence of any period of the system in any direction[25], the spectrum of possible values of k is continuous rather than discrete. As a result, the sum Eq. (2.84) now becomes an integral over k; and with our current eigenfunctions $X_k(x)$ and $Y_k(y)$, it is

$$\phi(x, y) = \int_0^\infty c_k \cos kx\, e^{-ky} dk. \qquad (*)$$

[24] Strictly speaking, complex values of k, with $\text{Re}k > 0$, are also acceptable, but as will be shown below the full solution of the problem may be constructed without involving imaginary parts of k, which would describe rather unphysical oscillations of the potential in the vertical direction.

[25] The problem analyzed in the lecture notes (see figure 2.13) may be considered as a periodic one, with the rectangular volume repeated infinitely in all three directions.

The continuous set of coefficients c_k (essentially, a function of k) may be found from the boundary condition at $y = 0$ and (due to problem's symmetry) $x \geqslant 0$:

$$\int_0^\infty c_k \cos kx \; dk = \phi(x, 0) = \begin{cases} V, & \text{for } 0 \leqslant x < w/2, \\ 0, & \text{for } w/2 < x < \infty. \end{cases} \qquad (**)$$

Since this is just the expansion of the function on the right-hand side into the usual Fourier integral, the way of calculating c_k is well known: we need to multiply both parts of Eq. (**) by $\cos k'x$, with an arbitrary k', and then integrate the result over x. Changing the order of integration on the left-hand side, we obtain

$$\begin{aligned} \int_0^\infty c_k dk \int_0^\infty \cos k'x \cos kx \; dx &= \int_0^\infty \phi(x, 0) \cos k'x \; dx \\ &\equiv V \int_0^{w/2} \cos k'x \; dx. \end{aligned} \qquad (***)$$

The integral over x on the left-hand side is just a delta-function[26],

$$\begin{aligned} \int_0^\infty \cos k'x \cos kx \; dx &= \frac{1}{4} \operatorname{Re} \int_{-\infty}^{+\infty} \exp\{-ik'x\} \exp\{ikx\} \; dx \\ &= \frac{1}{4} \operatorname{Re} \int_{-\infty}^{+\infty} \exp\{i(k - k')x\} \; dx = \frac{\pi}{2} \delta(k - k'), \end{aligned}$$

while that on the right-hand side is elementary:

$$\int_0^{w/2} \cos k'x \; dx = \frac{1}{k'} \sin \frac{k'w}{2},$$

so that Eq. (**) is reduced to

$$\frac{\pi}{2} c_{k'} = \frac{V}{k'} \sin \frac{k'w}{2},$$

Plugging the resulting expression for $c_{k'}$ (with index k' replaced with k) into Eq. (*), we obtain

$$\phi(x, y) = \frac{2V}{\pi} \int_0^\infty \sin \frac{kw}{2} \cos kx \; e^{-ky} \frac{dk}{k}. \qquad (****)$$

In this particular (simple!) case the integral over k may be worked out analytically[27], but generally such an integral form of the result is typical for the variable-separation solutions. Plots in the figure below show the potential ϕ as a function of the horizontal coordinate x, for several values of the vertical coordinate y. Close to the conductor's surface, the voltage-induced potential bump is almost rectangular, with a width $\Delta x \approx w$ and a height approaching V, but

[26] See, e.g. Eq. (A.88).

[27] For the result, see the model solution of problem 2.36.

at large distances from the surface the bump is lower and more spread out, with a width Δx of the order of $y \gg w$.

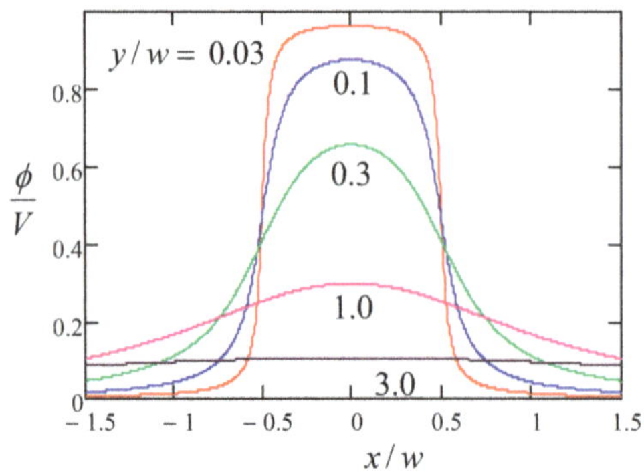

***Problem* 2.19.** The previous problem is now slightly modified: the cut-out and voltage-biased part of the conducting plane is now not a strip, but a square with side w. Calculate the potential distribution above the conductor's surface.

Solution: This problem is conceptually very similar to the previous one, except that now the distribution is three-dimensional—just like the problem solved in section 2.5 of the lecture notes, so that its solution is essentially a synthesis of the two known solutions, and the explanations below will be very short. In the Cartesian coordinates with the axes x and y in the plane of the conductor's surface, parallel to the sides of the cut square, the axis z directed normally to the plane (toward the free space), and with the origin in its center, the variable separation yields

$$\phi(x, y, z) = \int_0^\infty d\alpha \int_0^\infty d\beta\, c_{\alpha\beta} \cos \alpha x\ \cos \beta y\ e^{-\gamma z},$$

where the separation constants are related by Eq. (2.88) of the lecture notes:

$$\gamma = (\alpha^2 + \beta^2)^{1/2} > 0.$$

The boundary condition at the conductor's surface is

$$\int_0^\infty d\alpha \int_0^\infty d\beta\, c_{\alpha\beta} \cos \alpha x\ \cos \beta y = \phi(x, y, 0) = \begin{cases} V, & \text{for } |x| < w/2 \text{ and } |y| < w/2, \\ 0, & \text{otherwise.} \end{cases}$$

It is clearly satisfied by the product $c_{\alpha\beta} = Vf(\alpha)f(\beta)$, where the function f satisfies the condition

$$\int_0^\infty f(\xi) \cos \xi x\ d\xi = \begin{cases} 1, \text{ for } 0 < x < w/2, \\ 0, \text{ for } w/2 < x < \infty. \end{cases}$$

But such function was already calculated in the previous problem:

$$f(\xi) = \frac{2}{\pi\xi} \sin \frac{\xi w}{2},$$

so that, finally, the solution (*) may be spelled out as

$$\phi(x, y, z) = \frac{4V}{\pi^2} \int_0^\infty \sin \frac{\alpha w}{2} \cos \alpha x \frac{d\alpha}{\alpha} \int_0^\infty \sin \frac{\beta w}{2} \cos \beta y \frac{d\beta}{\beta} \exp\{-(\alpha^2 + \beta^2)^{1/2} z\}.$$

This result is plotted, in the figure below, as a function of one of horizontal coordinates (x), for the counterpart horizontal coordinate (y) fixed at zero, and the same values of the vertical coordinate z as were used for the similar plots in problem 2.14. (Note the difference in the vertical coordinate notation.)

The comparison of these two plot series shows that, quite naturally, the voltage-induced potential bump decays faster (at $z \gg w$, much faster) with height in our current case of the voltage-biased square than in the case of a biased infinite strip.

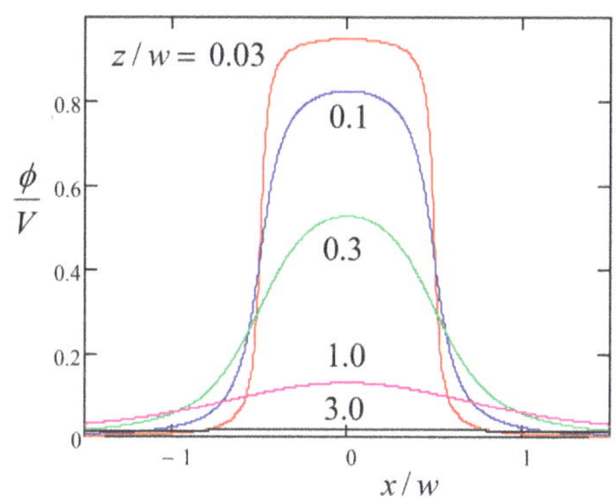

***Problem* 2.20.** Each electrode of a large plane capacitor is cut into long strips of equal width w, with very narrow gaps between them. These strips are kept at the alternating potentials, as shown in the figure below. Use the variable separation method to calculate the electrostatic potential distribution in space, and explore the limit $w \ll d$.

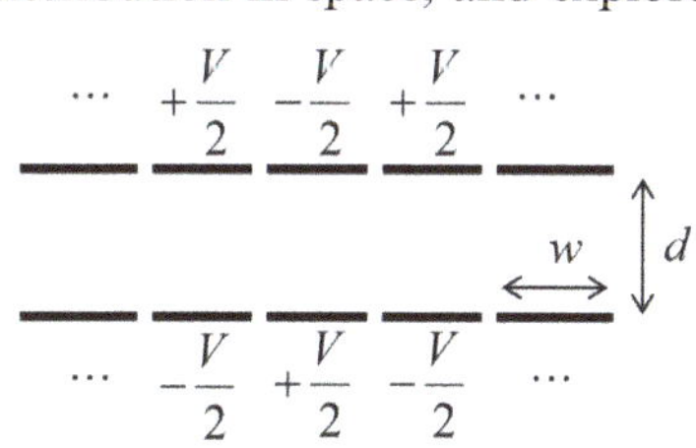

Solution: Due to the symmetry of the problem, the electrostatic potential should vanish at:

(i) the horizontal symmetry plane passing in the middle between the planes, and
(ii) any vertical plane passing between the strips.

Selecting these planes (or rather their traces on the plane of drawing) for the coordinate axes, we see that the both functions $X(x)$ and $Y(y)$, forming any particular solution $\phi_k = XY$, should be odd. Since the functions $X(x)$ have to be periodic, with period $\Delta x = 2w$, they may be taken in the form of sin αx with $\alpha = \pi n/w$. Hence, in order to satisfy the Laplace equation, the functions $Y(y)$ have to equal sinhαy, so that the full solution takes the form

$$\phi(x, y) = \sum_{n=1}^{\infty} \phi_n \sin\frac{\pi n x}{w} \sinh\frac{\pi n y}{w}. \qquad (*)$$

The coefficients ϕ_n should be found from the boundary conditions on the electrode surface. Since Eq. (*) already ensures the proper symmetry and periodicity of the solution, it is sufficient to require that it fits the boundary value on just one strip (say, $0 < x < l$, at $y = +d/2$, where $\phi = +V/2$):

$$\frac{V}{2} = \sum_{n=1}^{\infty} \phi_n \sin\frac{\pi n x}{w} \sinh\frac{\pi n d}{2w}.$$

Applying the reciprocal Fourier transform formula (or just multiplying the both parts of the last equation by $\sin(\pi x n'/w)$ and integrating the result over x from 0 to w), we readily obtain

$$\phi_n = \begin{cases} \dfrac{2V}{\pi n}\left(\sinh\dfrac{\pi n d}{2w}\right)^{-1}, & \text{for } n \text{ odd,} \\ 0, & \text{for } n \text{ even.} \end{cases}$$

The plots in the figure below show the potential distribution between the plates for two values of the ratio w/d, and for several distances y from the central plane.

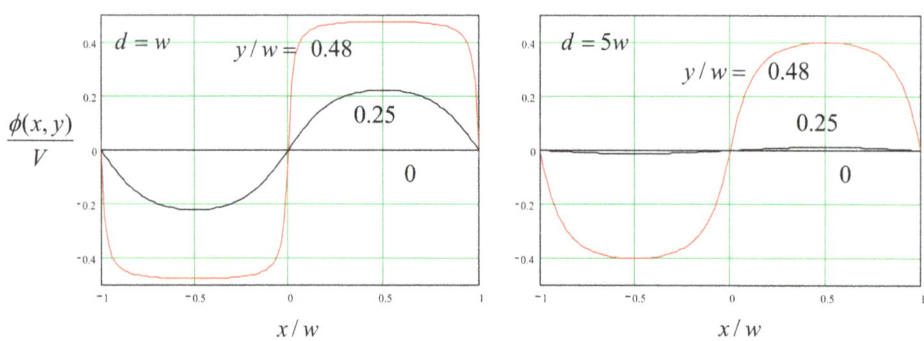

If $w \ll d$, the potential follows the square-wave profile, dictated by the electrodes, only very close (at distances $\Delta y \sim w$) to their surface, while in most of the volume it is close to the simple sine function given by the first term of the series (*):

$$\phi(x, y) \approx \frac{2V}{\pi} \frac{\sin(\pi x/w)\sinh(\pi y/w)}{\sinh(\pi d/2w)}.$$

The solution in the outer regions ($|y| > d/2$) is very similar, with the multiplier $\sinh(\pi n y/w)$ replaced with $\operatorname{sgn}(y)\exp\{-\pi n|y|/w\}$.

***Problem* 2.21.** Complete the cylinder problem started in section 2.7 of the lecture notes (see figure 2.17, reproduced below), for the cases when the top lid's voltage is fixed as follows:

(i) $V = V_0 J_1(\xi_{11}\rho/R)\sin\varphi$, where $\xi_{11} \approx 3.832$ is the first root of the Bessel function $J_1(\xi)$;
(ii) $V = V_0 = \text{const}$.

For both cases, calculate the electric field at the centers of the lower and upper lids. (For task (ii), an answer including series and/or integrals is acceptable.)

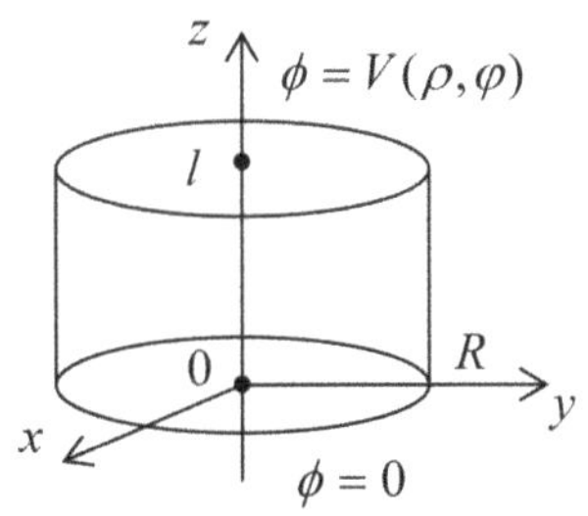

Solution: The general solution to this problem was found in section 2.7 of the lecture notes—see Eq. (2.139):

$$\phi(\rho, \varphi, z) = \sum_{n=0}^{\infty}\sum_{m=1}^{\infty} J_n\left(\frac{\xi_{nm}\rho}{R}\right)(c_{nm}\cos n\varphi + s_{nm}\sin n\varphi)\sinh\frac{\xi_{nm}z}{R}, \qquad (*)$$

where ξ_{nm} is the mth root of the Bessel function $J_n(\xi)$. This solution already satisfies not only the Laplace equation inside the cylinder, but also the boundary conditions on its side wall and bottom lid; hence the coefficients c_{nm} and s_{nm} have to be found from the boundary condition on the top lid:

$$\phi(\rho, \varphi, l) = V(\rho, \varphi). \qquad (**)$$

In task (i), this function has only one azimuthal harmonic proportional to $\sin\varphi$, and only one radial harmonic, proportional to $J_1(\xi_{11}\rho/R)$, so that only one of the coefficients c_{nm}, s_{nm} is not vanishing,

$$s_{11}\sinh\frac{\xi_{11}l}{R} = V_0,$$

and Eq. (*) is reduced to a very simple analytical solution

$$\phi = V_0 J_1\left(\frac{\xi_{11}\rho}{R}\right)\frac{\sinh(\xi_{11}z/R)}{\sinh(\xi_{11}l/R)}\sin\varphi.$$

Since $J_1(0) = 0$, the potential equals zero along all the symmetry axis of the system, so that in this case there is no normal electric field $E_n = -\partial\phi/\partial z$ in the center of either lid. However, since the potential changes within the top lid plane,

$$V = V_0 J_1\left(\frac{\xi_{11}\rho}{R}\right)\sin\varphi \to_{\rho\to 0} V_0\frac{\xi_{11}\rho}{2R}\sin\varphi = V_0\frac{\xi_{11}}{2R}y,$$

the field does have a horizontal component directed along axis y:

$$E_y\Big|_{\rho=0} = -V_0\frac{\xi_{11}}{2R} \approx -1.916\frac{V_0}{R}.$$

In task (ii), the top lid's potential does not depend on the angle φ, and hence only one angular function (with $n = 0$) fits this boundary condition, so that Eq. (*) is reduced to an axially symmetric form:

$$V_0 = \sum_{m=1}^{\infty} c_{0m} J_0\left(\frac{\xi_{0m}\rho}{R}\right)\sinh\frac{\xi_{0m}l}{R}.$$

Multiplying both parts of this equation by $\rho J_0(x_{0m}\rho/R)$, integrating the result over ρ from 0 to R, and using Eq. (2.141) of the lecture notes, we obtain

$$c_{0m} = \frac{2V_0}{R^2 J_1^2(\xi_{0m})\sinh\dfrac{\xi_{0m}l}{R}}\int_0^R J_0\left(\xi_{0m}\frac{\rho}{R}\right)\rho\, d\rho.$$

(The last integral may be worked out analytically, giving a more explicit result:

$$c_{0m} = \frac{2V_0}{\xi_{0m}J_1(\xi_{0m})\sinh\dfrac{\xi_{0m}l}{R}},$$

but I did not require my students to accomplish this astonishing feat.)

According to Eq. (*), and due to the fact that $J_n(0) = 0$ unless $n = 0$ (see, e.g. figure 2.18 of the lecture notes), the potential distribution along the symmetry axis of the cylinder is contributed only by terms with $n = 0$:

$$\phi(z) = \sum_{m=1}^{\infty} c_{0m}\sinh\frac{\xi_{0m}z}{R},$$

so that the (vertical) field in the center of the bottom lid ($z = 0$) is

$$E_n\big|_{\text{bottom}} = E_z\big|_{z=0} = -\frac{d\phi(z)}{dz}\bigg|_{z=0} = -\sum_{m=1}^{\infty} c_{0m}\frac{\xi_{0m}}{R}, \qquad (***)$$

while the field at the top lid ($z = l$) is

$$E_n\big|_{\text{top}} = -E_z\big|_{z=l} = \frac{d\phi(z)}{dz}\bigg|_{z=l} = \sum_{m=1}^{\infty} c_{0m}\frac{\xi_{0m}}{R}\cosh\frac{\xi_{0m}l}{R}.$$

For any $l > 0$, each term of the last series is larger than the corresponding term of Eq. (***), so that the field's magnitude at the bottom lid is always lower (physically, this fact is due to the screening effects of the side wall of the cylinder), although at $l/R \to 0$, $\cosh(\xi_{0m}l/R) \to 1$ for all leading terms in the series, and this difference is negligible.

***Problem* 2.22.** Solve the problem shown in figure 2.21 of the lecture notes (reproduced below). In particular:

(i) calculate and sketch the distribution of the electrostatic potential inside the system for various values of the ratio R/h, and
(ii) simplify the results for the limit $R/h \to 0$.

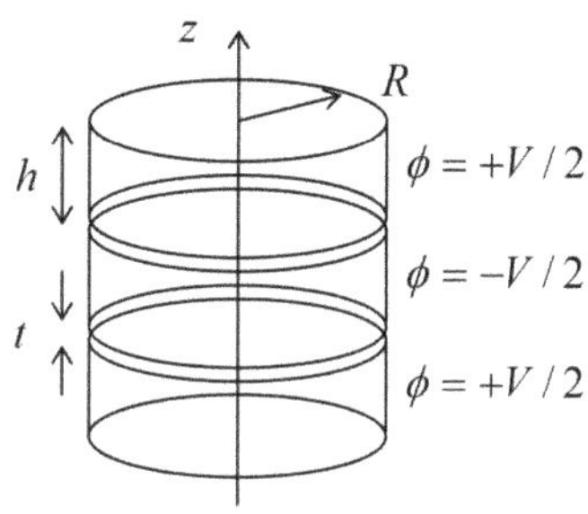

Solutions:

(i) Due to the problem's $2h$-periodicity along axis z, its particular solutions should be proportional to linear combinations of $\sin kz$ and $\cos kz$, with $2kh = 2\pi m$, where m are integers, so that the spectrum of possible values of the parameter k is restricted to $k = \pi m/h$. Moreover, if the origin of the axis z is selected at one of the gaps between the rings, the solution should be an odd function of z, so that for the internal problem ($\rho \leqslant R$) we may write

$$\phi(\rho, z) = \sum_{m=1}^{\infty} c_m I_0\left(\frac{\pi m \rho}{h}\right) \sin \frac{\pi m z}{h}.$$

The particular modified Bessel function of the first kind $I_\nu(\xi)$, used in this expression, has zero index ν, due to the evident axial symmetry of the problem—see Eq. (2.128); the function's argument results from the discreteness of the variable separation constant k: $\xi = k\rho = \pi m\rho/h$. Finally, we had to drop the modified Bessel functions of the second kind, $K_0(\xi)$, from our solution, because they diverge at $\rho \to 0$ (see the right panel of figure 2.22) and thus cannot be used to represent the finite potential inside the rings[28].

Coefficients c_m in the above series should be found from the boundary condition on the conducting rings; since the proper periodicity is already incorporated in our solution, it is sufficient to require that

$$\phi(R, z) \equiv \sum_{m=1}^{\infty} c_m I_0\left(\frac{\pi m R}{h}\right) \sin \frac{\pi m z}{h} = \frac{V}{2}, \quad \text{for } 0 \leqslant z \leqslant h.$$

[28] Note that these functions have to be used in the solution of the *external* problem ($\rho \geqslant R$), which is absolutely similar to that discussed above, except for the replacement of $I_0(\pi m\rho/h)$ with $K_0(\pi m\rho/h)$.

Multiplying both sides of this expression by $\sin(\pi m' z/2h)$, and integrating the result from 0 to h, we obtain

$$c_m = \frac{V}{\pi}\left[I_0\left(\frac{\pi m R}{h}\right)\right]^{-1} \times \begin{cases} 1/m, & \text{for } m \text{ odd}, \\ 0, & \text{for } m \text{ even}, \end{cases}$$

so that with the substitution $m = 2n - 1$ (where $n = 1, 2, \ldots$ are positive integers), we obtain

$$\phi(\rho, z) = \frac{V}{\pi}\sum_{n=1}^{\infty}\frac{1}{2n-1}\sin\frac{\pi(2n-1)z}{h}\frac{I_0[\pi(2n-1)\rho/h]}{I_0[\pi(2n-1)R/h]}. \qquad (*)$$

The figure below shows the plots of this result for cylinder's axis ($\rho = 0$) on one period of its z-dependence, for several values of the ratio R/h. At $R \gg h$, the result is dominated by the first term of the series (with $m = 1$), which is proportional to $1/I_0(\pi R/2h)$, so that at $R \gg h$, according to Eq. (2.158), the potential variation amplitude equals $2\pi V(R/2h)^{1/2}\exp\{-\pi R/2h\} \ll V$.

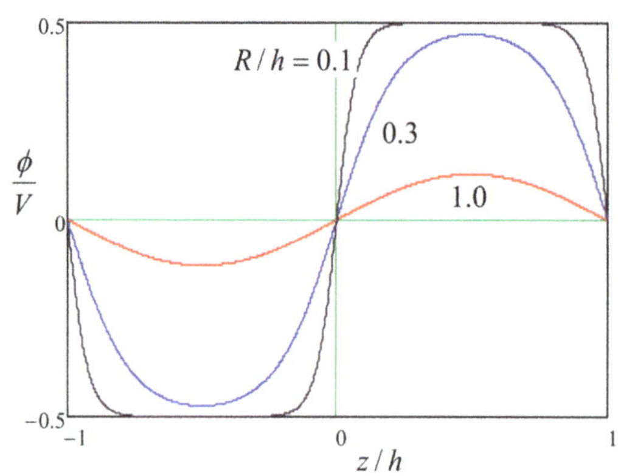

In contrast, at $R \ll h$, the potential follows that of the closest ring electrode at most of the axis, except for narrow ($\Delta z \sim R$) intervals centered on gaps between the electrodes. The exact behavior of the potential in these intervals (for example, in the close vicinity of the origin, $|z| \sim R \ll h$) is of considerable practical interest, because it is also pertinent to a system of just two long coaxial cylinders, with voltage V between them, which may serve as an effective electron lens[29].

(ii) In order to simplify Eq. (*) in this limit, we may notice that in the range $|z| \sim R$, each term of the sum is much smaller than 1, so that it may be well approximated by an integral:

[29] A convenience of such lens in comparison with one discussed in problem 1.8 is that in it, the accelerating electric field is confined to a limited length $\Delta z \sim R$.

$$\phi(\rho, z) \to \frac{V}{\pi}\int_0^\infty \frac{1}{2n-1} \sin\frac{\pi(2n-1)z}{h}\frac{I_0(\pi(2n-1)\rho/h)}{I_0(\pi(2n-1)R/h)}dn$$
$$\equiv \frac{V}{2\pi}\int_0^\infty \frac{\sin kz}{k}\frac{I_0(k\rho)}{I_0(kR)}dk, \quad \text{for } |z| \sim R \ll h,$$

where $k \equiv \pi(2n-1)/h$. This integral converges both at the lower limit (because at $k \to 0$, $\sin(kz)/k \to z$), and at the upper limit (because at $kR \to \infty$, the function $I_0(kR)$ grows exponentially—see, e.g. Eq. (2.158) and/or the left panel of figure 2.22), and is convenient for numerical calculations (for example, see the figure below) and for asymptotic behavior analyses, because it does not depend on h.

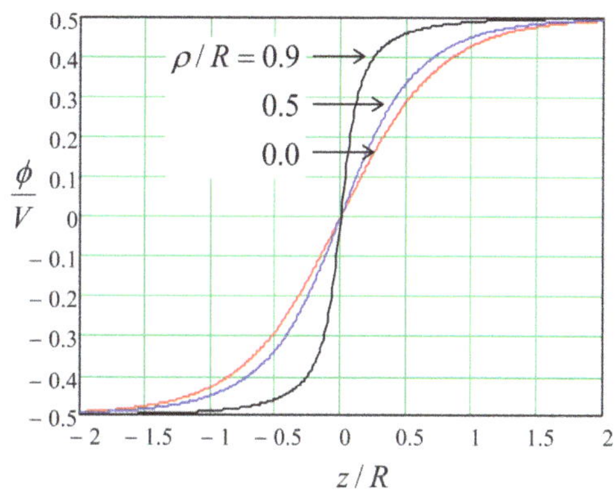

***Problem* 2.23.** Use the variable separation method to find the potential distribution inside and outside a thin spherical shell of radius R, with a fixed potential distribution: $\phi(R, \theta, \varphi) = V_0 \sin\theta\cos\varphi$.

Solution: From Eq. (2.182) of the lecture notes, we see that the surface potential is proportional to the product $\mathcal{P}_1^1(\cos\theta)\mathcal{F}_1(\phi)$, i.e. to a single angular function with $l = 1$ and $n = 1$. According to Eq. (2.184), inside the sphere the particular solution of the Laplace equation, proportional to such function, can use only radial functions $a_l r^l$, which do not diverge at $r \to 0$, while for the outer problem we may only use functions b_l/r^{l+1}, which do not diverge at $r \to \infty$. As a result, the general solution of the Laplace equation in spherical coordinates is reduced, for our case, to

$$\phi(r, \theta, \varphi) = \sin\theta\cos\varphi \times \begin{cases} a_1 r, & \text{for } r \leqslant R, \\ \dfrac{b_1}{r^2}, & \text{for } R \leqslant r. \end{cases}$$

Finding the constants a_1 and b_1 from the boundary conditions on the shell ($r = R$), we finally obtain

$$\phi(r, \theta, \varphi) = V_0 \sin\theta\cos\varphi \times \begin{cases} (r/R), & \text{for } r \leqslant R, \\ (R/r)^2, & \text{for } R \leqslant r. \end{cases}$$

Note that the first of these results may be simply expressed in Cartesian coordinates,

$$\phi = \frac{V_0}{R}x, \quad \text{for } r \leqslant R,$$

so that the field inside the sphere is uniform: $\mathbf{E} = -(V_0/R)\mathbf{n}_x$.

***Problem* 2.24.** A thin spherical shell carries charge with areal density $\sigma = \sigma_0 \cos\theta$. Calculate the spatial distribution of the electrostatic potential and the electric field, both inside and outside the shell.

Solution: Taking into account the axial symmetry of the problem, the potential distributions both outside (ϕ_{out}) and inside (ϕ_{in}) the shell may be represented as the series given by Eq. (2.172), retaining only the terms that do not diverge in the corresponding space region:

$$\phi_{\text{in}} = \sum_{l=0}^{\infty} a_l r^l \mathcal{P}_l(\cos\theta), \qquad \phi_{\text{out}} = a_0' + \sum_{l=0}^{\infty} \frac{b_l}{r^{l+1}} \mathcal{P}_l(\cos\theta).$$

We may select the arbitrary constants a_0 and a_0' to equal zero (conveniently corresponding to $\phi \to 0$ at $r \to \infty$), while the coefficient b_0 equals zero, because the corresponding term would describe the Coulomb field of shell's net charge $Q = \int \sigma d^2r$ (which our particular charge distribution does not have), so that

$$\phi_{\text{in}} = \sum_{l=1}^{\infty} a_l r^l \mathcal{P}_l(\cos\theta), \qquad \phi_{\text{out}} = \sum_{l=1}^{\infty} \frac{b_l}{r^{l+1}} \mathcal{P}_l(\cos\theta).$$

The internal and external fields are related by two boundary conditions. The first of them is the potential continuity (necessary to avoid infinite electric field), giving a set of homogenous linear relations

$$a_l R^l = \frac{b_l}{R^{l+1}}, \quad \text{for } l = 1, 2, \ldots \tag{*}$$

The second boundary condition may be obtained by applying the Gauss law to a small, flat pillbox drawn around a small part of the shell:

$$\left[\frac{\partial \phi_{\text{out}}}{\partial r} - \frac{\partial \phi_{\text{in}}}{\partial r}\right]_{r=R} = -\frac{\sigma}{\varepsilon_0} \equiv -\frac{\sigma_0}{\varepsilon_0}\cos\theta.$$

For all Legendre harmonics but the first one (with $l = 1$), this condition also gives homogeneous linear relations between a_l and b_l. Such system of two homogeneous equations is satisfied by the trivial solution $a_l = b_l = 0$, and since the solution of the electrostatic problem is unique, this is *the* real solution. The only exception is the relation for $l = 1$, with $\mathcal{P}_l(\cos\theta) = \cos\theta$, that does have a non-zero right-hand side:

$$-2\frac{b_1}{R^3} - a_1 = -\frac{\sigma_0}{\varepsilon_0}.$$

Solving this equation together with Eq. (*) for $l = 1$, namely $a_1 R = b_1/R^2$, we obtain

$$a_1 = \frac{\sigma_0}{3\varepsilon_0}, \qquad b_1 = \frac{\sigma_0 R^3}{3\varepsilon_0}$$

so that, finally,

$$\phi_{\text{in}} = \frac{\sigma_0}{3\varepsilon_0} r\cos\theta, \qquad \phi_{\text{out}} = \frac{\sigma_0 R}{3\varepsilon_0 r^2}\cos\theta.$$

This result corresponds to a uniform electric field,

$$\mathbf{E}_{\text{in}} = -\frac{\sigma_0}{3\varepsilon_0}\mathbf{n}_z,$$

inside the shell and a purely dipole field corresponding to the dipole moment $\mathbf{p} = \mathbf{n}_z (4\pi/3)\sigma_0 R$ outside it:

$$\mathbf{E}_{\mathbf{out}} = \frac{\sigma_0}{3\varepsilon_0}\left(\mathbf{n}_r\frac{2\cos\theta}{r^3} + \mathbf{n}_\theta\frac{\sin\theta}{r^3}\right).$$

As will be shown in section 3.1, these results are particular manifestations of the general relations (3.24) and (3.13), correspondingly.

***Problem* 2.25.** Use the variable separation method to calculate the potential distribution both inside and outside of a thin spherical shell of radius R, separated with a very thin cut, along plane $z = 0$, into two halves, with voltage V applied between them—see the figure below. Analyze the solution; in particular, compare the compare the field at the axis z, for $z > R$, with Eq. (2.218).

Hint: You may like to use the following integral of a Legendre polynomial with odd index $l = 1, 3, 5, \ldots = 2n - 1$:[30]

$$I_n \equiv \int_0^1 \mathcal{P}_{2n-1}(\xi)d\xi = \frac{1}{n!}\cdot\left(\frac{1}{2}\right)\cdot\left(-\frac{3}{2}\right)\cdot\left(-\frac{5}{2}\right)\ldots\left(\frac{3}{2} - n\right) \equiv (-1)^{n-1}\frac{(2n-3)!!}{2n(2n-2)!!}.$$

[30] As a reminder, the *double factorial* (also called 'semifactorial') operator (!!) is similar to the usual factorial operator (!), but with the product limited to numbers of the same parity as its argument (in our particular case, of odd numbers in the nominator, and even numbers in the denominator).

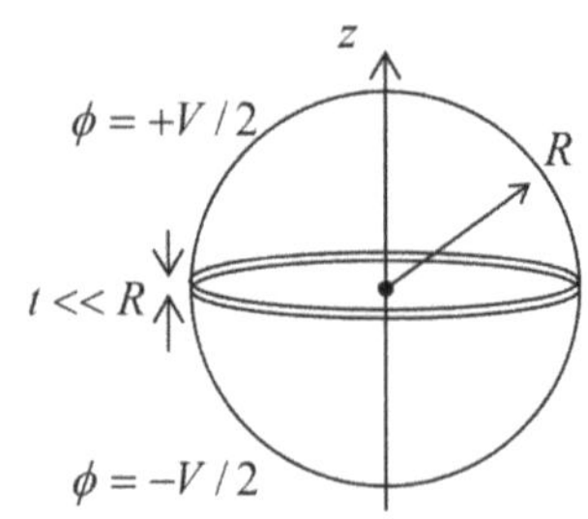

Solution: Due to the axial symmetry of the problem, we may look for the solution of the Laplace equation in the form given by Eq. (2.172) of the lecture notes. Moreover, due to the symmetry of the applied potentials, in that series we may take $a_0 = b_0 = 0$, so the term with $l = 0$ may be dropped, giving

$$\phi(r, \theta) = \sum_{l=1}^{\infty}\left(a_l r^l + \frac{b_l}{r^{l+1}}\right)P_l(\cos\theta).$$

For the internal problem (potential at $r \leqslant R$) we have to set all coefficients b_l to zero, because otherwise the corresponding terms would diverge at $r \to 0$. For the coefficients a_n, the boundary conditions at $r = R$ yield the following system of equations:

$$\sum_{l=1}^{\infty} a_l R^l P_l(\cos\theta) = \phi(R, \theta) \equiv \frac{V}{2}\,\text{sgn}\left(\frac{\pi}{2} - \theta\right), \qquad \text{for } r \leqslant R. \qquad (*)$$

In contrast, for the external problem ($r \geqslant R$) these are coefficients a_n that have to vanish to avoid the divergence at $r \to \infty$, while the coefficients b_n have to be found from a system of equations similar to Eq. (*):

$$\sum_{l=1}^{\infty} \frac{b_l}{R^{l+1}} \mathcal{P}_l(\cos\theta) = \phi(R, \theta) \equiv \frac{V}{2}\,\text{sgn}\left(\frac{\pi}{2} - \theta\right), \qquad \text{for } r \geqslant R. \qquad (**)$$

Due to this similarity, both systems may be solved similarly: by multiplying their both parts by the product $\sin\theta\,\mathcal{P}_{l'}(\cos\theta)$, with an arbitrary index $l' > 0$, integrating the results over the segment $0 \leqslant \theta \leqslant \pi$, and then using the polynomial orthonormality condition Eq. (2.171). The results are, naturally, also similar:

$$a_l R^l\big|_{r \leqslant R} = \frac{b_l}{R^{l+1}}\bigg|_{r \geqslant R} = \frac{V}{2}\frac{2l+1}{2}\int_{-1}^{+1} \text{sgn}\,(\xi)\,\mathcal{P}_l(\xi)d\xi,$$

where $\xi \equiv \cos\theta$. Since, according to Eq. (2.169) (see also figure 2.23) of the lecture notes, the Legendre polynomials with an even index are symmetric functions of their argument, while those with odd index are asymmetric, the above integral vanishes for even l, while for odd $l = 2n - 1$ (where $n = 1, 2, \ldots$) it may be rewritten as

$$a_{2n-1}R^{2n-1}\big|_{r \leqslant R} = \frac{b_{2n-1}}{R^{2n}}\bigg|_{r \geqslant R} = V\left(2n - \frac{1}{2}\right)\int_0^{+1} \mathcal{P}_{2n-1}(\xi)d\xi \equiv V\left(2n - \frac{1}{2}\right)I_{2n-1},$$

where I_{2n-1} are the numbers specified in the hint. Since these numbers drop very rapidly with the growth of n, and also alternate their signs, the series giving the potential distribution inside the shell

$$\begin{aligned}\phi(r, \theta) &= \sum_{n=1}^{\infty} a_{2n-1} r^{2n-1} \mathcal{P}_{2n-1}(\cos\theta) \\ &= V\sum_{n=1}^{\infty}\left(2n - \frac{1}{2}\right) I_{2n-1}\left(\frac{r}{R}\right)^{2n-1} \mathcal{P}_{2n-1}(\cos\theta), \quad \text{for } r \leqslant R,\end{aligned} \tag{***}$$

and outside it

$$\begin{aligned}\phi(r, \theta) &= \sum_{n=1}^{\infty} \frac{b_{2n-1}}{r^{2n}} \mathcal{P}_l(\cos\theta) \\ &= V\sum_{n=1}^{\infty}\left(2n - \frac{1}{2}\right) I_{2n-1}\left(\frac{R}{r}\right)^{2n} \mathcal{P}_{2n-1}(\cos\theta), \quad \text{for } r \geqslant R,\end{aligned} \tag{****}$$

converge reasonably fast even for the points rather close to the shell ($r \approx R$).

For example, the figure below shows the resulting potential distribution along the radius r for two values of the polar angle θ. For $\theta = 0$ (i.e. along the symmetry axis z), and $r \geqslant R$, the result exactly coincides with that given by Eq. (2.218) of the lecture notes, obtained by the Green's function method.

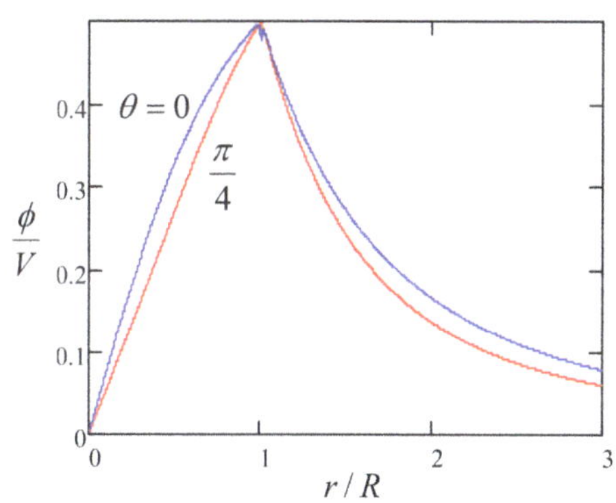

Apart from giving the result for any point of the system, Eqs. (***) and (****) are very convenient for obtaining analytical asymptotic expressions for important cases $r \ll R$ (the shell center's vicinity) and $r \gg R$ (points far from the shell). Indeed, in both cases, good approximations are given by the first terms of these series (with $n =$ 1), so that since $I_1 = ½$, we obtain

$$\phi(r, \theta) \to \frac{3}{4} V \cos\theta \times \begin{cases} (r/R), & \text{for } r \ll R, \\ (R/r)^2, & \text{for } r \gg R. \end{cases}$$

The former expression yields the electric field near the shell's center,

$$\mathbf{E}|_{r=0} = -\nabla\phi|_{r<R} = -\frac{3}{4}\frac{V}{R}\mathbf{n}_z,$$

while the latter one generalizes Eq. (2.219) for arbitrary values of the polar angle.

***Problem* 2.26.** Calculate, up to the terms $O(1/r^2)$, the long-range electric field induced by a cut and voltage-biased conducting sphere—similar to that discussed in section 2.7 of the lecture notes (see figure 2.32) and in the previous problem, but with the cut's plane at an arbitrary distance $d < R$ from the center—see the figure below.

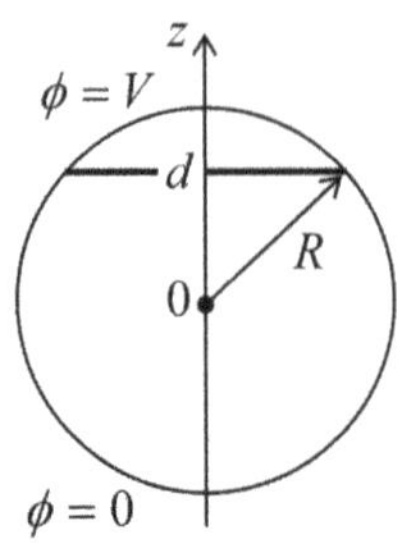

Solution: Applying the variable separation method just as was done in the model solution of problem 2.25, for the field outside the sphere[31] we may write the similar expression:

$$\phi(r, \theta) = \sum_{l=0}^{\infty} \frac{b_l}{r^{l+1}} \mathcal{P}_l(\cos\theta), \quad \text{for } r \geqslant R, \tag{*}$$

and calculate the expansion coefficients b_l in a similar way, from the following evident generalization of Eq. (**) of that solution:

$$\sum_{l=0}^{\infty} \frac{b_l}{R^{l+1}} \mathcal{P}_l(\cos\theta) = \phi(R, \theta) = V \times \begin{cases} 0, \text{ for } & -1 < \cos\theta < d/R, \\ 1, \text{ for } & d/R < \cos\theta < +1. \end{cases}$$

As in that problem, multiplying both parts of this equation by $\mathcal{P}_{l'}(\cos\theta)$ and integrating the result over the surface of the sphere, we may use the orthonormality condition (2.171) to obtain

$$\frac{b_l}{R^{l+1}} \frac{2}{2l+1} = \int_{-1}^{+1} \mathcal{P}_l(\xi)\phi(R, \cos^{-1}\xi)d\xi = V\int_{d/R}^{+1} \mathcal{P}_l(\xi)d\xi. \tag{**}$$

(Here $\xi \equiv \cos\theta$, so that $d/R = \cos\theta_0$, where θ_0 is the polar angle of the sphere surface's cut line.)

Using the Rodrigues' formula (2.169), the integral on the right-hand side of Eq. (**) may be expressed via the Legendre polynomials of d/R. However, we are only asked about the field at large distances, $r \gg R$. This field, with the requested accuracy $O(1/r^2)$, is given by the first two terms of the expansion (*):

$$\phi(r, \theta) \to \frac{b_0}{r}\mathcal{P}_0(\cos\theta) + \frac{b_1}{r^2}P_1(\cos\theta) \equiv \frac{b_0}{r} + \frac{b_1}{r^2}\cos\theta, \quad \text{for } r \gg R, \tag{***}$$

so that we only need to calculate two coefficients, b_0 and b_1. For them, Eq. (**) yields simple results:

[31] For the field in this region, it does not matter whether the conducting sphere is solid or it is a spherical shell.

$$b_0 = \frac{1}{2}RV\int_{d/R}^{+1} d\xi = \frac{1}{2}V(R - d), \qquad b_1 = \frac{3}{2}R^2V\int_{d/R}^{+1} \xi d\xi = \frac{3}{4}V(R^2 - d^2),$$

so that Eq. (***) gives

$$\phi(r, \theta) \to \frac{1}{2}V\frac{R - d}{r} + \frac{3}{4}V\frac{R^2 - d^2}{r^2}\cos\theta, \quad \text{for } r \gg R.$$

As a sanity check, at $d = 0$ (i.e. at the cut's plane passing through the sphere's center), the first term of this expression, $(1/2)VR/r$, is, very appropriately, half that of an uncut sphere kept at potential V, while the second term coincides with (2.219) of the lecture notes, and with the solution of problem 2.25. On the other hand, at $d \to R$ (i.e. in the limit on an uncut, grounded sphere), both terms vanish—as they should.

Note also that all the above calculations, and hence their results, are valid for any sign of d, with the sole limitation $-R \leqslant d \leqslant +R$.

***Problem* 2.27.** A small conductor (in this context, called the *single-electron island*) is placed between two conducting electrodes, with voltage V applied between them. The gap between the island and one of the electrodes is so narrow that electrons may tunnel quantum-mechanically through this 'junction'—see the figure below. Neglecting thermal excitations, calculate the equilibrium charge of the island as a function of V.

Hint: To solve this problem, you do not need to know much about the quantum-mechanical tunneling between conductors, except that such tunneling of an electron, followed by energy relaxation of the resulting excitations, may be considered as a single inelastic (energy-dissipating) event. At negligible thermal excitations, such an event takes place only when it decreases the total potential energy of the system[32].

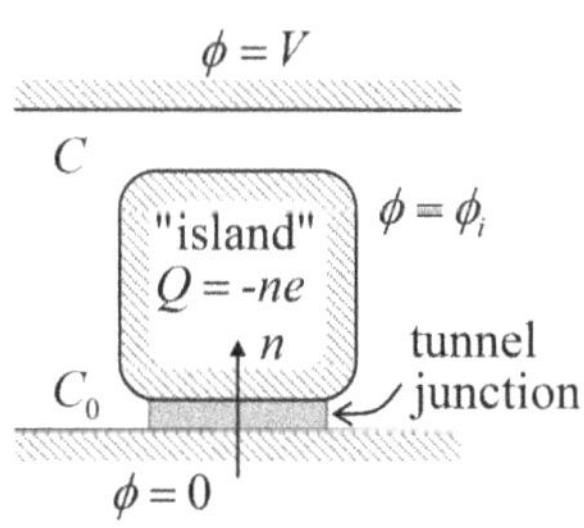

Solution: Let us assume that the electric field, induced by the island's charge Q, does not extend beyond the space between the two electrodes. Then, regardless of the island's geometry, according the linear superposition principle, the electrostatic

[32] Strictly speaking, this model, implying negligible quantum-mechanical coherence of tunneling events, is correct only if the junction transparency is sufficiently low, so that its effective electric resistance is much higher than the quantum unit of resistance (see, e.g. *Part QM* section 3.2), $R_Q \equiv \pi\hbar/2e^2 \approx 6.5$ kΩ—which is usually the case.

potential ϕ_i of the island has to be a linear function of Q and the applied voltage V. For the convenience of what follows, we may write this linear relation in the form

$$Q = C_0\phi_i + C(\phi_i - V), \quad \text{i.e.} \quad \phi_i = \frac{Q}{C_\Sigma} + \phi_{\text{ext}},$$
$$\text{with} \quad \phi_{\text{ext}} \equiv \frac{C}{C_\Sigma}V, \quad \text{and} \quad C_\Sigma \equiv C + C_0, \qquad (*)$$

where the coefficients C_0 and C have the physical sense of the mutual capacitances between the island and the system's electrodes (see the figure above), so that C_Σ of the full self-capacitance of the island at $V = 0$. (Note that Eq. (*) is valid even if the electric fields associated with the capacitances C_0 and C are not spatially separated in full, and hence the lumped capacitor model cannot be used from the very beginning—cf. figure 2.5 of the lecture notes)

Since the first term in the potential ϕ_i given by Eq. (*) is proportional to Q, the corresponding electrostatic energy has to be calculated using Eq. (1.61) of the lecture notes. On the other hand, the energy associated with the externally induced potential ϕ_{ext}, which does not depend on Q, is described by Eq. (1.54). As the result, the total electrostatic energy of the system (or rather its part dependent on Q) is

$$U(Q) = \frac{Q^2}{2C_\Sigma} + Q\phi_{\text{ext}}. \qquad (**)$$

In order to use this result, it is convenient to use the notion of the 'external charge',

$$Q_{ext} \equiv \phi_{\text{ext}}C_\Sigma \equiv CV,$$

because with it, Eq. (**) takes a very simple form:

$$U(Q) = \frac{Q^2}{2C_\Sigma} + \frac{QQ_{\text{ext}}}{C_\Sigma} \equiv \frac{(Q + Q_{\text{ext}})^2}{2C_\Sigma} + \text{const.} \qquad (***)$$

The plot of this expression as a function of Q is a quadratic parabola with the minimum at $Q = -Q_{\text{ext}}$ (see the left panel of the figure below) and if Q could take any real value the system's static equilibrium would follow this minimum. However, in reality Q is a multiple of the fundamental charge $e \approx 1.6 \times 10^{-19}$ C: $Q = -ne$, where n is the number of electrons that has tunneled into the island since it was electrically neutral[33]. As a result, the system's equilibrium corresponds to the value $Q = -en$ closest to $-Q_{\text{ext}}$, and the change of n by ± 1 (via tunneling of a single electron into or from the island) happens when Q_{ext} crosses any of the critical points $e(n + \frac{1}{2})$—see the right panel in the figure below.

[33] It is crucial that this discreteness does *not* pertain to Q_{ext}, which is just a normalized external voltage, and hence is *not* quantized—at least on this scale. In other words, Q_{ext} is just the charge of polarization of the island by voltage V, which may take any real values—see the discussion in section 2.9 of the lecture notes.

The simplest interpretation of this *Coulomb staircase* pattern is that at small C_Σ, those n extra electrons in the island, whose negative charge is not compensated by the positive charge of the atomic lattice, repel each other strongly, so that the island has only as many of them as necessary to compensate the externally induced polarization charge Q_{ext}.

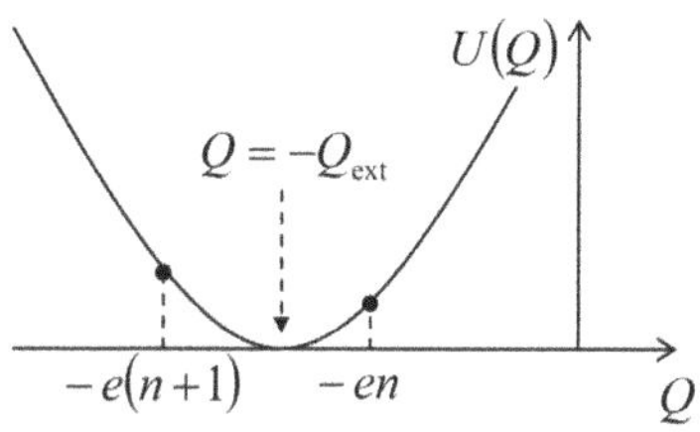

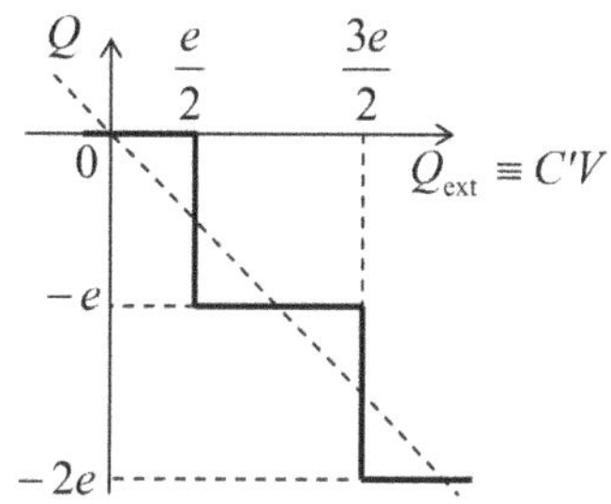

The most counter-intuitive aspect of this effect is that it may take place in rather macroscopic conducting islands—with the size up to a few microns, with zillions (e.g. $\sim 10^{10}$) of background electrons. However, for that the temperature has to be low enough, to make the energy scale of the masking thermal fluctuations, k_BT, much smaller than the relevant scale of U, namely $e^2/2C_\Sigma$. As a result, the Coulomb staircase (or rather its footprint on the rf impedance of the system) was observed experimentally and explained theoretically only in the 1970s.[34] Currently, this system, called the *single-electron island*, is considered a generic device, from which quite a few other single-electron devices have been derived—see, e.g. the next problem.

***Problem* 2.28.** The system discussed in the previous problem is now generalized as shown in the figure below. If the voltage V' applied between two bottom electrodes is sufficiently large, electrons can successively tunnel through two junctions of this system (called the *single-electron transistor*), carrying some dc current between these electrodes. Neglecting thermal excitations, calculate the region of voltages V and V' where such a current is fully suppressed (*Coulomb-blocked*).

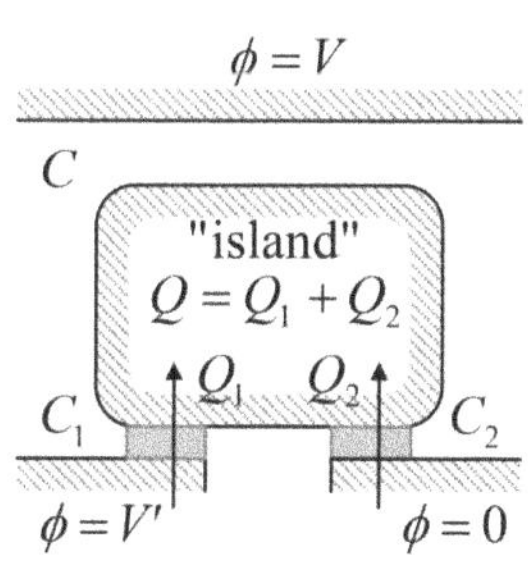

Solution: Repeating the argumentation of the previous problem, for the island's electrostatic potential we may write a similar expression:

[34] See [1].

$$\phi_i = \frac{Q}{C_\Sigma} + \phi_{\text{ext}} + \phi_{\text{ext}}', \quad \text{with} \quad \phi_{\text{ext}} \equiv \frac{C}{C_\Sigma}V,$$
$$\phi_{\text{ext}}' \equiv \frac{C_1}{C_\Sigma}V', \quad \text{and} \quad C_\Sigma \equiv C_1 + C_2 + C, \qquad (*)$$

where C_1, C_2, and C have the sense of mutual capacitances between the island and the electrodes of the system—see the figure above.

Now, writing the expression for the system's potential energy, we may follow the argumentation of the model solution of the previous problem, but should take into account that Eq. (1.54) of the lecture notes, used in that solution, is valid only if the charge q_k has been moved into the system under analysis from a region with negligible potential. If that initial potential $\phi_{k\text{–ini}}$ is different from zero, then we should generalize that relation as

$$U_{\text{ext}} = \sum_k q_k\left[\phi_{\text{ext}}(\mathbf{r}_k) - \phi_{k-\text{ini}}\right].$$

Indeed, the expression in square brackets is just the work of external forces, and hence the potential energy's increase, at the move of a unit charge between the points with potentials $\phi_{k\text{–ini}}$ and $\phi_{\text{ext}}(\mathbf{r}_k)$. In our current problem, ϕ_{ext} is actually the sum $\phi_{\text{ext}} + \phi_{\text{ext}}'$ defined by Eq. (2.62), $\phi_{k\text{–ini}} = V'$ for the charge Q_1 brought into the island by the electrons tunneling through the left junction, and $\phi_{k\text{–ini}} = 0$ for the charge Q_2 brought in through the right junction. (The total charge of the island is evidently $Q = Q_1 + Q_2$, see the figure above.) As a result, for the total electrostatic energy of the system we obtain

$$U(Q_1, Q_2) = \frac{Q^2}{2C_\Sigma} + Q_1(\phi_{\text{ext}} + \phi_{\text{ext}}' - V') + Q_2(\phi_{\text{ext}} + \phi_{\text{ext}}'),$$

Now plugging in Eqs. (*) for ϕ_{ext} and ϕ_{ext}', we may rewrite this expression in the form which is the generalization of Eq. (**) of the model solution of the previous problem:

$$U(Q_1, Q_2) = \frac{(Q_1 + Q_2 + Q_{\text{ext}})^2}{2C_\Sigma} + \left(\frac{C_1}{C_\Sigma}Q_2 - \frac{C_2 + C}{C_\Sigma}Q_1\right)V' + \text{const}, \qquad (**)$$
$$\text{where } Q_{\text{ext}} \equiv CV.$$

In order to use this expression to find the dc current threshold, we should take into account the discreteness of the electron charges passed through the junctions,

$$Q_1 = -n_1 e, \quad Q_2 = -n_2 e,$$

and compare $U(-en_1, -en_2)$ with all four values of energy, $U(-en_1 \pm e, -en_2)$ and $U(-en_1, -en_2 \pm e)$, resulting from tunneling of an additional electron through either junction, in either direction. (Only if any of these events leads to an increase of U, the current state is stable, and dc current cannot flow through the system.) An elementary calculation yields four stability conditions, which may be summarized as follows:

$$\left(n - \frac{1}{2}\right)e < \left\{ \begin{matrix} Q_{\text{ext}} + C_1 V' \\ Q_{\text{ext}} - (C_2 + C)V' \end{matrix} \right\} < \left(n + \frac{1}{2}\right)e,$$

where $n \equiv n_1 + n_2$, i.e. the total number of charge-uncompensated electrons in the island. This result may be understood more easily by plotting the boundaries for these conditions on the plane of applied voltages V and V'—see the figure below. Only if the point $\{V, V'\}$ is inside one of the 'Coulomb diamonds' of this periodic diagram (which continues infinitely to both sides of the horizontal axis), with the indicated value of n, the charge state is stable, with no persistent sequential tunneling of electrons. This stable state is called the *Coulomb blockade* of current. Note that the largest magnitude of the blockade threshold is e/C_Σ, clearly indicating its single-electron nature.

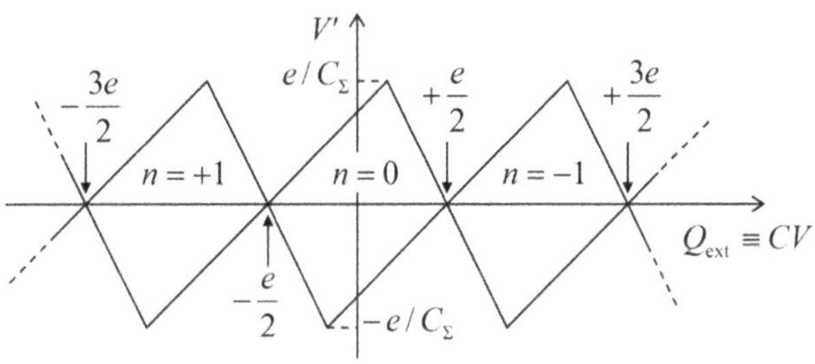

This result means that the gate voltage V may control static ('dc') current in this system, so that it may be indeed used as a transistor, working just on the Coulomb repulsion of electrons, rather than on any specific material properties. Invented in 1985, and experimentally demonstrated for the first time in 1987, the single-electron transistors are used in experimental physics, mostly for sensitive electric measurements at low temperatures. The transistor may be further modified to obtain other Coulomb-blockade devices, in particular fundamental dc current standards, and non-volatile memory cells, with digital bits stored in the form of single electrons. Unfortunately, the tough temperature-to-size trade-offs, and uncontrollable blockade threshold offsets by random charged impurities have so far prevented a practical use of these devices in digital electronics[35].

Note also that similar devices with superconducting electrodes enable controllable transfer of not only discrete single electrons, but also discrete single Cooper pairs between two or more superconducting Bose–Einstein condensates, and also an electrostatic control of supercurrents. However, due to the quantum coherence of

[35] For more discussion of these issues, see the literature cited in section 2.9 of the lecture notes.

the condensates, any quantitative discussion of these devices requires quantum mechanics[36].

***Problem* 2.29.** Use the image charge method to calculate the surface charges induced in the plates of a very broad plane capacitor of thickness D by a point charge q separated from one of the electrodes by distance d.

Solution: The system of image charges that satisfies the corresponding Poisson equation, and boundary conditions for this problem have already been discussed in section 2.9 of the lecture notes—see figure 2.28c, reproduced below in an extended version. Here the red balls denote charges $+q$ (including the real one), while the blue balls denote charges $-q$.

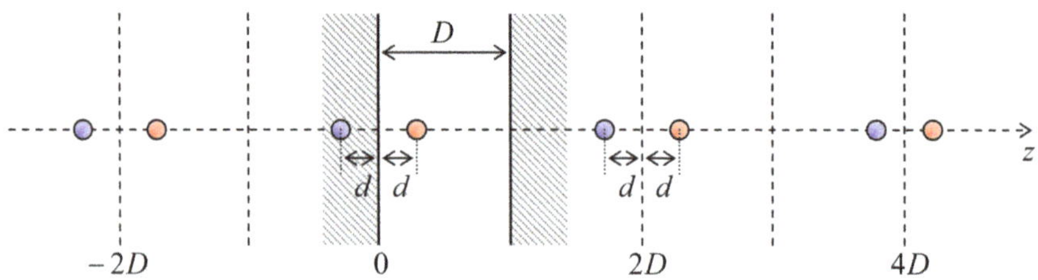

Conceptually, the total charge induced in any electrode surface (say, the one located at $z = 0$) may be readily calculated as was discussed in section 2.9:

$$Q = \int_A \sigma d^2r = -\varepsilon_0 \int_A \frac{\partial \phi}{\partial z}\Big|_{z=0} d^2r$$
$$= -\varepsilon_0 \int_A \frac{\partial}{\partial z} \sum_j \phi_j\Big|_{z=0} d^2r = -\varepsilon_0 \int_0^\infty \frac{\partial}{\partial z} \sum_{j,\pm} \frac{\pm q}{4\pi\varepsilon_0 \left[\rho^2 + z_j^2\right]^{1/2}}\Big|_{z=0} 2\pi\rho d\rho, \qquad (*)$$

where ρ is the distance from axis z, and z_j is the position of an image belonging to the jth pair of adjacent charges, centered at $2Dj$. According to Eq. (2.194) (but also as evident from the figure above),

$$z_j = 2Dj \pm d.$$

The differentiation in Eq. (*) is elementary, and the summation and integration may be swapped at will; however, neither operation is too pleasing. Perhaps the simplest way to obtain a simple analytical result is to apply the Gauss law first to the set of *only the 'positive'*[37] *charges* ($+q$) inside a round cylinder with the axis coinciding with axis z, a very large radius $R \gg D$, and the following special choices for the positions of the lids (both parallel to capacitor planes): one coinciding with the capacitor plate of our interest, $z = 0$, and the other located exactly in the middle

[36] See, e.g. *Part QM* section 2.8.

[37] I am using the quote signs because the actual point charge q may be of any sign.

between any two charges ($+q$), i.e. at $z = L_+ \equiv (2N + 1)D + d$. Due to the condition $R \gg D$, the electric field E_+ of the 'positive' charges at the lateral (curved) wall of the Gaussian cylinder is virtually uniform, perpendicular to axis z, and is the same as of a continuous charge line with uniform linear density $\lambda_+ = q/2D$. Hence, according to the (very easy) solution of problem 1.1, $E_+ = \lambda_+/2\pi\varepsilon_0$, so that its flux through the wall is

$$\int_{\rho=R\gg D}(E_+)_n d^2r = \frac{\lambda L_+}{\varepsilon_0}.$$

The electric flux through cylinder's lids should be calculated more carefully, because parts of them are at distances of the order of $d \sim D$ from the nearest charges. Fortunately, with our (smart :-) choice of lid positions, the flux through the lid at $z = L_+$ equals zero due to the symmetry, while, according to Eq. (2.3), the flux through the lid with $z = 0$ is proportional to the surface charge Q_+ of the electrode, induced by the 'positive' charges:

$$\int_{z=0}(E_+)_n d^2r = \frac{1}{\varepsilon_0}\int_{z=0}(\sigma_+)_n d^2r = \frac{Q_+}{\varepsilon_0}.$$

Now taking into account that the number of the 'positive' charges inside the cylinder is $(N + 1)$, the Gauss law reads

$$\oint(E_+)_n d^2r \equiv \frac{q}{2D\varepsilon_0}[(2N + 1)D + d] + \frac{Q_+}{\varepsilon_0} = \frac{N+1}{\varepsilon_0}q,$$

giving

$$Q_+ = q\left[N + 1 - \frac{(2N+1)D + d}{2D}\right] \equiv \frac{q}{2}\left(1 - \frac{d}{D}\right).$$

Now repeating these calculations for the system of 'negative' charges ($-q$), with the evident replacements of L_+ with $L_- = (2N + 1)D - d$, and λ_+ with $\lambda_- = -q/2D$, we obtain

$$Q_- = -\frac{q}{2}\left(1 + \frac{d}{D}\right),$$

so that the total surface charge of the electrode is[38]

$$Q \equiv Q_+ + Q_- = -q\frac{d}{D}. \qquad (**)$$

[38] For the surface charge of the counter-electrode located at $z = D$, we can readily obtain $Q' = -q(D - d)/D$. This result may be obtained either by reproducing the above calculations, or (more simply) by applying the Gauss law to a large volume with the real (point + surface) charges inside, giving $Q + Q' = q$, or just from the mirror symmetry of the problem about plane $z = D/2$.

This very simple result may be verified by a numerical summation and integration in Eq. (*), and also by the following simple reasoning. Let us assume that our plane capacitor, with the point charge inside, is connected to a battery fixing the voltage V across the plates, independent on the charge position—see the figure below. Then the uniform external electric field $E = V/D$ creates a position-independent force $F = qE = qV/D$, applied to the charge and directed perpendicular to electrode surfaces. On a small displacement Δd, this force performs work $\Delta\mathscr{W} = F\Delta d = qV\Delta d/D$. On the other hand ('from the point of view of the battery'), the same work $\Delta\mathscr{W}$ has to be equal to $V\Delta Q_V$, where ΔQ_V is the charge that has flown through the battery as a result of this displacement. But this charge has nowhere to come from other than the change of the surface charges of the capacitor plates: $\Delta Q_V = -\Delta Q$. Comparing these two expressions for $\Delta\mathscr{W}$, we obtain $\Delta Q = -q(\Delta d/D)$, i.e. $Q = -qd/D + \text{const}$. Since the surface charge should tend to zero when the point charge approaches the surface ($d \to 0$), the constant in the last expression has to equal zero, so that we return to Eq. (**).

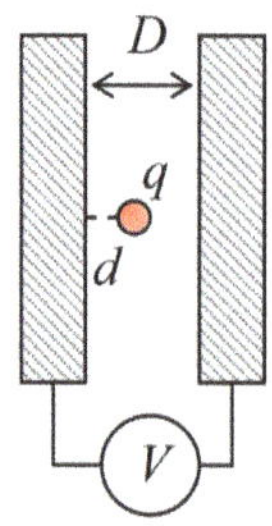

***Problem* 2.30.** Prove the statement, made in section 2.9 of the lecture notes, that the 2D boundary problem, shown in the figure below, can be solved using a finite number of image charges only if the angle β equals π/n, where $n = 1, 2,\ldots$

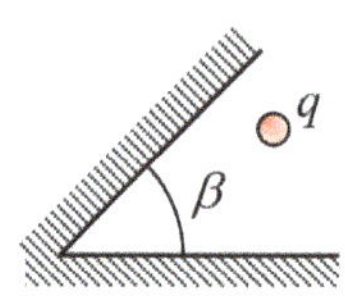

Solution: Let the angle between the direction to the charge and the nearest wall be equal to φ (see the figure below), so that its angular distance from the opposite wall is $(\beta - \varphi)$. In order to satisfy the boundary condition ($\phi = 0$) at the nearest wall, let us add the charge's reflection in the nearest wall: a negative image charge at the same distance from the corner, and at angle $(\beta + \varphi)$.

Still, the boundary condition at the opposite wall is not satisfied. In an attempt to correct that, let us reflect in that wall's surface the pair we already have, i.e. add a charge $+q$ at angle $(-\beta - \varphi)$ and a charge $-q$ at $(-\beta + \varphi)$—see the figure below. We see that the resulting two pairs of charges have the same angular direction of their dipole moments[39], and their centers are separated by angle 2β.

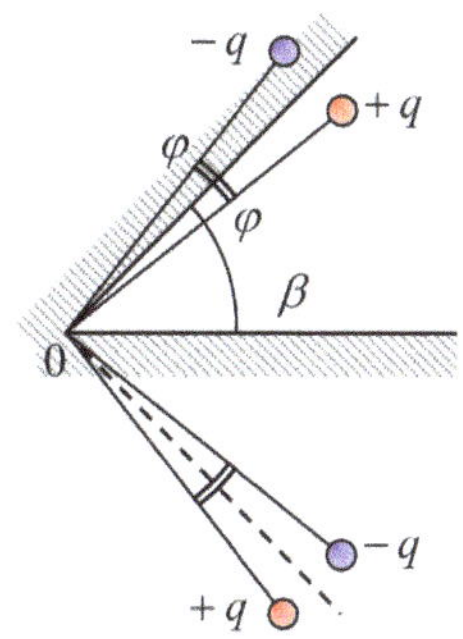

Now, the boundary condition at the first wall's surface (nearest to the real charge) is not satisfied again. In order to correct that, let us repeat the reflection procedure again, now in this wall. At this new reflection, the first pair does not change, but the second pair gives a new one (the third pair overall), with the same angular direction of the dipole moment. Now the pair centers occupy angle 3β, but (for the general case) the boundary conditions on the second wall's surface are not satisfied again.

Repeating this charge system reflection procedure n times overall, we obtain n charge pairs, with their centers within a sector $2n\beta$ wide. In order to have the process stopped (i.e. the boundary conditions satisfied on both surfaces), we need the last generated pair to overlap one of the previously generated pairs. This is only possible if $2n\beta$ equals 2π, i.e. $\beta = \pi/n$, with $n = 1, 2,\ldots$ (For $n = 1$, the situation is reduced to the generic image charge situation shown in figure 2.26 of the lecture notes, while the cases $n = 2$ and $n = 4$ are shown on panels (a) and (b) of figure 2.28. A useful exercise: draw the situation for $n = 3$.)

Superficially, the set of possible β is broader, and is given by expression $(m/n)\pi$, with m also an arbitrary integer. Indeed, after n pair reflections the pairs would cover angle $2n\beta = 2m\pi$, i.e. an integer number of 2π, and start overlapping with old ones. However, at $m > 1$ at least one of the image charges will be in the free-space (observable) region, so that the field in that region would have more than one point-charge singularity, and thus would not satisfy our initial problem.

***Problem* 2.31.** Use the image charge method to calculate the potential energy of the electrostatic interaction between a point charge placed in the center of a spherical cavity that was carved inside a grounded conductor, and the cavity's walls. Looking at the result, could it be obtained in a simpler way (or ways)?

[39] OK, OK, I know we have not discussed the dipole moments in this course yet (see chapter 3 of the lecture notes). If you insist you do not know what it is, in this context you may understand the moment just as the vector connecting the charges $-q$ and $+q$ of the pair.

Solution: First, let us calculate the force **F** exerted by the conductor on a point charge q' shifted by some distance $d' < R$ from cavity's center—see the figure below. As follows from the discussion of an example in section 2.9 of the lecture notes (see figure 2.29a and its discussion), both the Poisson equation inside the cavity and the boundary conditions at the conductor's surface may be satisfied by adding, to the unperturbed Coulomb field of the real charge, that of an image charge q placed at distance $d > R$ from the center, with q and d related to q' and d' by Eqs. (2.198). Solving these equations for q and d (now functions of the given q' and d'), we obtain

$$d = \frac{R^2}{d'}, \quad q = -q'\frac{d}{R} = -q'\frac{R}{d'}.$$

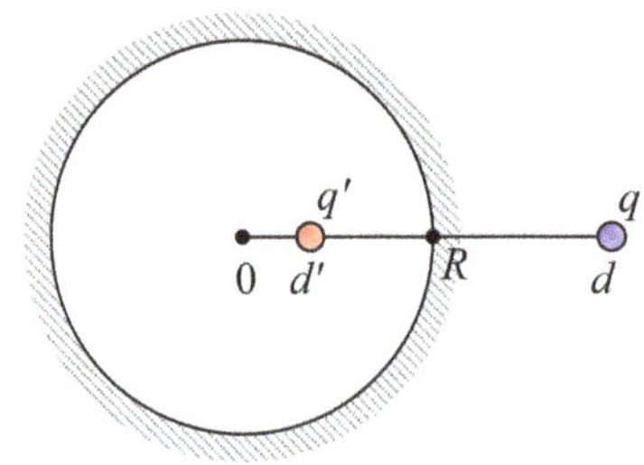

From here, the magnitude of force **F** (directed toward the image charge, i.e. the nearest point of the conducting surface) is

$$F(d') = \frac{|qq'|}{4\pi\varepsilon_0}\frac{1}{(d-d')^2} = \frac{q'^2}{4\pi\varepsilon_0}\frac{R}{d'}\frac{1}{(R^2/d' - d')^2} = \frac{q'^2}{4\pi\varepsilon_0}\frac{Rd'}{(R^2 - d'^2)^2}.$$

As a sanity check, the force vanishes at $d' \to 0$:

$$F(d') \to \frac{q'^2}{4\pi\varepsilon_0}\frac{d'}{R^3} \to 0, \quad \text{at } \frac{d'}{R} \to 0,$$

in accord with problem's symmetry, while at $d' \to R$ it tends to the expression,

$$F_{\text{plane}}(R - \varepsilon) = \frac{q'^2}{4\pi\varepsilon_0}\frac{1}{(2\delta)^2}, \qquad \delta \equiv R - d',$$

which we would get for the charge at distance δ from a conducting plane.

Now the required interaction energy U may be defined as the difference between the energy of the (charge + sphere) system and the total energy of these objects when they are far from each other. Hence, U may be calculated as the (minus) integral of the interaction force on a charge's path from infinity to the cavity center. Our results for the force **F** are only valid inside the sphere, but we may circumvent this problem by extending the integration from the center to distance $\varepsilon \ll R$ from the surface, and subtracting from the result the potential energy (2.191) provided by at that point by a conducting plane (reader, explain why):

$$U(0) = -\int_0^{R-\varepsilon} F(x)dx - U_{\text{plane}}(R-\varepsilon)$$
$$= -\frac{q'^2}{4\pi\varepsilon_0}\left[\int_0^{R-\varepsilon} \frac{R^2 x}{(R^2-x)^2}dx - \frac{1}{4\varepsilon}\right] = -\frac{q'^2}{8\pi\varepsilon_0 R}. \qquad (*)$$

According to Eq. (2.192), the magnitude of this energy is twice larger than what a conducting plane would give at distance R from the charge. The reason for such difference is clear: for a spherical cavity, all induced surface charges are at the same distance (R) from the initial charge, while only the closest charges on the conducting plane are at such distance, so that the effect of their attraction is smaller.

A much simpler way to obtain Eq. (*) is to use Eq. (1.65) for the electric field energy, limiting the integral to the cavity interior, $r < R$ (since there is no field in the conductor). The divergence of the integral at $r \to 0$ disappears if U is understood as above, i.e. as the difference between the charge's field in the cavity and that in the free space:

$$U \equiv \frac{\varepsilon_0}{2}\left(\int_{\text{cavity}} E^2_{\text{in cavity}}\, d^3r - \int_{\text{all space}} E^2_{\text{in free space}}\, d^3r\right)$$
$$= \frac{\varepsilon_0}{2}\left(\int_{r<R}\left(E^2_{\text{in cavity}} - E^2_{\text{in free space}}\right)d^3r - \int_{r>R} E^2_{\text{in free space}}\, d^3r\right).$$

According to the Gauss law, if charge is in the center, its field at $r < R$ is not affected by the conductor, so that the first integral in the last expression vanishes and we obtain

$$U = -\frac{\varepsilon_0}{2}\int_{r>R} E^2_{\text{in free space}}\, d^3r = -\frac{\varepsilon_0}{2}\int_R^{\infty}\left(\frac{q'}{4\pi\varepsilon_0 r^2}\right)^2 4\pi r^2 dr = -\frac{q'^2}{8\pi\varepsilon_0 R},$$

i.e. the same result as given by Eq. (*).

One more way to obtain the same result is to first calculate the small energy increment δU due to motion of a small addition $\delta q \ll q$ to the charge, moving it from infinity to the charge location, i.e. the center of the sphere, replacing the conducting sphere with a thin spherical shell with a uniformly distributed charge q. (Physically, such replacement of the conductor by the charge of its external surface is equivalent to the subtraction made at the previous solution method.) Since, due to its spherical symmetry, such a charge does not create an electric field inside the sphere,

$$\delta U = -\int_0^{\infty} \delta F dr = -\delta q \int_0^{\infty} E dr$$
$$= -\delta q \int_R^{\infty} E dr = -\delta q \int_R^{\infty} \frac{q}{4\pi\varepsilon_0 r^2} dr = -\frac{q\delta q}{4\pi\varepsilon_0 R} = \delta\left(-\frac{q^2}{8\pi\varepsilon_0 R}\right).$$

Integrating the result from $q = 0$ to $q = q'$, we again arrive at Eq. (*).

***Problem* 2.32.** Use the method of images to find the Green's function of the system shown in the figure below, where the bulge on the conducting plane has the shape of a semi-sphere of radius R.

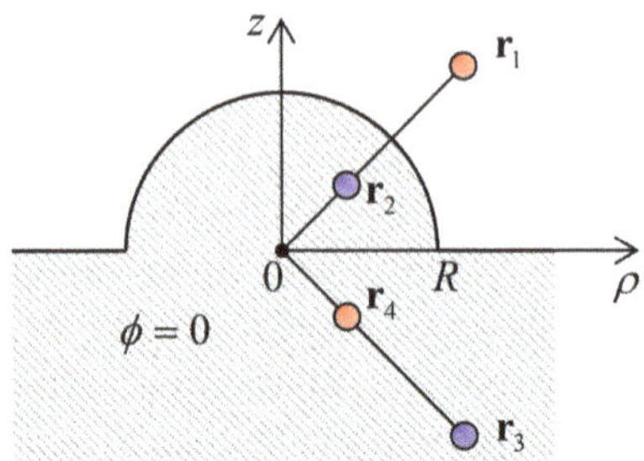

Solution: Let a (real) charge q_1 be at point $\mathbf{r}_1$ in the free-space part of the system. Then all the boundary conditions may be satisfied using the three charge images shown in the figure above, with

$$q_2 = -q_1\frac{R}{r_1}, \quad \mathbf{r}_2 = \left(\frac{R}{r_1}\right)^2\mathbf{r}_1;$$
$$q_3 = -q_1, \quad \boldsymbol{\rho}_3 = \boldsymbol{\rho}_1, \quad z_3 = -z_1;$$
$$q_4 = -q_3\frac{R}{r_3}, \quad \mathbf{r}_4 = \left(\frac{R}{r_3}\right)^2\mathbf{r}_3;$$

where $\boldsymbol{\rho}_j$ are the horizontal components of the radius-vectors $\mathbf{r}_j$. As a result, the Green's function may be represented simply as

$$G(\mathbf{r}, \mathbf{r}_1) = \sum_{j=1}^{4}\frac{q_j/q_1}{|\mathbf{r} - \mathbf{r}_j|}.$$

However, when expressed explicitly in the apparently most suitable, cylindrical coordinates, in which $r_1 = (\rho_1^2 + z_1^2)^{1/2}$, this compact expression becomes somewhat bulky, because each denominator looks like

$$|\mathbf{r} - \mathbf{r}_j| = \left[\rho^2 + \rho_j^2 - 2\rho\rho_j\cos(\varphi - \varphi_j) + (z - z_j)^2\right]^{1/2}.$$

Problem* 2.33. Use the spherical inversion, expressed by Eq. (2.198) of the lecture notes, to develop an iterative method for more and more precise calculation of the mutual capacitance between two similar metallic spheres of radius R, with centers separated by distance $d > 2R$.

Solution: There are several ways to develop the requested iterative process; the following one uses perhaps the clearest logic. At each step of the process, let us keep the total charge of the right sphere equal to $+Q$, and that of the left sphere equal to $-Q$, and keep the surface of each sphere equipotential (at potentials ϕ_L and ϕ_R, correspondingly), while calculating more and more exact values of these potentials, and hence that of the mutual capacitance

$$C \equiv \frac{Q}{V} = \frac{Q}{\phi_R - \phi_L}, \qquad (*)$$

by replacing the genuine distribution of the charge on the surface of each sphere with more and more complex systems of point image charges q, q', q'', …—see the figure below.

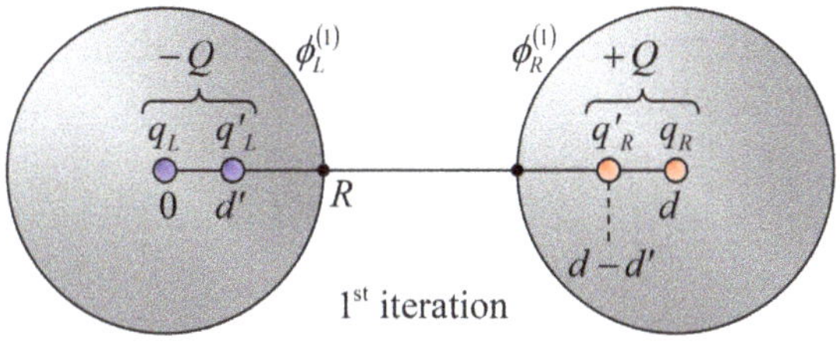

As the zeroth iteration, the distributed charges of the sphere surfaces are approximated with point charges $q_R^{(0)} = +Q$ and $q_L^{(0)} = -Q$, located at their centers. (As we know, this approximation correctly describes the field distribution in space at very large distances, i.e. it is asymptotically correct at $d/R \gg 1$.)

In order to make the first iteration, let us calculate the value of ϕ_L using the fact proved in section 2.6: if the field is induced by a point charge $q_R^{(0)}$, located at distance $d > R$ from sphere's center, then the potential of sphere's surface equals zero if its genuine surface charge is replaced with a point charge $q'_L = -(R/d)q_R^{(0)} = -(R/d)Q$, located at distance $d' = R^2/d$ from the center—see Eqs. (2.198) of the lecture notes. Now, following our general commitment, let us correct the central charge q_L to such a value $q_L^{(1)}$ that the sum of all charges inside the left sphere is equal to $-Q$:

$$q_L^{(1)} + q'_L = -Q, \quad \text{giving} \quad q_L^{(1)} = -Q - q'_L = -Q + \frac{R}{d}Q = -Q\left(1 - \frac{R}{d}\right).$$

This charge still keeps the surface potential of the left sphere constant, but changes its value to

$$\phi_L^{(1)} = \frac{q_L}{4\pi\varepsilon_0 R} = -\frac{Q}{4\pi\varepsilon_0 R}\left(1 - \frac{R}{d}\right).$$

Generally, we would need to carry out a similar adjustment of charge q_R and potential ϕ_R, but due to the symmetry of our problem, the results are evident:

$$q_R^{(1)} = -q_L^{(1)} = Q\left(1 - \frac{R}{d}\right), \qquad \phi_R^{(1)} = -\phi_L^{(1)} = \frac{Q}{4\pi\varepsilon_0 R}\left(1 - \frac{R}{d}\right).$$

Now we may calculate the mutual capacitance in the first approximation:

$$C^{(1)} = \frac{Q}{\phi_R^{(1)} - \phi_L^{(1)}} = 2\pi\varepsilon_0 R \frac{1}{1 - R/d};$$

this expression shows that C increases as the distance d between the spheres is reduced—as it should.

For the next, second iteration, we have to use the same approach to correct the values of ϕ_L and ϕ_R, taking into account the corrected values of the fields of the two central point charges, $q_L{}^{(1)}$ and $q_R{}^{(1)}$, as well as the image charges q'_L and q'_R. According to Eq. (2.198), the former goal may be achieved just by the adjustment of charges q' and q'_R to new values,

$$q'^{(2)}_L = -\frac{R}{d} q_R^{(1)} = -\frac{R}{d}\left(1 - \frac{R}{d}\right)Q, \qquad q'^{(2)}_R = -q'^{(2)}_L = \frac{R}{d}\left(1 - \frac{R}{d}\right)Q,$$

but the latter step requires the introduction of new image charges,

$$q''_L = -\frac{R}{d - d'} q'_R = \frac{R}{d - d'} q'_L = -\frac{R}{d - R^2/d}\frac{R}{d}Q = -\frac{R^2}{d^2 - R^2}Q,$$

$$q''_R = -q''_L = \frac{R^2}{d^2 - R^2}Q,$$

located at a different distance, $d'' = R^2/(d - d') = R^2/(d - R^2/d) > d'$, from the centers of the respective spheres—see the figure below.

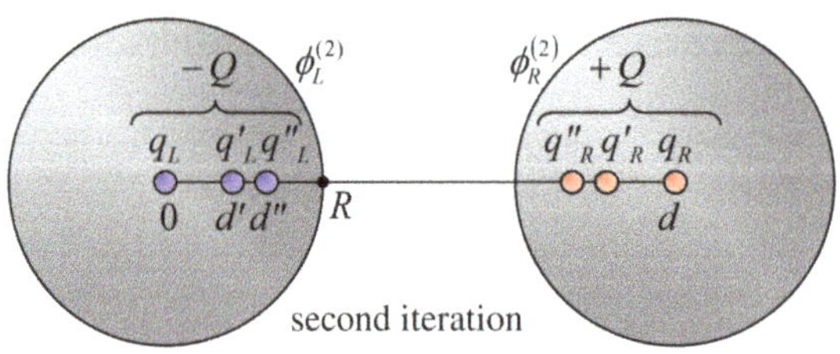

Now again, just at the first iteration, we need to adjust the centrally positioned charges to keep the total charge inside each sphere equal to $\pm Q$; for example,

$$q_R^{(2)} + q'^{(2)}_R + q''_R = +Q,$$

obtaining

$$q_R^{(2)} = Q - q'^{(2)}_R - q''_R = Q - \frac{R}{d}\left(1 - \frac{R}{d}\right)Q - \frac{R^2}{d^2 - R^2}Q$$

$$= \frac{d^4 - Rd^3 - R^2d^2 - R^3d - R^4}{d^2(d^2 - R^2)}Q.$$

Since the net effect of all other charges on ϕ_R vanishes, in the second iteration this potential is:

$$\begin{aligned}\phi_R^{(2)} &= \frac{q_R^{(2)}}{4\pi\varepsilon_0 R}\\ &= \frac{Q}{4\pi\varepsilon_0 R}\frac{d^4 - Rd^3 - R^2d^2 + R^3d - R^4}{d^2(d^2 - R^2)}\\ &\equiv \frac{Q}{4\pi\varepsilon_0 R}\frac{1 - (R/d) - (R/d)^2 + (R/d)^3 - (R/d)^4}{1 - (R/d)^2}.\end{aligned}$$

Due to the problem's symmetry, $\phi_L = -\phi_R$ at all iterations, so that the second iteration gives the following mutual capacitance value:

$$C^{(2)} = \frac{Q}{\phi_R^{(2)} - \phi_L^{(2)}} = 2\pi\varepsilon_0 R\frac{1 - (R/d)^2}{1 - (R/d) - (R/d)^2 + (R/d)^3 - (R/d)^4}.$$

This process may be continued similarly, in each iteration adding one more image charge inside each sphere, and then readjusting its prior charges and the surface potential. However, even the initial iterations give surprisingly accurate results if the ratio R/d is not too close to the convergence limit $(R/d)_{\max} = 1/2$; for example, for d as small as $4R$ (i.e. the gap between the spheres as small as one diameter) the accuracy of the first iteration is already better than 1%.

Problem* 2.34. A metallic sphere of radius R_1, carrying electric charge Q, is placed inside a spherical cavity of radius $R_2 > R_1$, cut inside another metal. Calculate the force exerted on the inner sphere if its center is displaced by a small distance $\delta \ll R_1$, $R_2 - R_1$ from that of the cavity—see the figure below.

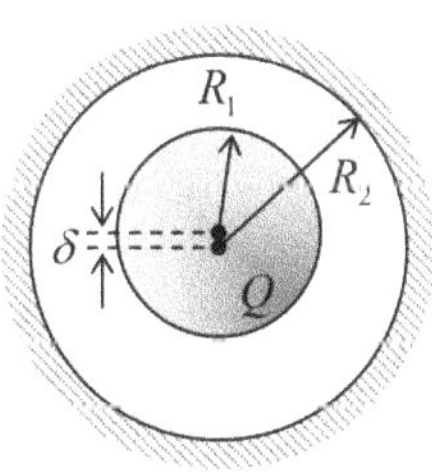

Solution: As follows from the solution of the previous problem, this problem could be solved by building up a sequence of point image charges at the following distances from cavity's center (where 'positive' charges means those with the same charge sign as Q):

$$\begin{array}{ll}\text{positive charges} & \text{negative charges}\\ d = \delta \text{ (real } Q), & D = R_2^2/\delta,\\ d' = \delta + R_1^2/D, & D' = R_2^2/d',\\ d'' = \delta + R_1^2/D', & D'' = R_2^2/d'', \ldots\end{array}$$

This is perhaps the most straightforward way toward an analytical solution, but it would be in the form of a series in the ratio R_1/R_2.

For a more compact solution, let us notice that in the reference frame whose origin coincides with the center of the cavity, and axis z directed along the shift δ, the surface of the inner sphere is described by the following relation:

$$x_1^2 + y_1^2 + (z_1 - \delta)^2 = R_1^2,$$

where the index '1' is added to distinguish a point on the sphere's surface from an arbitrary point in space. In the spherical coordinates, we may rewrite this relation as

$$(r_1 \sin\theta \cos\varphi)^2 + (r_1 \sin\theta \sin\varphi)^2 + (r_1 \cos\theta - \delta)^2 = R_1^2,$$
$$\text{giving } r_1^2 \sin^2\theta + (r_1 \cos\theta - \delta)^2 = R_1^2.$$

In the limit $\delta/R_1 \to 0$, the last expression reduces to the first-order approximation

$$r_1 \approx R_1 + \delta \cos\theta. \qquad (*)$$

Now we will calculate the potential distribution $\phi(r, \theta)$ between the spheres, taking into account its initial (radially symmetric) part, plus the first non-vanishing perturbation proportional to $\delta\cos\theta$. This means that in the general axial-symmetric solution (2.172) of the Laplace equation we may drop all terms with $l > 1$:

$$\phi(r, \theta) \approx a_0 + \frac{b_0}{r} + \left(a_1 r + \frac{b_1}{r^2}\right)\cos\theta, \quad \text{with} \quad a_1, b_1 \propto \delta. \qquad (**)$$

Plugging this distribution into the appropriate boundary conditions,

$$\phi(r_1, \theta) \approx \phi(R_1 + \delta\cos\theta, \theta) = \phi_0, \quad \phi(R_2, \theta) = 0,$$

(where ϕ_0 is the so-far unknown potential of the inner sphere, and the potential of the outer metal taken for zero), we obtain two equations[40],

$$a_0 + \frac{b_0}{R_1 + \delta\cos\theta} + \left(a_1 R_1 + \frac{b_1}{R_1^2}\right)\cos\theta$$
$$\approx a_0 + \frac{b_0}{R_1} - \frac{b_0\delta}{R_1^2}\cos\phi + \left(a_1 R_1 + \frac{b_1}{R_1^2}\right)\cos\theta = \phi_0,$$
$$a_0 + \frac{b_0}{R_2} + \left(a_1 R_2 + \frac{b_1}{R_2^2}\right)\cos\theta = 0,$$

which should be satisfied at all θ. This requirement gives us a system of four equations for four unknown coefficients a and b:

$$a_0 + \frac{b_0}{R_1} = \phi_0, \quad a_1 R_1 + \frac{b_1}{R_1^2} = \frac{b_0\delta}{R_1^2}; \quad a_0 + \frac{b_0}{R_2} = 0, \quad a_1 R_2 + \frac{b_1}{R_2^2} = 0. \quad (*)$$

[40] Taking into account the variation, $\delta\cos\theta$; of radius r_1 in the denominators of terms with a_1 and b_1 would give us additional terms of the order of $(\delta\cos\theta)^2$ and higher, which have to be ignored in our linear approximation.

This system may be easily solved, because the equations for a_0 and b_0, and hence their solution,

$$a_0 = -\phi_0 \frac{R_1}{R_2 - R_1}, \qquad b_0 = \phi_0 \frac{R_1 R_2}{R_2 - R_1},$$

do not depend on a_1 and b_1.

At this stage we may interrupt the solution of the system (*) for a minute, to express the potential ϕ_0 of the inner sphere via its charge Q, by applying the Gauss law to a spherical surface of some radius r within the range $r_1 < r < R_2$:

$$\oint_S E_n d^2r \equiv -\oint_S \frac{\partial \phi}{\partial r} d^2r = \frac{Q}{\varepsilon_0}.$$

For the potential distribution given by Eq. (2.69), the surface integral equals

$$\oint_S \left[\frac{b_0}{r^2} - \left(a_1 - 2\frac{b_1}{r^3}\right)\cos\theta\right] d^2r = 2\pi r^2 \int_0^{2\pi} \left[\frac{b_0}{r^2} - \left(a_1 - 2\frac{b_1}{r^3}\right)\cos\theta\right] \sin\theta d\theta$$
$$= 4\pi b_0 = 4\pi\phi_0 \frac{R_2 - R_1}{R_1 R_2}$$

so that the relation between ϕ_0 and Q (in the linear approximation in δ only!) is the same as at $\delta = 0$:

$$\phi_0 = \frac{Q}{C}, \qquad \text{where } C = 4\pi\varepsilon_0 \frac{R_1 R_2}{R_2 - R_1} \equiv 4\pi\varepsilon_0 \left(\frac{1}{R_1} - \frac{1}{R_2}\right)$$

—cf. Eq. (2.56) of the lecture notes. Currently we need this result only to express the coefficient b_0 via the given charge Q:

$$b_0 = \phi_0 \frac{R_1 R_2}{R_2 - R_1} = \frac{Q}{4\pi\varepsilon_0}.$$

Using this relation and solving the remaining system (*) of equations for a_1 and b_1, we obtain

$$a_1 = -b_0 \frac{1}{R_2^3 - R_2^3}\delta = -\frac{Q}{4\pi\varepsilon_0}\frac{1}{R_2^3 - R_1^3}\delta, \quad b_1 = b_0 \frac{R_2^3}{R_2^3 - R_2^3}\delta = \frac{Q}{4\pi\varepsilon_0}\frac{R_2^3}{R_2^3 - R_1^3}\delta,$$

so that the spatial distribution of the radial electric field is

$$E_r = -\frac{\partial \phi}{\partial r} = \frac{b_0}{r^2} + \left(-a_1 + 2\frac{b_1}{r^3}\right)\cos\theta = \frac{Q}{4\pi\varepsilon_0}\left[\frac{1}{r^2} + \frac{1}{R_2^3 - R_1^3}\left(1 + 2\frac{R_2^3}{r^3}\right)\delta\cos\theta\right].$$

We will need to know E_r at the shifted surface of the inner sphere, i.e. at $r = r_1 = R_1 + \delta\cos\theta$. The small difference between r_1 and R_1 has to be taken into account only in the largest, first term of the above expression for E_r, and only in the first, linear approximation:

$$\frac{1}{r_1^2} = \frac{1}{(R_1 + \delta \cos\theta)^2} = \frac{1}{R_1^2[1 + 2(\delta/R_1)\cos\theta]}$$
$$\approx \frac{1 - 2(\delta/R_1)\cos\theta}{R_1^2} = \frac{1}{R_1^2} - 2\frac{1}{R_1^3}\delta\cos\theta.$$

In all other terms, the difference between r_1 and R_1 may be ignored, because they are already proportional to δ. As a result, we obtain

$$E_r|_{r=r_1} \approx \frac{Q}{4\pi\varepsilon_0}\left[\frac{1}{R_1^2} - 2\frac{1}{R_1^3}\delta\cos\theta + \frac{1}{R_2^3 - R_1^3}\left(1 + 2\frac{R_2^3}{R_1^3}\right)\delta\cos\theta\right]$$
$$= \frac{Q}{4\pi\varepsilon_0}\left(\frac{1}{R_1^2} + \frac{3\delta\cos\theta}{R_2^3 - R_1^3}\right).$$

According to the solution of problem 2.1, the normal force f_n per unit area of the inner sphere may be calculated as

$$f_n = \frac{\varepsilon_0}{2}E_r^2\Big|_{r=r_1} = \frac{\varepsilon_0}{2}\left(\frac{Q}{4\pi\varepsilon_0}\right)^2\left(\frac{1}{R_1^2} + \frac{3\delta\cos\theta}{R_2^3 - R_1^3}\right)^2.$$

In the first, linear approximation in small $\delta\cos\theta$:

$$f_n = \frac{\varepsilon_0}{2}\left(\frac{Q}{4\pi\varepsilon_0 R_1^2}\right)^2\left(1 + \frac{3R_1^2\delta\cos\theta}{R_2^3 - R_1^3}\right)^2 \approx \frac{\varepsilon_0}{2}\left(\frac{Q}{4\pi\varepsilon_0 R_1^2}\right)^2\left(1 + \frac{6R_1^2\delta\cos\theta}{R_2^3 - R_1^3}\right).$$

Due to the axial symmetry of the problem, the net force may have only one, vertical component:

$$F = F_z = \oint_{r_1} f_z d^2r = \oint_{r_1} f_n \cos\theta d^2r = 2\pi R_1^2\int_0^\pi f_n \cos\theta \sin\theta d\theta.$$

At the integration, the net contribution of the unperturbed, radially symmetric normal forces vanishes, while the first-perturbation term yields a very simple final result:

$$F = 2\pi R_1^2\ \frac{\varepsilon_0}{2}\left(\frac{Q}{4\pi\varepsilon_0 R_1^2}\right)^2\frac{6R_1^2\delta}{R_2^3 - R_1^3}\int_0^\pi \cos^2\theta\sin\theta d\theta = 2\pi\frac{\varepsilon_0}{2}\left(\frac{Q}{4\pi\varepsilon_0}\right)^2\frac{6\delta}{R_2^3 - R_1^3}\frac{2}{3}$$
$$= \frac{Q^2}{4\pi\varepsilon_0}\frac{\delta}{R_2^3 - R_1^3}.$$

As the first sanity check, at $R_2 \to \infty$ the net force vanishes—as it has to for a charged sphere in free space. More substantially, at $R_1 \to 0$, when the inner sphere turns into a point charge, the result is reduced exactly to the corresponding limit of the by-product of the model solution of problem 2.31:

$$F = \frac{Q^2}{4\pi\varepsilon_0}\frac{\delta}{R_2^3},$$

obtained using the charge image method.

***Problem* 2.35.** Within the simple models of the electric field screening in conductors, discussed in section 2.1 of the lecture notes, analyze the partial screening of the electric field of a point charge q by a plane, uniform conducting film of thickness $t \ll \lambda$, where λ is (depending on the charge carrier statistics) either the Debye or the Thomas–Fermi screening length—see, respectively, Eqs. (2.8) or (2.10). Assume that the distance d between the charge and the film is much larger than t.

Solution: Due to the condition $d \gg t$, the gradient of the electrostatic potential in the film is dominated by its component along the axis (say, z) normal to the film's plane, so we still may use the 1D equation (2.7):

$$\frac{\partial^2 \phi}{\partial z^2} = \frac{1}{\lambda^2}\phi,$$

even if ϕ is relatively slowly (at distances of the order of d) changing in the plane of the film. Integrating both sides of this equation over the film's thickness t, we obtain

$$\frac{\partial \phi_+}{\partial z} - \frac{\partial \phi_-}{\partial z} = \frac{1}{\lambda^2}\int_{-t/2}^{+t/2} \phi dz, \qquad (*)$$

where indices $\pm$ refer to two film surfaces. This equation shows that if $t \ll \lambda$, the film can create only a small difference (of the order of $\phi t/\lambda^2$) of the derivatives, and hence just a negligible (of the order of $\phi t^2/\lambda^2$) variation of the potential inside it. Hence Eq. (*) yields two boundary conditions:

$$\phi_+ = \phi_- \equiv \phi, \qquad \frac{\partial \phi_+}{\partial z} - \frac{\partial \phi_-}{\partial z} = \frac{t}{\lambda^2}\phi. \qquad (**)$$

Now, inspired by the charge-image analysis of the basic problem shown in figure 2.26 of the lecture notes, we may try to describe the fields at $z > 0$ as a superposition of the fields by the real charge q, and by its image q' located at the same distance d, but the opposite side of the film—see the figure below. However, now q' is not necessarily equal to $(-q)$; indeed, if $t \to 0$, all the film's effects, in particular the charge q', should vanish.

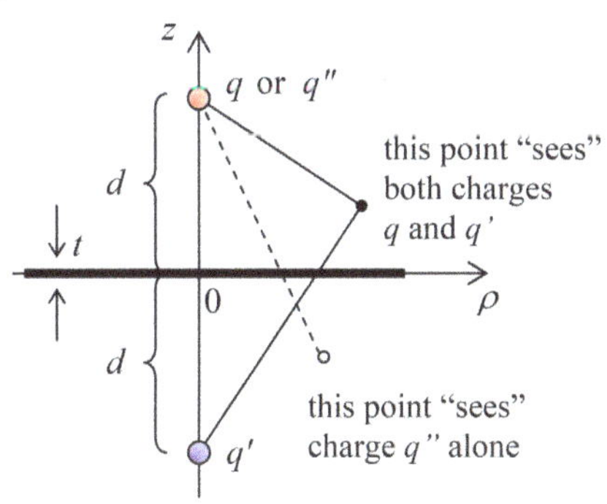

In addition, in contrast to the conducting half-space case, we should expect to find some field's remnants at the other side of the film, at $z < 0$. This field cannot be contributed by the image charge q', because it would provide a potential divergence at its location. Thus, in that half-space we should try to use the real point source only, but with a re-normalized charge q'' rather than the genuine charge q—see the figure above. As a result, we may look for the potential distribution in the form[41]

$$\phi(\rho, z) = \frac{1}{4\pi\varepsilon_0} \times \begin{cases} \dfrac{q}{[\rho^2 + (z-d)^2]^{1/2}} + \dfrac{q'}{[\rho^2 + (z+d)^2]^{1/2}}, & \text{for } z > 0, \\ \dfrac{q''}{[\rho^2 + (z-d)^2]^{1/2}}, & \text{for } z < 0. \end{cases}$$

(In this 'coarse-grain' picture, with the size scale given by $d \gg t$, the film thickness is negligible.)

Plugging this solution into the boundary conditions (**), we see that they are indeed satisfied (so that this is the unique solution of our boundary problem), if the effective charges q' and q'' obey the following two relations:

$$q + q' = q'', \qquad (q - q') - q'' = \frac{td}{\lambda^2} q''.$$

Solving this simple system of two linear equations, we obtain, in particular, the following expression for ratio q''/q—which is evidently the most relevant measure of field's penetration behind the film:

$$\frac{q''}{q} = \frac{1}{1 + td/2\lambda^2}.$$

As the sanity check, q'' tends to q at $t \to 0$, and to zero at $t \to \infty$—as it should. However, the most interesting feature of this result is that the ratio q''/q starts to drop as soon as t is increased to $\sim\lambda^2/d \ll \lambda$, i.e. the field behind the film may be well (though not exponentially well) screened even by a film *much thinner than the screening length*! Very unfortunately, such an enhancement of the electric field's screening by thin conducting films (and the similar enhancement of the magnetic field's shielding by thin ferromagnetic and superconducting films[42]) is overlooked even in some popular textbooks.

***Problem* 2.36.** Suggest a convenient definition of the Green's function for 2D electrostatic problems, and calculate it for:

(i) the unlimited free space, and
(ii) the free space above a conducting plane.

Use the latter result to re-solve problem 2.18.

Solution: As was discussed in the beginning of section 2.10 of the lecture notes, the Green's function $G(\mathbf{r}, \mathbf{r}')$ for 3D electrostatic problems is defined as the partial

[41] As is discussed in section 3.3 of the lecture notes, absolutely the same solution (although with different constants q' and q'') describes the electric field penetration into a half-space filled with a linear dielectric.
[42] See, e.g. sections 5.6, 6.3, and 6.4 of the lecture notes.

solution (in the specific geometry of the problem) of the Poisson equation (1.41) for the electrostatic potential at point **r**, provided that the right-hand side of the equation is proportional to the 3D delta-function $\delta(\mathbf{r} - \mathbf{r}')$ - see Eq. (2.208). If the field is two-dimensional (say, a function of only coordinates x and y), its potential is described by the Poisson equation with $\partial/\partial z = 0$, so that the charge distribution is also z-independent:

$$\left(\frac{\partial^2}{\partial x^2} + \frac{\partial^2}{\partial y^2}\right)\phi = -\frac{\rho(x, y)}{\varepsilon_0}.$$

Hence the most reasonable definition of the 2D Green's function, $G(\boldsymbol{\rho}, \boldsymbol{\rho}')$, where $\boldsymbol{\rho} = \{x, y\}$ is the 2D radius-vector, is the solution of this equation with the similar but 2D delta-functional right-hand side:

$$\nabla_\rho^2 G(\boldsymbol{\rho}, \boldsymbol{\rho}') = -4\pi\delta(\boldsymbol{\rho} - \boldsymbol{\rho}').$$

Physically, this solution is the potential created by a thin straight line, perpendicular to plane $[x, y]$, with the 'unit' linear charge density (in the sense $\lambda/4\pi\varepsilon_0 = 1$). This formulation allows us to readily calculate the Green's function for simple geometries.

(i) In unlimited free space, we know the electric field induced by a thin line of charges with linear density λ—see problem 1.1:

$$E_r = \frac{\lambda}{2\pi\varepsilon_0\rho},$$

where ρ is the shortest distance from the field observation point to the line of charges, i.e. the 2D distance between the observation point and the point (taken for the origin) where the charged line, parallel to axis z, pierces the $[x, y]$ plane. From here, the electrostatic potential is

$$\phi = -\int E_r d\rho = -\frac{\lambda}{2\pi\varepsilon_0}\ln\rho + \text{const} \equiv \frac{\lambda}{4\pi\varepsilon_0} \times (-2\ln\rho + \text{const}).$$

Hence the 2D Green's function may be represented as

$$G(\boldsymbol{\rho}, \boldsymbol{\rho}') = -2\ln|\boldsymbol{\rho} - \boldsymbol{\rho}'| + \text{const} \equiv -\ln|\boldsymbol{\rho} - \boldsymbol{\rho}'|^2 + \text{const},$$

where $\boldsymbol{\rho}$ and $\boldsymbol{\rho}'$ are the 2D radius-vectors of the observation point and charged line's trace ('cross-section'), respectively.

(ii) For a charged line parallel to a grounded conducting plane $y = 0$, crossing the plane of drawing at point $\boldsymbol{\rho}' = \{x', y'\}$, both the Poisson equation and the boundary condition, $\phi(x, 0) = 0$, may be satisfied by the introduction of an image line with the charge density $-\lambda$, crossing the plane of drawing at point $\boldsymbol{\rho}'' = \{x', -y'\}$. Hence the Green's function of the system is

$$\begin{aligned} G(\boldsymbol{\rho}, \boldsymbol{\rho}') &= -\ln|\boldsymbol{\rho} - \boldsymbol{\rho}'|^2 + \ln|\boldsymbol{\rho} - \boldsymbol{\rho}''|^2 \\ &= -\ln[(x - x')^2 + (y - y')^2] + \ln[(x - x')^2 + (y + y')^2]. \end{aligned} \qquad (*)$$

For the particular system that was the subject of problem 2.18 (see the figure below), all we need from the Green's function is the normal component of its gradient at the surface.

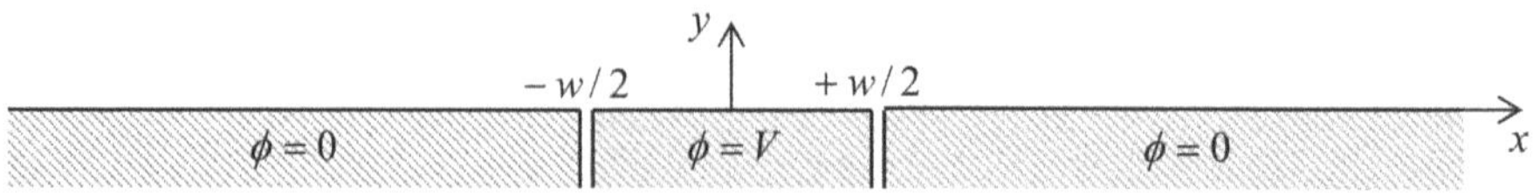

A straightforward differentiation of Eq. (*) yields

$$\left.\frac{\partial G}{\partial n'}\right|_S \equiv \left.\frac{\partial G}{\partial y'}\right|_{y'=0} = \frac{4y}{(x - x')^2 + y^2}.$$

Plugging this expression into the 2D version of the main formula (2.210) of the Green's function theory (with no first term, describing the spatially distributed charge), we obtain

$$\phi(\boldsymbol{\rho}) = \frac{1}{4\pi}\sum_k \phi_k \oint_{l_k} \frac{\partial G(\boldsymbol{\rho}, \boldsymbol{\rho}')}{\partial n'} dr' = \frac{V}{4\pi}\int_{-w/2}^{+w/2} \frac{4y}{(x - x')^2 + y^2} dx'.$$

This integral may be readily worked out using the substitution $\xi \equiv (x - x')/y$, giving[43]

$$\phi = -\frac{V}{\pi}\int_{(x+w/2)/y}^{(x-w/2)/y} \frac{d\xi}{\xi^2 + 1} = \frac{V}{\pi}\left[\tan^{-1}\frac{x + w/2}{y} - \tan^{-1}\frac{x - w/2}{y}\right]. \qquad (**)$$

This formula gives the exactly the same result as the integral (****) in the model solution of problem 2.18. (See the plots in that solution.) Actually, Eq. (**) above may be obtained by working out that integral, but the above Green's-function solution is much simpler. Note also that alternatively the problem could be solved using the 3D Green's function given by Eq. (2.211), but this would cost us one more integration (along the z axis).

***Problem* 2.37.** Calculate the 2D Green's function for the free spaces:

(i) outside a round conducting cylinder, and
(ii) inside a round cylindrical hole in a conductor.

Solutions:

(i) Taking into account the solution of task (i) of the previous problem,

$$G(\boldsymbol{\rho}, \boldsymbol{\rho}')|_{\text{unlimited space}} = -2\ln|\boldsymbol{\rho} - \boldsymbol{\rho}'| + \text{const}$$

(where $\boldsymbol{\rho} = \{\rho, \varphi\}$ is the 2D radius-vector of a point), and inspired by the solution of the similar 3D problem (see Eqs. (2.197) and (2.198) of the lecture notes), let us try to look for the solution of our current problem in a similar form:

[43] See, e.g. Eq. (A.30d).

$$G(\boldsymbol{\rho}, \boldsymbol{\rho}') = -2\ln|\boldsymbol{\rho} - \boldsymbol{\rho}'| + 2a\ln|\boldsymbol{\rho} - \boldsymbol{\rho}''| + b \\ \equiv -\ln\ |\boldsymbol{\rho} - \boldsymbol{\rho}'|^2 + a\ln\ |\boldsymbol{\rho} - \boldsymbol{\rho}''|^2 + b, \qquad (*)$$

where $\boldsymbol{\rho}''$ is the charge image point, whose distance from the center is related to that of the actual source point $\boldsymbol{\rho}'$ (see the figure below) by the inversion relation similar to the first of Eqs. (2.198):

$$\rho'' = \frac{R^2}{\rho'},$$

and coefficients a (physically, the relative magnitude of the image charge density) and b are still to be determined.

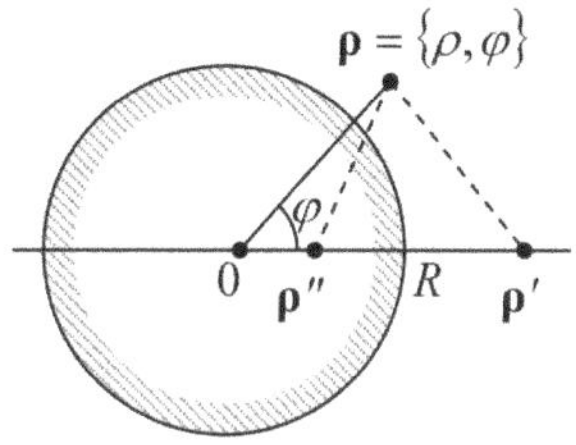

Since Eq. (*) satisfies the 2D Laplace equation for any a and b, these coefficients may be found (and thus the solution (*) proved) by requiring the Dirichlet boundary condition

$$G(\boldsymbol{\rho}, \boldsymbol{\rho}') = 0, \quad \text{for } \boldsymbol{\rho} \equiv \{R, \varphi\},$$

to be satisfied at any point at the boundary, i.e. for any polar angle φ. Referring φ to the direction from the center toward points $\boldsymbol{\rho}'$ and $\boldsymbol{\rho}''$ (so that $\boldsymbol{\rho}' = \{\rho', 0\}$ and $\boldsymbol{\rho}'' = \{\rho'', 0\}$), and applying the basic trigonometry to the figure above, we may spell out this condition as

$$-\ln(\rho'^2 + R^2 - 2R\rho'\cos\varphi) + a\ln(\rho''^2 + R^2 - 2R\rho''\cos\varphi) + b$$
$$\equiv -\ln(\rho'^2 + R^2 - 2R\rho'\cos\varphi) + a\ln\left[\left(\frac{R^2}{\rho'}\right)^2 + R^2 - 2R\left(\frac{R^2}{\rho'}\right)\cos\varphi\right] + b$$
$$\equiv -\ln(\rho'^2 + R^2 - 2R\rho'\cos\varphi) + a\ln\left[\frac{R^2}{\rho'^2}(R^2 + \rho'^2 - 2R\rho'\cos\varphi)\right] + b = 0\,.$$

It is evident that this condition is satisfied if $a = 1$ and $b = -\ln(R^2/\rho'^2) \equiv 2\ln(\rho'/R)$, so that, finally, the Green's function is

$$G(\boldsymbol{\rho}, \boldsymbol{\rho}') = -\ln\ |\boldsymbol{\rho} - \boldsymbol{\rho}'|^2 + \ln\ |\boldsymbol{\rho} - \boldsymbol{\rho}''|^2 - \ln\frac{R^2}{\rho'^2} \equiv 2\ln\frac{\rho'}{R}\frac{|\boldsymbol{\rho} - \boldsymbol{\rho}''|}{|\boldsymbol{\rho} - \boldsymbol{\rho}'|}. \qquad (**)$$

In terms of electrostatics, the result $a = 1$ may be interpreted as the image charge (or rather its linear density) having, in contrast to the 3D cases, the same magnitude as the real charge density, and the opposite charge. This equality kills the logarithmic divergence of the Green's function (pertinent to the unlimited free space) at large distances.

(ii) Repeating the solution of part (i) for this geometry, we see that Eq. (**) is valid for the 'inner' problem as well, but now with $\rho' < R$ (see the figure below). In particular, at the center of the circle (at $\boldsymbol{\rho} = 0$),

$$G(0, \boldsymbol{\rho}') = 2 \ln \frac{\rho''}{R} = 2 \ln \frac{R}{\rho'} > 0.$$

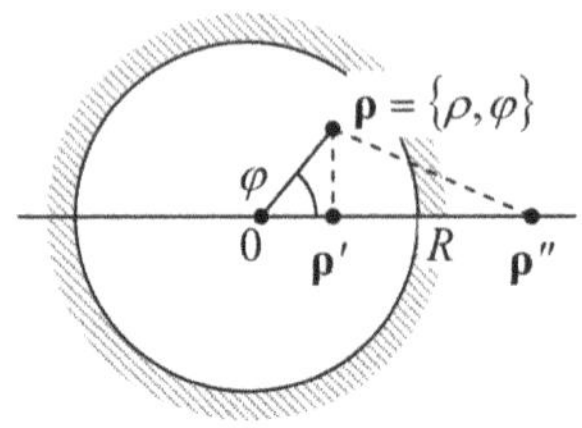

***Problem* 2.38.** Solve task (i) of problem 2.16 (see also problem 2.17), using the Green's function method.

Solution: For our problem, the 2D form of Eq. (2.210), with no space charge, and the half-pipe potentials specified in the problem's assignment, yields

$$\phi(\boldsymbol{\rho}) = \frac{1}{4\pi}\frac{V}{2} \operatorname{sgn}(\rho - R)\left(\int_0^{\pi} - \int_{-\pi}^{0}\right)\frac{\partial G(\boldsymbol{\rho}, \boldsymbol{\rho}')}{\partial \rho'}\bigg|_{\rho'=R} R d\varphi', \qquad (*)$$

where $G(\boldsymbol{\rho}, \boldsymbol{\rho}')$ is the 2D Green's function of the thin, hollow, grounded pipe, and the sign factor reflects the different signs of the scalar product of the vector $\boldsymbol{\rho}'$ by the external normal $\mathbf{n}$ to the surface for the external problem ($\rho > R$, $\boldsymbol{\rho}' \cdot \mathbf{n} > 0$, see the left panel of the figure below) and the internal problem ($\rho < R$, $\boldsymbol{\rho}' \cdot \mathbf{n} < 0$, shown in the right-hand panel of that figure).

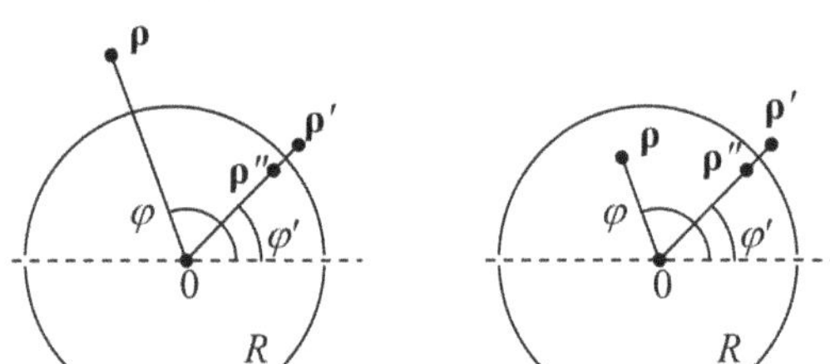

As was shown in the solution of the previous problem, the Green's function has the same form for the inner and outer case:

$$G(\boldsymbol{\rho}, \boldsymbol{\rho}') = 2 \ln \frac{\rho'}{R}\frac{|\boldsymbol{\rho} - \boldsymbol{\rho}''|}{|\boldsymbol{\rho} - \boldsymbol{\rho}'|},$$

where the charge image point $\boldsymbol{\rho}''$ sits on the same radius as the source point $\boldsymbol{\rho}'$, at distance $\rho'' = R^2/\rho'$ from the center. In the notation of the figure above,

$$G(\boldsymbol{\rho}, \boldsymbol{\rho}') = 2\ln\rho' + \ln[\rho^2 + \rho''^2 - 2\rho\rho''\cos(\varphi' - \varphi)] \\ - \ln[\rho^2 + \rho'^2 - 2\rho\rho'\cos(\varphi' - \varphi)] + \text{const}.$$

Taking into account that at $\rho' = R$, the derivative $d\rho''/d\rho' = -(R/\rho')^2$ equals (-1), this expression yields

$$\left.\frac{\partial G(\boldsymbol{\rho}, \boldsymbol{\rho}')}{\partial \rho'}\right|_{\rho'=R} = \frac{2}{R}\left[1 - \frac{2R^2 - 2\rho R\cos(\varphi' - \varphi)}{\rho^2 + R^2 - 2\rho R\cos(\varphi' - \varphi)}\right] \\ = \frac{2}{R}\frac{\rho^2 - R^2}{\rho^2 + R^2 - 2\rho R\cos(\varphi' - \varphi)}$$

Now using Eq. (A.40)[44], with

$$a - b = (\rho^2 + R^2) + 2R\rho \equiv (\rho + R)^2, \\ (a^2 - b^2)^{1/2} = [(\rho^2 + R^2)^2 - (2\rho R)^2]^{1/2} \equiv |\rho^2 - R^2|,$$

we obtain

$$\int \left.\frac{\partial G(\boldsymbol{\rho}, \boldsymbol{\rho}')}{\partial \rho'}\right|_{\rho'=R} d\varphi' = \frac{4}{R}\,\mathrm{sgn}(\rho - R)\tan^{-1}\left[\frac{\rho + R}{|\rho - R|}\tan\frac{\varphi' - \varphi}{2}\right].$$

At the substitution of this expression into Eq. (*), the product of two front sign functions yields 1, and we obtain

$$\phi(\boldsymbol{\rho}) = \frac{V}{2\pi}\left\{\tan^{-1}\left[\frac{\rho + R}{|\rho - R|}\tan\frac{\varphi' - \varphi}{2}\right]\Bigg|_{\varphi' = 0}^{\varphi' = \pi} \right. \\ \left. - \tan^{-1}\left[\frac{\rho + R}{|\rho - R|}\tan\frac{\varphi' - \varphi}{2}\right]\Bigg|_{\varphi' = -\pi}^{\varphi' = 0}\right\} \\ = \frac{V}{2\pi}\left\{2\tan^{-1}\left[\frac{\rho + R}{|\rho - R|}\mathrm{cotan}\frac{\varphi}{2}\right] + 2\tan^{-1}\left[\frac{\rho + R}{|\rho - R|}\tan\frac{\varphi}{2}\right] - \pi\,\mathrm{sgn}(\sin\varphi)\right\},$$

This is already the final answer, and its numerical plots coincide with those shown in the model solution of problem 2.16. However, for the analytical comparison of the results we may rewrite it as

$$\phi(\boldsymbol{\rho}) = \frac{V}{\pi}\left(\alpha + \alpha' - \frac{\pi}{2}\mathrm{sgn}(\sin\varphi)\right),$$

$$\text{with}\quad \tan\alpha = \frac{\rho + R}{|\rho + R|}\mathrm{cotan}\frac{\varphi}{2}, \qquad \tan\alpha' = \frac{\rho + R}{|\rho + R|}\tan\frac{\varphi}{2},$$

[44] It is applicable to our problem, because in it, $a^2 - b^2 = (R^2 + \rho^2)^2 - 4R^2\rho^2 = (R^2 - \rho^2)^2 \geqslant 0$.

so that using the well-known formula for the tangent of a sum, we obtain

$$\tan(\alpha + \alpha') \equiv \frac{\tan\alpha + \tan\alpha'}{1 - \tan\alpha\tan\alpha'}$$
$$= \frac{\rho + R}{|\rho + R|}\frac{1}{1 - (R+\rho)^2/(R-\rho)^2}\left(\text{cotan}\frac{\varphi}{2} + \tan\frac{\varphi}{2}\right)$$
$$= -\frac{|\rho^2 - R^2|}{4\rho R}\frac{\sin^2(\varphi/2) + \cos^2(\varphi/2)}{\sin(\varphi/2)\cos(\varphi/2)} = -\frac{|\rho^2 - R^2|}{2\rho R\sin\varphi},$$

and our result may be rewritten in the form derived in problem 2.16:

$$\phi(\boldsymbol{\rho}) = \frac{V}{\pi}\left\{\tan^{-1}\left[-\frac{|\rho^2 - R^2|}{2\rho R\sin\varphi}\right] + \frac{\pi}{2}\,\text{sgn}(\sin\varphi)\right\} \equiv \frac{V}{\pi}\tan^{-1}\frac{2\rho R\sin\varphi}{|\rho^2 - R^2|}.$$

An advantage of the Green's function approach, as well as the variable separation method, over the conformal mapping is that they may be more readily used in less symmetric situations—for example, applied to a similar problem with the pipe cut into two fragments as shown in the figure below, where β is an arbitrary angle rather than exactly π as in the problem solved above. (Carrying out such a calculation, using both methods, is a very useful additional exercise, highly recommended to the reader.)

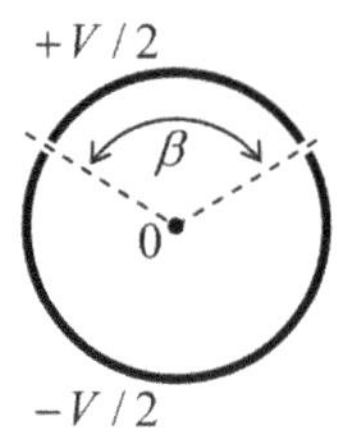

***Problem* 2.39.** Solve the 2D boundary problem that was discussed in section 2.11 of the lecture notes (figure 2. 34) using:

(i) the finite difference method, with a finer square mesh, $h = a/3$, and
(ii) the variable separation method.

Compare the results at the mesh points, and comment.

Solutions:

(i) Due to problem's symmetry, there are only two types of internal nodes in this mesh: A and B (see the figure below), so we need only two finite difference equations, each describing a five-point 'cross' (see figure 2.33b) with its center at each of these points. For these two cases, Eq. (2.221) yields:

$$\text{–point A:} \quad 0 + \phi_A + V + \phi_B - 4\phi_A = 0,$$
$$\text{–point B:} \quad 0 + \phi_B + \phi_A + 0 - 4\phi_B = 0.$$

Solving this simple system of two linear equations, we obtain

$$\phi_A = \frac{3}{8}V = 0.375\ V, \quad \phi_B = \frac{1}{8} = 0.125\ V.$$

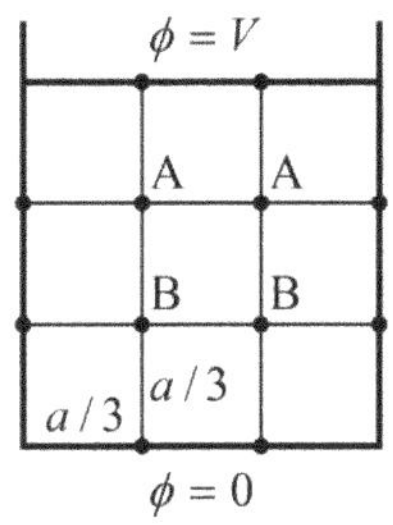

(ii) In full analogy with the solution (2.95) of the similar 3D problem solved by the variable separation method in section 2.5, the solution of this 2D problem is

$$\phi(x, y) = \sum_{n=1}^{\infty} c_n \sin\frac{\pi n x}{a} \sinh\frac{\pi n y}{a},$$

where the coefficients c_n have to be found from the boundary condition on the top lid ($y = a$):

$$V = \sum_{n=0}^{\infty} c_n \sin\frac{\pi n x}{a} \sinh \pi n.$$

Performing the reciprocal Fourier transform (i.e. multiplying both parts of this equation by $\sin(\pi n' x/a)$, and integrating each of them over x from 0 to a), we obtain

$$\begin{aligned} c_n &= \frac{1}{(a/2)\sinh \pi n} V \int_0^a \sin\frac{\pi n x}{a} dx \\ &= \begin{cases} 4V/\pi n \sinh \pi n, & \text{for} \quad n = 2j + 1, \\ 0, & \text{for} \quad n = 2j, \end{cases} \qquad \text{with } j = 0, 1, 2 \ldots \end{aligned}$$

In the $a/3$ mesh nodes, these formulas yield

$$\phi_A = \frac{4}{\pi} V \sum_{j=0}^{\infty} \frac{\sinh[2\pi(2j+1)/3]}{(2j+1)\sinh[\pi(2j+1)]} \approx 0.381\ V,$$

$$\phi_B = \frac{4}{\pi} V \sum_{j=0}^{\infty} \frac{\sinh[\pi(2j+1)/3]}{(2j+1)\sinh[\pi(2j+1)]} \approx 0.119\ V.$$

Note that though the sums are formally infinite, virtually all their values are provided by the first terms, because the factor $(2j + 1)$ in the denominator and the exponential character of the sinh functions at large values of its arguments function ensure a fast convergence of the series.

Comparing the results provided by the two methods, we see that the finite difference method's error, even with this rather coarse mesh, is just a few percent.

Reference

[1] Lambe J and Jaklevic R 1976 *Phys. Rev. Lett.* **22** 1371

IOP Publishing

Chapter 3

Dipoles and dielectrics

***Problem* 3.1.** Prove Eqs. (3.3) and (3.4) of the lecture notes, starting from Eqs. (1.38) and (3.2).

Solution: In section 3.1 of the lecture notes, the first two terms of the multipole expansion (3.3) have already been derived from the general Eq. (1.38),

$$\phi(\mathbf{r}) = \frac{1}{4\pi\varepsilon_0}\int \frac{\rho(\mathbf{r}')}{|\mathbf{r} - \mathbf{r}'|} d^3r',$$

so that we need only calculate the third, quadrupole term using the third term on the right-hand side of Eq. (3.2):

$$\phi_q(\mathbf{r}) = \frac{1}{4\pi\varepsilon_0}\int \rho(\mathbf{r}')\frac{1}{2!}\sum_{j,j'=1}^{3} \left.\frac{\partial^2 f(R)}{\partial r_j \partial r_{j'}}\right|_{\mathbf{r}'=0} r'_j r'_{j'}\, d^3r', \qquad (*)$$

with $f(R) = 1/R$, where $R = |\mathbf{R}|$, and $\mathbf{R} \equiv \mathbf{r} - \mathbf{r}'$. Spelling out the function f via the Cartesian coordinates $R_j = r_j - r_j'$ of the vector $\mathbf{R}$,

$$f(R) = \left[\sum_{j=1}^{3}(r_j - r'_j)^2\right]^{-1/2},$$

makes its first differentiation in Eq. (*) elementary:

$$\frac{\partial f}{\partial r_j} = -\frac{1}{2}\left[\sum_{j=1}^{3}(r_j - r'_j)^2\right]^{-3/2} 2(r_j - r'_j) = -\left[\sum_{j=1}^{3}(r_j - r'_j)^2\right]^{-3/2}(r_j - r'_j).$$

However, the result of the second differentiation depends on whether indices j and j' coincide, because if they do, we should differentiate the second multiplier $(r_j - r'_j)$ in the first derivative as well:

doi:10.1088/978-0-7503-1407-7ch3

$$\frac{\partial^2 f}{\partial r_j \partial r_{j'}} \equiv \frac{\partial}{\partial r_{j'}}\left(\frac{\partial f}{\partial r_j}\right) = \frac{\partial}{\partial r_{j'}}\left\{-\left[\sum_{j=1}^{3}(r_j - r'_j)^2\right]^{-3/2}(r_j - r'_j)\right\}$$

$$= \frac{3}{2}\left[\sum_{j=1}^{3}(r_j - r'_j)^2\right]^{-5/2} 2(r_j - r'_j)(r_{j'} - r'_{j'}) - \delta_{jj'}\left[\sum_{j=1}^{3}(r_j - r'_j)^2\right]^{-3/2}$$

$$= \frac{3R_jR_{j'}}{R^5} - \frac{\delta_{jj'}}{R^3},$$

so that the value participating in Eq. (*) is

$$\left.\frac{\partial^2 f}{\partial r_j \partial r_{j'}}\right|_{\mathbf{r}'=0} = \frac{3r_jr_{j'}}{r^5} - \frac{\delta_{jj'}}{r^3} = \frac{3r_jr_{j'} - \delta_{jj'}r^2}{r^5},$$

and may be moved out of the integral over the $\mathbf{r}'$ space. The resulting expression for the quadrupole potential,

$$\phi_q(\mathbf{r}) = \frac{1}{4\pi\varepsilon_0}\frac{1}{2r^5}\sum_{j,j'=1}^{3}\left(3r_jr_{j'} - \delta_{jj'}r^2\right)\int \rho(\mathbf{r}')\, r'_j r'_{j'} d^3r', \qquad (**)$$

although absolutely correct, is still not quite convenient for most applications: it is beneficial to move the account for the special character of the diagonal components of the sum into the integration over r'.

The standard way of doing that is shown in Eq. (3.4); plugging that expression for $\mathcal{Q}_{jj'}$ into the last term of Eq. (3.3), we may rewrite the latter as

$$\phi_q(\mathbf{r}) = \frac{1}{4\pi\varepsilon_0}\frac{1}{2r^5}\sum_{j,j'=1}^{3} r_jr_{j'}\int \rho(\mathbf{r}')\left(3r'_jr'_{j'} - \delta_{jj'}r'^2\right) d^3r'; \qquad (***)$$

let us prove that Eqs. (***) and the already proved Eq. (**) are equivalent. The off-diagonal terms under both sums are clearly similar (they differ only by the location of the factor 3), so that the sums over off-diagonal elements are equal as well. Hence we need to analyze only the sums of diagonal elements:

$$\text{in Eq. (**):}\quad \sum_{j=j'=1}^{3}\left(3r_jr_{j'} - \delta_{jj'}r^2\right)\int \rho(\mathbf{r}')\, r'_jr'_{j'}d^3r' = \sum_{j=1}^{3}\left(3r_j^2 - r^2\right)\int \rho(\mathbf{r}')\, r'^2_j d^3r'$$

$$= 3\sum_{j=1}^{3} r_j^2 \int \rho(\mathbf{r}')\, r'^2_j d^3r' - r^2\int \rho(\mathbf{r}')r'^2 d^3r';$$

$$\text{in Eq. (***):}\quad \sum_{j=j'=1}^{3} r_jr_{j'}\int \rho(\mathbf{r}')\left(3r_jr_{j'} - \delta_{jj'}r^2\right) d^3r' = \sum_{j=1}^{3} r_j^2\int \rho(\mathbf{r}')\left(3r'^2_j - r'^2\right) d^3r'$$

$$= 3\sum_{j=1}^{3} r_j^2 \int \rho(\mathbf{r}')\, r'^2_j d^3r' - r^2\int \rho(\mathbf{r}')r'^2 d^3r'\ .$$

We see that the results are identical, thus completing the required proof.

***Problem* 3.2.** A plane thin ring of radius R is charged with a constant linear density λ. Calculate the exact electrostatic potential distribution along the symmetry axis of the ring, and prove that at large distances, $r \gg R$, the three leading terms of its multipole expansion are indeed correctly described by Eqs. (3.3)–(3.4) of the lecture notes.

Solution: Due to the axial symmetry of the system, the exact calculation of its electrostatic potential from Eq. (1.38), integrated over the cross-section of the ring, is elementary (see the figure below):

$$\phi = \frac{\lambda}{4\pi\varepsilon_0}\int_{\text{ring}} \frac{dr'}{|r - r'|} = \frac{\lambda}{4\pi\varepsilon_0}\int_0^{2\pi} \frac{Rd\varphi}{(R^2 + z^2)^{1/2}} = \frac{\lambda}{4\pi\varepsilon_0}\frac{2\pi R}{(R^2 + z^2)^{1/2}}$$
$$\equiv \frac{\lambda R}{2\varepsilon_0 z} f(\xi),$$

where $\xi \equiv \left(\frac{R}{z}\right)^2$, and $f(\xi) \equiv (1 + \xi)^{-1/2}$.

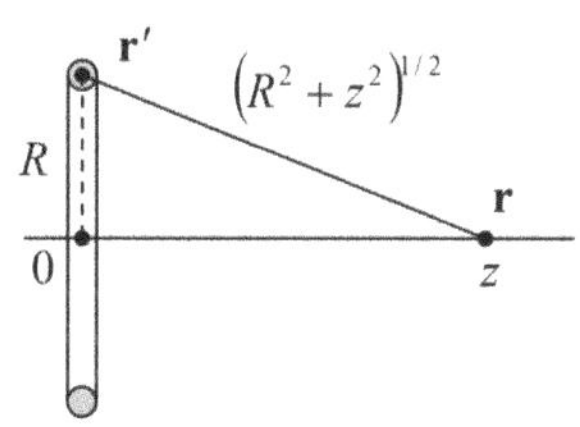

Let us expand this function $f(\xi)$ into the Taylor series in ξ at point $\xi = 0$:

$$f(\xi) = (1 + \xi)^{-1/2}|_{\xi=0} + \frac{d}{d\xi}(1 + \xi)^{-1/2}|_{\xi=0}\xi + \frac{1}{2}\frac{d^2}{d\xi^2}(1 + \xi)^{-1/2}|_{\xi=0}\xi^2 + \cdots$$
$$= 1 - \frac{1}{2}\xi + \frac{3}{8}\xi^2 + \cdots.$$

At large distances from the ring, where ξ is small, the function $f(\xi)$ is well approximated by these three leading terms, so that

$$\phi \approx \frac{\lambda}{2\varepsilon_0}\left[\frac{R}{z} - \frac{1}{2}\left(\frac{R}{z}\right)^3 + \frac{3}{8}\left(\frac{R}{z}\right)^5\right], \quad \text{at } z \gg R. \qquad (**)$$

This expression should be compared with the first three terms of the quadrupole expansion (3.3):

$$\phi(\mathbf{r}) \approx \frac{1}{4\pi\varepsilon_0}\left(\frac{1}{r}Q + \frac{1}{r^3}\sum_{j=1}^{3} r_j p_j + \frac{1}{2r^5}\sum_{j,j'=1}^{3} r_j r_{j'} Q_{jj'}\right), \qquad (***)$$

where

$$Q = \int \rho(\mathbf{r}')d^3r', \quad p_j = \int \rho(\mathbf{r}')r'_j d^3r', \quad \mathscr{Q}_{jj'} = \int \rho(\mathbf{r}')\left(3r'_j r'_{j'} - r'^2\delta_{jj'}\right)d^3r',$$

and $\rho(\mathbf{r}')$ is the electric charge density (per unit volume). In the standard polar coordinates, with the origin in the ring's center (see the figure above), the integration over the ring yields

$$Q = \lambda \int_0^{2\pi} R\, d\varphi' = 2\pi R\lambda;$$

$$p_x = \lambda \int_0^{2\pi} R \cos\varphi' \; R\, d\varphi' = 0,$$

$$p_y = \lambda \int_0^{2\pi} R \sin\varphi' \; R\, d\varphi' = 0,$$

$$p_z = \lambda \int_0^{2\pi} 0 \cdot R\, d\varphi' = 0;$$

$$\mathscr{Q}_{xx} = \lambda \int_0^{2\pi} (3R^2\cos^2\varphi' - R^2)R\, d\varphi' = \pi\lambda R^3,$$

$$\mathscr{Q}_{yy} = \lambda \int_0^{2\pi} (3R^2\sin^2\varphi' - R^2)R\, d\varphi' = \pi\lambda R^3,$$

$$\mathscr{Q}_{zz} = \lambda \int_0^{2\pi} (0 - R^2)R\, d\varphi' = -2\pi\lambda R^3,$$

$$\mathscr{Q}_{xy} = \mathscr{Q}_{yx} = \lambda \int_0^{2\pi} 3R^2 \sin\varphi' \cos\varphi' R\, d\varphi' = 0,$$

$$\mathscr{Q}_{xz} = \mathscr{Q}_{zx} = \lambda \int_0^{2\pi} 3 \cdot 0 \cdot R\cos\varphi' R\, d\varphi' = 0,$$

$$\mathscr{Q}_{yz} = \mathscr{Q}_{zy} = \lambda \int_0^{2\pi} 3 \cdot 0 \cdot R\sin\varphi' R\, d\varphi' = 0\,.$$

(As a sanity check, the sum of the diagonal components of the quadrupole tensor equals 0—as it should—see the model solution of problem 3.1.) Plugging these results into Eq. (*) written for the z axis, i.e. for $r_1 = r_2 = 0$, and $r = r_3 = z$,

$$\phi(z) \approx \frac{1}{4\pi\varepsilon_0}\left(\frac{1}{z}Q + \frac{1}{z^2}p_z + \frac{1}{2z^3}\mathscr{Q}_{zz}\right),$$

we obtain

$$\phi(z) \approx \frac{1}{4\pi\varepsilon_0}\left(\frac{1}{z}2\pi R\lambda + \frac{1}{z^2}\cdot 0 - \frac{1}{2z^3}2\pi\lambda R^3\right) = \frac{1}{4\pi\varepsilon_0}\left(\frac{2\pi R\lambda}{z} - \frac{\pi\lambda R^3}{z^3}\right).$$

Comparing this expression with Eq. (**), we see that the quadrupole approximation exactly describes the three leading (two non-zero) terms of that expansion.

Evidently, for this simple geometry the exact calculation was much simpler, but in more complex cases the multipole expansion may be the only way to carry out an (approximate) analytical calculation.

***Problem* 3.3.** In suitable reference frames, calculate the dipole and quadrupole moments of the following systems (see the figures below):

(i) four point charges of the same magnitude, but alternating signs, placed in the corners of a square;
(ii) a similar system, but with a pair charge sign alternation; and
(iii) a point charge in the center of a thin ring carrying a similar but opposite charge, uniformly distributed along its circumference.

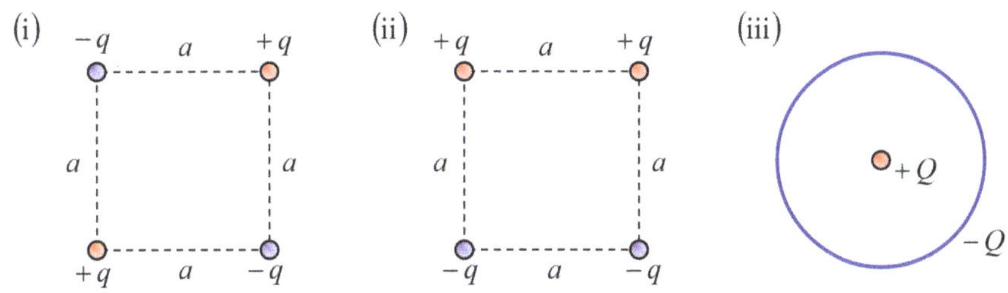

Solutions: For point charges, the definitions of the dipole and quadrupole moments of a system (see Eq. (3.4) of the lecture notes) take the following forms

$$p_j = \sum_k q_k (r_j)_k, \qquad (*)$$

$$\mathscr{Q}_{jj'} = \sum_k q_k \left(3r_j r_{j'} - r^2\delta_{jj'}\right)_k, \qquad (**)$$

where the index k numbers the charges, while the indices $j, j' = 1, 2, 3$ mark the Cartesian coordinates. Let us apply these formulas to systems (i) and (ii).

(i) For this system, in the reference frame and charge numbering shown in the figure below, Eq. (*) yields

$$p_x = q\frac{a}{2}(1 + 1 - 1 - 1) = 0, \quad p_y = q\frac{a}{2}(1 - 1 - 1 + 1) = 0, \quad p_z = 0,$$

where the order of terms in the parentheses follows the particle numbers. Since $p_z = 0$ as well (because $z = 0$ for all charges), the dipole moment $\mathbf{p}$ completely vanishes. Similarly, Eq. (**) yields

$$\mathscr{Q}_{xx} = q\left(\frac{a}{2}\right)^2(1 - 1 + 1 - 1) = 0, \quad \mathscr{Q}_{yy} = q\left(\frac{a}{2}\right)^2(1 - 1 + 1 - 1) = 0,$$

$$\mathscr{Q}_{zz} = q\left(\frac{a}{2}\right)^2(-2 + 2 - 2 + 2) = 0, \quad \mathscr{Q}_{xy} = \mathscr{Q}_{yx} = q\left(\frac{a}{2}\right)^2(3 + 3 + 3 + 3) = 3qa^2,$$

with all off-diagonal components involving coordinate z being equal zero as well, so that quadrupole moment may be represented by the following off-diagonal matrix:

$$\mathscr{Q} = qa^2 \begin{pmatrix} 0 & 3 & 0 \\ 3 & 0 & 0 \\ 0 & 0 & 0 \end{pmatrix}.$$

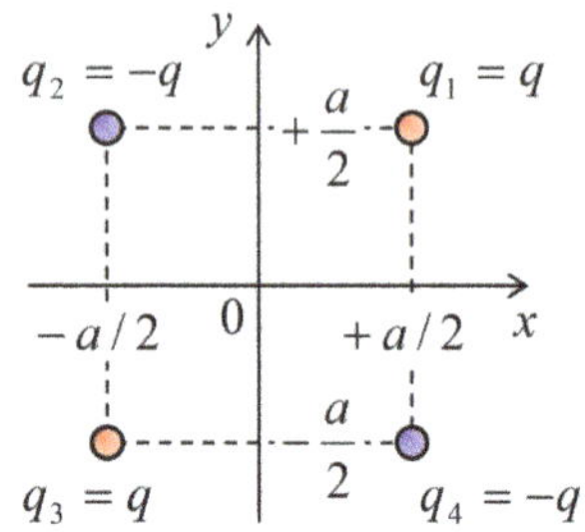

(ii) For this system, in a similar reference frame (see the figure below):

$$p_x = q\frac{a}{2}(1 - 1 + 1 - 1) = 0, \quad p_y = q\frac{a}{2}(1 + 1 + 1 + 1) = 2qa, \quad p_z = 0,$$

so that the system has a non-vanishing dipole moment $\mathbf{p} = 2qa\mathbf{n}_y$, and

$$\mathscr{Q}_{xx} = q\left(\frac{a}{2}\right)^2(1 + 1 - 1 - 1) = 0, \quad \mathscr{Q}_{yy} = q\left(\frac{a}{2}\right)^2(1 + 1 - 1 - 1) = 0,$$

$$\mathscr{Q}_{zz} = q\left(\frac{a}{2}\right)^2(-2 - 2 + 2 + 2) = 0, \quad \mathscr{Q}_{xy} = \mathscr{Q}_{yx} = q\left(\frac{a}{2}\right)^2(3 - 3 - 3 + 3) = 0,$$

with all other components involving coordinate z being equal zero as well, so that now it is the quadrupole moment that completely vanishes.

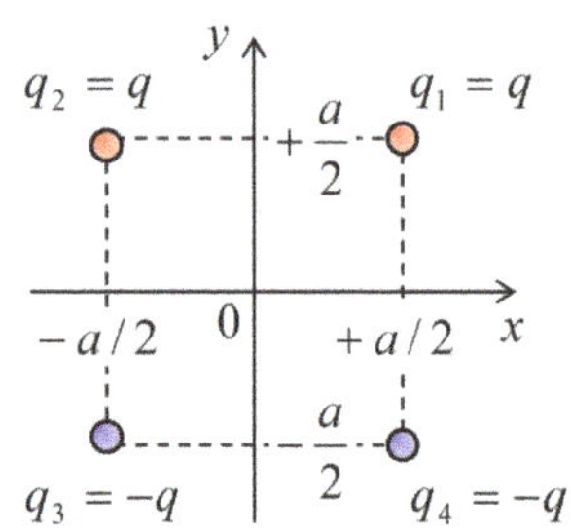

(iii) Generally speaking, for this system we need to combine contributions from both the central point charge $+Q$ and the distributed charge of the ring, with the linear density $\lambda = -Q/2\pi R$. However, selecting a reference frame with its origin in the center of the ring, the former charge (now with $\mathbf{r} = 0$) yields no contribution to either p_j or $\mathscr{Q}_{jj'}$, and we may focus on the ring alone. Directing the axes so that

$x = R\cos\varphi$, $y = R\sin\varphi$, and $z = 0$, so that $r^2 = x^2 + y^2 = R^2$ for all points of the ring, we obtain $\rho d^3r = \lambda R\, d\varphi$, so that

$$p_x = \lambda R \int_0^{2\pi} R \cos\varphi d\varphi = 0, \quad p_x = \lambda R \int_0^{2\pi} R \sin\varphi d\varphi = 0, \quad p_z = 0,$$

(i.e. the dipole moment $\mathbf{p}$ is completely vanishing), and

$$\mathcal{Q}_{xx} = \lambda R \int_0^{2\pi} (3x^2 - r^2)\, d\varphi = \lambda R \int_0^{2\pi} (3R^2 \cos^2\varphi - R^2)\, d\varphi$$
$$= \lambda R^3 \left(\frac{3}{2} - 1\right) 2\pi = -\frac{1}{2} QR^2,$$
$$\mathcal{Q}_{yy} = \lambda R \int_0^{2\pi} (3y^2 - r^2)\, d\varphi = \lambda R \int_0^{2\pi} (3R^2 \sin^2\varphi - R^2)\, d\varphi$$
$$= \lambda R^3 \left(\frac{3}{2} - 1\right) 2\pi = -\frac{1}{2} QR^2,$$

$$\mathcal{Q}_{zz} = \lambda R \int_0^{2\pi} (0 - r^2)\, d\varphi = -\lambda R \int_0^{2\pi} R^2 d\varphi = QR^2,$$
$$\mathcal{Q}_{xy} = \mathcal{Q}_{yx} = \lambda R \int_0^{2\pi} 3xy d\varphi = \lambda R \int_0^{2\pi} R^2 \sin\varphi \cos\varphi d\varphi = 0,$$

with all other components involving coordinate z being equal zero as well, so that quadrupole tensor has the following diagonal matrix:

$$\mathcal{Q} = qR^2 \begin{pmatrix} -1/2 & 0 & 0 \\ 0 & -1/2 & 0 \\ 0 & 0 & 1 \end{pmatrix}.$$

As a sanity check, all the quadrupole moment matrices calculated above have zero trace (the sum of their diagonal elements), as they should, for any system, by the very definition of the tensor $\mathcal{Q}_{jj'}$:

$$\mathrm{Tr}(\mathcal{Q}) \equiv \sum_{j=1}^{3} \mathcal{Q}_{jj} \equiv \sum_{j=1}^{3} \int \rho(\mathbf{r}) \left(3r_j^2 - r^3\right) d^3r = \int \rho(\mathbf{r}) \sum_{j=1}^{3} \left(3r_j^2 - r^2\right) d^3r$$
$$= \int \rho(\mathbf{r})(3r^2 - 3r^2)\, d^3r = 0.$$

***Problem* 3.4.** Without carrying out an exact calculation, can you predict the spatial dependence of the interaction between various electric multipoles, including point charges (in this context, frequently called *monopoles*), dipoles, and quadrupoles? Based on these predictions, what is the functional dependence of the interaction between dumbbell-shaped diatomic molecules such as H_2, N_2, O_2, etc, on the distance between them, if the distance is much larger than the molecular size?

Solution: According to Eq. (1.35) of the lecture notes, the electrostatic potential ϕ_m of a monopole is proportional to $1/r$. Since, according to Eq. (1.31), the potential energy U of a monopole in an external potential is proportional to the potential itself, the potential energy of monopole–monopole interaction energy is inversely proportional to the distance between them:

$$U_{m-m} \propto \frac{1}{r}.$$

This conclusion is of course elementary, but the above reasoning establishes the framework which may be readily extended to higher multipoles.

For example, according to Eq. (3.7), potential ϕ_d of a dipole scales as $1/r^2$. Hence its interaction with a monopole

$$U_{d-m} \propto \frac{1}{r^2}.$$

This conclusion is confirmed by Eq. (3.15) with $\mathbf{E}_{ext} = \mathbf{E}_m = -\nabla\phi_m \propto 1/r^2$.

The same Eq. (3.15) may be used to calculate the dipole–dipole interaction, now taking $\mathbf{E}_{ext} = \mathbf{E}_d = -\nabla\phi_d \propto 1/r^3$, so that

$$U_{d-d} \propto \frac{1}{r^3},$$

the result detailed by Eq. (3.16).

The general trend is perhaps already clear to the reader, but let us follow it one step further. According to the multipole expansion whose proof was the subject of problem 3.1, the quadrupole potential ϕ_q decreases with distance as $1/r^3$, so its interaction with a monopole scales as

$$U_{q-m} \propto \frac{1}{r^3}.$$

Again using Eq. (3.15), we may conclude that its interaction with a dipole is proportional to $\mathbf{E}_q = -\nabla\phi_q \propto 1/r^4$, i.e.

$$U_{q-d} \propto \frac{1}{r^4}.$$

Now we may formulate the general rule, which covers all the above cases: the interaction between two multipoles as a function of the distance between them acquires one more factor $1/r$ each time when the order of one of the interacting multipoles is increased by 1. On this basis, we may safely predict that the interaction between two quadrupoles drops with distance as

$$U_{q-q} \propto \frac{1}{r^5}.$$

The formal proof of this statement, and the exact expression for U_{q-q} may be obtained by the corresponding extension of Eq. (3.14)—an exercise highly recommended to the reader.

Due to their symmetry, dumbbell-shaped molecules do not have a dipole moment, but may have a non-vanishing quadrupole moment. Indeed, a very crude but fair model of their charge distribution is given by a non-uniform but symmetric linear density $\lambda(z) = \lambda(-z)$, where z is the symmetry axis, with the origin in molecule's center. For such a density, the dipole moment vanishes, because its only significant component,

$$p_z = \int \lambda(z) z dz,$$

is proportional to an integral of an odd function in symmetric limits, i.e. is equal to zero. Using the above results of the previous problem, one may conclude that the long-range intermolecular interaction is proportional to $1/r^5$.

This is true, however, only for rather large distances, and molecules' quadrupole moments are typically so small that this interaction is rather insignificant. Another important factor diminishing the importance of such interaction is that the quadrupole potential has a zero angular average, so that if the mutual orientation of the interacting molecules is random (as it is, for example, in the gas phase), the quadrupole interaction is averaged out. This is why another long-range interaction (called the *London dispersion force*), with the potential energy proportional to $1/r^6$, is usually more important. As was mentioned in section 3.1, this effect results from the interaction of two mutually induced (rather than spontaneous) dipole moments, due to their quantum fluctuations[1]. Due to this mutual correlation of the induced dipole moments, this effect is not averaged out at the random orientation of the molecules, and always has the same sign, corresponding to mutual attraction of the molecules.

***Problem* 3.5.** Two similar electric dipoles of fixed magnitude p, located at a fixed distance r from each other, are free to rotate, changing their directions. What stable equilibrium position(s) may they take as a result of their electrostatic interaction?

Solution: Directing the axis z along the line connecting the dipoles, we may use the last form of Eq. (3.16) of the lecture notes for the potential energy of the dipole–dipole interaction:

$$U_{\text{int}} = \frac{1}{4\pi\varepsilon_0} \frac{p_{1x} \cdot p_{2x} + p_{1y} \cdot p_{2y} - 2p_{1z} \cdot p_{2z}}{r^3}.$$

Plugging into this relation the expression of Cartesian components of both dipole moments via the polar and azimuthal angles of their orientation,

$$p_{jx} = p \sin\theta_j \cos\varphi_j, \quad p_{jy} = p \sin\theta_j \sin\varphi_j, \quad p_{jz} = p \cos\theta_j, \quad \text{where } j = 1, 2,$$

we may rewrite the interaction potential as $U_{\text{int}} = Cf$, where $C \equiv (1/4\pi\varepsilon_0)p^2/r^3$ is just a constant, and f is a function of angles θ_j and φ_j:

$$\begin{aligned} f &= \sin\theta_1 \cos\varphi_1 \sin\theta_2 \cos\varphi_2 + \sin\theta_1 \sin\varphi_1 \sin\theta_2 \sin\varphi_2 - 2\cos\theta_1 \cos\theta_2 \\ &\equiv \sin\theta_1 \sin\theta_2 \cos(\varphi_1 - \varphi_2) - 2\cos\theta_1 \cos\theta_2 \\ &\equiv \sin\theta_1 \sin\theta_2 \cos\varphi - 2\cos\theta_1 \cos\theta_2, \end{aligned}$$

[1] A quantitative discussion of this interaction may be found, e.g. in *Part QM*, chapters 3, 5, and 6.

where $\varphi \equiv \varphi_1 - \varphi_2$ is the mutual shift of the azimuthal angles of the dipole moment vectors. (The last form of the function f reflects the fact that it is invariant to any simultaneous turn of both vectors around the line connecting them.)

The stable equilibrium orientations of the dipoles correspond to the minima of this function of three arguments. In order to find them, let us use symmetry arguments. Since the dipoles are similar, their equilibrium angles θ_1 and θ_2 should be either equal or π-complementary, giving us two options:

$$\text{Option A: } \theta_1 = \theta_2 \equiv \theta, \quad \text{giving } f = f_0(\theta, \varphi) \equiv \sin^2\theta \cos\varphi - 2\cos^2\theta = \cos\varphi - (2 + \cos\varphi)\cos^2\theta,$$

$$\text{Option B: } \theta_1 = \pi - \theta_2 \equiv \theta, \quad \text{giving } f = -f_0\,(\theta, \varphi).$$

On the relevant segment $0 \leqslant \theta \leqslant \pi$, the function $f_0(\theta, \varphi)$ has two equal minima, $f_0(0, \varphi) = f_0(\pi, \varphi) = -2$, and only one maximum, $f_0(\pi/2, \varphi) = \cos\varphi$. Since $|\cos\varphi|$ is always less than 2, this means that the function has two minima, at $\theta = 0$ and $\theta = \pi$ (when the azimuthal angle φ is inconsequential). So, the system has two similar stable equilibrium states, in which both dipoles are aligned with each other ($\mathbf{p}_1 = \mathbf{p}_2$), and are together oriented in either direction of the line connecting them.

***Problem* 3.6.** An electric dipole is located above an infinite, grounded conducting plane. Calculate:

(i) the distribution of the induced charge in the conductor,
(ii) the dipole-to-plane interaction energy, and
(iii) the force and the torque acting on the dipole.

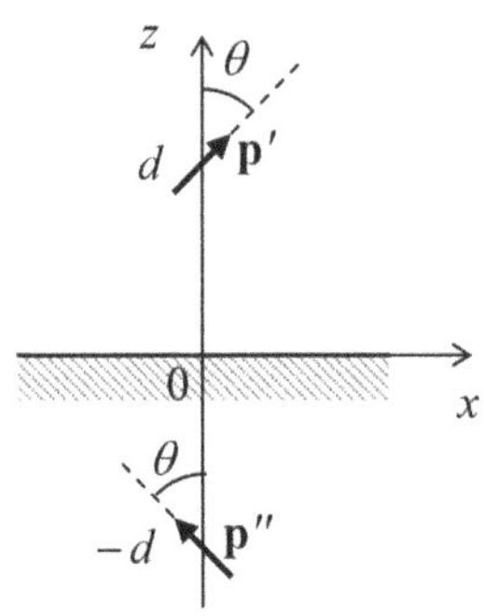

Solutions:

(i) The problem may be solved by the introduction of a dipole image, at the same distance d below the plane, and with the same dipole moment magnitude p as the original dipole, but reflected in the vertical plane perpendicular to that containing the dipole moment vector (see the figure above). The simplest way to understand this fact is to represent the dipole in the approximate form of two point charges, $(+q)$ and $(-q)$, slightly displaced along the direction of the dipole moment vector, and then construct the dipole image from the mirror images of these point charges in the conducting plane. However, this is just a guess, not as a proof; let us prove it.

The net field of these two dipoles evidently satisfies the Poisson equation in the upper half-space, so that the only thing we have to prove is that it also satisfies the boundary condition ($\phi = 0$) on the plane surface. Let us generalize Eq. (3.7) of the lecture notes to the system of two dipoles (calling them $\mathbf{p}'$ and $\mathbf{p}''$, see the figure above), with Cartesian components

$$p_x{}' = -p_x{}'' = p\sin\theta, \quad p_y{}' = p_y{}'' = 0, \quad p_z{}' = p_z{}'' = p\cos\theta, \qquad (*)$$

located, respectively, at points $\mathbf{r}'$ and $\mathbf{r}''$ with the following Cartesian coordinates:

$$x' = x'' = 0, \qquad y' = y'' = 0, \qquad z' = -z'' = d.$$

(Here x is the coordinate within the vertical plane that contains the vectors $\mathbf{p}'$ and $\mathbf{p}''$, i.e. in the plane of our drawing, while the axis y is perpendicular to the plane.) In these coordinates, the generalization yields

$$\begin{aligned}\phi &= \frac{1}{4\pi\varepsilon_0}\left[\frac{(\mathbf{r}-\mathbf{r}')\cdot\mathbf{p}'}{|\mathbf{r}-\mathbf{r}'|^3} + \frac{(\mathbf{r}-\mathbf{r}'')\cdot\mathbf{p}''}{|\mathbf{r}-\mathbf{r}''|^3}\right] \\ &= \frac{p}{4\pi\varepsilon_0}\left[\frac{(z-d)\cos\theta + x\sin\theta}{(x^2+y^2+(z-d)^2)^{3/2}} + \frac{(z+d)\cos\theta - x\sin\theta}{(x^2+y^2+(z+d)^2)^{3/2}}\right].\end{aligned}$$

This equation shows that potential indeed vanishes at an arbitrary point of the surface ($z = 0$), thus proving our guess.

Now the induced surface charge density may be calculated from Eq. (2.3) of the lecture notes as

$$\sigma = -\varepsilon_0 \frac{\partial\phi}{\partial z}\bigg|_{z=0},$$

giving

$$\sigma = \frac{p}{2\pi}\frac{(2d^2 - x^2 - y^2)\cos\theta - 3dx\sin\theta}{(x^2+y^2+d^2)^{5/2}}.$$

(ii) The potential energy of interaction between the actual dipole and its image (i.e. the conducting plane) may be calculated using Eqs. (3.16) of the lecture notes, with the additional factor ½, because the image dipole is induced by the actual one—see the discussion following that formula in the lecture notes. Taking $r = 2d$, and using Eqs. (*), we obtain

$$U_{\text{int}} = -\frac{1}{8\pi\varepsilon_0}\frac{p^2}{(2d)^3}(1+\cos^2\theta). \qquad (**)$$

Note that for any angle θ, the interaction energy is negative, i.e. the dipole is always attracted to a conductor.

(iii) Now we can use Eq. (**) to calculate the force and the torque acting of the dipole. As should be clear from the symmetry of that expression (namely, its independence on the horizontal position of the dipole), the force has only one non-vanishing component,

$$F_z = -\frac{\partial U_{\text{int}}}{\partial d} = -\frac{1}{4\pi\varepsilon_0}\frac{3p^2}{16d^4}(1 + \cos^2\theta) < 0,$$

so that the force is directed toward the plane. The torque vector also has only one Cartesian component, perpendicular to the plane of drawing:

$$\tau_y = -\frac{\partial U_{\text{int}}}{\partial \theta} = -\frac{1}{4\pi\varepsilon_0}\frac{p^2}{16d^3}\sin 2\theta.$$

(Alternatively, this result may be obtained from Eq. (3.17) of the lecture notes, again with the additional factor 1/2). It is interesting that the torque disappears at $\theta = 0$, π, and $\pm\pi/2$, i.e. in all positions in which the dipole moment is aligned with the field created by its image. Of these configurations, the former two, $\theta = 0$ (both dipoles up), and $\theta = \pi$ (both dipoles down), are stable with respect to dipole rotation, because they correspond to the minima of the interaction energy (**)[2].

***Problem* 3.7.** Calculate the net charge Q induced in a grounded conducting sphere of radius R by a dipole $\mathbf{p}$ located at point $\mathbf{r}$ outside the sphere—see the figure below.

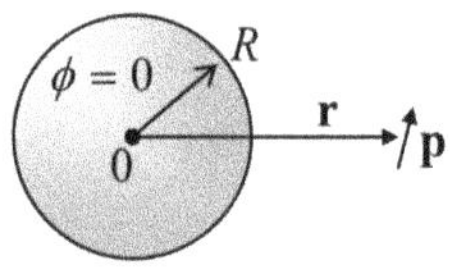

Solution: According to Eq. (3.9) of the lecture notes, the dipole $\mathbf{p}$ may be represented as a limiting case of a couple of equal but opposite charges $\pm q$, displaced by $\mathbf{a} = \mathbf{p}/q$, at $a \to 0$, but $q \to \infty$, so that $p = qa = \text{const}$. For our problem's geometry, in the first approximation in the small parameter $a/r \ll 1$, the distances of the components of such a pair from sphere's center are

$$d_\pm = r \pm \frac{a}{2}\cos\theta,$$

where θ is the angle between the vectors $\mathbf{r}$ and $\mathbf{p}$—see the figure below. Now applying Eq. (2.199) to each of the charges of the pair, we obtain

$$Q = -\frac{R}{d_+}q + \frac{R}{d_-}q = Rq\left[-\frac{1}{r + (a/2)\cos\theta} + \frac{1}{r + (a/2)\cos\theta}\right]$$
$$= \frac{Rq}{r}\left[-\frac{1}{1 + (a/2r)\cos\theta} + \frac{1}{1 + (a/2r)\cos\theta}\right].$$

[2] In principle, such a bistable system may be used as a binary memory cell, potentially scalable to atomic size, using spontaneous molecular dipoles. However, to be practicable, such cells would need atomic-scale devices for their state readout.

The limit of this expression at $a/r \to 0$,

$$Q \to \frac{Rqa}{r^2}\cos\theta = \frac{Rp}{r^2}\cos\theta = \frac{R\mathbf{r}\cdot\mathbf{p}}{r^3},$$

gives the solution of our problem.

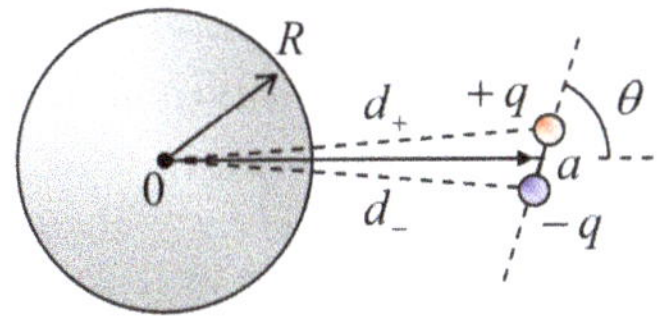

***Problem* 3.8.** Use two different approaches to calculate the energy of interaction between a grounded conductor and an electric dipole **p**, placed in the center of a spherical cavity of radius R, carved in this conductor.

Solution: Let us use spherical coordinates $\{r, \theta, \varphi\}$, with the axis z directed along the dipole moment vector **p**.

Approach 1. As we know from the solution of problem 2.24, if a thin spherical layer of radius R is charged with areal density

$$\sigma = \sigma_0 \cos\theta, \qquad (*)$$

it creates, outside it, a purely dipole field corresponding to dipole moment $\mathbf{p}_\sigma = \mathbf{n}_z(4\pi/3)\sigma_0 R^3$. Hence, if we select σ_0 to satisfy the condition $\mathbf{p} + \mathbf{p}_\sigma = 0$, i.e. take

$$\sigma_0 = -\frac{3}{4\pi R^3}p, \quad \text{so that } \sigma = -\frac{3}{4\pi R^3}p\cos\theta, \qquad (**)$$

the net field of these two dipoles at $r \geqslant R$ vanishes, thus satisfying the boundary condition $\phi(R) = 0$.

Now let us use one more result of the solution of problem 2.24, namely that the internal electric field induced by the distribution (*) is uniform, and equal to

$$\mathbf{E}_{\text{in}} = -\frac{\sigma_0}{3\varepsilon_0}\mathbf{n}_z.$$

With the σ_0 given by Eq. (**), this field of the dipole-induced surface charges is

$$\mathbf{E}_{\text{in}} = \frac{\mathbf{p}}{4\pi\varepsilon_0 R^3}.$$

According to Eq. (3.15), with the usual additional factor ½ (due to the self-induced character of this field), the dipole–cavity interaction energy is

$$U = -\frac{p^2}{8\pi\varepsilon_0 R^3}.$$

As a sanity check, according to this expression the interaction energy rapidly vanishes at $R \to \infty$, as it should.

Approach 2. The model solution of problem 2.31 had the following by-product: the force **F** of attraction of a point charge q, displaced from the center of a spherical cavity by distance $z \ll R$, is directed to the nearest point of the surface and is equal to

$$F(z) = \frac{q^2}{4\pi\varepsilon_0 R^3}z. \qquad (***)$$

Let us build up our dipole **p** by a gradual separation of two point charges $+q$ and $-q$, moving each of them, from the center of the cavity along axis z, to the final positions $\pm a/2$, where $a = p/q$—see Eq. (3.9) of the lecture notes. (The charge magnitudes q should be sufficiently large to ensure that the final distance p/q between them is still much smaller than R, so that the field of this pair of charges, felt by the sphere, is well approximated by that of a dipole of magnitude p.) Physically, this force is exerted by the surface charge induced by the point charge. In our case, there are two such distributions (induced by each of the two charges), and each of them is exerted on both charges, so that the full work

$$\mathscr{W} = 4\int_0^{a/2} F(z)dz = 4\frac{q^2}{4\pi\varepsilon_0 R^3}\int_0^{p/2q} zdz = 4\frac{q^2}{4\pi\varepsilon_0 R^3}\frac{1}{2}\left(\frac{p}{2q}\right)^2 = \frac{p^2}{8\pi\varepsilon_0 R^3}.$$

(Note that we should not take into account the forces of the direct interaction of the charges, because they account for the built-in internal energy of the dipole, and do not depend on R.) Since in the beginning of the build-up the dipole did not exist, the potential energy of the interaction in the end of the build-up interaction is just minus the calculated work, giving the same result as in approach 1.

***Problem* 3.9.** A plane separating two parts of otherwise free space is densely and uniformly (with a constant areal density n) filled with dipoles, with similar dipole moments **p** oriented in a direction normal to the plane.

(i) Calculate the boundary conditions for the electrostatic potential on both sides of the plane.
(ii) Use the result of task (i) to calculate the potential distribution created in space by a spherical surface, with radius R, densely and uniformly filled with radially oriented dipoles.
(iii) What condition should be imposed on the dipole density n for your results to be qualitatively valid?

Solutions:

(i) Let the dipole density be so high that they may be treated as a 2D continuum, with the polarization

$$\mathbf{P} = n\mathbf{p}\delta(z) = \mathbf{n}_z np\delta(z), \qquad (*)$$

where z is the direction normal to the plane, whose position is taken for $z = 0$. Due to the 1D character of this problem, the electrostatic potential may depend only on the same coordinate z, so that the Poisson equation for the macroscopic dipole potential takes the form

$$\frac{d^2\phi}{dz^2} = -\frac{\rho_{\text{ef}}}{\varepsilon_0},$$

where ρ_{ef} in given by Eq. (3.30), which in our case is reduced to

$$\rho_{\text{ef}} = -\frac{dP_z}{dz}.$$

Integrating the Poisson equation once, we obtain

$$\frac{d\phi}{dz} = -\frac{1}{\varepsilon_0}\int \rho(z)dz = \frac{P(z)}{\varepsilon_0} + c_1 = \frac{np}{\varepsilon_0}\delta(z) + c_1, \qquad (**)$$

where the physical sense of the constant c_1 is the (minus) electric field outside of the interface, which the dipole layer cannot create or affect (because its net charge density σ equals zero). Now integrating the last equation again over a negligibly small interval across the plane (so that the contribution of the integral of c_1 is negligible), we obtain

$$\Delta\phi \equiv \phi(+0) - \phi(-0) = \frac{np}{\varepsilon_0}. \qquad (***)$$

The physical sense of this potential leap becomes clear from the simple cartoon in which each dipole $\mathbf{p}$ is represented by two equal and opposite charges $\pm q$, separated by a small distance vector $\mathbf{a} = \mathbf{p}/q$—see Eq. (3.9) of the lecture notes, in our case normal to the dipole sheet. Then the dipole layer we are discussing is represented by two, slightly separated charge layers, with the equal and opposite areal densities $\pm\sigma = \pm nq$. As we know from the numerous discussions of such a system in chapters 1 and 2, the electric field induced by the charges between these layers equals $E_z = -nq/\varepsilon_0$, so that the potential difference it creates is $\Delta\phi = -E_z a = nqa/\varepsilon_0 = np/\varepsilon_0$, thus returning us to Eq. (***).

(ii) Since the sphere carries no net charge, the electric field not only inside it, but also outside it should be zero, so that the electrostatic potential should be constant in each of these regions:

$$\phi(\mathbf{r}) = \begin{cases} \phi_{\text{in}} = \text{const}, & \text{for } r < R, \\ \phi_{\text{out}} = \text{const}, & \text{for } r > R, \end{cases}$$

with the origin of r at the center of the sphere. Very close to any smooth surface, it may be well approximated by a plane, so that the boundary condition (***) is valid for two points very close to sphere's surface, giving

$$\phi_{\text{out}} - \phi_{\text{in}} = \frac{np}{\varepsilon_0}.$$

With the usual convention $\phi(\infty) = 0$, this result gives

$$\phi(\mathbf{r}) = \begin{cases} -np/\varepsilon_0, & \text{for } r < R, \\ 0, & \text{for } r > R. \end{cases} \qquad (****)$$

(iii) The above analyses, based on the approximation of the dipole system with a 2D continuum, are valid if the average distance $d \equiv n^{-1/2}$ between the dipoles is much smaller than any lateral size of the problem. For the case of the sphere in task (ii), this role is played by sphere's radius R, so that Eqs. (***) and (****) are strictly valid if $d \ll R$, i.e. if

$$nR^2 \gg 1,$$

meaning just that the total number of the dipoles on the surface should be much larger than 1. Actually, the above condition of validity of Eq. (***) is applicable to any interface, provided that R is understood as the smallest of its two curvature radii.

***Problem* 3.10.** Prove the Clausius–Mossotti relation (see Eq. (3.52) of the lecture notes) for the case of a cubic lattice of similar dipoles obeying Eq. (3.48): $\mathbf{p} = \alpha\mathbf{E}_\text{m}$, where $\mathbf{E}_\text{m}$ is the microscopic electric field at the dipole's location point.

Hint: Use Eq. (3.65) of the lecture notes to account for the difference between the external field and the macroscopic field.

Solution: At a substantial dipole density, the microscopic field $\mathbf{E}_\text{m}$ acting on each dipole may include not only the external field $\mathbf{E}_0$, but also a 'molecular field' $\mathbf{E}_\text{mol}$ of the surrounding dipoles:

$$\mathbf{E}_\text{m} = \mathbf{E}_0 + \mathbf{E}_\text{mol}(0), \qquad (*)$$

where the position of the particular dipole under consideration is taken for $\mathbf{r} = 0$. Let us calculate $\mathbf{E}_\text{mol}(0)$, using a very simple model: a regular cubic lattice of identical dipoles—see the figure below. In a Cartesian coordinate system with axes directed along the lattice vectors, coordinates of the dipoles are

$$x_{jkl} = aj, \quad y_{jkl} = ak, \quad z_{jkl} = al,$$

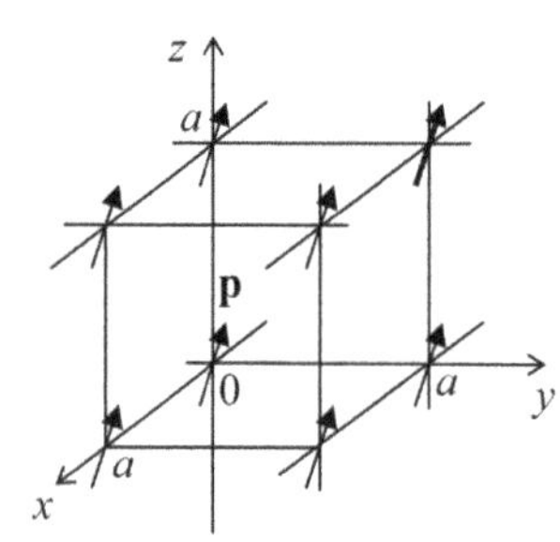

where j, k, and l are integers numbering the dipoles. Now we may use the last form of Eq. (3.13), and the linear superposition principle, to calculate one of the Cartesian components (say, along axis x) of the molecular field induced by all other dipoles of the lattice:

$$(E_{\text{mol}})_x(0) = \frac{1}{4\pi\varepsilon_0 a^3} \sum_{j,k,l=-\infty}^{+\infty} \frac{3j(jp_x + kp_y + lp_z) - p_x(j^2 + k^2 + l^2)}{(j^2 + k^2 + l^2)^{5/2}}, \qquad (**)$$

with the term $j = k = l = 0$ excluded. The sums of all cross-terms, proportional to jk and jl, vanish due to the system's symmetry, so that we obtain

$$(E_{\text{mol}})_x(0) = \frac{1}{4\pi\varepsilon_0 a^3} \sum_{j,k,l=-\infty}^{+\infty} \frac{[3j^2 - (j^2 + k^2 + l^2)]}{(j^2 + k^2 + l^2)^{5/2}} p_x.$$

Since all the sums participating in this expression are equal,

$$\sum_{j,k,l=-\infty}^{+\infty} \frac{j^2}{(j^2 + k^2 + l^2)^{5/2}} = \sum_{j,k,l=-\infty}^{+\infty} \frac{k^2}{(j^2 + k^2 + l^2)^{5/2}} = \sum_{j,k,l=-\infty}^{+\infty} \frac{l^2}{(j^2 + k^2 + l^2)^{5/2}},$$

we obtain $(E_{\text{mol}})_x(0) = 0$. But, due to the system symmetry, the same result is valid for all other components of the dipole field. Hence, a bit counter-intuitively, $\mathbf{E}_{\text{mol}}(0) = 0$, so that (due to the equivalence of all the dipoles of the system), the molecular field vanishes at the location of each dipole, Eq. (*) is reduced to $\mathbf{E}_{\text{m}} = \mathbf{E}_0$, and Eq. (3.48) yields

$$\mathbf{p} = \alpha\mathbf{E}_{\text{m}} = \alpha\mathbf{E}_0, \qquad \text{so that } \mathbf{P} \equiv n\mathbf{p} = \alpha n\mathbf{E}_0. \qquad (***)$$

In order to relate the external field $\mathbf{E}_0$ and the macroscopic field $\mathbf{E}$ in the medium, we may follow the hint and use the first of Eqs. (3.65) of the lecture notes,

$$\mathbf{E} = \frac{3}{\kappa + 2}\mathbf{E}_0,$$

for the macroscopic field inside a uniform, macroscopic sphere with a radius R much larger than the inter-dipole distance a. (The arguments typically used for this model are that the spherical geometry, due to its isotropy, most fairly represents the relation between $\mathbf{E}$ and $\mathbf{E}_0$, and that due to the condition $R \gg a$, the limitation of the volume filled by the dipoles does not affect the above calculation of the rapidly converging sum (**) much.) Plugging this field into the general formula Eq. (3.45) for linear dielectrics:

$$\mathbf{P} = (\kappa - 1)\varepsilon_0\mathbf{E},$$

we obtain

$$\mathbf{P} = 3\frac{\kappa - 1}{\kappa + 2}\varepsilon_0\mathbf{E}.$$

Now requiring this relation to give the same result as Eq. (***), we finally obtain the equality

$$3\frac{\kappa - 1}{\kappa + 2}\varepsilon_0 = \alpha n,$$

which is just another form of the Clausius–Mossotti relation (3.52).

This proof shows clearly why this formula is not valid for most ferroelectric materials—and more generally for other materials in which the size of each molecule is comparable with the distance between them. Indeed, for each molecule the derivation explicitly uses Eq. (3.13), which is invalid at distances comparable with the molecular size.

***Problem* 3.11.** A sphere of radius R is made of a material with a uniform, fixed polarization $\mathbf{P}_0$.

(i) Calculate the electric field everywhere in space—both inside and outside the sphere.
(ii) Compare the result for the internal field with Eq. (3.24) of the lecture notes.

Solutions:

(i) According to the general Eq. (3.31), the effective charges, with density $\rho_{\text{ef}} \equiv -\nabla \cdot \mathbf{P}$, affect the electric field distribution exactly as the stand-alone charges (which are absent in this problem). Since in our current problem the polarization $\mathbf{P}$ is uniform both inside the sphere ($\mathbf{P} = \mathbf{P}_0$) and outside it ($\mathbf{P} = 0$), ρ_{ef} is different from zero only on sphere's surface, forming the effective surface charge density σ_{ef}, which may be found by the integration of the above relation for ρ_{ef} through the infinitesimal interval of the outer normal to the surface:

$$\sigma_{\text{ef}} = \int_{n=-0}^{n=+0} \rho_{\text{ef}} dn = -\int_{n=-0}^{n=+0} \nabla \cdot \mathbf{P} dn = P_0 \cos\theta,$$

where θ is the angle between the vector $\mathbf{P}_0$ and the local normal to the surface. With the axis z taken along the vector $\mathbf{P}_0$, θ coincides with the usual polar angle of spherical coordinates (with the origin in sphere's center). But this means that the spatial distribution of the electrostatic potential ϕ, and the electric field $\mathbf{E} = -\nabla\phi$, in this problem is exactly the same as was calculated in the solution of problem 2.24, with the replacement $\sigma_0 \to P_0$:

$$\phi_{\text{in}} = \frac{P_0}{3\varepsilon_0} r \cos\theta, \qquad \phi_{\text{out}} = \frac{P_0 R^3}{3\varepsilon_0 r^2} \cos\theta. \qquad (*)$$

Comparing the second of these expressions with Eq. (3.7) of the lecture notes, we see that the electric field outside the sphere exactly coincides with that of an electric dipole with the moment

$$\mathbf{p} = \frac{4\pi}{3}R^3\mathbf{P}_0 = V\mathbf{P}_0, \qquad (**)$$

equal to the sum of all elementary dipoles within the sphere—the result that could be conjectured even without the formal solution of the problem.

On the other hand, the first of Eqs. (*) describes a uniform electric field inside the sphere, with

$$\mathbf{E}_{\text{in}} = -\frac{\mathbf{P}_0}{3\varepsilon_0}, \qquad (***)$$

and $\mathbf{D}_{\text{in}} \equiv \varepsilon_0\mathbf{E}_{\text{in}} + \mathbf{P}_0 = (2/3)\mathbf{P}_0$. (Note that the vectors **E** and **D** have *different* directions, one again demonstrating a radical difference between these two macroscopic fields.)

Note also another interesting way to derive Eq. (***). In section 3.4 of the lecture notes, we have obtained the following result for the internal field and polarization of a sphere made of a *linear* dielectric (with $\mathbf{P} \propto \mathbf{E}$) placed into a uniform external field $\mathbf{E}_0$—see Eqs. (3.65):

$$\mathbf{E}_{\text{in}} = \frac{3}{\kappa + 2}\mathbf{E}_0, \qquad \mathbf{P} = 3\varepsilon_0\frac{\kappa - 1}{\kappa + 2}\mathbf{E}_0. \qquad (****)$$

The field $\mathbf{E}_{\text{in}}$ may be considered as the sum of the external field $\mathbf{E}_0$ and the field $\mathbf{E}_{\text{self}}$ created by the polarization of the sphere itself. For the latter field, Eqs. (****) yield

$$\mathbf{E}_{\text{self}} \equiv \mathbf{E}_{\text{in}} - \mathbf{E}_0 = \frac{3}{\kappa + 2}\mathbf{E}_0 - \mathbf{E}_0 \equiv -\frac{\kappa - 1}{\kappa + 2}\mathbf{E}_0 \equiv -\frac{\mathbf{P}}{3\varepsilon_0},$$

i.e. the equality similar to Eq. (***). This is natural, because the field induced by a uniform polarization of the sphere should not depend on the origin of this polarization.

(ii) Using Eq. (**) for the dipole moment **p** of the sphere, Eq. (***) may be recast as

$$\mathbf{E}_{\text{in}} = -\frac{\mathbf{p}}{3\varepsilon_0}\frac{1}{V}, \qquad \text{so that} \quad \int_V \mathbf{E}_{\text{in}}d^3r = -\frac{\mathbf{p}}{3\varepsilon_0},$$

in agreement with the general equation (3.24). So, our calculations confirm this general expression at least for one particular case.

***Problem* 3.12.** Calculate the electric field at the center of a cube with side *a*, made of material with the uniform spontaneous polarization vector $\mathbf{P}_0$ parallel to one of the cube's sides.

Solution: As was discussed in the previous problem, the effect of a uniform polarization of a volume is equivalent to its surface's charging with the areal density

$$\sigma_{\text{ef}} = P_0 \cos\theta,$$

where θ is the angle between the vector $\mathbf{P}_0$ of the polarization and the outer normal to the surface. In our current case this means that two cube's faces have equal but opposite charges, with constant densities $\sigma_{ef} = \pm P_0$, while other four faces have none. Evidently, the two faces with non-zero surface charges give equal contributions to the electric field in the center of the cube, directed normally to those faces—from the face with $\sigma_{ef} = +P_0$ toward the face with $\sigma_{ef} = -P_0$, i.e. opposite to vector $\mathbf{P}_0$.

The calculation of the magnitude of the field induced by one cube's face carrying a constant charge density σ was one of the tasks of problem 1.6; the result was $|E| = \sigma/6\varepsilon_0$. Replacing σ with σ_{ef}, doubling the result (to reflect the equal contributions of both charged faces), and taking into account the field direction, we finally obtain

$$\mathbf{E} = -\frac{\mathbf{P}_0}{3\varepsilon_0}, \qquad (*)$$

i.e. exactly the same result as was obtained in the solution of the previous problem for the spontaneously polarized sphere.

Note also that multiplying both sides of Eq. (*) by the volume V of the cube, and taking into account that its total dipole moment of the cube is $\mathbf{p} = \mathbf{P}_0 V$, we may write the result in the form of Eq. (3.24) of the lecture notes,

$$\int_V \mathbf{E}_{in} d^3r = -\frac{\mathbf{p}}{3\varepsilon_0},$$

thus providing a confirmation of this general result for one more particular case.

***Problem* 3.13.** A stand-alone charge Q is distributed, in some way, in the volume of a body made of a uniform linear dielectric with a dielectric constant κ. Calculate the polarization charge Q_{ef} residing on the surface of the body, provided that it is surrounded by free space.

Solution: Let us apply the *macroscopic* Gauss law (see Eq. (3.34) of the lecture notes) to two closed surfaces—one (S_{in}) immediately inside the body's surface and one (S_{out}) immediately outside it, so that the stand-alone charge inside them is the same (Q):

$$\oint_{S_{in}} D_n d^2r = Q, \qquad \oint_{S_{out}} D_n d^2r = Q. \qquad (*)$$

The inner surface is inside the dielectric, so that at all its points $\mathbf{D} = \varepsilon\mathbf{E}$, i.e. $D_n = \varepsilon E_n$ (where $\varepsilon \equiv \kappa\varepsilon_0$ is dielectric's permittivity), while at the outer surface $\mathbf{D} = \varepsilon_0\mathbf{E}$, and $D_n = \varepsilon_0 E_n$. Plugging these relations into Eqs. (*), we obtain

$$\oint_{S_{in}} E_n d^2r = \frac{Q}{\varepsilon}, \qquad \oint_{S_{out}} E_n d^2r = \frac{Q}{\varepsilon_0}. \qquad (**)$$

But according to the 'microscopic' Gauss law (see Eq. (1.16) of the lecture notes), applied to the slim volume between two surfaces (which, by the problem's conditions, does not contain any stand-alone surface charge, but may have a surface polarization charge Q_{ef}),

$$\oint_{S_{\text{out}}} E_n d^2r - \oint_{S_{\text{in}}} E_n d^2r = \frac{Q_{\text{ef}}}{\varepsilon_0}.$$

Plugging in the integrals from Eqs. (**), we obtain

$$\frac{Q_{\text{ef}}}{\varepsilon_0} = \frac{Q}{\varepsilon_0} - \frac{Q}{\varepsilon}, \qquad \text{i.e. } Q_{\text{ef}} = Q\left(1 - \frac{\varepsilon_0}{\varepsilon}\right) \equiv Q\left(1 - \frac{1}{\kappa}\right).$$

Note that this result does not describe the part of the polarization ('effective') charge $\rho_{\text{ef}} = -\nabla \cdot \mathbf{P}$ that may be distributed in the volume of the body. (This part vanishes only if the polarization inside the body is uniform: $\mathbf{P} = \text{const}$.)

***Problem* 3.14.** In two separate experiments, a thin, planar sheet of a linear dielectric with $\kappa = \text{const}$ is placed into a uniform external electric field $\mathbf{E}_0$:

(i) with the sheet's surfaces parallel to the electric field, and
(ii) the surfaces normal to the field.

For each case, find the electric field $\mathbf{E}$, the electric displacement $\mathbf{D}$, and the polarization $\mathbf{P}$ inside the dielectric (far from the sheet's edges).

Solution: The same reasoning that was used in section 3.4 to analyze vacuum slits in dielectrics, based on the continuity of E_τ and D_n at the major surfaces of the sheet, gives the following results:

(i) at the sheet parallel to the electric field[3]:

$$\mathbf{E} = \mathbf{E}_0, \qquad \text{and hence } \mathbf{D} \equiv \varepsilon\mathbf{E} = \kappa\varepsilon_0\mathbf{E}_0, \quad \mathbf{P} = (\kappa - 1)\varepsilon_0\mathbf{E}_0;$$

(ii) at the sheet normal to the field:

$$\mathbf{D} = \mathbf{D}_0 \equiv \varepsilon_0\mathbf{E}_0, \qquad \text{and hence } \mathbf{E} \equiv \frac{\mathbf{D}}{\kappa\varepsilon_0} = \frac{\mathbf{E}_0}{\kappa}, \quad \mathbf{P} = \frac{\kappa - 1}{\kappa}\varepsilon_0\mathbf{E}_0.$$

We see that in the latter case *all* the fields are κ times weaker than in the former one.

One may wonder why the problem's assignment mentions that the sheet is 'thin', while the solution does not use this condition. The reason is that in the case (ii), fixing a constant external field $\mathbf{E}_0$ is easy (may be done, for example, using a large plane capacitor) only if the sheet's thickness t is much smaller than $\sim[(\kappa - 1)/(\kappa + 2)]a$, where a is the lateral size of the sheet; otherwise, the sheet-induced deformation of the external field (similar to that shown, for example, in figure 3.11 for the sphere) would require a compensation using quite an elaborate setup. So, the assignment without the 'thin' qualifier would describe a hardly practicable experiment.

---

[3] Note that this result was also derived in section 3.5 from the condition of the minimum of the Gibbs potential energy (3.78).

***Problem* 3.15.** A point charge q is located at a distance $r \gg R$ from the center of a uniform sphere of radius R, made of a uniform linear dielectric. In the first non-vanishing approximation in small parameter R/r, calculate the interaction force, and the energy of interaction between the sphere and the charge.

Solution: The point charge's field $E = (1/4\pi\varepsilon_0)\, q/r^2$, and at the sphere's location, is nearly uniform on the scale of its radius $R \ll r$. From the problem solved in section 3.4 of the lecture notes, we know that such a uniform field induces in a dielectric sphere a dipole moment of magnitude $p = 4\pi\varepsilon_0 ER^3(\kappa - 1)/(\kappa + 2)$, directed along the initial electric field—see Eq. (3.64). Hence we can use the general formula for the radial component of the dipole field, $E_r = 2p\cos\theta/4\pi\varepsilon_0 r^3$ (see the first form of Eq. (3.13) of the lecture notes), with $\theta = 0$, to calculate the magnitude of the interaction force:

$$F = qE_r = q\frac{2}{4\pi\varepsilon_0 r^3}p = q\frac{2}{4\pi\varepsilon_0 r^3}4\pi\varepsilon_0 R^3\frac{\kappa - 1}{\kappa + 2}E = q\frac{2}{r^3}R^3\frac{\kappa - 1}{\kappa + 2}\frac{1}{4\pi\varepsilon_0}\frac{q}{r^2}$$
$$= \frac{1}{4\pi\varepsilon_0}2\frac{\kappa - 1}{\kappa + 2}\frac{q^2R^3}{r^5}.$$

Integrating the result from ∞ to r, we obtain the potential energy of the charge–sphere interaction (attraction, for any sign of q):

$$U = -\frac{1}{4\pi\varepsilon_0}\frac{1}{2}\frac{\kappa - 1}{\kappa + 2}\frac{q^2R^3}{r^4}.$$

Another way to obtain the last result is to use Eq. (3.15b) of the lecture notes for the interaction energy of a dipole with an external electric field, accounting for the fact that in our case the dipole moment is induced by the external field itself. In that approach, the interaction force's magnitude may be obtained at the second step, by calculating the (minus) gradient of U over r, i.e. its derivative over r.

Finally, the same results may be obtained by the proper generalization of the image charge method. (My strong recommendation to the reader is to explore this route as well.)

***Problem* 3.16.** A fixed dipole $\mathbf{p}$ is placed in the center of a spherical cavity of radius R, cut inside a uniform, linear dielectric. Calculate the electric field distribution everywhere in the system (both at $r < R$ and $r > R$).

Hint: You may start with the assumption that the field at $r > R$ has a distribution typical for a dipole (but be ready for surprises).

Solution: Following the hint, let us look for the electric potential's distribution in the form

$$\phi = \frac{1}{4\pi\varepsilon_0} \times \begin{cases} \dfrac{p'\cos\theta}{r^2}, & \text{for } r \geqslant R, \\ \dfrac{p\cos\theta}{r^2} + \varsigma(r, \theta), & \text{for } r \leqslant R, \end{cases}$$

where $\varsigma(r,\theta)$ is an axially symmetric function that satisfies the Laplace equation. With this assumption, the boundary conditions at $r = R$ take the form

$$\phi = \text{const}: \quad \frac{p' \cos\theta}{R^2} = \frac{p\cos\theta}{R^2} + \varsigma(R, \theta),$$

$$\varepsilon\frac{\partial\phi}{\partial n} = \text{const}: \quad -2\kappa\frac{p' \cos\theta}{R^3} = -2\frac{p\cos\theta}{R^3} + \frac{\partial\varsigma}{\partial r}(R, \theta).$$

Representing the function $\varsigma(r, \theta)$ as a series over the Legendre polynomials (see, e.g. Eq. (2.172) of the lecture notes), we see from the boundary conditions that only the first term of that expansion, $a_1 r\cos\theta$, may have a non-vanishing magnitude. For a_1, the above boundary conditions yield:

$$\frac{p'}{R^2} = \frac{p}{R^2} + a_1 R, \qquad -2\kappa\frac{p'}{R^3} = -2\frac{p}{R^3} + a_1,$$

giving

$$p' = \frac{3p}{2\kappa + 1}, \qquad a_1 = -\frac{2(\kappa - 1)p}{(2\kappa + 1)R^3}.$$

Hence the electric field outside the spherical cavity indeed corresponds to a single dipole, but (here is the promised surprise) with a different dipole moment $\mathbf{p}' = 3\mathbf{p}/(2\kappa + 1)$, while that inside the sphere is the sum of the original dipole field and an additional *uniform* field

$$\mathbf{E}_{\text{in}} = \frac{1}{4\pi\varepsilon_0}\frac{2(\kappa - 1)}{2\kappa + 1}\frac{\mathbf{p}}{R^3}.$$

As a sanity check, in the limit $\kappa \to \infty$, this field approaches the κ-independent value that was calculated in the model solution of problem 3.8. On the other hand, in the limit $\kappa \to 1$ (no dielectric at all), the field vanishes.

***Problem* 3.17.** A spherical capacitor (see the figure below) is filled with a linear dielectric whose permittivity ε depends on spherical angles θ and φ, but not on the distance r from the system's center. Derive an explicit expression for its capacitance C.

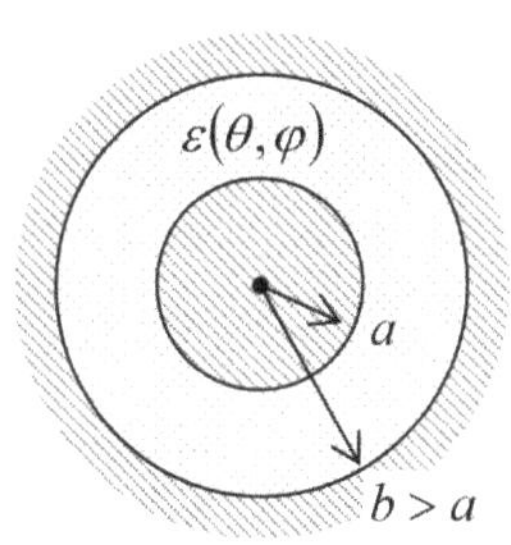

Solution: Let us prove that the boundary problem for the field distribution inside the capacitor is satisfied by the following radially directed fields:

$$\mathbf{E} = \mathbf{n}_r E(r), \qquad \mathbf{D} = \mathbf{n}_r \varepsilon(\theta, \phi) E(r),$$

Since the curl of such a vector E equals zero[4], the first macroscopic Maxwell equation for $\mathbf{E}$, given by Eq. (3.36) of the lecture notes, is satisfied. On the other hand, the macroscopic Maxwell equation (3.32) for $\mathbf{D}$ is satisfied if $\nabla \cdot \mathbf{D} = 0$. For the selected form of $\mathbf{D}$, this equation is reduced to[5]

$$\frac{d}{dr}[r^2 E(r)] = 0.$$

Its straightforward integration yields the same result,

$$E(r) = \frac{c_1}{r^2},$$

as for the capacitor without the dielectric (i.e. with $\varepsilon = \varepsilon_0 = \text{const}$), which was analyzed in section 2.3(iii) of the lecture notes, so that the boundary conditions for the corresponding electrostatic potential,

$$\phi(r) = \phi(a) - \int_a^r E(r)dr = \phi(a) + \frac{c_1}{r} + c_2,$$

may be satisfied by the same choice of constants $c_{1,2}$ just as in that case.

Hence, our proof is accomplished, and now we may use Eq. (2.53) to write the expression for the electric field on the surface of any conductor—say, the inner one:

$$E(a) = \frac{V}{a^2}\left(\frac{1}{a} - \frac{1}{b}\right)^{-1}.$$

Only at this stage does the difference with the dielectric-free case cut in, because the areal density of stand-alone charges on the surface of the conductor has to be calculated using the integral of Eq. (3.54) rather than of Eq. (2.3):

$$\sigma_a = D(a) = \varepsilon(\theta, \varphi)E(a) = \varepsilon(\theta, \varphi)\frac{V}{a^2}\left(\frac{1}{a} - \frac{1}{b}\right)^{-1},$$

so that the mutual capacitance is

$$C \equiv \frac{Q_a}{V} = \frac{1}{V}\int_{r=a} \sigma_a d^2r = \frac{a^2}{V}\int_{r=a} \sigma_a d\Omega = \left(\frac{1}{a} - \frac{1}{b}\right)^{-1}(\theta, \varphi)d\Omega$$

$$\equiv \left(\frac{1}{a} - \frac{1}{b}\right)^{-1}\int_0^\pi \sin\theta d\theta \int_0^{2\pi} d\varphi\, \varepsilon(\theta, \varphi).$$

[4] See, e.g. Eq. (A.69) with $f_\theta = f_\varphi = 0$ and $\partial/\partial\theta = \partial/\partial\varphi = 0$.

[5] See, e.g. Eq. (A.68) with $f_\theta = f_\varphi = 0$.

For the particular case ε = const, this double integral is equal to $4\pi\varepsilon$, and we immediately get the evident generalization of the (easy) problem 2.8(i):

$$C = 4\pi\varepsilon\left(\frac{1}{a} - \frac{1}{b}\right)^{-1} \equiv 4\pi\varepsilon\frac{ab}{b - a}.$$

***Problem* 3.18.** For each of the two capacitors shown in figure 3.10 of the lecture notes, calculate the electric forces (per unit area) exerted on the interface between the two different dielectrics, in terms of fields in the system.

Solution: By the definition of the potential energy U, the force may be calculated as $-\partial U/\partial x$, where x is the boundary's position, provided that no other forces perform work at its displacement. In our case, this condition means that the capacitors should be disconnected from any external circuit (say, a battery), so that U, before the differentiation, could be expressed via the of the constant charge $Q = CV$ of the capacitor rather than the applied voltage V or the electric field(s) in it. As a sanity check, for the force F exerted on the lower plate of a simple uniform, plane capacitor (see the figure below), with $C = \varepsilon A/d$, this procedure, with the appropriate choice x = const − d (where the constant denotes the reference point) yields the correct result:

$$U = \frac{Q^2}{2C} = \frac{Q^2 d}{2\varepsilon A}, \qquad F = -\frac{\partial U}{\partial x} = \frac{\partial U}{\partial d} = \frac{Q^2}{2\varepsilon A} = \frac{(\sigma A)^2}{2\varepsilon A} = \frac{(DA)^2}{2\varepsilon A} = \frac{D^2}{2\varepsilon}A,$$

$$\text{i.e. } \frac{F}{A} = \frac{D^2}{2\varepsilon} = \frac{\varepsilon E^2}{2}.$$

This result coincides with the solution of problem 2.1, with the natural replacement $\varepsilon_0 \to \varepsilon$.

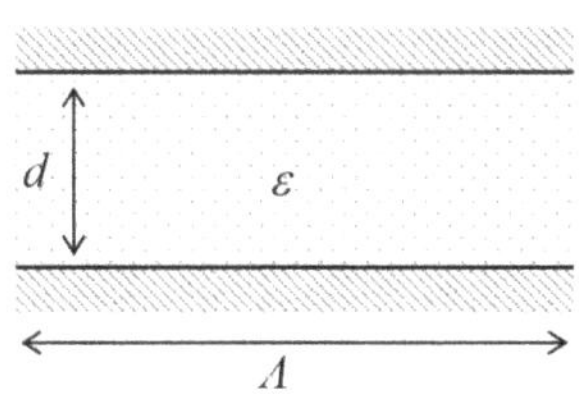

Applying this recipe to the capacitor with the dielectric interface perpendicular to the electrode planes (see the figure below), whose capacitance is given by Eq. (3.57) of the lecture notes, we obtain

$$U = \frac{Q^2}{2C} = \frac{Q^2 d}{2(\varepsilon_1 A_1 + \varepsilon_2 A_2)} = \frac{Q^2 d}{2b(\varepsilon_1 x_1 + \varepsilon_2 x_2)},$$

where b is the linear size of the capacitor in the direction perpendicular to the plane of drawing. Now differentiating this expression over $x \equiv x_1$ = const − x_2, we obtain

$$F = -\frac{\partial U}{\partial x} = \frac{\partial U}{\partial x_2} - \frac{\partial U}{\partial x_1} = \frac{Q^2 d(\varepsilon_1 - \varepsilon_2)}{2b(\varepsilon_1 x_1 + \varepsilon_2 x_2)^2} = \frac{(\varepsilon_1 - \varepsilon_2)E^2}{2}bd,$$

$$\text{i.e. } \frac{F}{A_{\text{interface}}} = \frac{F}{bd} = \frac{(\varepsilon_1 - \varepsilon_2)E^2}{2},$$

where $E = V/d = Q/Cd$ is the electric field—common to both parts of the capacitor.

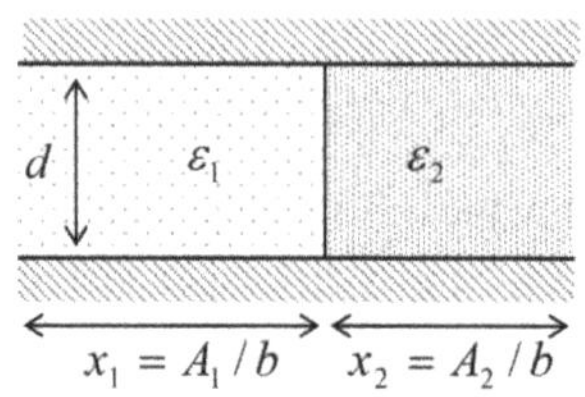

In contrast (as was already discussed in section 3.4 of the lecture notes), in a capacitor with the dielectric boundary parallel to the electrodes (see the figure below), the field common to both dielectrics is $D = \sigma = Q/A$, so it is more natural to express the final result via this field. Using Eq. (3.59) of the lecture notes to write

$$U = \frac{Q^2}{2C} = \frac{Q^2}{2A}\left(\frac{d_1}{\varepsilon_1} + \frac{d_2}{\varepsilon_2}\right),$$

and taking $x \equiv d_2 = \text{const} - d_1$, we obtain

$$F = -\frac{\partial U}{\partial x} = \frac{\partial U}{\partial d_1} - \frac{\partial U}{\partial d_2} = \frac{Q^2}{2A}\left(\frac{1}{\varepsilon_1} - \frac{1}{\varepsilon_2}\right) = \frac{D^2}{2}\left(\frac{1}{\varepsilon_1} - \frac{1}{\varepsilon_2}\right)A,$$

$$\text{i.e. } \frac{F}{A} = \frac{D^2}{2}\left(\frac{1}{\varepsilon_1} - \frac{1}{\varepsilon_2}\right).$$

In both cases the force is directed toward the dielectric with the lowest permittivity, i.e. electric field 'tries' to pull in the material with the higher dielectric constant, pushing out that with the lower ε.

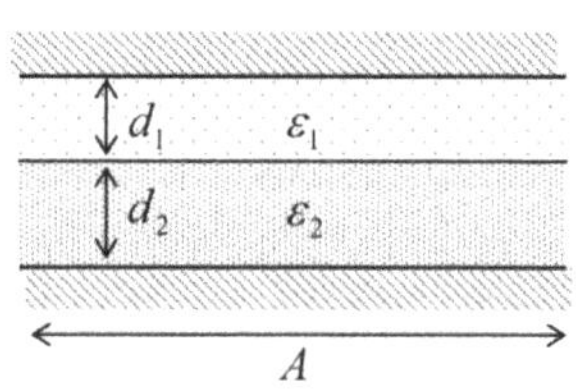

***Problem* 3.19.** A uniform electric field $\mathbf{E}_0$ has been created (by external sources) inside a uniform linear dielectric. Find the change of the electric field, created by cutting out a cavity in the shape of a round cylinder of radius R, with the axis perpendicular to the external field—see the figure below.

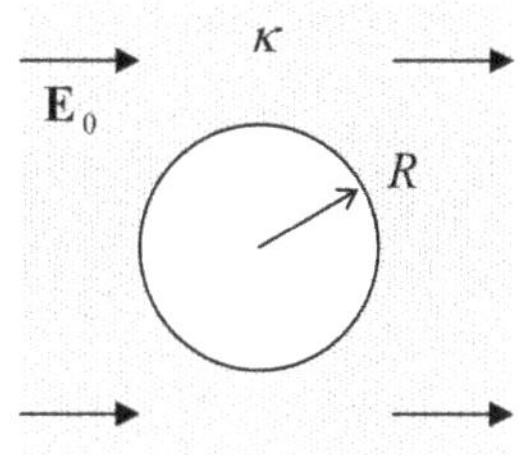

Solution: Introducing the usual polar coordinates, we can use the general solution (2.112) of the Laplace equation, and our experience with using it for the problem shown in figure 2.15 of the lecture notes, as the guidance for looking for the electrostatic potential ϕ in the following form:

$$\phi|_{\rho\leqslant R} = -a_1\rho\cos\varphi, \qquad \phi|_{\rho\geqslant R} = \left(-E_0\rho + \frac{b_1}{\rho}\right)\cos\varphi,$$

where the coefficient a_1 has the sense of the uniform field inside the cavity. Using the boundary conditions of continuity of ϕ and $\varepsilon\partial\phi/\partial n$ (and hence $\kappa\partial\phi/\partial\rho$) on the cavity surface ($\rho = R$), we obtain the following system of two linear equations,

$$a_1R = -E_0R + \frac{b_1}{R}, \qquad a_1 = \kappa\left(-E_0 - \frac{b_1}{R^2}\right),$$

for two unknown coefficients, a_1 and b_1. Solving the system, we obtain:

$$a_1 = -\frac{2\kappa}{\kappa+1}E_0, \qquad b_1 = -\frac{\kappa-1}{\kappa+1}E_0R^2.$$

As a result, the electrostatic potential distribution may be represented as

$$\phi|_{\rho<R} = -\frac{2\kappa}{\kappa+1}E_0\rho\cos\varphi,$$

$$\phi|_{\rho>R} = -E_0\left(\rho + \frac{\kappa-1}{\kappa+1}\frac{R^2}{\rho}\right)\cos\varphi = -E_0x\left(1 + \frac{\kappa-1}{\kappa+1}\frac{R^2}{x^2+y^2}\right),$$

where x is the Cartesian coordinate along the initial field.

As a (necessary for us all :-) sanity check, at $\kappa = 1$ (uniform space with no dielectric), the potential distribution is the same at both $\rho > R$ and $\rho < R$:

$$\phi = \phi_0 = -E_0\rho\cos\theta,$$

and corresponds to the uniform field $E_0\mathbf{n}_x$. In the general case, $\kappa \neq 1$, the electric field, $\mathbf{E} = -\nabla\phi$, is:

$$\mathbf{E}|_{\rho<R} = \frac{2\kappa}{\kappa+1}E_0\mathbf{n}_x,$$

$$\mathbf{E}|_{\rho>R} = E_0\mathbf{n}_x + \frac{\kappa-1}{\kappa+1}E_0\left[\mathbf{n}_x\frac{R^2(y^2-x^2)}{(x^2+y^2)^2} - \mathbf{n}_y\frac{2R^2xy}{(x^2+y^2)^2}\right].$$

From here, the electric field change from its original value $E_0\mathbf{n}_x$ is:

$$\Delta\mathbf{E}|_{\rho<R} = \left(\frac{2\kappa}{\kappa+1} - 1\right)E_0\mathbf{n}_x \equiv \frac{\kappa-1}{\kappa+1}E_0\mathbf{n}_x,$$

$$\Delta\mathbf{E}|_{\rho>R} = \frac{\kappa-1}{\kappa+1}E_0\left[\mathbf{n}_x\frac{R^2(y^2-x^2)}{(x^2+y^2)^2} - \mathbf{n}_y\frac{2R^2xy}{(x^2+y^2)^2}\right].$$

It is curious that in the limit $\kappa \to \infty$, the internal electric field *increases* by exactly $E_0\mathbf{n}_x$, i.e. by 100% of its initial value, while $\mathbf{D}$ *drops* dramatically (by a factor of κ). The reader is invited to interpret this fact in the light of the thin-gap 'experiments' discussed in section 3.4 of the lecture notes—see figure 3.9.

***Problem* 3.20.** Similar small spherical particles, made of a linear dielectric, are dispersed in free space with a low concentration $n \ll 1/R^3$, where R is particle's radius. Calculate the average dielectric constant of such a medium. Compare the result with the apparent, but wrong answer

$$\bar{\kappa} - 1 = (\kappa - 1)\,nV,$$

(where κ is the dielectric constant of the particle's material and $V = (4\pi/3)R^3$ is its volume), and explain the origin of the difference.

Solution: Due to the given condition $nR^3 \ll 1$, we may neglect the mutual electrostatic interaction between the dipole moments $\mathbf{p}$ of particles, induced by an external electric field $\mathbf{E}_0$, so that for each $\mathbf{p}$ we may use Eq. (3.64) of the lecture notes:

$$\mathbf{p} = 4\pi\varepsilon_0 R^3\frac{\kappa-1}{\kappa+2}\mathbf{E}_0.$$

Now we may calculate the average polarization of the medium by just summing all moments of a unit volume. This yields

$$\overline{\mathbf{P}} = n\mathbf{p} = 4\pi nR^3\frac{\kappa-1}{\kappa+2}\varepsilon_0\mathbf{E}_0,$$

so that applying Eq. (3.45) to the average values of $\mathbf{P}$ and κ, we obtain

$$\bar{\kappa} = 1 + 4\pi nR^3\frac{\kappa-1}{\kappa+2} \equiv 1 + (\kappa-1)\,nV\frac{3}{\kappa+2}. \qquad (*)$$

The formula shows that the apparent result cited in the assignment is valid only asymptotically, if the last fraction on its right-hand side approaches 1, i.e. in the limit $\kappa \to 1$, when the particle material's polarization is very weak. The reason of this difference is clear from the solution of the single sphere polarization problem in section 3.4, and in particular from figure 3.11b and Eq. (3.65): due to the non-uniform distribution of the electric field in the vicinity of the sphere (caused by its polarization), the field inside the sphere (albeit uniform) differs, by exactly the same factor $3/(\kappa + 2)$, from the external field far from the sphere. The comparison implies that generally, the average dielectric constant depends not only on the concentration, volume, and material of the dispersed particles, but also on their shape and orientation in space.

Note also that in the limit $\kappa \to \infty$, the correct result (*) approaches Eq. (3.51) for a diluted medium of metallic spheres, derived in the lecture notes in a similar way.

Problem* 3.21. Calculate the spatial distribution of the electrostatic potential induced by a point charge q placed at distance d from a very wide parallel plate, of thickness D, made of a linear dielectric—see the figure below.

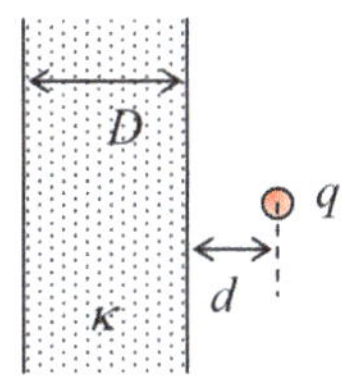

Solution: Let us direct the axis z normally to the plate, so that the radius-vector of the charge is $\mathbf{r} = \{0, 0, d\}$, as shown in the figure below, and examine what system of charge images may describe the field everywhere in the system, satisfying the boundary conditions (3.35) and (3.37) on both surfaces of the plate.

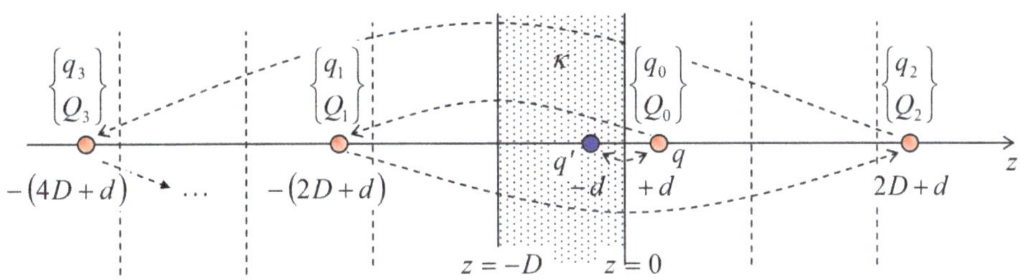

From section 3.4 of the lecture notes (see figure 3.12 and its discussion) we know that the conditions on the right surface ($z = 0$) may be satisfied by using, in addition to the real charge q, just two its images: the charge

$$q' = -\frac{\kappa - 1}{\kappa + 1}q, \tag{*}$$

located at $z = -d$ and describing the additional field in the free space (at $z \geqslant 0$), and the charge

$$Q_0 = \frac{2}{\kappa + 1}q, \tag{**}$$

(in section 3.4, called q''), describing the field inside the dielectric (at $z \leqslant 0$).

Now let us try to satisfy the boundary conditions at the opposite surface of the plate, located at $z = -D$, introducing an image charge Q_1, located at point $z = -(2D + d)$, i.e. at the same distance $(D + d)$ from this surface as the charge Q_0 (see the figure above), and describing the additional field inside the dielectric. The most important point here (and in this solution as a whole) is that the value (*) of the charge Q_0 should not be adjusted, because the charge Q_1 (as well as all other image charges we will introduce later) gives a different spatial dependence of the potential inside the dielectric, so that their net potential is

$$\phi_1(\mathbf{r}) = \frac{1}{4\pi\varepsilon_0}\left\{\frac{Q_0}{[\rho^2 + (z - d)^2]^{1/2}} + \frac{Q_1}{[\rho^2 + (z + 2D + d)^2]^{1/2}}\right\}, \quad \text{for } -D \leqslant z \leqslant 0, \qquad (***)$$

(where $\rho \equiv (x^2 + y^2)^{1/2}$ is the distance of the field observation point $\mathbf{r}$ from the axis z), so that any adjustment of Q_0 would irreparably violate the boundary conditions at $z = 0$.

However, in order to balance the field (***) in the dielectric, and hence at the right of the border $z = -D$, with that in the free space left of the border (i.e. at $z < -D$), we need a new contribution which would have, at the border, the same dependence on ρ. Such contribution may be provided by another charge image sitting at the same distance $(D + d)$ from the border as Q_0 and Q_1. This charge cannot be co-located with Q_1, because it would violate the Laplace equation at that point, $z = -(2D + d)$, which does not have a real charge. Hence, this new image charge (let us call it q_0) has to be co-located with the charge Q_0, i.e. sit at the same point $z = d$ (see the figure above), creating the contribution

$$\phi_1(\mathbf{r}) = \frac{1}{4\pi\varepsilon_0}\frac{q_0}{[\rho^2 + (z - d)^2]^{1/2}}, \qquad \text{for } z \leqslant -D, \qquad (****)$$

to the potential[6]. Now applying the boundary conditions (3.35) and (3.37) at $z = -D$ to Eqs. (***) and (****), we obtain a system of equations very similar to Eqs. (3.60),

$$Q_0 + Q_1 = q_0, \qquad \kappa(Q_0 - Q_1) = q_0,$$

which allows us to express the new image charges via the already determined Q_0:

$$Q_1 = \frac{\kappa - 1}{\kappa + 1}Q_0, \qquad q_0 = \frac{2\kappa}{\kappa + 1}Q_0.$$

(As an intermediate sanity check, at $\kappa = 1$, i.e. with no dielectric plate at all, $Q_1 = 0$, and $q_0 = Q_0 = q$—as they should be, because in this case the field everywhere is just that of the real point charge q.)

[6] Here and later, I will use lower-case letters q to denote any charge image describing a contribution to the field in a free space region (necessarily on the opposite side of the dielectric plate from this charge's location), while upper-case letters Q are reserved for the charges giving contributions into the field inside the plate.

Now we should revisit the surface $z = 0$, and compensate the field contribution by charge Q_1 in the dielectric (see Eq. (***) above) by an absolutely similar introduction of two new charge images, sitting at the same distance, $2D + d$, from that boundary: charge Q_2, located at $z = 2D + d$, and describing a new contribution to the field in the dielectric, and charge q_1, co-located with Q_1, which describes a new contribution (besides those from q and q') to the field in the free space outside the dielectric. The boundary conditions relating Q_1, Q_2, and q_1 are exactly the same as the above relations for Q_0, Q_1, and q_0, and lead to similar expressions:

$$Q_2 = \frac{\kappa - 1}{\kappa + 1} Q_1 = \left(\frac{\kappa - 1}{\kappa + 1}\right)^2 Q_0, \qquad q_1 = \frac{2\kappa}{\kappa + 1} Q_1 = \frac{2\kappa}{\kappa + 1}\frac{\kappa - 1}{\kappa + 1} Q_0.$$

Absolutely similarly returning to the surface $z = -D$, then to surface $z = 0$, etc (see the arrows in the figure above, showing the sequence of new charge image generation), we obtain an infinite system of image charges

$$Q_n = \left(\frac{\kappa - 1}{\kappa + 1}\right)^n Q_0, \text{ and } q_n = \frac{2\kappa}{\kappa + 1} Q_n = \frac{2\kappa}{\kappa + 1}\left(\frac{\kappa - 1}{\kappa + 1}\right)^n Q_0, \qquad \text{for } n \geqslant 0,$$

located at $z = nD + d = 2mD + d$ for even $n = 2m$, and at $z = -(n + 1)D - d = -2mD - d$ for odd $n = 2m - 1$ (with $m = 0, 1, 2,\ldots$). From this series, using Eq. (**) for Q_0, and taking into account the contributions from the actual charge q and the 'irregular' image charge q' given by Eq. (*)[7], we may readily construct the potential distribution in the system:

$$\phi(\mathbf{r}) = \frac{q}{4\pi\varepsilon_0}\frac{4\kappa}{(\kappa + 1)^2}\sum_{m=0}^{\infty}\left(\frac{\kappa - 1}{\kappa + 1}\right)^{2m}\frac{1}{[\rho^2 + (z - 2mD - d)^2]^{1/2}}, \qquad \text{for } z < -D,$$

$$\phi(\mathbf{r}) = \frac{q}{4\pi\varepsilon_0}\frac{2}{\kappa+1}\sum_{m=0}^{\infty}\left(\frac{\kappa-1}{\kappa+1}\right)^{2m-1} \times \left\{\begin{array}{l} \dfrac{\kappa-1}{\kappa+1}\dfrac{1}{[\rho^2 + (z - 2mD - d)^2]^{1/2}} \\ +\dfrac{(1 - \delta_{m0})}{[\rho^2 + (z + 2mD + d)^2]^{1/2}} \end{array}\right\}, \qquad \text{for } -D \leqslant z < 0,$$

$$\phi(\mathbf{r}) = \frac{q}{4\pi\varepsilon_0}\left\{\begin{array}{l} \dfrac{1}{[\rho^2 + (z - d)^2]^{1/2}} - \dfrac{\kappa - 1}{\kappa + 1}\dfrac{1}{[\rho^2 + (z + d)^2]^{1/2}} \\ + \dfrac{4\kappa}{(\kappa + 1)^2}\displaystyle\sum_{m=1}^{\infty}\left(\frac{\kappa - 1}{\kappa + 1}\right)^{2m-1}\frac{1}{[\rho^2 + (z + 2mD + d)^2]^{1/2}} \end{array}\right\}, \qquad \text{for } 0 \leqslant z.$$

Plots of this potential as a function of z, for the particular values $\rho = d = D$, and three representative values of the dielectric constant κ are shown in the figure below. (Necessary sanity checks are that at both borders, the potential is continuous, while its

[7] This charge (the only one with the sign opposite to that of the actual charge q) drops out of the regular sequence $\{Q_n, q_n\}$, because this is the only charge that may be located inside the dielectric plate, and hence cannot have a co-located partner (say, Q') which would give a contribution into the field inside the plate.

derivative jumps by a factor of κ—as they should.) The plots show that the increase of κ suppresses the electric field not only behind the dielectric plate, but also inside and in front of it, i.e. in the vicinity of the actual point charge. (According to Eqs. (3.66) and (3.68) of the lecture notes, the last effect is even stronger for a dielectric half-space, which is the particular case of our current problem, with $D \to \infty$.)

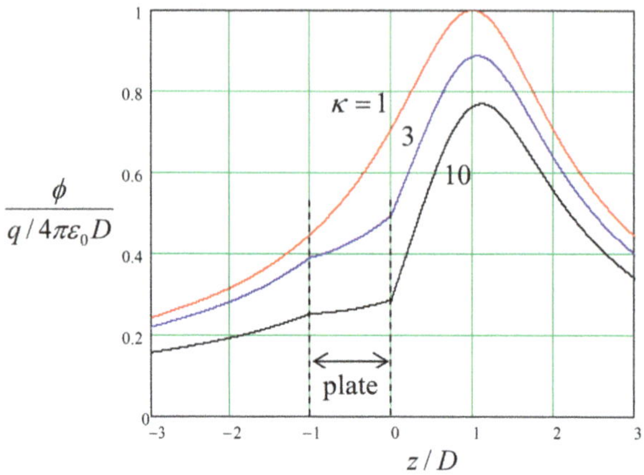

As an additional exercise, I can recommend the reader also solve this problem using the variable separation method, with the following (generally useful) expansion:

$$\frac{1}{(\rho^2 + z^2)^{1/2}} = \int_0^\infty J_0(k\rho)\, e^{-k|z|} dk,$$

where $J_0(\xi)$ is the Bessel function of the first kind—see section 2.7 of the lecture notes.

***Problem* 3.22.** Discuss the physical nature of Eq. (3.76) of the lecture notes. Apply your conclusions to a material with a fixed (field-independent) polarization $\mathbf{P}_0$, and calculate the electric field's energy of the uniformly polarized sphere (see problem 3.11).

Solution: In a dipole media, with a constant number n of elementary dipole moments $\mathbf{p}$ per unit volume, $\delta\mathbf{P} = n\delta\mathbf{p}$, so that the electric energy variation, given by Eq. (3.76), equals

$$\delta u = \mathbf{E} \cdot \delta(\varepsilon_0 \mathbf{E} + \mathbf{P}) \equiv \varepsilon_0 \mathbf{E} \cdot \delta\mathbf{E} + \mathbf{E} \cdot \delta\mathbf{P} = \delta u_0 + n\mathbf{E} \cdot \delta\mathbf{p}, \quad \text{with } u_0 \equiv \varepsilon_0 \frac{E^2}{2},$$

where u_0 is the energy density of the electric field as such—see Eq. (1.65) of the lecture notes. But the product $\mathbf{E} \cdot \delta\mathbf{p}$ is just the energy transfer from the electric field to the internal degrees of freedom of the elementary dipole, resisting its polarization[8]. (For example, in simplest model of the dipole, a point charge q deviating from its equilibrium position by distance $\mathbf{r}$, with $\mathbf{p} = q\mathbf{r}$, this product is $\mathbf{E} \cdot \delta\mathbf{p} = \mathbf{E} \cdot q\delta\mathbf{r} = \mathbf{F} \cdot \delta\mathbf{r}$, where $\mathbf{F} = q\mathbf{E}$ is the force exerted by the field.) Hence, the deviation of the electric field energy U from the free-space value (1.65) is due to the additional energy, possibly of non-electric origin, stored inside elementary dipoles—for example, polarized atoms or molecules.

For a medium with a fixed polarization $\mathbf{P}_0$ (say, an electret or a saturated ferroelectric), the internal energy of the elementary dipoles does not depend on the electric field ($\delta\mathbf{P} = 0$), so that it is reasonable to associate with the field only the energy density u_0. Hence, the full electric field energy may be calculated as an integral

$$U = U_0 = \frac{\varepsilon_0}{2}\int E^2 d^3r,$$

taken over the whole space where the electric field is significant—both inside and outside the volume filled with the polarized medium. However, in some cases it may be technically easier to calculate the energy using the equivalent Eq. (1.60), with the charge density replaced by the effective charge (3.30):

$$U = \frac{1}{2}\int \rho_{\text{ef}}(\mathbf{r})\phi(\mathbf{r})d^3r = -\frac{1}{2}\int \nabla \cdot \mathbf{P}_0(\mathbf{r})\phi(\mathbf{r})d^3r,$$

because this integration may be limited to the polarized medium—moreover, if $\mathbf{P}_0 =$ const inside it, just to its surface.

For example, for problem 3.11 (the uniformly polarized sphere), in which the effective charge resides on the sphere's surface S, with the areal density $\sigma_{\text{ef}} = P_0\cos\theta$, this expression yields

$$U = \frac{1}{2}\int_S \sigma_{\text{ef}}\phi|_{r=R}d^2r = \frac{R^2P_0}{2}\int_0^{2\pi} d\phi \int_0^{\pi} \cos\theta\phi|_{r=R} \sin\theta d\theta.$$

Plugging in the potential distribution over the surface of the sphere, calculated in the solution of that problem,

$$\phi|_{r=R} = \frac{P_0R}{3\varepsilon_0}\cos\theta,$$

we obtain

$$\begin{aligned} U &= \frac{R^2P_0}{2}\frac{P_0R}{3\varepsilon_0}\int_0^{\pi} \cos^2\theta \sin\theta d\theta = \frac{P_0^2R^3}{6\varepsilon_0}\int_0^{2\pi} d\phi\int_0^{\pi} \cos^2\theta \sin\theta d\theta \\ &= \frac{P_0^2R^3}{6\varepsilon_0}2\pi\frac{2}{3} = \frac{2\pi P_0^2R^3}{9\varepsilon_0}. \end{aligned}$$

[8] This expression is in agreement with Eq. (3.15), despite its apparently different sign. For example, if the dipole moment *increases* by $\delta\mathbf{p}$ in a fixed external field $\mathbf{E}_{\text{ext}}$, according to Eq. (3.15) its energy in the field *decreases* by $\delta\mathbf{p} \cdot \mathbf{E}_{\text{ext}}$, and hence its internal energy *increases* by the same amount.

***Problem* 3.23.** Use Eqs. (3.73) and (3.81) to calculate the force of attraction of the plane capacitor's plates (per unit area), for two cases:

(i) the capacitor is charged to a voltage V, and then disconnected from the battery, and
(ii) the capacitor remains connected to the battery.

Solution: The attraction force may be calculated as

$$F = \frac{\partial U}{\partial d},$$

where d is the distance between the plates, and U is the appropriate potential energy. For the case (i), when the capacitor is disconnected from the battery, the only relevant energy is that of the capacitor itself, so that we should use Eq. (3.73). In this particular case (of a uniform field $E = V/d$ inside the volume Ad between the plates, and a negligible field outside it), this formula gives

$$U = \frac{\varepsilon}{2}E^2 Ad = \frac{\varepsilon}{2}\left(\frac{V}{d}\right)^2 Ad \equiv \frac{\varepsilon A V^2}{2d},$$

where A is the capacitor's area. Before the partial differentiation over d, however, we have to take into account that as d is varied, so is the voltage V. In order to spell out this dependence, we may use the fact that since the capacitor is disconnected from the battery, its electric charge Q is not affected by d, so that using the well-known expression for the capacitance C (see Eq. (3.55) of the lecture notes), we obtain

$$V = \frac{Q}{C} = \frac{Qd}{\varepsilon A}, \qquad \text{and hence } U = \frac{\varepsilon A}{2d}\left(\frac{Qd}{\varepsilon A}\right)^2 \equiv \frac{Q^2}{2\varepsilon A}d.$$

(Actually, for the particular case of no dielectric filling, i.e. $\varepsilon = \varepsilon_0$, this expression, in the form

$$\frac{U}{A} = \frac{\sigma^2}{2\varepsilon_0}d,$$

where $\sigma = Q/A$ is the areal density of the plate's charge, was already obtained in the solution of problem 1.14.) Now the differentiation over d yields

$$F = \frac{\partial}{\partial d}\left(\frac{Q^2}{2\varepsilon A}d\right) = \frac{Q^2}{2\varepsilon A} \equiv \frac{(CV)^2}{2\varepsilon A} \equiv \frac{\varepsilon A V^2}{2d^2}. \qquad (*)$$

This result may be readily verified by a direct calculation of the force per unit area, $F/A = \sigma E/2$, with the factor of ½ resulting from the fact that the force exerted on one plate is due to the electric field $E/2$, produced by the other plate alone.

However, in the case (ii), when the voltage V is fixed by the battery, the use of the 'ordinary' potential energy (3.73) would give an incorrect result:

$$F = \frac{\partial}{\partial d}\left(\frac{\varepsilon A V^2}{2d}\right) = -\frac{\varepsilon A V^2}{2d^2}, \qquad \text{(WRONG!)} \qquad (**)$$

which differs from the correct result (*) by the sign. The reason for such an error is that if the capacitor stays connected with the battery during the virtual change of d, we have to take into account the total energy of the (capacitor + battery) system, or at least of its part depending on d. This is exactly the definition of the Gibbs potential energy U_G, given (for the particular case of the electrostatic field) by Eq. (3.81). Since according to that expression (besides an inconsequential constant), $U_G = -U$, there is no need to perform the calculations again, and we may just revert the sign in the incorrect Eq. (**), turning it into the correct Eq. (*). (Of course, the attraction force between the plates, at the same voltage between them, does not depend on whether the capacitor is still connected to the battery or not.)

Let me hope that this simple example illustrates the difference between the potential energies U and U_G very clearly.

Chapter 4

DC current

***Problem* 4.1.** A dc voltage V_0 is applied to the end of a semi-infinite chain of lumped Ohmic resistors, shown in the figure below. Calculate the voltage across the *j*th link of the chain.

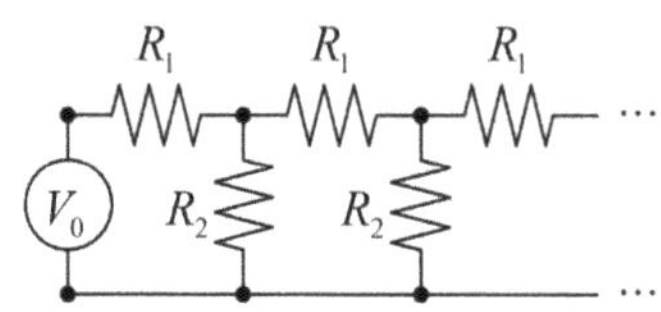

Solution: This problem is a close analog of problem 2.4 for a similar chain of lumped capacitances. Again, since all links of the chain are similar, it is sufficient to analyze just a couple of them, say with numbers j and $(j + 1)$, where j is the 'distance' from the left end. Denoting the currents through the 'series' resistors R_1, and the voltages across the 'parallel' resistors R_2 as shown in the figure below, the first Kirchhoff law (4.7a) for the left of the displayed two nodes gives

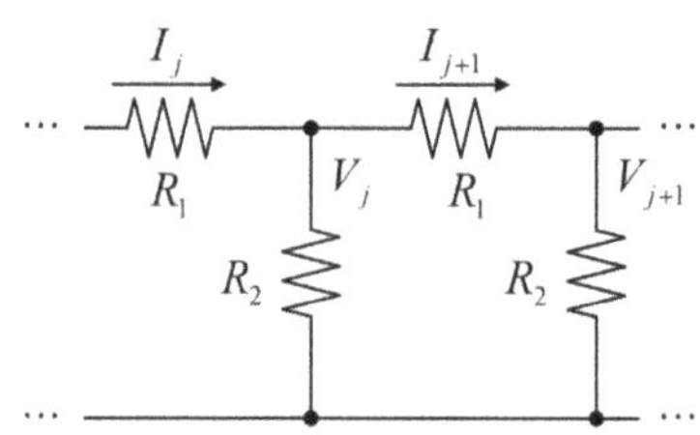

$$I_j - I_{j+1} = \frac{V_j}{R_2}, \qquad (*)$$

while the second Kirchhoff law (4.7b) for the displayed loop reads

$$V_j - V_{j+1} = R_1 I_{j+1}. \qquad (**)$$

These relations are absolutely similar to Eqs. (*) of the model solution of problem 2.4 (with the natural replacements $Q \to I$, and $C \to 1/R$), so that we may use the same arguments to look for the solution of the infinite system of equations (*) and (**), for $j = 1, 2, \ldots$, in the similar form:

$$V_j \propto I_j \propto e^{-\lambda j}, \qquad \text{so that } \frac{V_{j+1}}{V_j} = \frac{I_{j+1}}{I_j} = \alpha \equiv e^{-\lambda} < 1.$$

Indeed, using the last relations to eliminate V_{j+1} and I_{j+1} from Eqs. (*) and (**), we obtain a system of two linear homogenous equations for the same two variables V_j and I_j:

$$(1 - \alpha)Q_j = \frac{V_j}{R_2}, \qquad (1 - \alpha)\, V_j = R_1 \alpha I_j.$$

These equations are compatible (and hence the assumed solution in valid for all j) if the determinant of this system equals zero:

$$\begin{vmatrix} 1 - \alpha & -\alpha R_1 \\ -1/R_2 & 1 - \alpha \end{vmatrix} = 0, \qquad \text{i.e. if } \alpha^2 - \left(2 + \frac{R_1}{R_2}\right)\alpha + 1 = 0.$$

The needed root $\alpha < 1$ of this quadratic equation is

$$\alpha = 1 + \frac{R_1}{2R_2} - \left(\frac{R_1}{R_2} + \frac{R_1^2}{4R_2^2}\right)^{1/2}. \qquad (***)$$

Since our definition of V_j (see the figure above) at $j = 0$ complies with our boundary condition V_0, the requested voltage across the jth link of the chain is

$$V_j = V_0 \alpha^j \equiv V_0 e^{-\lambda j},$$

where α is given by Eq. (***). In the limit $R_1/R_2 \to \infty$, this relation yields

$$\alpha \to \frac{R_1}{2R_2} \gg 1.$$

This result means that the applied voltage does not propagate noticeably beyond the first resistance capacitor R_1 of the chain, because the next, relatively small resistance R_2 effectively 'grounds' it—see the figure above.

In the opposite limit, $R_1/R_2 \to 0$, Eq. (***) yields

$$\alpha \to 1 - \left(\frac{R_1}{R_2}\right)^{1/2} \to 1, \qquad \text{so that } \lambda \equiv \ln\frac{1}{\alpha} \to \left(\frac{R_1}{R_2}\right)^{1/2} \ll 1.$$

According to the definition of the parameter λ, this means that the applied voltage penetrates deep into the chain, decaying at the 'distance' of the order of $N \equiv 1/\lambda \approx (R_2/R_1)^{1/2} \gg 1$.

***Problem* 4.2.** It is well known that properties of many dc current sources (e.g. batteries) may be reasonably well represented as a connection in series of a *perfect voltage source* and an Ohmic *internal resistance*. Discuss the option, and possible advantages, of using a different equivalent circuit that would include a *perfect current source.*

Solution: The well-known representation mentioned in the assignment is shown in the figure below. Here the circle denotes an imaginary two-terminal circuit element that sustains a fixed voltage $\mathscr{V}$,[1] regardless of what is connected to it. (Naturally, it is impossible to implement such ideal element separately from the battery's internal resistance r; indeed, short-circuiting its terminals, we would obtain infinite current at the non-zero voltage $\mathscr{V}$, i.e. an infinite power.) We may apply the second Kirchhoff law (4.7b) to the loop formed by this equivalent circuit and the (unspecified) external circuit to the write the relation between the voltage V at the battery's terminals and the current I flowing through the battery[2]:

$$V = \mathscr{V} - rI. \qquad (*)$$

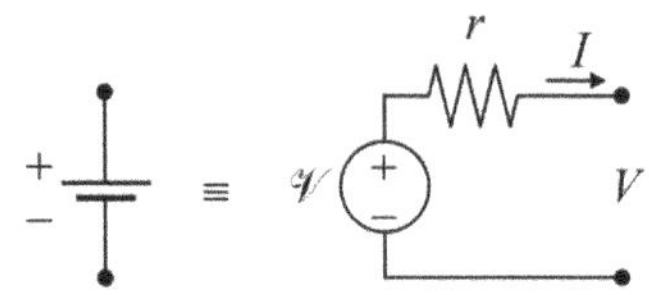

This is just an analytical representation of the equivalent circuit shown above (and hence of the battery properties), and is well known to most undergraduate students.

The fact which attracts much less attention is that the following equivalent form of Eq. (*),

$$I = \mathscr{I} - \frac{V}{r}, \qquad \text{with } \mathscr{I} \equiv \frac{\mathscr{V}}{r}, \qquad (**)$$

may be graphically represented with another equivalent circuit—see the figure below. Here the circle with an arrow is the common notation of a *prefect current source*, defined as a two-terminal circuit element that sustains a certain current $\mathscr{I}$ (in our case, equal to $\mathscr{V}/r$), regardless of the circuit it is connected to. Evidently, such ideal element is also impossible to implement separately, without the parallel connection of a finite resistance r, because it would push the same current $\mathscr{I}$ even through the empty space, creating infinite voltage and hence supplying infinite power.

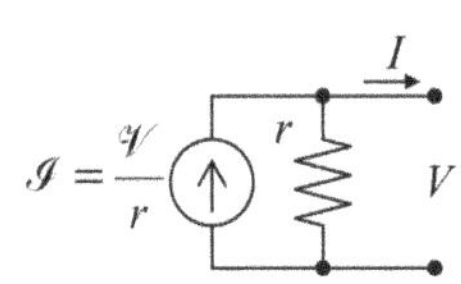

[1] This is exactly the formal definition of the electromotive force $\mathscr{V}$. (Note that in the circuit theory it is frequently denoted as either E or $\mathscr{E}$.)

[2] This relation, in the limit $R_i \to 0$, explains why the perfect voltage source is sometimes called the 'battery with zero internal resistance'.

Due to the mathematical equivalency of Eqs. (*) and (**), the two equivalent circuits shown above always give the same final result, but each of them may be more or less convenient, depending on the problem. For example, if we need to calculate the current in the circuit shown in the figure below, it is more convenient to use the first equivalent circuit of each battery, with the perfect voltage source. Indeed, as the right panel of the figure shows, its application reduces the full circuit to a connection, in series, of the net e.m.f. ($\mathscr{V}_1 + \mathscr{V}_2$) to the total resistance ($r_1 + r_2 + R$), immediately giving the final result:

$$I = \frac{\mathscr{V}_1 + \mathscr{V}_2}{r_1 + r_2 + R}.$$

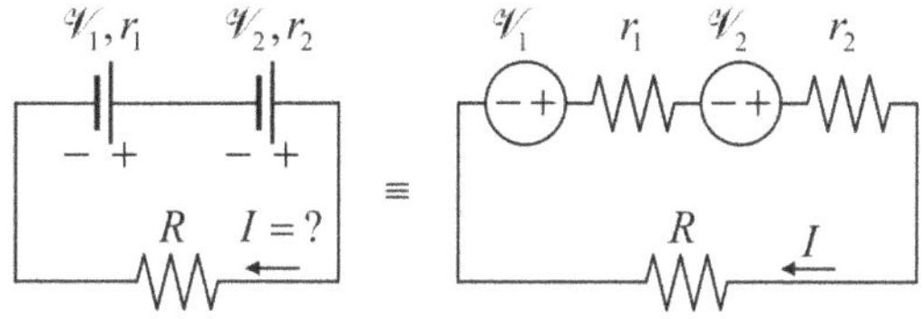

However, using this equivalent circuit for the solution of the (infamous :-) problem shown in the figure below is much less convenient, and requires the heavy artillery of the Kirchhoff laws (4.7). On the other hand, using the perfect-current-source alternative immediately reduces the full circuit to a simple connection of an ideal source of the net current ($\mathscr{I}_1 + \mathscr{I}_2$) to the net Ohmic conductance ($1/r_1 + 1/r_2 + 1/R$) of three resistors connected in parallel, so that the requested voltage across the circuit is calculated very similarly to the current in the first problem:

$$V = \frac{\mathscr{I}_1 + \mathscr{I}_2}{1/r_1 + 1/r_2 + 1/R} \equiv \frac{\mathscr{V}_1/r_1 + \mathscr{V}_2/r_2}{1/r_1 + 1/r_2 + 1/R}.$$

(The current through the 'load' R is of course just V/R.)

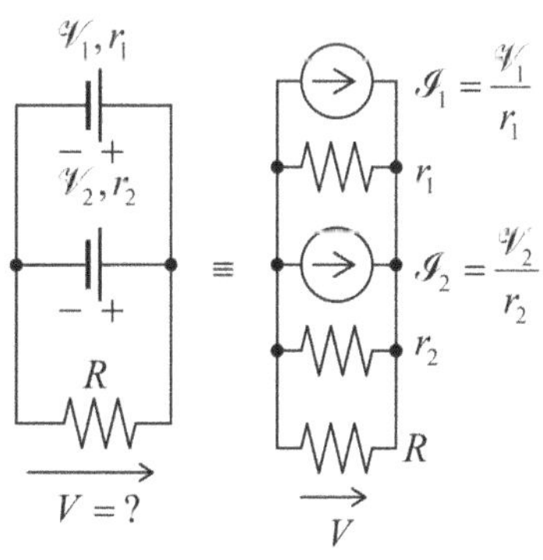

***Problem* 4.3.** Calculate the resistance between two large, uniform Ohmic conductors separated with a very thin, planar, insulating partition, with a circular hole of radius R in it—see the figure below.

Hint: You may like to use the degenerate ellipsoidal coordinates, which had been used in section 2.4 of the lecture notes.

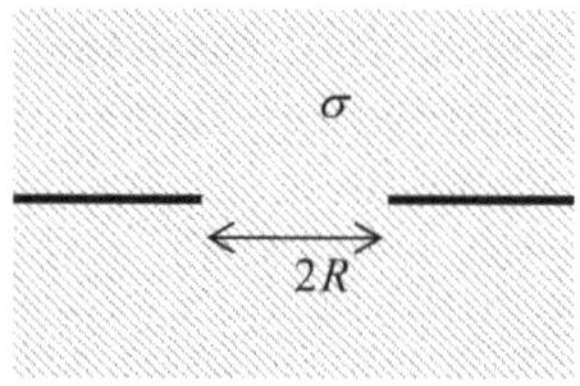

Solution: Selecting the parameter R of the degenerate ellipsoidal coordinates defined by Eqs. (2.59) of the lecture notes,

$$x = R\cosh\alpha\sin\beta\cos\varphi,$$
$$y = R\cosh\alpha\sin\beta\sin\varphi,$$
$$z = R\sinh\alpha\cos\beta.$$

to be equal to the hole radius, and placing the coordinate origin into its center, we see that the intact part of the partition corresponds to $\beta = \pi/2 = \text{const}$, while for the conductor points slightly above the partition, $z \approx R\sinh\alpha\,(\pi/2 - \beta)$, so that

$$\left.\frac{\partial\phi}{\partial n}\right|_{\substack{z=0\\ r>R}} = \left.\frac{\partial\phi}{\partial z}\right|_{\substack{z=0\\ r>R}} = -\frac{1}{R\sinh\alpha}\frac{\partial\phi}{\partial\beta}.$$

This means that if the potential is a function of α alone, the boundary condition of having no current flow through the partition is automatically satisfied. (Due to the mirror symmetry of the problem about the partition plane, the same conclusion is valid for the lower semi-space as well.) From the disk capacitance problem's solution in section 2.4 of the lecture notes, we already know that the Laplace equation may be also satisfied with the potential distribution (2.64):

$$\phi = c_1\tan^{-1}(\sinh\alpha) + c_2. \qquad (*)$$

For the sake of notation simplicity, let us accept the (anti-)symmetric boundary conditions at infinity:

$$\phi \to \mp\frac{V}{2}\,\text{sgn}(z) = \mp\frac{V}{2}\,\text{sgn}(\alpha), \quad \text{at } |z| \to \infty, \quad \text{i.e. at } |\alpha| \to \infty,$$

where V is the voltage applied between the distant points of the conductors; in this case the solution (*) takes the simple form

$$\phi = -\frac{V}{\pi}\tan^{-1}(\sinh\alpha).$$

Note that according to this formula, the contribution of distant parts of the conductor (with $r \gg R$) into the total voltage drop is negligible; this is why the exact shape of external electrodes is not important, provided that they are not too small and not too close to the hole.

What remains is to calculate the total current

$$I = \int_S j_n d^2r = -\sigma \int_S \frac{\partial \phi}{\partial n} d^2r,$$

where S is an arbitrary surface crossed by the whole current flow. (Due to the current continuity, the result does not depend on the choice of the surface.) The easiest choice of such a surface is evidently the plane $z = 0$—more exactly, its fraction inside the hole, $r < R$. Near the plane, the coordinate relation for z is reduced to $z \approx R\alpha \cos \beta$, while our solution is just $\phi \approx -(V/\pi)\, \alpha$, so that $\phi \approx -(V/\pi R \cos \beta)\, \alpha$, and $-\partial\phi/\partial z = V/R\pi \cos \beta$. Since at this plane, the distance of a point from the center is $\rho \equiv (x^2 + y^2)^{1/2} = R \sin \beta$, we obtain

$$I = -\sigma \int_{\substack{z=0 \\ \rho<R}} \frac{\partial \phi}{\partial z} d^2\rho = -\, 2\pi\sigma \int_0^R \frac{\partial \phi}{\partial z} \rho d\rho = 2\pi\sigma \frac{V}{\pi} \int_0^R \frac{\rho d\rho}{R \cos \beta}$$

$$= \sigma V \int_{\rho=0}^{\rho=R} \frac{d(\rho^2)}{(R^2 - \rho^2)^{1/2}} = 2\sigma V R,$$

so that the hole's resistance V/I is just $1/2\sigma R$.

This *Maxwell* (or 'Maxwell–Rayleigh') *formula* for resistance is frequently remembered as the ratio of the material resistivity $1/\sigma$ to the hole diameter $2R$, without any numerical coefficients. The result may be interpreted as a longitudinal resistance of a round cylinder with the cross-section $A = \pi R^2$, and the effective length $l = 2\pi R$, defined by the gradual spreading of the current into the electrodes.

***Problem* 4.4.** Calculate the effective (average) conductivity σ_{ef} of a medium with many empty spherical cavities of radius R, carved at random positions in a uniform Ohmic conductor (see the figure below), in the limit of low density $n \ll R^{-3}$ of the spheres.

Hint: Try to use the analogy with a dipole medium—see, e.g. section 3.2 of the lecture notes.

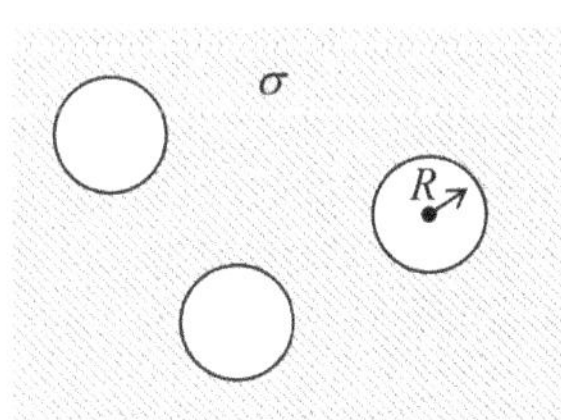

Solution: As was calculated in section 4.3 of the lecture notes, a single spherical cavity, cut in a conductor with a uniform current flow of density $\mathbf{j}_0$, creates outside it an additional dipole field, described by the second term in Eq. (4.26)

$$\phi_{\text{d}} = -\frac{j_0}{\sigma} \frac{R^3}{2r^2} \cos \theta.$$

According to Eq. (3.7), such a field corresponds to the following dipole moment:

$$\mathbf{p} = -4\pi\varepsilon_0\frac{R^3}{2\sigma}\mathbf{j}_0.$$

If the concentration of spheres is low ($nR^3 \ll 1$), these dipole fields add up independently, creating the average electric polarization

$$\mathbf{P} = n\mathbf{p} = -4\pi\varepsilon_0\frac{nR^3}{2\sigma}\mathbf{j}_0.$$

Our analysis of electrically polarized media in section 3.2 has shown that the polarization (divided by ε_0) is effectively subtracted from the field created by external charges. In particular, Eq. (3.33) may be rewritten as

$$\mathbf{E} = \mathbf{E}_0 - \frac{\mathbf{P}}{\varepsilon_0}, \qquad \text{with } \mathbf{E}_0 \equiv \frac{\mathbf{D}}{\varepsilon_0}, \tag{*}$$

where **E** is the macroscopic electric field, and vector **D** is defined by the external charge density ρ via Eq. (3.32). Since the section 3.2 arguments concerning the effect of polarization did not use any assumptions about the media, Eq. (*) may be used for conductors as well. In this case, $\mathbf{E}_0$ is the field, created by an external e.m.f., which would be equal to **E** at $\mathbf{P} = 0$, i.e. $\mathbf{E}_0 = \mathbf{j}_0/\sigma$. This means that in our case the average field is

$$\mathbf{E} = \mathbf{E}_0 - \frac{\mathbf{P}}{\varepsilon_0} = \frac{\mathbf{j}_0}{\sigma} + 4\pi\frac{nR^3}{2\sigma}\mathbf{j}_0 \equiv \frac{\mathbf{j}_0}{\sigma_{\text{ef}}},$$

where σ_{ef} is the effective conductance we were looking for:

$$\sigma_{\text{ef}} = \frac{\sigma}{1 + 2\pi nR^3} \approx \sigma(1 - 2\pi nR^3) \leqslant \sigma.$$

The last (approximate) equality is valid because our analysis is limited to the small cavity concentrations, $nR^3 \ll 1$, when the correction to σ is small. Of course, the very fact of the conductivity reduction, i.e. the negative sign of the difference $\Delta\sigma \equiv \sigma_{\text{ef}} - \sigma$ due to carving out the cavities could be predicted from handwaving arguments, and so could be the proportionality of the ratio $\Delta\sigma/\sigma \ll 1$ to the dimensionless product nR^3, so that the main result of our calculation is the particular (and rather large) numerical coefficient 2π before the product.

***Problem* 4.5.** In two separate experiments, a narrow gap, possibly of an irregular width, between two close metallic electrodes is filled with some material—in the first case, with a uniform linear insulator with an electric permittivity ε, and in the second case, with a uniform conducting material with an Ohmic conductivity σ. Neglecting the fringe effects, calculate the relation between the mutual capacitance

C between the electrodes (in the first case) and the dc resistance R between them (in the second case).

Solution: According to Eqs. (3.53) and (4.14), the electrostatic potential distribution $\phi(\mathbf{r})$ inside both materials satisfies the same (Laplace) differential equation. The boundary conditions on the electrode surfaces ($\phi(\mathbf{r}_1) = \phi_1 = \text{const}$, and $\phi(\mathbf{r}_1) = \phi_2 = \text{const}$) are also the same. (If fringe effects are negligible, the boundary conditions on the lateral surfaces of the materials are not important.) Thus the distribution $\phi(\mathbf{r})$ in both cases is the same, and defines in the first case the electrode surface charge density (3.54):

$$\sigma_s = -\varepsilon \frac{\partial \phi}{\partial n},$$

and in the second case the current density (4.8) on the surfaces:

$$j_n = -\sigma \frac{\partial \phi}{\partial n}.$$

As the result, the ratio σ_s/j_n, and hence the ratio of the full charge Q to the full dc current I, and hence the ratio of the mutual capacitance $C \equiv Q/V$ to dc conductance $G \equiv 1/R \equiv I/V$, all are independent of the exact potential distribution (and hence of the electrode shape):

$$\frac{C}{G} = \frac{Q}{I} = \frac{\sigma_s}{j_n} = \frac{\varepsilon}{\sigma}.$$

According to this formula, the product RC (= C/G) depends only on the parameters of the materials:

$$RC = \frac{\varepsilon}{\sigma} \equiv \frac{\varepsilon_0 \kappa}{\sigma}.$$

Note that both effects (of the surface charge accumulation and Ohmic conductivity) may coexist in the same material, and in this case the RC constant equals to its relaxation time constant τ_r—see Eq. (4.10) and its discussion.

***Problem* 4.6.** Calculate the voltage V across a uniform, wide resistive slab of thickness t, at distance l from the points of the injection/pickup of the dc current I passed across the slab—see the figure below.

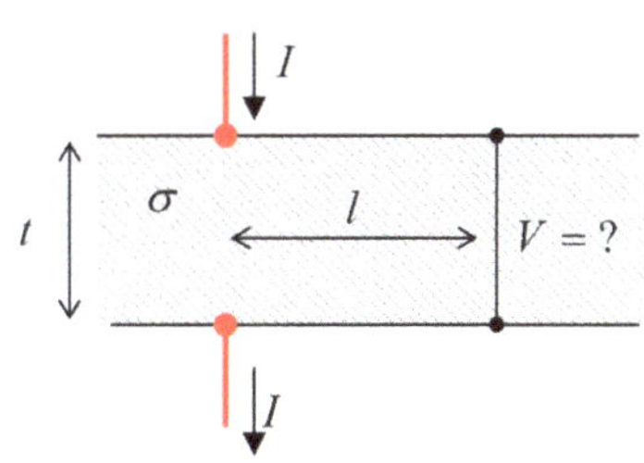

Solution: Following the discussion in section 4.3 of the lecture notes, both the Laplace equation for the electrostatic potential ϕ inside the conductor (related to the dc current density as $\mathbf{j} = \sigma\mathbf{E} = -\sigma\nabla\phi$), and boundary conditions outside the injection/pickup points ($j_n = -\sigma\partial\phi/\partial n = 0$) at its surface may be satisfied by complementing the actual slab with an infinite stack of imaginary slabs, with a periodic pattern of mirror images of the injection/pickup points—see the figure below, where the point colors code the current sign.

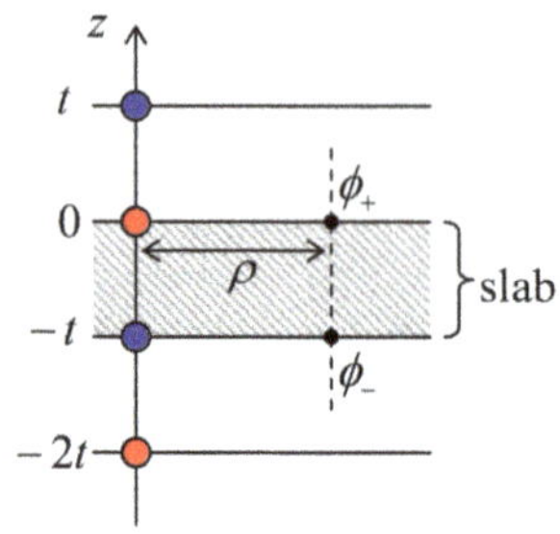

With the coordinates selected as shown in this figure, points of current injection (shown red) are located at $z_k = 2kt$, while points of current pickup (shown blue) are at $z_k' = (2k - 1)t$, where k are integers. Since the magnitude of current at each injection/ejection point is $2I$ (as follows, e.g. from figure 4.8 of the lecture notes, with $d \to 0$), it creates, in the slab stack, a spherically symmetric distribution of current density and electric field,

$$\mathbf{j}_k(\mathbf{r}) = \frac{2(-1)^k I}{4\pi\,|\mathbf{r}_k - \mathbf{r}|^3}(\mathbf{r}_k - \mathbf{r}), \qquad \mathbf{E}_k(\mathbf{r}) = \frac{\mathbf{j}_k}{\sigma} = \frac{(-1)^k I}{2\pi\sigma\,|\mathbf{r}_k - \mathbf{r}|^3}(\mathbf{r}_k - \mathbf{r}),$$

and hence a spherically symmetric potential distribution:

$$\phi_k(z) = (-1)^k \frac{I}{2\pi\sigma\,|\mathbf{r}_k - \mathbf{r}|} = (-1)^k \frac{I}{2\pi\sigma[(kt - z)^2 + \rho^2]^{1/2}}.$$

Summing up these contributions for all k, for the top point of voltage pick-up ($z = 0$) we have

$$\phi_+ = \frac{I}{2\pi\sigma}\sum_{k=-\infty}^{+\infty}\frac{(-1)^k}{(k^2 + \rho^2)^{1/2}}.$$

From the figure above, it is evident that in this symmetric situation, the potential at the lower pick-up point is equal and opposite, $\phi_- = -\phi_+$, so that the voltage drop V is

$$V = \phi_+ - \phi_- = 2\phi_+ = \frac{I}{\pi\sigma}\sum_{k=-\infty}^{+\infty}\frac{(-1)^k}{(k^2t^2 + \rho^2)^{1/2}}.$$

This result is shown (in the appropriate log–log scale) in the figure below. At close distances, $\rho/t \to 0$, the sum is dominated by the term with $k = 0$, so that the voltage drop does not depend on t and increases as $I/\pi\sigma\rho$, reflecting the potential divergence near the current's injection and pickup points. (This divergence would naturally disappear at an account of a non-zero size of the electrodes.) In the opposite limit $\rho/t \gg 1$, the voltage drop decreases with distance exponentially, as $\exp\{-\pi\rho/t\}$.

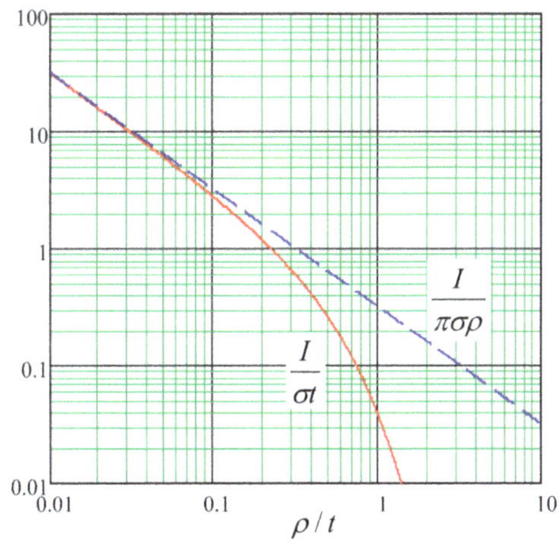

***Problem* 4.7.** Calculate the voltage V between two corners of a square cut from a uniform, resistive sheet of a very small thickness t, induced by the dc current I that is passed between its two other corners—see the figure below.

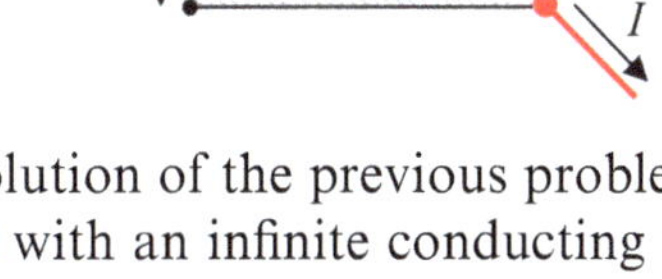

Solution: Extending the solution of the previous problem, we see that the original problem may be replaced with an infinite conducting sheet with a periodic mesh of the current injection/ejection points shown in the figure below. Each red dot means the injection of current $4I$, while a blue point means the pickup of a similar current. (The value $4I$ is due to the merger of four equal currents, due to the same-sign reflection in both sides of the square—cf. figure 4.8 of the lecture notes.)

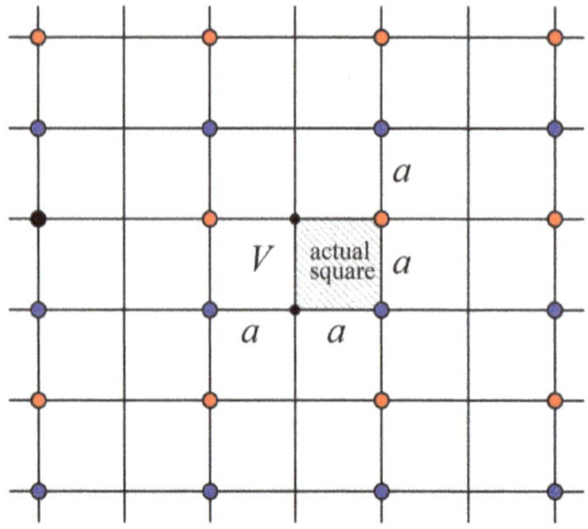

Indeed, one can readily check that the net current, which is a vector sum of the currents radially spreading from/into each injection/pickup point, does not cross any border of the actual sheet square (shown by hatching). In our 2D case, each of these currents creates the current density

$$\mathbf{j}_k(\boldsymbol{\rho}) = \frac{I_k}{2\pi t}\frac{(\boldsymbol{\rho} - \boldsymbol{\rho}_k)}{|\boldsymbol{\rho} - \boldsymbol{\rho}_k|^2} = \pm\frac{2I}{\pi t}\frac{(\boldsymbol{\rho} - \boldsymbol{\rho}_k)}{|\boldsymbol{\rho} - \boldsymbol{\rho}_k|^2},$$

(where t is the sheet thickness), and hence the following electric field,

$$\mathbf{E}_k(\boldsymbol{\rho}) = \frac{\mathbf{j}_k(\boldsymbol{\rho})}{\sigma} = \pm\frac{2I}{\pi t\sigma}\frac{(\boldsymbol{\rho} - \boldsymbol{\rho}_k)}{|\boldsymbol{\rho} - \boldsymbol{\rho}_k|^2},$$

which may be represented as the (minus) gradient of the following electrostatic potential:

$$\phi_k(\boldsymbol{\rho}) = \mp\frac{2I}{\pi t\sigma}\ln|\boldsymbol{\rho} - \boldsymbol{\rho}_k| + \text{const} = \mp\frac{I}{\pi t\sigma}\ln[|\boldsymbol{\rho} - \boldsymbol{\rho}_k|^2] + \text{const}.$$

The integration constant in the last expression is unimportant, because we are interested only in the voltage V (i.e. the potential difference) between the two points, and we can take it for zero.

For our points of voltage measurement (at square corners),

$$|\boldsymbol{\rho} - \boldsymbol{\rho}_k|^2 = a^2[(2k + 1)^2 + k'^2],$$

where k and k' are the integers listing, respectively, the horizontal and vertical distances (in the units of a) between the corner and the current injection/pickup points. Now using the linear superposition principle to sum up the contributions of all the points (see the figure above), we obtain

$$V = \frac{2I}{\pi t\sigma}\sum_{k,k'=-\infty}^{+\infty}\ln\frac{(2k + 1)^2 + (2k' + 1)^2}{(2k + 1)^2 + (2k')^2} \approx 0.72\frac{I}{t\sigma}.$$

This result shows that the voltage drop is somewhat lower that the value $I/t\sigma$ that we would obtain from the passing a uniformly distributed current between the opposite sides of the square sheet. This is natural, because at the corner injection/pickup of the current, most of it passes closer to the side connecting the current electrodes (its right side in the figures above), thus reducing the voltage across the opposite (left) side.

Also, note that the calculated voltage is directly proportional to the so-called *sheet resistance* $R_\square \equiv 1/t\sigma$. This notion is very popular in experimental and engineering practice, because it characterizes the average lateral conductive properties of a thin sheet or a thin film, even if it is non-uniform at the distances of the order of t. The standard notation of this constant (and its alternative name, 'resistance per square') is due to that, according to Eq. (4.21), $R_\square$ is the resistance of a square of any size, cut of the film, if the current is uniformly distributed in it—say, flows between two straight electrodes contacting the opposite sides of the square.

***Problem* 4.8.** Calculate the distribution of the dc current's density in a thin, round, uniform resistive disk, if the current is inserted into a point at its rim, and picked up at the center.

Solution: We can solve analytically even a more general problem, when the current I is inserted into an arbitrary point of the disk, at distance d from the center—see the figure below. Following the discussion in the end of section 4.3 of the lecture notes, let us look for the solution of the problem in the form of the current in an unlimited thin layer made of the same material as the disk, with an additional current I' inserted into a point at a certain distance d' from the center, on the same radius. According to Eq. (4.14), the distribution of the electric potential in the layer outside the current injection and pickup points has to satisfy the same 2D Laplace equation as in 2D problems of electrostatics, including problem 2.37. Guided by the model solution of that problem, we may guess that the additional injection point is the inversion point, with $d' = R^2/d$, and that the image current I' exactly equals the actual injection current I.

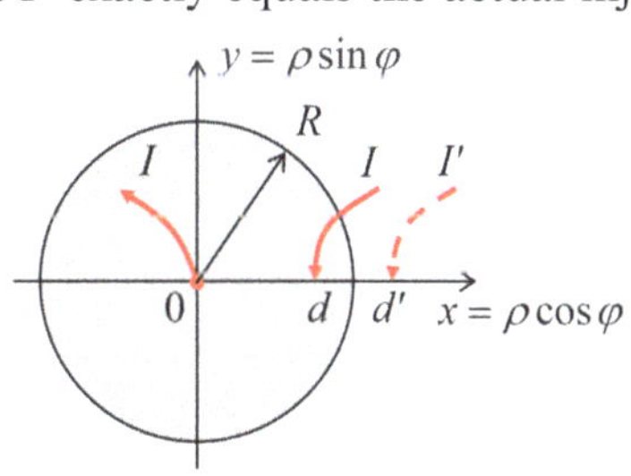

Taking into account that, according to the solution of problem 2.36(i), the Green's function of the 2D Poisson equation for the free space is proportional to $\ln|\boldsymbol{\rho} - \boldsymbol{\rho}'|$, we may look for the solution in the form

$$\begin{aligned}\phi(\boldsymbol{\rho}) = C_0 \ln\rho + C_d \ln\left(\rho^2 + d^2 - 2\rho d\cos\varphi\right)^{1/2} \\ + C_{d'} \ln\left(\rho^2 + d'^2 - 2\rho d'\cos\varphi\right)^{1/2},\end{aligned} \qquad (*)$$

where the argument of each log function is the distance of the observation point, with polar coordinates $\{\rho, \varphi\}$, from one of current injection/pickup point—see the figure above. The constants C should be selected so that the total current flowing out of each contact would be equal to the insertion/pickup current. For example, the first term, describing the current pickup point from the center of the disk, gives the linear current density

$$\mathbf{J} = t\mathbf{j} = t\sigma\mathbf{E} = -t\sigma\nabla\phi = -t\sigma\frac{\partial\phi}{\partial\rho}\mathbf{n}_\rho = -\frac{t\sigma C_0}{\rho}\mathbf{n}_\rho,$$

where t is disk's thickness, so that the above requirement,

$$\oint_{\text{around }\rho=0} J_n dr \equiv 2\pi\rho J_\rho = -I,$$

yields $C = I/2\pi t\sigma$. The constants in two other terms of Eq. (*) are evidently similar by magnitude but opposite by sign, $C_d = C_{d'} = -C_0$, because they describe the injection of similar currents I.

With these constants, Eq. (*) satisfies the 2D Poisson equation for our problem and the boundary conditions at the current injection/pickup points, and the last step is to prove that it also satisfies the boundary condition at the disk's rim:

$$J_n\bigg|_{\rho=R} \equiv -t\sigma\frac{\partial\phi}{\partial\rho}\bigg|_{\rho=R} = 0.$$

The proof is straightforward; indeed, according to Eq. (*), at an arbitrary point of the disk,

$$\begin{aligned}\frac{1}{C_0}\frac{\partial\phi}{\partial\rho} &= \frac{\partial}{\partial\rho}\left[\ln\rho - \frac{1}{2}\ln(\rho^2 + d^2 - 2\rho d\cos\varphi) - \frac{1}{2}\ln(\rho^2 + d'^2 - 2\rho d'\cos\varphi)^{1/2}\right]\\ &= \frac{1}{\rho} - \frac{\rho - d\cos\varphi}{\rho^2 + d^2 - 2\rho d\cos\varphi} - \frac{\rho - d'\cos\varphi}{\rho^2 + d'^2 - 2\rho d'\cos\varphi}.\end{aligned}$$

Now taking $\rho = R$, and using the inversion rule $d' = R^2/d$, we obtain

$$\begin{aligned}\frac{1}{C_0}\frac{\partial\phi}{\partial\rho}\bigg|_{\rho=R} &= \frac{1}{R} - \frac{R - d\cos\varphi}{R^2 + d^2 - 2Rd\cos\varphi} - \frac{R - (R^2/d)\cos\varphi}{R^2 + (R^2/d)^2 - 2R(R^2/d)\cos\varphi}\\ &= \frac{1}{R}\left[1 - \frac{R^2 - Rd\cos\varphi}{R^2 + d^2 - 2Rd\cos\varphi} - \frac{d^2 - Rd\cos\varphi}{R^2 + d^2 - 2Rd\cos\varphi}\right] = 0.\end{aligned}$$

So, Eq. (*), with proper constants C_0, C_d, and $C_{d'}$, is indeed the correct electric potential distribution. We may simplify it for our case of current injection at the disk's rim, $d = d' = R$, obtaining a very simple expression:

$$\begin{aligned}\phi(\boldsymbol{\rho}) &= \frac{I}{2\pi t\sigma}[\ln\rho - \ln(\rho^2 + R^2 - 2\rho R\cos\varphi)]\\ &\equiv \frac{I}{2\pi t\sigma}\ln\frac{\rho}{\rho^2 + R^2 - 2\rho R\cos\varphi}.\end{aligned} \qquad (**)$$

From here, the current density components are

$$J_\rho = -t\sigma\frac{\partial\phi}{\partial\rho} = \frac{I}{2\pi}\frac{\rho^2 - R^2}{\rho(\rho^2 + R^2 - 2\rho R\cos\varphi)},$$
$$J_\varphi = -t\sigma\frac{1}{\rho}\frac{\partial\phi}{\partial\varphi} = \frac{I}{2\pi}\frac{2R\sin\varphi}{\rho^2 + R^2 - 2\rho R\cos\varphi}.$$

The figure below shows the color-coded contour plots of the potential distribution (**), so that the boundaries between the colors show the equipotential surfaces, with the current density vectors normal to them. The results show that the current flows mostly directly from the injection point to the pickup point, somewhat spreading out on the way.

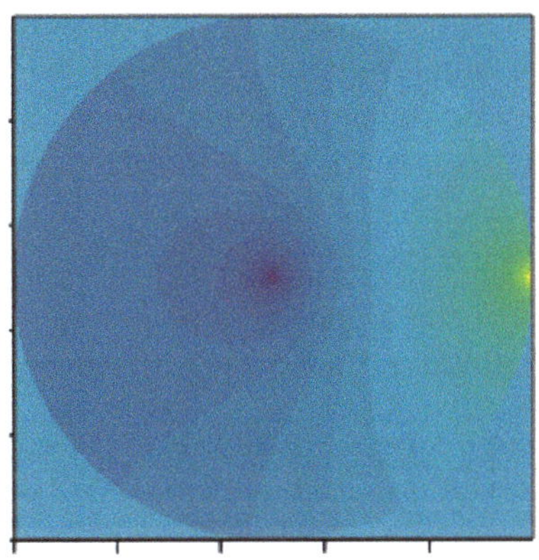

Problem* 4.9. The simplest model of a vacuum diode consists of two plane, parallel metallic electrodes of area A, separated by a gap of thickness $d \ll A^{1/2}$: a 'cathode' which emits electrons to vacuum, and an 'anode' which absorbs the electrons arriving at its surface. Calculate the dc I–V curve of the diode, i.e. the stationary relation between the current I flowing between the electrodes and the voltage V applied between them, using the following simplifying assumptions:

(i) due to the effect of the negative space charge of the emitted electrons, the current I is much smaller than the emission ability of the cathode,
(ii) the initial velocity of the emitted electrons is negligible, and
(iii) the direct Coulomb interaction of electrons (besides the space charge effect) is negligible.

Solution: Due to system's symmetry, the current is uniformly distributed over the electrode area and directed along the axis (say, x) normal to their surfaces, so that we may integrate the relation $\mathbf{j} = \rho\mathbf{v}$, where ρ is the space charge density, to calculate the total current:

$$I = Aj_x = A\rho(x)v_x(x). \qquad (*)$$

Due to the same symmetry, the electrostatic potential ϕ also depends only on the same coordinate: $\phi = \phi(x)$. Since the direct electron–electron interaction is negligible, the total energy of each electron, consisting of its electrostatic energy $q\phi(x) = -e\phi(x)$ and the kinetic energy $m_e v^2/2$, is conserved during its motion from the cathode to the anode, i.e. does not depend on x. Since the initial velocity of electrons is also negligible, $v = v_x$ everywhere in the gap, and taking cathode's electrostatic potential for zero, we may write this energy conservation law as

$$\frac{m_e v_x^2(x)}{2} - e\phi(x) = 0. \qquad (**)$$

Now we may use Eqs. (*) and (**) to eliminate v_x, and hence express the local charge density as a function of the local value of the electrostatic potential: $\rho = I/Av_x(x) = I/A(2e/m_e)^{1/2}\phi^{1/2}$. Plugging this expression into the Poisson equation (1.41), with $\nabla^2 = d^2/dx^2$ for our 1D geometry, we get the following equation for the distribution of the electrostatic potential along axis x:

$$\frac{d^2\phi}{dx^2} = \frac{C}{\phi^{1/2}}, \qquad \text{with } C \equiv \frac{I}{\varepsilon_0 A}\left(\frac{m_e}{2e}\right)^{1/2}.$$

For the first integration of the equation, we may use the very common[3] trick of multiplying and dividing the second derivative by $d\phi$:

$$\frac{d^2\phi}{dx^2} \equiv \frac{d}{dx}\left(\frac{d\phi}{dx}\right) = \frac{d}{d\phi}\left(\frac{d\phi}{dx}\right)\frac{d\phi}{dx} = \frac{1}{2}\frac{d}{d\phi}\left[\left(\frac{d\phi}{dx}\right)^2\right],$$

so that the equation takes the form

$$d\left[\left(\frac{d\phi}{dx}\right)^2\right] = 2C\frac{d\phi}{\phi^{1/2}}.$$

After such separation of variables $d\phi/dx$ and ϕ, we may integrate both parts of this relation, obtaining

$$\left(\frac{d\phi}{dx}\right)^2 = 2C\int\frac{d\phi}{\phi^{1/2}} = 4C\phi^{1/2} + \text{const.} \qquad (***)$$

Now comes the (only :-) non-trivial point of the solution. According to assumption (i), if the electric field $E = -d\phi/dx$ on the cathode's surface was substantially negative (with the force $\mathbf{F} = -e\mathbf{E}$, acting on each electron, directed towards the vacuum gap), it would extract a much larger current than we are calculating. On the other hand, if the field was substantially positive, it would suppress the emission altogether, chasing the emitted electrons back to the cathode. Hence E has to be at the very emission's threshold, and a good approximation may

[3] For example, such trick is used in classical mechanics for the simplest derivation of the work–energy principle—see, e.g., *Part CM*, Eq. (1.20).

be achieved by taking $E = 0$ on the cathode's surface. In this approximation, the integration constant in Eq. (***) equals zero. Extracting the root of both parts of the resulting relation (with the physically justified positive sign),

$$\frac{d\phi}{dx} = 2C^{1/2}\phi^{1/4}, \qquad \text{i.e.} \quad \frac{d\phi}{\phi^{1/4}} = 2C^{1/2}dx,$$

we may readily integrate it again, taking into account the second boundary condition, on the anode's surface: $\phi(d) = V$, where the cathode's surface it taken for $x = 0$. The integration yields[4]

$$\int_0^V \frac{d\phi}{\phi^{1/4}} \equiv \frac{4}{3}V^{3/4} = 2C^{1/2}\int_0^d dx \equiv 2C^{1/2}d.$$

After squaring both sides of this relation and using the definition of the constant C,

$$\frac{16}{9}V^{3/2} = 4Cd^2 \equiv 4\frac{I}{\varepsilon_0 A}\left(\frac{m_e}{2e}\right)^{1/2} d^2,$$

we obtain the requested dc I–V curve:

$$I = \frac{4}{9}\varepsilon_0\left(\frac{2e}{m_e}\right)^{1/2}\frac{A}{d^2}V^{3/2}.$$

This is the famous *Child–Langmuir equation* (frequently called the '$V^{3/2}$ law')[5], one of the foundations of all vacuum electronics. Notice how drastically it differs from Ohm's law $I = V/R$; physically, this difference is due to the effect of the negative space charge of propagating electrons, which suppresses the electric field on cathode's surface to $E \approx 0$—and hence the current to the values much lower than the cathode's emission ability. The $V^{3/2}$ law is very well obeyed by many vacuum electronic devices until the applied voltage becomes so large that is dissolves the electron 'cloud'.

Note that very similar space-charge effects take place in semiconductor devices, but because of strong scattering of electrons by impurities, the charge carriers (electrons and holes) they do not move ballistically (as was assumed in the above calculation), but rather drift through the material, dissipating their energy at the collisions. Because of that, the quantitative result of the (conceptually, similar) theory is different—see the next problem.

Problem* 4.10. Calculate the space-charge-limited current in a system with the same geometry, and using the same assumptions as in the previous problem, besides assuming that now the emitted charge carriers move not ballistically, but drift in

---

[4] The same integration, but with an arbitrary upper limit, yields the spatial distribution of the electrostatic potential: $\phi \propto x^{4/3}$.

[5] It was first derived in 1911 by C Child for (much lower) ionic currents, and in 1913 by I Langmuir for electrons.

accordance with the Ohm law, with the conductivity given by Eq. (4.13) of the lecture notes: $\sigma = q^2\mu n$, with a constant mobility μ.

Hint: In order to obtain a realistic result, assume that the medium in which the charge carriers move[6] has a certain dielectric constant κ, unrelated to the carriers.

Solution: Let us combine the Ohm law for the current, with the conductivity σ taken in the suggested form,

$$I = Aj = A\sigma(x)E(x) = Aq\mu\rho(x)E(x), \tag{*}$$

where $\rho(x) = qn(x)$ is the space charge density, with the 1D version of the Maxwell equation (3.53):

$$\frac{dE}{dx} = \frac{\rho(x)}{\kappa\varepsilon_0}. \tag{**}$$

For that, we may, for example, express $\rho(x)$ from Eq. (*) and plug it into Eq. (**), obtaining the following differential equation for the electric field's distribution:

$$E\frac{dE}{dx} = \frac{I}{Aq\mu\kappa\varepsilon_0}.$$

Integrating this equation, we obtain

$$\frac{E^2}{2} = \frac{I}{Aq\mu\varepsilon_r\varepsilon_0}x + \text{const}.$$

Now using exactly the same argumentation as in the previous problem to conclude that the proper boundary condition at the emission surface ($x = 0$) is $E = 0$, so that the integration constant equals zero, and taking the appropriate (negative) sign before the square root[7], we obtain

$$\frac{d\phi}{dx} = -E = \left(\frac{2I}{Aq\mu\kappa\varepsilon_0}x\right)^{1/2}.$$

Integrating this equation, with the boundary conditions $\phi = 0$ at $x = 0$, and $\phi = V$ at $x = d$, we obtain

$$V = \left(\frac{2I}{Aq\mu\kappa\varepsilon_0}\right)^{1/2}\int_0^d x^{1/2}\,dx = \left(\frac{2I}{Aq\mu\kappa\varepsilon_0}\right)^{1/2}\frac{2}{3}d^{3/2}.$$

This is the final result, which is usually represented in the reciprocal form,

$$I = \frac{9}{8}\,\varepsilon_0\kappa\mu q\frac{A}{d^3}V^2,$$

[6] As was mentioned in section 4.2 of the lecture notes, the assumption of a constant (charge-density-independent) mobility is most suitable for semiconductors.

[7] In our system with, say, positive anode voltage V, the electric field should be directed from the anode to the cathode, i.e. (for our coordinate choice) should be negative.

and called the *Mott–Gurney law* (first derived in 1948). Note that it significantly differs from both the integral form of the Ohm law, $I \propto V$ (valid in semiconductors only at a negligible space charge accumulation) and the Child–Langmuir formula, $I \propto V^{3/2}$ (valid only at the ballistic motion of charge carriers).

***Problem* 4.11.** Prove that the distribution of dc currents in a uniform Ohmic conductor, at a fixed voltage applied at its boundaries, corresponds to the minimum of the total power dissipation ('Joule heat').

Solution: According to Eq. (4.40) of the lecture notes, the total dissipation power in a uniform Ohmic conductor may be expressed as

$$\mathscr{P} = \sigma \int_V (\nabla\phi)^2 d^3r. \qquad (*)$$

According to Eq. (4.14), the distribution of the scalar potential ϕ in a uniform Ohmic conductor (with σ = const) satisfies the Laplace equation, just as in a free space. Hence, its distribution is the same as in a similar empty volume V with the same boundary conditions $\phi|_S$. But as we know from chapter 1 (see the model solution of problem 1.16), the latter distribution corresponds to the minimum of electrostatic energy

$$U = \frac{\varepsilon_0}{2}\int_V E^2 d^3r = \frac{\varepsilon_0}{2}\int_V (\nabla\phi)^2 d^3r,$$

i.e. to the minimum of the same integral as in Eq. (*).

IOP Publishing

Classical Electrodynamics
Problems with solutions
Konstantin K Likharev

Chapter 5

Magnetism

***Problem* 5.1.** A circular wire loop, carrying a fixed dc current, has been placed inside a similar but larger loop, carrying a fixed current in the same direction—see the figure below. Use semi-quantitative arguments to analyze the mechanical stability of the coaxial, coplanar position of the inner loop with respect to its possible angular, axial, and lateral displacements relative to the outer loop.

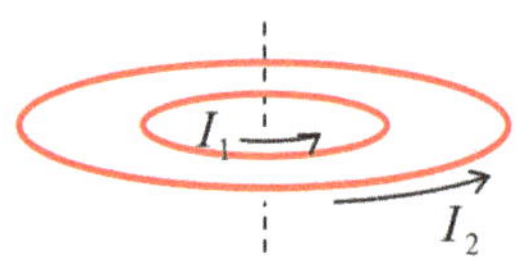

Solution: The stability may be analyzed using semi-quantitative force arguments. Since according to the basic Eq. (5.1) of the lecture notes, the currents flowing in the same directions attract each other, and this force decreases with the distance d between them, the interaction of the closest parts of two loops plays the most important role. Hence, a small axial displacement of the inner ring from the outer ring's plane results in a rise of a net force **F** dragging it back—see the figure below, where colors code the current direction—in or out of the plane of drawing.

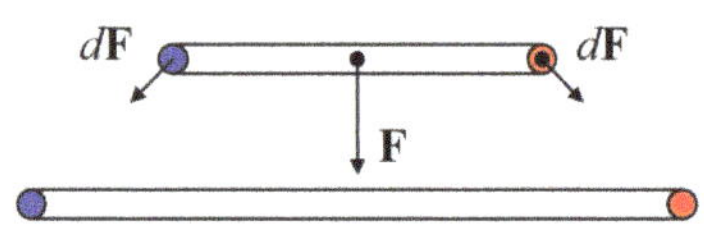

Very similarly, an angular displacement of the inner loop causes the elementary forces $d\mathbf{F}$ creating a mechanical torque pulling it back to the coplanar position—see the figure below.

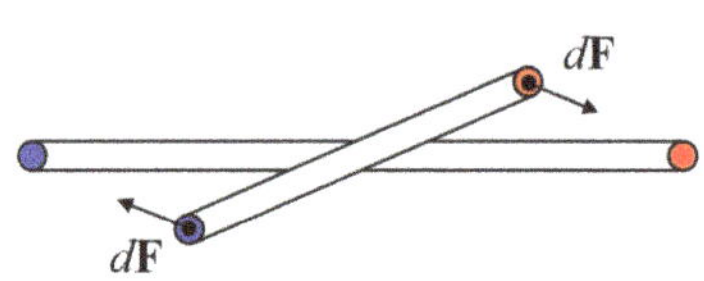

doi:10.1088/978-0-7503-1407-7ch5

The analysis of the lateral displacement is a bit more subtle. According Eq. (5.7) of the lecture notes, the attraction of two currents (per unit length) scales *approximately*[1] as $1/d$, where d is the distance between them. At a lateral displacement of the loop by small distance $\Delta \ll d$ (see the figure below), the nearest distances d for one half of the ring decrease by $\sim\Delta$, and for the other half increase by the same amount. However, the function $f(d) = 1/d$ grows faster at the argument decrease than it drops at its increase:

$$f(d \pm \Delta) = \frac{1}{d \pm \Delta} = \frac{1}{d(1 \pm \Delta/d)} \approx \frac{1}{d}\left(1 \mp \frac{\Delta}{d} + \frac{\Delta^2}{d^2}\right),$$

so that we may expect the net force to be directed toward the nearest part of the outer ring.

Hence, the coaxial and coplanar position of the inner loop is *stable* with respect to its axial and angular displacements[2], but is *unstable* with respect to its lateral displacements.

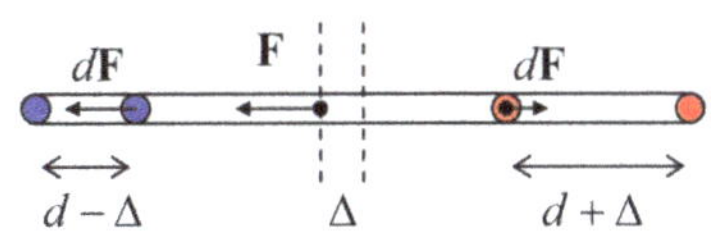

***Problem* 5.2.** Two straight, plane, parallel, long, thin conducting strips of width w, separated by distance d, carry equal but oppositely directed currents I—see the figure below. Calculate the magnetic field in the plane located at the middle between the strips, assuming that the flowing currents are uniformly distributed across the strip widths.

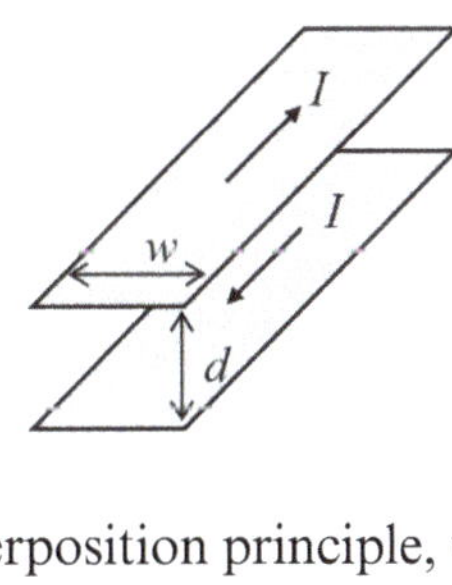

Solution: Due to the linear superposition principle, we may calculate the total field **B** as a vector sum of elementary fields $d\mathbf{B}$ induced by elementary currents $dI = (I/w)dw$, flowing in each elementary segment dw of the strip widths, considering it as a thin wire. The magnetic field of such a wire was calculated in section 5.1 of the lecture

[1] As a reminder, Eq. (5.7) is strictly valid only for infinite, straight wires, but as we know, for example from Eq. (5.24), the results for other geometries scale almost similarly.

[2] Eq. (5.53) of the lecture notes allows an easy confirmation of these two conclusions using potential energy arguments, but I will defer this task until the conclusion of a new discussion of the magnetic field energy (or rather energies) at the beginning of the next chapter.

notes (see Eq. (5.20) and its derivation), and we just need to rewrite it in a form more convenient for our current purposes. Directing the axis z along the current (see the figure below), we obtain

$$dB_x = dB\sin\varphi = \frac{\mu_0}{2\pi d}dI\frac{\Delta y}{d} = \frac{\mu_0}{2\pi}\frac{y - y'}{(x - x')^2 + (y - y')^2}dI,$$

$$dB_y = -\,dB\cos\varphi = -\frac{\mu_0}{2\pi d}dI\frac{\Delta x}{d} = -\frac{\mu_0}{2\pi}\frac{x - x'}{(x - x')^2 + (y - y')^2}dI,$$

where $\boldsymbol{\rho} = \{x, y\}$ is the 2D radius-vector of the field observation point, while $\boldsymbol{\rho}' = \{x', y'\}$ is that of the source (wire segment).

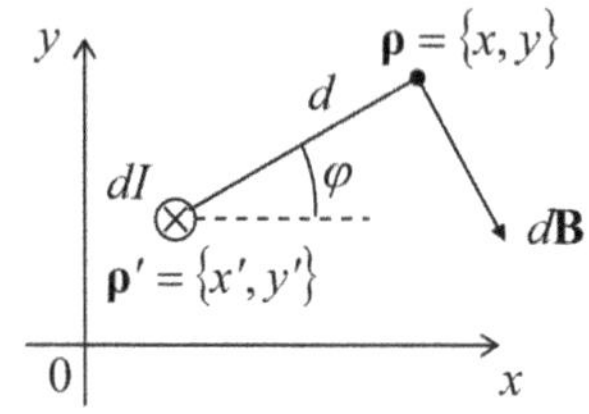

Now we should integrate this expression over both current-carrying strips. Selecting the reference frame in the natural way shown in the figure below, so that the mid-plane in question corresponds to $y = 0$, we see that due to problem's symmetry, the integral of dB_y vanishes, while the 'horizontal' component of the field is twice that of a single strip:

$$\begin{aligned} B_x &= 2\int_{x'=-w/2}^{x'=w/2} dB_x = -\frac{\mu_0}{\pi}\int_{x'=-w/2}^{x'=w/2}\frac{d/2}{(x - x')^2 + (d/2)^2}dI \\ &= -\frac{\mu_0}{\pi}\frac{I}{w}\frac{d}{2}\int_{-w/2}^{+w/2}\frac{dx'}{(x - x')^2 + (d/2)^2} \\ &= -\frac{\mu_0 I}{\pi w}\left(\tan^{-1}\frac{x - w/2}{d/2} - \tan^{-1}\frac{x + w/2}{d/2}\right). \end{aligned}$$

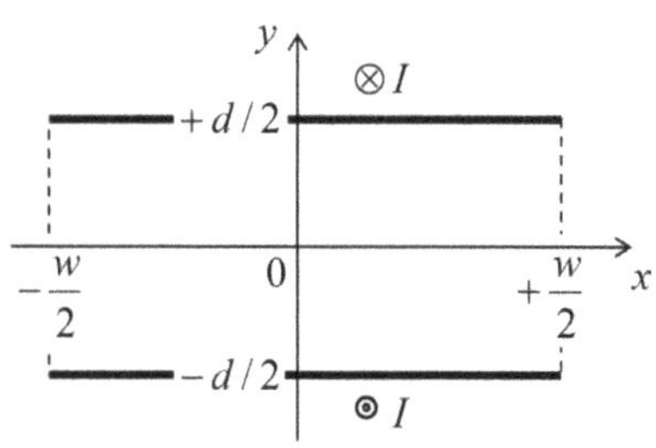

The figure below shows the plots of B_x as a function of x, for several values of the d/w ratio. If $d \gg w$, the field is relatively low, distributed along axis x over a broad interval $\Delta x \sim d \gg w$, and may be well approximated by the sum of fields of two thin wires.

On the other hand, as the distance between the strips is reduced, the field becomes localized within the gap between the strips. (At $d/w \to 0$, the so-called 'fringe fields' at $|x| > w/2$ become negligible.) Moreover, the field becomes uniform, and tends to the d-independent value

$$B_x = -\frac{\mu_0 I}{w}.$$

A simple explanation, and the discussion of consequences of this very important result are the subjects of the next problem.

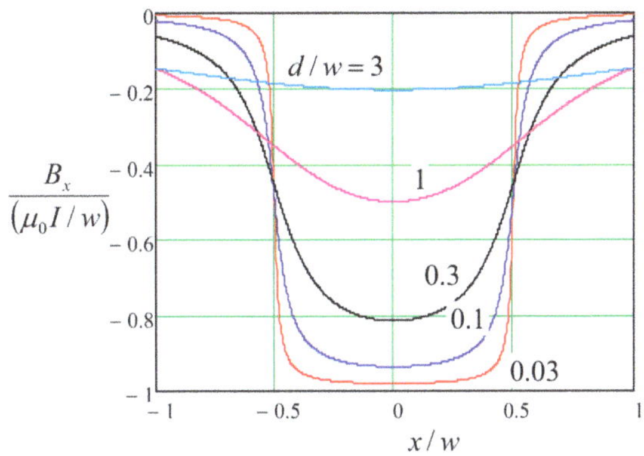

***Problem* 5.3.** For the system studied in the previous problem, but now only in the limit $d \ll w$, calculate:

(i) the distribution of the magnetic field (in the simplest possible way),
(ii) the vector-potential of the field,
(iii) the magnetic force (per unit length) exerted on each strip, and
(iv) the magnetic energy and self-inductance of the loop formed by the strips (per unit length).

Solutions:

(i) As was shown in the previous problem, in the limit $d/w \to 0$, the magnetic field is localized in the gap between the strips, and uniform. Applying the Ampère law, given by Eq. (5.37) of the lecture notes, to the contour shown with the dashed line in the figure below (which shows the cross-section of the system), we get the same result as was obtained in the previous problem,

$$B = -B_x = \mu_0 \frac{I}{w}, \qquad (*)$$

in a much simpler way.

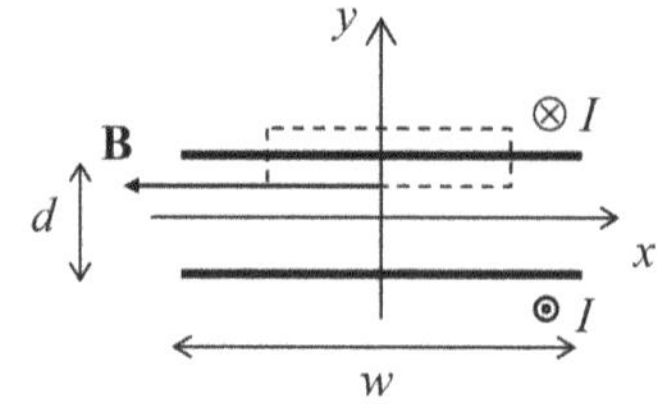

(ii) According to Eq. (5.28) of the lecture notes, the vector-potential has to be directed along the axis z, i.e. along the current, and be independent of z. From the structure of that equation, it is also clear that at $d \ll w$, well inside the gap, **A** should not depend on x either (because the strip edges are not 'visible' from those points). Hence we may look for the vector-potential in the form

$$\mathbf{A} = A(y)\mathbf{n}_z.$$

Calculating the curl of such a vector,

$$\nabla \times \mathbf{A} = \mathbf{n}_x \frac{\partial A}{\partial y},$$

and requiring it to be equal to the vector **B** described by Eq. (*), we obtain

$$\mathbf{A} = -\mu_0 \frac{I}{w} y \mathbf{n}_z + \text{const}.$$

This result could be also obtained by the direct integration of Eq. (5.28) along x' and z'. (Alternatively, we can integrate Eq. (5.51) written for each current component $dI = Idx'/w$, along axis x').

Note, however, that a uniform field, like our $\mathbf{B} = -B\mathbf{n}_x$, cannot 'tell' one transverse coordinate (say, y) from another one (z), and hence may be the same for different vector-potential distributions, for example[3]

$$\mathbf{A} = \mu_0 \frac{I}{w} z \mathbf{n}_y + \text{const},$$

or a linear combination of these two functions. This is one more manifestation of the gauge invariance of the magnetic field with respect to any transformation $\mathbf{A} \to \mathbf{A} + \nabla\chi$—see Eq. (5.46).

(iii) As follows from Eq. (5.1), the total magnetic force acting on each strip is directed along the axis y, and corresponds to strips' repulsion. In order to calculate its magnitude, it would be wrong to plug Eq. (*) into Eq. (5.15), because an elementary current (like a point electric charge) does exert force on itself[4]. As we can readily check using the Ampère law, the field created by a single strip is twice lower:

[3] Notice that this distribution, of course, does *not* satisfy Eq. (5.28).

[4] Of course, each component jdx of the current distributed in a strip does exerts a non-vanishing force on each other component jdx' of current in the same strip, but this force is directed along axis x, rather than y.

$$B_1 = \mu_0 \frac{I}{2w},$$

so that, according Eq. (5.15), the force magnitude (per unit length)

$$\frac{F}{l} = \frac{1}{l} \int_{\text{strip}} jB_1 d^3r = IB_1 = \mu_0 \frac{I^2}{2w}.$$

This expression may be also represented as the integral, over the strip area, of the positive pressure of the magnetic field:

$$\mathcal{P} \equiv \frac{F}{lw} = \frac{\mu_0 I^2}{2w^2} = \frac{B^2}{2\mu_0} = u,$$

where $u = U/lwd = U/V_{\text{gap}}$ is the magnetic energy density—see Eq. (5.57b).

(iv) Since the full magnetic field is uniform inside the gap (with the cross-section area dw) and vanishes outside of it, the magnetic energy per unit length is just[5]

$$\frac{U}{l} = \frac{1}{l} \int_{\text{gap}} \frac{B^2}{2\mu_0} d^3r = \frac{B^2}{2\mu_0} dw = \mu_0 \frac{d}{w} \frac{I^2}{2} = u \frac{V_{\text{gap}}}{l}.$$

From this result and Eq. (5.72) we immediately obtain the following expression for the self-inductance (also per unit length):

$$\frac{L}{l} = \mu_0 \frac{d}{w}.$$

The same result may be also calculated as the ratio $(\Phi/l)/I$, where Φ is the magnetic flux in the gap between the strips:

$$\frac{\Phi}{l} = \frac{1}{l} \int_{\text{gap}} B_n d^2r = Bt = \mu_0 \frac{I}{w} t.$$

This simple formula, which shows a clear way for the reduction of current loop inductances (bring the counterpart conductors as close to each other as possible!), is very important for applications in electronics, because self-inductance frequently plays a negative role in high-speed integrated circuits, reducing the interconnect bandwidth.

***Problem* 5.4.** Calculate the magnetic field distribution near the center of the system of two similar, planar, round, coaxial wire coils, fed by equal but oppositely directed currents—see the figure below.

[5] This result poses an additional question for the reader: how can the strip *repulsion* (i.e. their 'desire' to *increase* d, and hence the volume V_{gap} and the product $U = uV_{\text{gap}}$) be compatible with the apparent general trend for any system to *decrease* its potential energy? As a reminder, for the *electrostatic* force exerted on the conductor this paradox does not exist, because the force direction corresponds to *negative* pressure—see, e.g. the solution of problem 2.1. Thinking about this issue may be a good primer for the general discussion, in section 6.2, of the relation between two magnetic energies given by Eqs. (5.53) and (5.54).

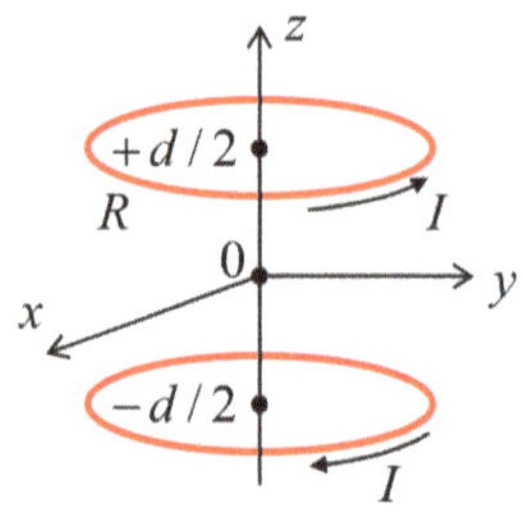

Solution: In order to find the field $\mathbf{B}(x, y, z)$ exactly on the system's axis, we may combine two versions of Eq. (5.23) of the lecture notes, applied to each of the loops:

$$\mathbf{B}(0, 0, z) = \frac{\mu_0 I}{2}\left\{-\frac{R^2}{[R^2 + (d/2 + z)^2]^{3/2}} + \frac{R^2}{[R^2 + (d/2 - z)^2]^{3/2}}\right\}\mathbf{n}_z,$$

where z is the axis directed along the system axis, and its center is taken for the reference frame origin (see the figure above). At the center of the system ($z = 0$), the field evidently vanishes, while for small ($|z| \ll R, d$) but non-vanishing deviations it may be found from the linear term of the Taylor expansion of the right-hand side:

$$\mathbf{B}(0, 0, z) \approx \frac{\mu_0 I}{2}\frac{\partial}{\partial z}\left\{\frac{R^2}{[R^2 + (d/2 + z)^2]^{3/2}} - \frac{R^2}{[R^2 + (d/2 - z)^2]^{3/2}}\right\}_{z=0} z\mathbf{n}_z$$
$$= \frac{3\mu_0 IdR}{2[R^2 + (d/2)^2]^{5/2}} z\mathbf{n}_z.$$

Evidently, this expression may be rewritten as

$$\left.\frac{\partial B_z}{\partial z}\right|_{\mathbf{r}=0} = \frac{3\mu_0 IdR}{2[R^2 + (d/2)^2]^{5/2}}.$$

Now, in the Cartesian coordinates (see the figure above), the general equation (5.29) reads

$$\frac{\partial B_x}{\partial x} + \frac{\partial B_y}{\partial y} + \frac{\partial B_z}{\partial z} = 0.$$

Due to the axial symmetry of the system, we may argue that at $z = 0$, the first two partial derivatives have to be equal to each other. Hence, the last two relations yield

$$\left.\frac{\partial B_x}{\partial x}\right|_{\mathbf{r}=0} = \left.\frac{\partial B_y}{\partial y}\right|_{\mathbf{r}=0} = -\frac{1}{2}\left.\frac{\partial B_z}{\partial z}\right|_{\mathbf{r}=0} = -\frac{3\mu_0 IdR}{4[R^2 + (d/2)^2]^{5/2}},$$

so that in the linear approximation in small x, y, and z we may write

$$\mathbf{B}(x, y, z) \approx \frac{3\mu_0 IdR}{4[R^2 + (d/2)^2]^{5/2}}(-x\mathbf{n}_x - y\mathbf{n}_y + 2z\mathbf{n}_z), \qquad \text{for } r \ll d, R.$$

Such *quadrupole magnetic field* pattern may provide trapping of atoms with zero net electric charge but a non-zero total spin (and hence non-vanishing magnetic moment), within a certain range of their initial velocities. A brief discussion of other magnetic traps, and references to additional literature on the subject, may be found in section 9.7 of the lecture notes.

***Problem* 5.5.** The two-coil-system, similar to that considered in the previous problem, now carries equal and similarly directed currents—see the figure below. Calculate what should be the ratio d/R for the second derivative $\partial^2 B_z/\partial z^2$ at $z = 0$ to vanish[6].

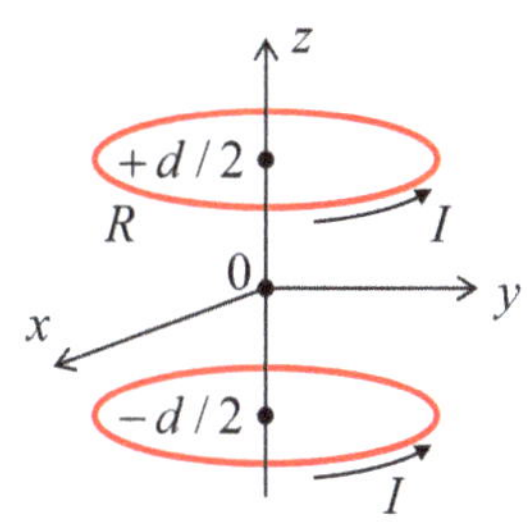

Solution: Just as in the model solution of the previous problem, using Eq. (5.23) for the field of each loop we obtain

$$B_z = \frac{\mu_0 I}{2}\sum_{\pm}\frac{R^2}{[R^2 + (d/2 \pm z)^2]^{3/2}} \equiv \frac{\mu_0 I R^2}{2}\sum_{\pm}\left[R^2 + \left(\frac{d}{2} \pm z\right)^2\right]^{-3/2}.$$

Note that at an arbitrary ratio d/R, the first derivative

$$\frac{\partial B_z}{\partial z} = \frac{\mu_0 I R^2}{2}\sum_{\pm}\left[\mp 3\left(\frac{d}{2} \pm z\right)\right]\left[R^2 + \left(\frac{d}{2} \pm z\right)^2\right]^{-5/2},$$

vanishes in the center of the system (at $z = 0$)—just as one might expect from system's symmetry. Now calculating the second derivative, we obtain

$$\left.\frac{\partial^2 B_z}{\partial z^2}\right|_{z=0} = \frac{\mu_0 I R^2}{2}\sum_{\pm}\left\{\left[R^2 + \left(\frac{d}{2} \pm z\right)^2\right] - 5\left(\frac{d}{2} \pm z\right)^2\right\}\left.\left[R^2 + \left(\frac{d}{2} \pm z\right)^2\right]^{-7/2}\right|_{z=0}$$

$$= \mu_0 I R^2\left[R^2 - 4\left(\frac{d}{2}\right)^2\right]\left[R^2 + \left(\frac{d}{2}\right)^2\right]^{-7/2}$$

This expression evidently turns to zero at

$$R^2 - 4\left(\frac{d}{2}\right)^2 = 0, \qquad \text{i.e. at } R = d.$$

[6] Such a system, producing a highly uniform magnetic field near its center, is called the *Helmholtz coils*, and is widely used in physics experiment.

This is exactly the relation used for the Helmholtz coil design. Note that due to system's symmetry, the third derivative, $\partial^3 B_z/\partial z^3$, also vanishes at $z = 0$ as well (this one for any d/R), so that the deviation of the field from its value at the center,

$$B_z\Big|_{\substack{z=0\\ d=R}} = \mu_0 I \frac{R^2}{[R^2 + (d/2)^2]^{3/2}}\Bigg|_{d=R} = \left(\frac{4}{5}\right)^{3/2}\frac{\mu_0 I}{R},$$

grows very slowly, as $(z/R)^4$, with the deviation from that point.

***Problem* 5.6.** Calculate the magnetic field's distribution along the axis of a straight solenoid (see figure 5.6a of the lecture notes, partly reproduced below) with a finite length l, and round cross-section of radius R. Assume that the solenoid has many ($N \gg 1$, l/R) wire turns, uniformly distributed along its length.

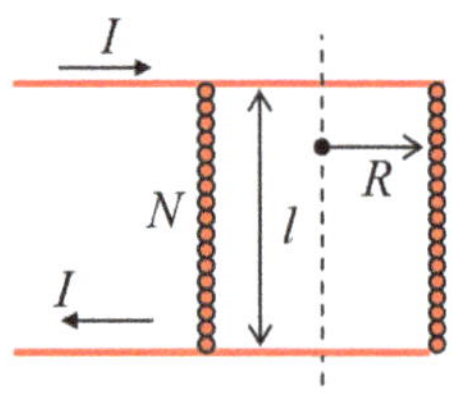

Solution: Because of the high wire turn density $n \equiv N/l \gg 1/R$, and hence the almost circular shape of each turn, the solenoid's field may be represented as a sum of fields of N circular current loops. The field induced by a single loop has been calculated in section 5.1 of the lecture notes—see Eq. (5.23). Generalizing that relation to the case of arbitrary position (z') of the loop, we obtain

$$B_1 = \frac{\mu_0 I}{2}\frac{R^2}{[R^2 + (z - z')^2]^{3/2}}.$$

Now we can use the linear superposition principle to find the field induced by the solenoid as a whole:

$$B_N = \frac{\mu_0 I}{2}\sum_{j=-N/2}^{N/2}\frac{R^2}{\left[R^2 + (z - z_j)^2\right]^{3/2}} = \frac{\mu_0 I}{2}\sum_{j=-N/2}^{+N/2}\frac{R^2}{[R^2 + (z - jl/N)^2]^{3/2}},$$

where the index j numbers wire turns (starting from the middle of the solenoid). For $N \gg 1$ this sum is well approximated by the integral

$$B_N = \frac{\mu_0 I}{2}\int_{-N/2}^{+N/2}\frac{R^2}{[R^2 + (z - jl/N)^2]^{3/2}}dj = \frac{\mu_0 I}{2}n\int_{-l/2}^{+l/2}\frac{R^2}{[R^2 + (z - z')^2]^{3/2}}dz'.$$

This is a table integral[7], giving finally:

$$B_N = \frac{\mu_0 In}{2} \frac{z - z'}{[R^2 + (z - z')^2]^{1/2}}\Bigg|_{z'=+l/2}^{z'=-l/2}$$

$$= \frac{\mu_0 In}{2}\left\{\frac{z + l/2}{[R^2 + (z + l/2)^2]^{1/2}} - \frac{z - l/2}{[R^2 + (z - l/2)^2]^{1/2}}\right\}.$$

The figure below shows the typical field profiles described by this formula. For points well inside a long solenoid ($l \gg |z|$, R), the expression in the figure brackets tends to 2, so that our result approaches the simple Eq. (5.40), $B_N = \mu_0 In$. Note, however, that a close approach of this value requires rather large values of the l/R ratio, i.e. long solenoids. (This handicap may be remedied by using either toroidal solenoids sketched in figure 5.6 of the lecture notes, or the straight solenoid's filling with a high-μ magnetic material—see section 5.6.)

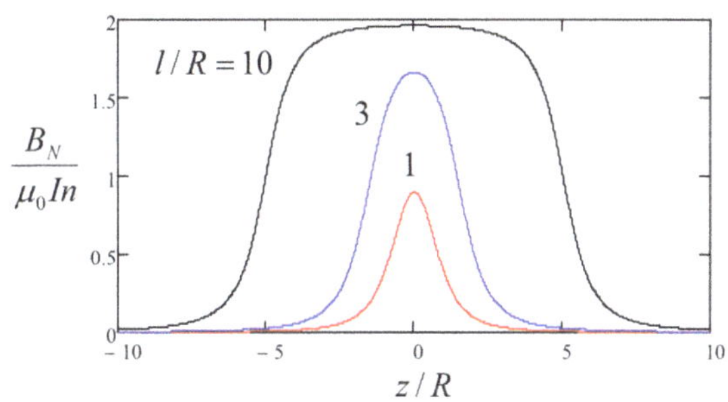

On the other hand, at $|z| \gg R$, l, our result yields

$$B_N \to \frac{\mu_0 INR^2}{2|z|^3} \equiv \frac{\mu_0}{4\pi}\frac{2m_N}{|z|^3}, \qquad \text{where } m_N \equiv \pi INR^2 = Nm_1,$$

i.e. the dipole field (5.99), with the dipole moment m_N, N times larger than that (m_1) of a single current loop—see Eq. (5.97).

***Problem* 5.7.** A thin round disk of radius R, carrying electric charge of a constant areal density σ, is being rotated around its axis with a constant angular velocity ω. Calculate:

(i) the induced magnetic field on the disk's axis,
(ii) the magnetic moment of the disk,

and relate these results.

Solutions:

(i) The rotating disk may be represented as a set of narrow rings of radius ρ and width $d\rho \ll \rho$, each carrying a ring current $dI = Jd\rho = \sigma v d\rho = \sigma\omega\rho d\rho$, each creating

[7] See, e.g. Eq. (A.29b).

the elementary field dB described by Eq. (5.23) of the lecture notes (with the notation replacements $B \to dB$, $I \to dI$, $R \to \rho$):

$$dB = \frac{\mu_0 dI}{2}\frac{\rho^2}{(\rho^2 + z^2)^{3/2}} = \frac{\mu_0 \sigma\omega}{2}\frac{\rho^3 d\rho}{(\rho^2 + z^2)^{3/2}},$$

where z is the distance of the field observation point from the ring's plane. Since the fields of all rings have the same direction (along the disk's axis), we may sum up their contributions to the total field as scalars:

$$B = \int_{\rho=0}^{\rho=R} dB = \frac{\mu_0\sigma\omega}{2}\int_0^R \frac{\rho^3 d\rho}{(\rho^2 + z^2)^{3/2}} = \frac{\mu_0\sigma\omega}{2}\frac{|z|}{2}\int_{\rho=0}^{\rho=R}\frac{\xi d\xi}{(\xi + 1)^{3/2}},$$

where $\xi \equiv \rho^2/z^2$. The last integral may be readily worked out as

$$\begin{aligned}\int\frac{\xi d\xi}{(\xi + 1)^{3/2}} &= \int\frac{(\xi + 1)d\xi}{(\xi + 1)^{3/2}} - \int\frac{d\xi}{(\xi + 1)^{3/2}}\\ &= \int\frac{d\xi}{(\xi + 1)^{1/2}} - \int\frac{d\xi}{(\xi + 1)^{3/2}}\\ &= 2(\xi + 1)^{1/2} + \frac{2}{(\xi + 1)^{1/2}} = \frac{2(\xi + 2)}{(\xi + 1)^{1/2}},\end{aligned}$$

and we finally obtain

$$B = \frac{\mu_0\sigma\omega}{2}\frac{|z|}{2}\left[\frac{2(\rho^2/z^2 + 2)}{(\rho^2/z^2 + 1)^{1/2}}\right]_{\rho=0}^{\rho=R} = \mu_0\sigma\omega\left[\frac{R^2 + 2z^2}{2(R^2 + z^2)^{1/2}} - |z|\right]. \qquad (*)$$

(ii) Very similarly, the magnitude of the magnetic moment $\mathbf{m}$ of the system (evidently, also directed along the symmetry axis z) may be calculated by summing up the elementary ring contributions given by Eq. (5.97):

$$dm = AdI = \pi\rho^2 dI = \pi\sigma\omega\rho^3 d\rho,$$

so that

$$m = \int_{\rho=0}^{\rho=R} dm = \pi\sigma\omega\int_0^R \rho^3 d\rho = \frac{\pi\sigma\omega R^4}{4}. \qquad (**)$$

In order to relate the above results, let us find the limit of Eq. (*) at large distances, $|z| \gg R$:

$$\begin{aligned}B &= \mu_0\sigma\omega|z|\left[\frac{1 + R^2/2z^2}{(1 + R^2/z^2)^{1/2}} - 1\right]\\ &\to \mu_0\sigma\omega|z|\left[\left(1 + \frac{R^2}{2z^2}\right)\left(1 - \frac{R^2}{2z^2} + \frac{3R^4}{8z^4}\right) - 1\right] \to \mu_0\sigma\omega\frac{R^4}{8|z|^3}.\end{aligned} \qquad (***)$$

On the other hand, the general expression for the magnetic dipole field, following from (5.99) of the lecture notes for the observation point on the magnetic dipole's axis ($\mathbf{r} = |z|\mathbf{m}/m$), is

$$B = \frac{\mu_0}{4\pi}\frac{2m}{|z|^3}.$$

With the m given by Eq. (**), this expression coincides with that Eq. (***), i.e. our two results do match.

***Problem* 5.8.** A thin spherical shell of radius R, with charge Q uniformly distributed over its surface, rotates about its axis with angular velocity ω. Calculate the distribution of the magnetic field everywhere in space.

Solution: As viewed from the lab reference frame, the rotation creates ring-like surface currents with the linear density $J = \sigma v$, where $\sigma = Q/4\pi R^2$ is the (constant) surface charge density, $v = \omega R\sin\theta'$ is the linear velocity, and θ' is the polar angle measured from the rotation axis—see the figure below, showing a part of the shell's cross-section. As was discussed in section 5.4 of the lecture notes, at large distances $r \gg R$, each elementary ring current

$$dI = JRd\theta' = \frac{Q}{4\pi R^2}\omega R\sin\theta' Rd\theta' = \frac{Q\omega}{4\pi}\sin\theta' d\theta',$$

creates a magnetic field with the same dipole distribution (5.99), with the elementary dipole moment (5.97):

$$dm = A'dI = \pi(R\sin\theta')^2\frac{Q\omega}{4\pi}\sin\theta' d\theta' = \frac{1}{4}Q\omega R^2\sin^3\theta' d\theta'.$$

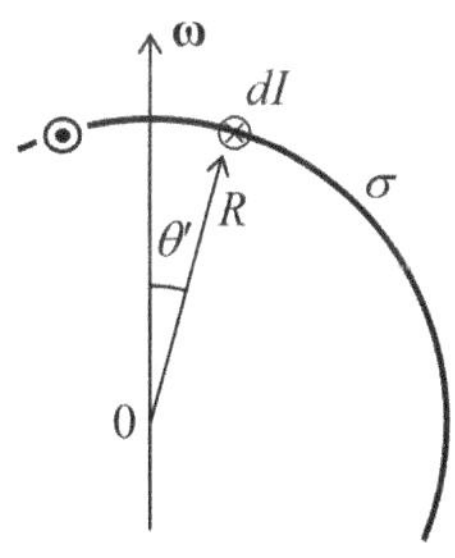

Due to the similarity of these distributions, the magnetic field of the whole shell is also similar, with the magnetic dipole moment m that may be calculated by summing dm of all elementary currents, i.e. integrating dm over the $0 \leqslant \theta' \leqslant \pi$—see the figure above[8]:

$$m = \int dm = \frac{1}{4}Q\omega R^2\int_0^\pi \sin^3\theta' d\theta' = \frac{1}{4}Q\omega R^2\frac{4}{3} = \frac{1}{3}Q\omega R^2.$$

---

[8] The integral over $d\theta'$ may be readily worked out using the following variable replacement: $\xi \equiv \cos\theta'$, so that $\sin^3\theta' d\theta' = -\sin^2\theta' d(\cos\theta') = -(1-\xi^2)d\xi$.

After solving so many problems on uniform spheres and thin spherical shells in this course, we may readily guess that the dipole field distribution (5.99),

$$\mathbf{H} = \frac{1}{4\pi}\frac{3\mathbf{r}(\mathbf{r}\cdot\mathbf{m}) - \mathbf{m}r^2}{r^5} \equiv \frac{m}{4\pi}\frac{2\mathbf{n}_r\cos\theta - \mathbf{n}_\theta\sin\theta}{r^3}, \qquad (*)$$

is valid not only at large distances, but for any $r > R$, and that the field inside the shell is uniform:

$$\mathbf{H} = \frac{\boldsymbol{\omega}}{\omega}H_0 = H_0(\mathbf{n}_r\cos\theta - \mathbf{n}_\theta\sin\theta), \qquad \text{for } r < R, \qquad (**)$$

where H_0 is a constant.

Indeed, as we already know, the fields (*) and (**) do satisfy the stationary Maxwell equations (5.36) in the current-free regions. Hence, due to the uniqueness of the solutions of linear boundary problems, it is sufficient to prove that, at a certain choice of the scalar constant H_0, these fields also satisfy the boundary conditions at the shell, i.e. at $r = R$. These conditions,

$$H_\theta|_{r=R+0} - H_\theta|_{r=R-0} = J, \qquad H_r|_{r=R+0} - H_r|_{r=R-0} = 0,$$

follow from, respectively, Eqs. (5.116) and (5.119) of the lecture notes. For our fields (*) and (**), and linear current density calculated above, $J = (Q/4\pi R^2)\omega R\sin\theta$, the conditions take the form

$$\left(-\frac{1}{4\pi}\frac{1}{3}Q\omega R^2\frac{\sin\theta}{R^3}\right) - (-H_0\sin\theta) = \frac{Q}{4\pi R^2}\omega R\sin\theta,$$

$$\left(\frac{1}{4\pi}\frac{1}{3}Q\omega R^2\frac{2\cos\theta}{R^3}\right) - (H_0\cos\theta) = 0.$$

One can readily see that if

$$H_0 = \frac{Q\omega}{6\pi R},$$

both boundary conditions are satisfied for any θ, so that we have found the (unique) solution of the problem.

***Problem* 5.9.** A sphere of radius R, made of an insulating material with a uniform electric charge density ρ, rotates about its diameter with angular velocity ω. Calculate the magnetic field distribution inside the sphere and outside it.

Solution: We may represent the sphere as a set of thin spherical shells, each with radius $R' \leqslant R$, thickness dR', and charge $dQ = \rho dV = \rho(4\pi R'^2 dR')$. As was discussed in the model solution of the previous problem, each shell creates outside it a purely dipole magnetic field, with the same spatial distribution

$$d\mathbf{H} = \frac{1}{4\pi}\frac{3\mathbf{r}(\mathbf{r}\cdot d\mathbf{m}) - d\mathbf{m}r^2}{r^5},$$

where $d\mathbf{m}$ is shell's magnetic dipole moment, directed along the axis of rotation, with magnitude

$$dm = \frac{1}{3}dQ\omega R'^2 = \frac{4\pi}{3}\rho\omega R'^4 dR',$$

and inside it a uniform field,

$$d\mathbf{H} = dH_0(\mathbf{n}_r \cos\theta - \mathbf{n}_\theta \sin\theta),$$

of magnitude

$$dH_0 = \frac{dQ\omega}{6\pi R'} = \frac{2}{3}\rho\omega R' dR'.$$

As the result of addition of shell fields, the total field *outside* the sphere (for $r \geqslant R$) has the same dipole distribution, with the total dipole moment $\mathbf{m}$ whose magnitude m that may be calculated by the scalar addition of elementary moments dm:

$$m = \int dm = \frac{4\pi}{3}\rho\omega \int_0^R R'^4 dR' = \frac{4\pi}{15}\rho\omega R^5.$$

In order to calculate the field *inside* the rotating sphere, i.e. at distances $r < R$ from its center, we should sum up the dipole field induced by shells with $R' < r$, giving the total dipole moment

$$m = \frac{4\pi}{15}\rho\omega r^5,$$

and the uniform magnetic field created by the thin shells with $R' \geqslant r$, whose magnitude is the scalar sum of all shell contributions:

$$H_0 = \int dH_0 = \frac{2}{3}\rho\omega \int_r^R R' dR' = \frac{1}{3}\rho\omega(R^2 - r^2).$$

Using an alternative form of the dipole field representation[9],

$$\mathbf{H} = \frac{1}{4\pi} m \frac{2\mathbf{n}_r \cos\theta - \mathbf{n}_\theta \sin\theta}{r^3},$$

these two contributions may be summed up, giving

$$\begin{aligned}\mathbf{H} &= \frac{1}{4\pi}\frac{4\pi}{15}\rho\omega r^5 \frac{2\mathbf{n}_r \cos\theta - \mathbf{n}_\theta \sin\theta}{r^3} + \frac{1}{3}\rho\omega(R^2 - r^2)(\mathbf{n}_r \cos\theta - \mathbf{n}_\theta \sin\theta) \\ &= \frac{\rho\omega r}{15}[\mathbf{n}_r \cos\theta(5R^2 - 3r^2) - \mathbf{n}_\theta \sin\theta(5R^2 - 4r^2)], \quad \text{for} \quad r \leqslant R.\end{aligned}$$

As a (useful) sanity check, on the sphere's surface (i.e. at $r = R$), the result gives a purely dipole field:

$$\mathbf{H}|_{r=R} = \frac{\rho\omega R^3}{15}(\mathbf{n}_r 2\cos\theta - \mathbf{n}_\theta \sin\theta),$$

which coincides with the external field calculated earlier.

[9] See, e.g. Eq. (3.13) of the lecture notes.

***Problem* 5.10.** The reader is (hopefully :-) familiar with the classical Hall effect when it takes place in the usual rectangular *Hall bar* geometry—see the left panel of the figure below. However, the effect takes a different form in the so-called *Corbino disk*—see the right panel below. (Dark shading shows electrodes, with no appreciable resistance.) Analyze the effect in both geometries, assuming that in both cases the conductors are thin, planar, have a constant Ohmic conductivity σ and a charge carrier density n, and that the applied magnetic field $\mathbf{B}$ is uniform and normal to the conductors' planes.

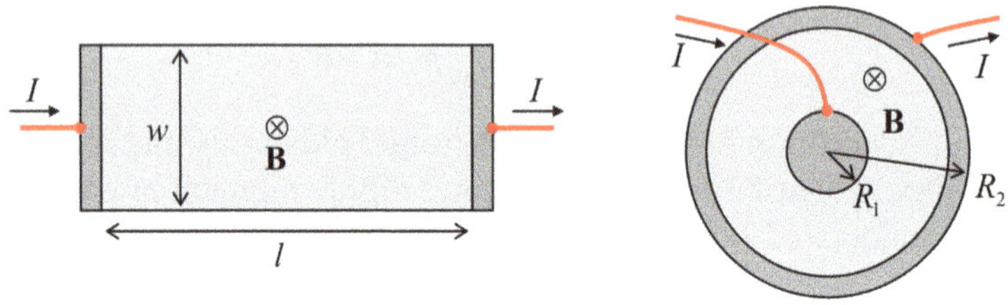

Solution: As we know from chapter 4 of the lecture notes, in the Hall bar geometry, the current I is uniformly distributed across the sample, with the density $j = I/wt$, where t is the bar's thickness. As was discussed section 4.2, in an Ohmic conductor the same density should be equal to $qn\langle v\rangle$, where $\langle v\rangle$ is the average, 'drift' velocity of the charge carriers (on top of their possible random thermal and/or quantum motion), so that $\langle v\rangle = I/wtqn$. This velocity gives rise to an average magnetic component of the Lorentz force (5.10), with magnitude $\langle F\rangle = q\langle v\rangle B$, and directed normally to the current and the magnetic field, i.e. across the Hall bar. Initially (as the current is turned on) this force causes the charge accumulation at the lateral sides of the bar, until their electric field $\mathbf{E}$ becomes large enough to compensate the magnetic force: $\mathbf{E} = -\langle\mathbf{v}\rangle \times \mathbf{B}$. So, at a stationary motion of the charge carriers along the current direction, the electric field has to be uniform, and hence resulting in the voltage of magnitude $V = R_H I$, where R_H is the so-called Hall resistance

$$R_H \equiv \frac{V}{I} = \frac{w|E|}{I} = \frac{w\langle v\rangle B}{I} = \frac{B}{qnt}.$$

Note the following important features of this well-known result: first, R_H does not depend on w and l, i.e. on the Hall bar's geometry, and, second, in contrast to the usual Ohmic resistance (see Eqs. (4.9) and (4.13) of the lecture notes) it *is* sensitive to the sign of q. These features make the measurement of R_H the standard technique for the measurement of sgn(q), as well as of the carrier density n in semiconductors.

Now addressing the Corbino disk, due to its axial symmetry of geometry, the radial component j_ρ of the current density has to be uniformly distributed over the angle, so that at distance ρ from the center,

$$j_\rho = \frac{I}{2\pi t\rho}, \qquad \text{and } \langle v_\rho\rangle = \frac{j_\rho}{qn} = \frac{I}{2\pi tqn\rho},$$

causing an azimuthally directed average magnetic force of magnitude

$$\langle F_\varphi \rangle = q\langle v_\rho \rangle B = \frac{IB}{2\pi tn\rho}.$$

Such field is equivalent, in its action on the carriers, to an azimuthal electric field

$$\langle E_\varphi \rangle = \frac{\langle F_\varphi \rangle}{q} = \frac{IB}{2\pi tqn\rho}.$$

However, in this axially symmetric geometry, the resulting azimuthal motion of carriers does not lead to their accumulation, and hence continues even in the stationary state, giving an azimuthal (circular) current with the density

$$j_\varphi = \sigma\langle E_\varphi \rangle = \frac{\sigma IB}{2\pi tqn\rho}.$$

Calculating the full circular current,

$$I_\varphi = t\int_{R_1}^{R_2} j_\varphi d\rho = I\frac{\sigma B}{2\pi qn}\ln\frac{R_2}{R_1},$$

we see that the I_φ/I ratio is independent of the disk's thickness t, and, according to Eq. (4.13), also of the charge carrier density:

$$\frac{I_\varphi}{I} = \frac{\omega_c \tau}{2\pi}\ln\frac{R_2}{R_1},$$

where τ is the effective scattering time participating in the Drude formula (4.13), and $\omega_c \equiv qB/m$ is the *cyclotron frequency* of the ballistically moving carriers in the same field. (For more about ω_c, see section 9.6 of the lecture notes.)

For typical conductors in practicable magnetic fields, $I_\varphi/I \ll 1$, and the magnetic field produced by the circular currents (which was ignored in the above solution) is typically much smaller than B.

Problem* 5.11. The simplest model of the famous *homopolar motor*[10] is a thin, round conducting disk, placed into a uniform magnetic field normal to its plane, and fed by a dc current flowing from the disk's center to a sliding electrode ('brush') on its rim—see the figure below.

(i) Express the torque, rotating the disk, via its radius R, the magnetic field $\mathbf{B}$, and the current I.
(ii) If the disk is allowed to rotate about its axis, and the motor is driven by a battery with e.m.f. $\mathscr{V}$, calculate its stationary angular velocity ω, neglecting friction and the electric circuit's resistance.
(iii) Now assuming that the current circuit (battery + wires + contacts + disk itself) has a non-zero resistance $\mathscr{R}$, derive and solve the equation for the time evolution of ω, and analyze the solution.

[10] It was invented by M Faraday in 1821, i.e. well before his celebrated work on electromagnetic induction. The adjective 'homopolar' refers to the constant 'polarity' (sign) of the current; the alternative term is 'unipolar'.

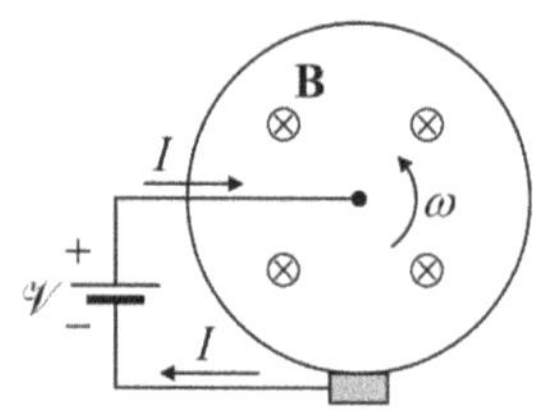

Solutions:

(i) Applying Eq. (5.8) of the lecture notes to an elementary area of the disk, we obtain the elementary Lorenz force

$$d\mathbf{F} = \mathbf{J}(\boldsymbol{\rho}) \times \mathbf{B}d^2\rho,$$

where $\mathbf{J}$ is the linear density (measured, e.g. in A m^{-1}) of the current in the disk, and $\boldsymbol{\rho}$ is the 2D radius-vector in disk's plane. Assuming that $\boldsymbol{\rho}$ is measured from the disk center (i.e. the rotation axis), the elementary force creates the following elementary rotating torque[11]

$$d\boldsymbol{\tau} = \boldsymbol{\rho} \times d\mathbf{F} = \boldsymbol{\rho} \times [\mathbf{J}(\boldsymbol{\rho}) \times \mathbf{B}]d^2\rho.$$

Since the vector $d\mathbf{F}$ is perpendicular to the vector $\mathbf{B}$, i.e. lies in disk's plane, and so is vector $\boldsymbol{\rho}$, all elementary torque vectors $d\boldsymbol{\tau}$ are directed along disk's axis, so that they may be summed as scalars. Moreover, the radial component of $d\mathbf{F}$, proportional to the angular component $J_\varphi \mathbf{n}_\varphi$ of the current, does not contribute to the net torque, so that its magnitude is[12]

$$\tau = \int_{\substack{\text{disk}\\ \text{area}}} \rho J_\rho(\boldsymbol{\rho}) B d^2\rho, \qquad (*)$$

where J_ρ is the radial component of the current density.

Superficially it may look like that in order to complete the calculation of the torque, we need to know in detail the distribution of the current density $\mathbf{J}$ (or at least its radial component) over the disk area—for example, the one calculated in problem 4.8 in the approximation of point current injection/pickup. However, it is not so. Indeed, recasting Eq. (*) in the form

$$\tau = B\int_0^R \rho d\rho \int_0^{2\pi} \rho d\varphi J_\rho(\boldsymbol{\rho}),$$

[11] See, e.g. Eq. (1.34).

[12] Formally, this result may be obtained by using the well-known 'bac minus cab rule'—see Eq. (A.47):

$$d\boldsymbol{\tau}/d^2\rho = \boldsymbol{\rho} \times (\mathbf{J} \times \mathbf{B}) = \mathbf{J}(\boldsymbol{\rho} \cdot \mathbf{B}) - \mathbf{B}(\boldsymbol{\rho} \cdot \mathbf{J}) = 0 - \mathbf{B}J_\rho.$$

we may notice that since $\rho d\varphi$ is an elementary arc of a circle of radius ρ, the internal integral is just the total current I flowing out from disk's center. Since due the charge conservation (see the discussion in section 4.1 of the lecture notes), the total dc current cannot depend on ρ, and we may complete the integration as

$$\tau = BI\int_0^R \rho d\rho = BI\frac{R^2}{2}. \qquad (**)$$

The background reason of this surprising simplicity will be discussed below.

(ii) The simplest, though somewhat formal way to solve this part of the problem is to apply the energy–work principle to an elementary time interval dt:

$$d\mathscr{W} \equiv \mathscr{V}dQ \equiv \mathscr{V}Idt = dT_{\text{rotation}} \equiv d\left(\frac{\mathscr{I}\omega^2}{2}\right),$$

where $\mathscr{I}$ is the disk's moment of inertia for rotation about its axis[13]. Dividing both sides of this equation by dt, and differentiating its right-hand side, we obtain

$$\mathscr{V}I = \mathscr{I}\omega\dot{\omega}.$$

But according to the basic equation of rotation dynamics[14], the product $\mathscr{I}\dot{\omega}$ (i.e. the time derivative of the disk's angular momentum) should be equal to the applied torque τ, so that using Eq. (**), we obtain

$$\mathscr{V}I = \omega\tau = \omega BI\frac{R^2}{2}, \qquad \text{giving } \omega = \omega_0 \equiv \frac{2\mathscr{V}}{BR^2}. \qquad (***)$$

At first glance this result is counter-intuitive. Since we have neglected the circuit's resistance, why could not the non-zero e.m.f. lead to an infinite growth of current I, and hence the torque, and hence an infinite speed-up of the disk's rotation? Some light on the underlying physics may be shed by the following simple model. Let the disk's conductivity be due to certain classical particles of charge q. Then the disk's rotation with angular velocity ω results in the particles' motion (in the lab reference frame, i.e. relative to the magnetic field's source) with the velocity $v = \rho\omega$, and hence to the radially directed Lorentz force (5.10), with magnitude $F = qvB = q\omega\rho B$. The ratio $F/q = \omega\rho B$ may be considered as the magnitude of some additional, rotation-induced electric field $\mathbf{E}_{\text{ind}}$; it is easy to check that this field is always (for any sign q) directed against the current I. As a result, the total additional e.m.f.,

$$\mathscr{V}_{\text{ind}} = -\int_0^R \mathbf{E}_{\text{ind}} \cdot d\boldsymbol{\rho} = \int_0^R \omega\rho B d\rho = \omega B\int_0^R \rho d\rho = \omega B\frac{R^2}{2}, \qquad (****)$$

[13] See, e.g. *Part CM* sections 4.2 and 4.3.

[14] See, e.g. *Part CM*, Eqs. (4.38).

induced by the rotation between the disk axis and the brush contact, has the polarity opposite to battery's e.m.f. $\mathscr{V}$. So, the physics of Eq. (***) is that at $\omega = \omega_0$, $\mathscr{V}_{\text{ind}}$ exactly compensates $\mathscr{V}$.[15]

(iii) The last discussion allows a ready generalization of the result to the case of non-vanishing resistance $\mathscr{R}$ of the current-supplying circuit. Indeed, writing the second Kirchhoff law (4.7b) with account of the rotation-induced e.m.f.,

$$\mathscr{V} - \mathscr{V}_{\text{ind}} = I\mathscr{R},$$

and plugging in the I expressed from Eq. (***) with $\tau = \mathscr{I}\dot{\omega}$, and $\mathscr{V}_{\text{ind}}$ from Eq. (****), we obtain

$$\mathscr{V} - \frac{\omega BR^2}{2} = \mathscr{I}\dot{\omega}\frac{2}{BR^2}\mathscr{R}.$$

This linear, first-order differential equation for ω may be rewritten in the standard form

$$\left(1 + t_0\frac{d}{dt}\right)\omega = \omega_0,$$

where t_0 is a time constant,

$$t_0 \equiv \frac{4\mathscr{I}\mathscr{R}}{B^2R^4},$$

and ω_0 is the stationary value of ω, given by Eq. (***). This differential equation may be readily solved, giving

$$\omega(t) = \omega(0)\, e^{-t/t_0} + \omega_0(1 - e^{-t/t_0}).$$

This formula describes a transient, of duration $\sim t_0$, from the initial rotation velocity $\omega(0)$ to its stationary value (***). The time constant is proportional to circuit's resistance $\mathscr{R}$, making it clear why in part (ii) of the problem (where $\mathscr{R}$ and hence t_0 were neglected) we obtained the value (***) as a constant, without any apparent transient process.

Finally note that the same device, driven by an external torque, and with the battery $\mathscr{V}$ replaced with an electric load, may serve as a *homopolar generator* of the dc e.m.f. (****)[16]. Since it has only one current loop, for most practical purposes it is less efficient than the usual multi-coil generators, but has certain advantages for producing short, very large pulses of current (in some cases, of several mega-

[15] This result is more general than the simple model used, and is actually one of manifestations of Faraday's law of electromagnetic induction, discussed in section 6.1. Indeed, it is easy to check that the product $\omega R^2/2$, which participates in Eq. (****), is just dA/dt, where A is the area swept by circle's radius, so that Eq. (****) may be rewritten as $\mathscr{V}_{\text{ind}} = d(BA)/dt$. Since, for a uniform magnetic field $\mathbf{B}$, perpendicular to the disk's plane and directed against axis z, BA is just minus magnetic flux Φ through area A, Eq. (****) is in agreement with the induction law in the well-known form of Eq. (6.1). (Note, however, that the exact specification of what exactly A is in this case, and hence the derivation of Eq. (****) from the Faraday law, require some care.) The most important philosophical result of this discussion is that the Faraday law is not something entirely new, and is at least consistent with the laws of magnetostatics. (More discussion of this fact will be offered in section 6.2.)

[16] It was also invented by M Faraday and is sometimes called the 'Faraday wheel'.

amperes!), using the gradually accumulated kinetic energy of massive flywheels, for such unique purposes as synchrotron experiments.

***Problem* 5.12.** Current I flows in a thin wire bent into a planar, round loop of radius R. Calculate the net magnetic flux through the plane in which the loop is located.

Solution: To solve this problem, an explicit calculation of the integral (5.65) is unnecessary. Indeed, applying the divergence theorem[17] to Eq. (5.29) and any volume limited by a closed surface S, we obtain

$$\oint_S B_n d^2r = 0. \qquad (*)$$

(Essentially, this is an analog of the Gauss law (1.16), but with no magnetic monopoles in sight.) Let us apply this relation to the axially-symmetric surface with the cross-section shown with the dashed line in the figure below. Its spherical 'cup' has a radius $a \gg R$, while the near-wire part follows a slim torus, drawn around the wire, with a much smaller radius $b \ll R$. Since at large distances the field of the loop follows Eq. (5.99), i.e. $B \propto 1/r^3$, the flux through the spherical part of the surface, of area $A \propto a^2$, tends to zero in the limit $a \to \infty$.

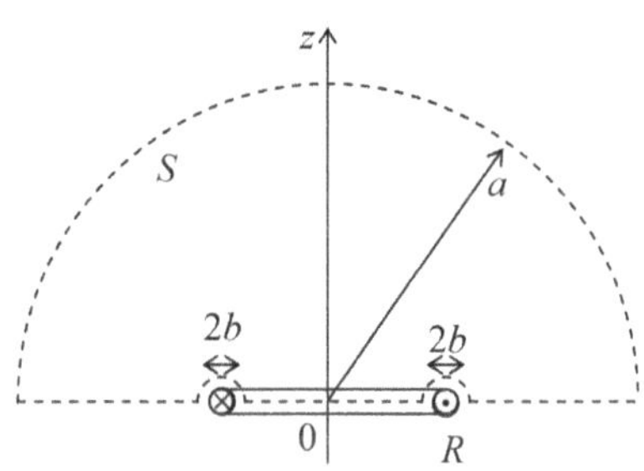

On the other hand, very close to the wire, the field lines are virtually circular, i.e. tangential to the slim torus' surface, also giving no contribution to the left-hand side of Eq. (*). Hence the flux we are interested in,

$$\Phi = \oint B_z|_{z=0} dxdy,$$

differs from the total flux (*) only by the integral of $B_z|_{z=0}$ over a narrow plane ring of radius R and width $2b$. However, since $B_z|_{z=0}$ near the wire is virtually the same as near a straight wire (see Eq. (5.20) of the lecture notes), i.e. an odd function of the deviation from the wire's center[18], this integral vanishes and we obtain $\Phi = 0$.

The above calculation is a strict proof of the following handwaving argument: due to the absence of magnetic charges (monopoles), the magnetic field lines do not start

[17] See, e.g. Eq. (A.79).

[18] Formally, for an infinitesimally thin wire, the field diverges, but this divergence is automatically removed in any realistic (physical) model of current distribution in a wire of non-vanishing thickness—see, e.g. Eq. (5.38).

or end at any point, but rather follow closed contours. Hence, in our particular case, the number of such lines crossing the loop's plane inside the loop (giving, say, a positive contribution to the net flux Φ through the plane) exactly equals to that crossing it outside of the loop—thus creating an exactly equal but negative contribution into Φ.

***Problem* 5.13.** Prove that:

(i) the self-inductance L of a current loop cannot be negative, and
(ii) each mutual inductance coefficient $L_{kk'}$, defined by Eq. (5.60) of the lecture notes, cannot be larger than $(L_{kk}L_{k'k'})^{1/2}$.

Solutions:

(i) Since, according to Eq. (5.57), the magnetic energy U of any system cannot be negative, Eq. (5.72) for a single current loop,

$$U = \frac{L}{2}I^2,$$

cannot give negative results for any current I, which is only possible if $L \geqslant 0$.

(ii) Let us apply a similar argument to the more general equation (5.59). It also cannot give a negative result for any combination of currents, in particular when only two of them[19], I_k and $I_{k'}$, are different from 0, and hence this relation (taking into account Eq. (5.61), $L_{kk'} = L_{k'k}$) is reduced to a form similar to Eq. (5.62):

$$U = \frac{L_{kk}}{2}I_{kk}^2 + L_{kk'}I_kI_{k'} + \frac{L_{k'k'}}{2}I_{k'}^2. \qquad (*)$$

This means, in particular, that if we fix one of the currents (say, $I_{k'}$), U as a function of the other current (I_k) cannot be negative even at its minimum, where

$$\frac{\partial U}{\partial I_k} \equiv L_{kk}I_k + L_{kk'}I_{k'} = 0.$$

Plugging the resulting value, $I_k = -L_{kk'}I_{k'}/L_{kk}$, back into Eq. (*), we obtain

$$U_{\min} = \frac{L_{kk}}{2}\left(\frac{L_{kk'}I_{k'}}{L_{kk}}\right)^2 - L_{kk'}\frac{L_{kk'}I_{k'}}{L_{kk}}I_{k'} + \frac{L_{k'k'}}{2}I_{k'k'}^2 \equiv \left(1 - \frac{L_{kk'}^2}{L_{kk}L_{k'k'}}\right)\frac{L_{k'k'}}{2}I_{k'k'}^2.$$

Now the requirement $U_{\min} \geqslant 0$ immediately gives the required proof of the inequality $L_{kk'} \leqslant (L_{kk}L_{k'k'})^{1/2}$.

Problem* 5.14. Estimate the values of magnetic susceptibility due to

(i) orbital diamagnetism, and
(ii) spin paramagnetism,

[19] If $k = k'$, the correctness of the statement (ii) is obvious.

for a medium with negligible interaction between the induced molecular dipoles. Compare the results.

Hints: For task (i), you may use the classical model described by Eq. (5.114) of the lecture notes—see figure 5.13. For task (ii), assume the mechanism of ordering of spontaneous magnetic dipoles $\mathbf{m}_0$, with a magnitude m_0 of the order of the Bohr magneton μ_B, similar to the one sketched for electric dipoles in figure 3.7a.

Solutions:

(i) According to Eq. (5.114), the angular velocity of the torque-induced precession of vector **L** (in this case called the *Larmor frequency*),

$$\omega_L = \frac{2\pi}{\mathcal{T}_L} = \frac{2\pi}{2\pi L \sin\theta/(dL/dt)} = \frac{qB}{2m_e} \equiv \frac{\mu_B}{\hbar}B,$$

does not depend on the angle θ between these two vectors[20]. This motion creates the average current

$$I = qf_L = q\frac{\omega_L}{2\pi} = \frac{q^2}{4\pi\, m_e}B = \frac{e^2}{4\pi\, m_e}B,$$

(note its independence of the sign of q) and hence the induced magnetic moment may be estimated as[21]

$$\Delta m = I\pi\,\langle\rho^2\rangle = \frac{e^2 B}{4m_e}\langle\rho^2\rangle = \frac{e^2 B}{4m_e}(\langle x^2\rangle + \langle y^2\rangle),$$

where ρ is the distance of the electron from the precession axis, taken for axis z. Now let us add such contributions from Z electrons of an atom, assuming that the initial orientations of their rotation planes are random, and hence the averages of all Cartesian coordinate squares are equal:

$$\langle x^2\rangle = \langle y^2\rangle = \langle z^2\rangle = \frac{\langle r^2\rangle}{3}, \quad \text{so that} \langle x^2\rangle + \langle y^2\rangle = \frac{2}{3}\langle r^2\rangle.$$

With this assumption, for the orbital molecular susceptibility we obtain the following *Langevin formula*:

$$|\chi_m| \equiv n\frac{\Delta m}{H} \approx n\frac{\Delta m}{B/\mu_0} = \frac{\mu_0 n Z e^2}{6m_e}\langle r^2\rangle,$$

where n is the number of atoms per unit volume.

Amazingly enough, this is *exactly* the result given by quantum mechanics[22], with the only difference that $\langle r^2\rangle$ should be understood as the simultaneously statistical *and* quantum-mechanical average, taking into account the spread of electrons'

[20] As a useful sanity check, $\hbar\Omega_L = \mu_B B$ is exactly the result given by quantum mechanics for the atomic energy level splitting by a magnetic field, due to the electron's orbital motion—see, e.g. *Part QM* section 6.4.

[21] The initial dipole moments **m** are assumed to be disordered, so their net contribution to the magnetization vanishes.

[22] See, e.g. *Part QM*, chapter 6.

wavefunctions. Thus the susceptibility may be crudely estimated by taking $\langle r^2 \rangle$ equal to the square of the Bohr radius $r_B \equiv \hbar^2/m_e(e^2/4\pi\varepsilon_0)$.[23] This substitution yields

$$|\chi_m| \sim \frac{\mu_0 n Z e^2 r_B^2}{6m_e} \equiv \frac{2\pi}{3} Z\left(n r_B^3\right) \alpha^2, \qquad (*)$$

where, in the last expression, parameter

$$\alpha \equiv \frac{e^2}{4\pi\varepsilon_0 \hbar c} \equiv \frac{e^2 \mu_0 c}{4\pi\hbar} \equiv \frac{e^2}{4\pi\hbar}\left(\frac{\mu_0}{\varepsilon_0}\right)^{1/2} \approx \frac{1}{137} \ll 1$$

is the *fine-structure constant*, which determines the relative strength of electromagnetic interactions on the quantum-relativistic scale[24]. It is evident that even for heaviest atoms ($Z \sim 10^2$) in their condensed matter form (where $n \sim r_B^{-3}$), this magnetic susceptibility is still small: $|\chi_m| \ll 1$, i.e. the orbital diamagnetism is always weak.

(ii) As was discussed in section 5.5 of the lecture notes, we may expect the full alignment of spontaneous magnetic dipoles to be achieved at $m_0 B \sim k_B T$, resulting in the induced magnetic moment $\Delta m = Z m_0$ per atom, and hence the saturation of the magnetization at the level $M_0 = nZm_0$. Hence it is reasonable to expect that in weak magnetic field, the magnetization scales as[25]

$$M \sim \frac{m_0 B}{k_B T} M_0 = \frac{nZm_0^2 B}{k_B T}, \qquad \text{for } M \ll M_0,$$

corresponding to the paramagnetic (positive) susceptibility

$$\chi_m \equiv \frac{M}{H} \approx \frac{M}{B/\mu_0} \approx \frac{\mu_0 n Z m_0^2}{k_B T}.$$

This result obeys the so-called *Curie law*, $\chi_m \propto 1/T$ (also valid for χ_e in paraelectrics).

In order to estimate this susceptibility, we may take $m_0 \sim \mu_B \equiv e\hbar/2m_e$, obtaining

$$\chi_m \sim \frac{\mu_0 n Z}{k_B T}\left(\frac{e\hbar}{2m_e}\right)^2 \equiv \pi Z(n r_B^3)\, \alpha^2 \frac{E_H}{k_B T}.$$

For room temperatures ($T \sim 300$K), the ratio $E_H/k_B T$ is close to 10^3, so that comparing this estimate with Eq. (*), we see that the spin paramagnetism is much stronger than the orbital diamagnetism[26], so that the latter effect prevails only in materials with zero net spins—see table 5.1 of the lecture notes.

---

[23] See, e.g. *Part QM* sections 1.1 and 3.6. As a mnemonic rule, the expression r_B may be obtained by equating the scales of the electrostatic ($e^2/4\pi\varepsilon_0 r_B$) and quantum-kinetic ($\hbar^2/m_e r_B^2$) energies in Bohr's model of the hydrogen atom, without numerical coefficients. (Note also that each of these fractions gives the atomic energy scale $E_H \equiv (e^2/4\pi\varepsilon_0)^2 m_e/\hbar^2 \approx 27.2$ eV, called the *Hartree energy*.)

[24] In this course, we will run into the fine-structure constant again in section 10.4; it will be repeatedly discussed in *Part QM* of the series. For its more exact value, see appendix B.

[25] This assumption is confirmed by results from statistical models based on the Gibbs distribution—see, e.g. *Part SM* section 2.4.

[26] This estimate is invalid in metals with their conduction electrons forming a degenerate Fermi 'sea'—see, e.g. *Part SM* section 3.3. In this case the diamagnetic susceptibility, calculated within a simple model of independent Fermi particles, called the *Landau diamagnetism*, equals exactly 1/3 of the so-called *Pauli paramagnetism* due to spin orientation of the same electrons, so that in metals with uncompensated spins (e.g. the alkali metals) the paramagnetism still prevails, but not by much. For more details see, e.g. [1].

Problem* 5.15. Use the classical picture of the orbital ('Larmor') diamagnetism, discussed in section 5.5 of the lecture notes, to calculate its (small) correction $\Delta\mathbf{B}(0)$ to the magnetic field $\mathbf{B}$, as felt by an atomic nucleus, modeling the electrons of the atom by a spherically symmetric cloud with an electric charge density $\rho(r)$. Express the result via the value $\phi(0)$ of the electrostatic potential of the electrons' cloud, and use this expression for a crude numerical estimate of the relative correction, $\Delta B(0)/B$, for the hydrogen atom.

Solution: As was discussed in section 5.5, a relatively low external magnetic field $\mathbf{B}$ causes the torque-induced precession of classical particle's orbit around field's direction, with the angular velocity

$$\boldsymbol{\Omega} = -\frac{q}{2m_0}\mathbf{B}.$$

In the (admittedly, somewhat eclectic, but very useful) model, in which the quantum-mechanical spread of electron positions at $\mathbf{B} = 0$ is represented by a time-independent classical charge density $\rho(\mathbf{r})$, the rotation induces loop currents (see the figure below) with density $\mathbf{j}(\mathbf{r}) = \mathbf{v}(\mathbf{r})\rho(\mathbf{r})$, where $\mathbf{v}(\mathbf{r}) = \boldsymbol{\Omega} \times \mathbf{r}$.[27] The magnetic field correction $\Delta\mathbf{B}$ due to these currents may be calculated using Eq. (5.14). For the origin-located nuclei (with $\mathbf{r} = 0$), and with $\mathbf{r}'$ replaced with $\mathbf{r}$ (just for notation simplicity), that formula yields

$$\Delta\mathbf{B}(0) = -\frac{\mu_0}{4\pi}\int \mathbf{j}(\mathbf{r}) \times \frac{\mathbf{r}}{r^3}d^3r = -\frac{\mu_0}{4\pi}\int (\boldsymbol{\Omega} \times \mathbf{r}) \times \mathbf{r}\frac{\rho(\mathbf{r})}{r^3}d^3r$$
$$= \frac{\mu_0}{4\pi}\frac{q}{2m_0}\int (\mathbf{B} \times \mathbf{r}) \times \mathbf{r}\frac{\rho(\mathbf{r})}{r^3}d^3r.$$

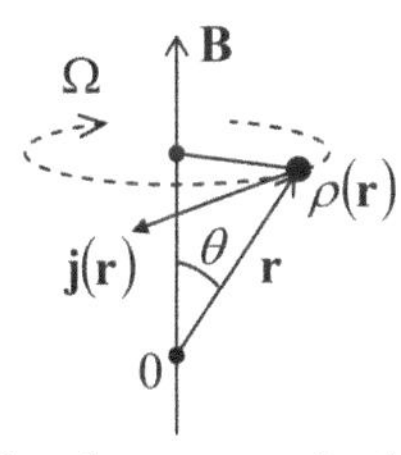

The vector $(\mathbf{B} \times \mathbf{r}) \times \mathbf{r}$ lies in the mutual plane of the vectors $\mathbf{r}$ and $\mathbf{B}$, has magnitude $Br^2|\sin\theta|$, and may be represented as a vector sum of two components: one, $-\mathbf{B}r^2\sin^2\theta$, directed against vector $\mathbf{B}$, and another one, normal to this vector. For any axially symmetric charge distribution $\rho(\mathbf{r})$, the integral of the latter component over the azimuthal angle vanishes. Hence, for any sign of the moving particles' charge (which controls the sign of not only q, but also ρ), the vector $\Delta\mathbf{B}(0)$ is directed exactly against vector $\mathbf{B}$, and hence describes the diamagnetic response of the orbital motion. Its relative magnitude,

$$\frac{\Delta B(0)}{B} = -\frac{\mu_0}{4\pi}\frac{q}{2m_0}\int \sin^2\theta\frac{\rho(\mathbf{r})}{r}d^3r = -\frac{\mu_0}{4\pi}\frac{q}{2m_0}2\pi\int_0^{\pi}\sin^3\theta d\theta\int_0^{\infty}\rho(\mathbf{r})\, r dr,$$

[27] If you are uncomfortable with the last relation, please revisit a classical mechanics course, e.g. the discussion in the beginning of *Part CM* section 4.1, and in particular Eq. (4.9).

where θ is the angle between vectors $\mathbf{B}$ are $\mathbf{r}$ (see the figure above), takes a particularly simple form if the charge distribution is spherically symmetric, i.e. $\rho(\mathbf{r}) = \rho(r)$, because in this case the integral over θ is may be readily worked out by the usual substitution $\xi \equiv \cos\theta$ (so that $d\xi = -\sin\theta d\theta$ and $\sin^2\theta = 1 - \xi^2$):

$$\int_0^{\pi} \sin^3\theta d\theta = \int_{-1}^{+1} (1 - \xi^2) d\xi = \frac{4}{3}.$$

The final result,

$$\frac{\Delta B(0)}{B} = -\frac{\mu_0}{4\pi}\frac{q}{m_0}\frac{4\pi}{3}\int_0^{\infty} \rho(r)\, r dr, \qquad (*)$$

may be readily related to the value (at the same central point) of the electrostatic potential $\phi(\mathbf{r})$ induced by the same electron charge distribution. Indeed, applying Eq. (1.38) of the lecture notes to the point $\mathbf{r} = 0$, and again replacing $\mathbf{r}'$ with $\mathbf{r}$, for the spherically symmetric charge density we obtain the same integral:

$$\phi(0) = \frac{1}{4\pi\varepsilon_0}\int \frac{\rho(r)}{r} d^3r = \frac{1}{4\pi\varepsilon_0} 4\pi \int_0^{\infty} \rho(r) r dr,$$

so that the comparison with Eq. (*) yields:

$$\frac{\Delta B(0)}{B} = -\frac{\varepsilon_0\mu_0}{3}\frac{q}{m_0}\phi(0) \equiv -\frac{1}{3}\frac{q}{m_0 c^2}\phi(0).$$

This equality enables a simple estimate of the diamagnetic correction in the hydrogen atom: for electrons, with $q = -e$ and $m_0c^2 \approx 0.51$ MeV, $|\phi(0)|$ may be crudely estimated as $e/4\pi\varepsilon_0 r_B \equiv E_H/e$, where r_B is the Bohr radius, and $E_H \approx 27$ eV is the Hartree energy unit[28], so that

$$\left|\frac{\Delta B(0)}{B}\right| \sim \frac{1}{3}\frac{E_H}{m_0c^2} \approx 2 \times 10^{-5}.$$

***Problem* 5.16.** Calculate the (self-) inductance of a toroidal solenoid (see figure 5.6b of the lecture notes) with the round cross-section of radius $r \sim R$ (see the figure below), with many ($N \gg 1$, R/r) wire turns uniformly distributed along the perimeter, and filled with a linear magnetic material of a magnetic permeability μ. Check your results by analyzing the limit $r \ll R$.

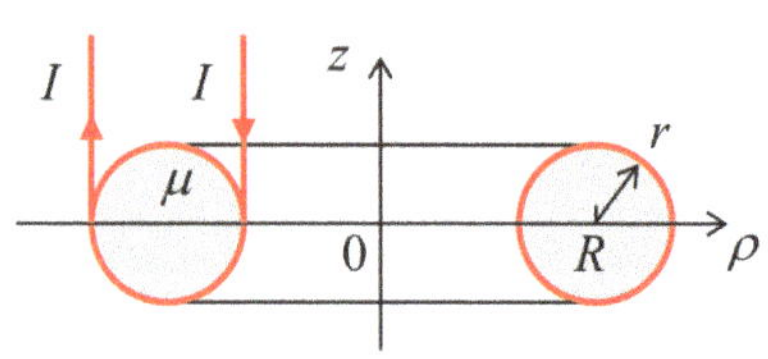

[28] See, e.g. either the solution of the previous problem, or *Part QM* section 1.1, in particular Eqs. (1.9) and (1.13).

Solution: In section 5.2 of the lecture notes, the magnetic field inside such a solenoid without magnetic filling was calculated—see Eq. (5.41). According to the discussion in the beginning of section 5.6, since all the field is concentrated inside the solenoid filled with a linear magnetic, we may just multiply that result by ratio μ/μ_0:

$$B = \frac{\mu NI}{2\pi\rho}.$$

Now we can calculate the magnetic flux piercing one wire loop:

$$\Phi_1 = \int_A B_n d^2r = \frac{\mu NI}{\pi}\int_0^r dz \int_{R-(r^2-z^2)^{1/2}}^{R+(r^2-z^2)^{1/2}} \frac{d\rho}{\rho}$$
$$= \frac{\mu NI}{\pi}\int_0^r \ln\frac{R+(r^2-z^2)^{1/2}}{R-(r^2-z^2)^{1/2}}dz = \frac{\mu NIr}{\pi}\int_0^1 \ln\frac{a+(1-\xi^2)^{1/2}}{a-(1-\xi^2)^{1/2}}d\xi,$$

where $a \equiv R/r > 1$, and $\xi \equiv z/r$. This is a table integral[29], equal to $\pi[a-(a^2-1)^{1/2}]$, so that, finally,

$$\Phi_1 = \mu NI\left[R-\left(R^2-r^2\right)^{1/2}\right].$$

Just as in the long solenoid discussed in section 5.2 of the lecture notes, the whole flux Φ piercing the whole wire is N times larger. As a result, the solenoid's inductance

$$L \equiv \frac{\Phi}{I} = \frac{\Phi_1 N}{I} = \mu N^2\left[R-\left(R^2-r^2\right)^{1/2}\right]. \qquad (*)$$

In the limit $r \ll R$, we may expand this expression into the Taylor series in small r/R, and in the first approximation obtain

$$L \approx \mu N^2\frac{r^2}{2R} = \mu n^2 lA, \quad \text{with } n \equiv \frac{N}{l}, \quad l \equiv 2\pi R, \quad \text{and } A \equiv \pi r^2.$$

We see that in this limit, the result coincides with the inductance of a long straight solenoid, calculated in section 5.6—see Eq. (5.143). It is interesting that in the opposite limit ($r = R$), Eq. (*) also acquires a very simple form,

$$L = \mu N^2 r.$$

Note that these results, as well as Eq. (5.142) of the lecture notes for the straight solenoid, may be used for relatively small values of N only if $\mu \gg \mu_0$, otherwise we have to account for 'stray' magnetic fields spilling out between separated wire turns.

[29] See, e.g. (A.40).

***Problem* 5.17.** A long straight, thin wire, carrying current I, passes parallel to a plane boundary between two uniform, linear magnetics—see the figure below. Calculate the magnetic field everywhere in the system, and the force (per unit length) exerted on the wire.

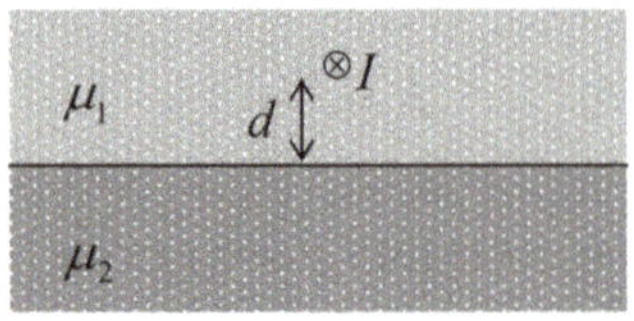

Solution: If the linear magnetic surrounding the wire was purely uniform ($\mu(\mathbf{r})$ = const), its field lines would be circular, with the magnitude described by Eq. (5.20) of the lecture notes, corrected as prescribed by Eq. (5.115). The result could be represented in the following vector form:

$$\mathbf{H}(\boldsymbol{\rho}) = \mathbf{n}_\varphi \frac{I}{2\pi|\boldsymbol{\rho} - \boldsymbol{\rho}_0|}, \qquad B(\boldsymbol{\rho}) = \mathbf{n}_\varphi \frac{\mu I}{2\pi|\boldsymbol{\rho} - \boldsymbol{\rho}_0|}, \qquad (*)$$

where $\boldsymbol{\rho}$ is the 2D radius-vector of the observation point within the plane perpendicular to the wire (i.e. the plane of the figure above), $\boldsymbol{\rho}_0$ is that of the wire, and $\mathbf{n}_\varphi$ is the unit azimuthal vector in this plane (normal to the vector $\boldsymbol{\rho} - \boldsymbol{\rho}_0$). However, in our current case, with $\mu_1 \neq \mu_2$, such a solution would not satisfy the second of the boundary conditions (5.118) and (5.119),

$$H_\tau = \text{const}, \quad B_n = \text{const}, \qquad (**)$$

at the interface between two magnetics.

Inspired by the solution of similar electrostatic and dc-current problems in chapters 2–4 of the lecture notes (see, in particular, the problem shown in figure 3.12), we may try to look for the solution in the form of a sum of two contributions of the type (*)—from the actual current I (for observation points in the lower half-space, replaced by another value, I'') and from an image current I' passing at the same distance d from the interface, but at its opposite side—see the figure below.

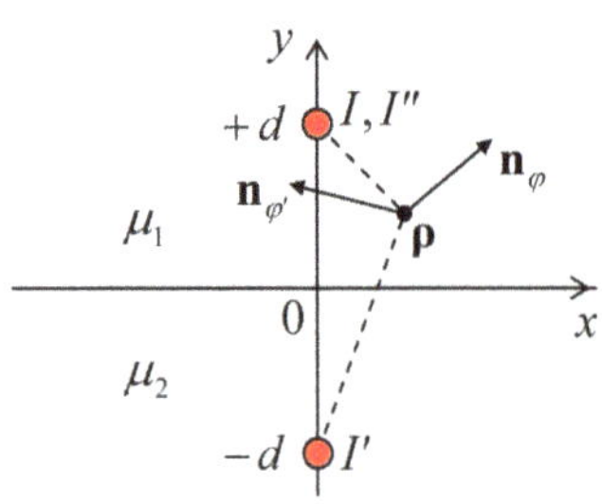

With the Cartesian coordinates selected as shown in this figure, the assumed solution (analogous to (3.66) of the similar electrostatics problem) is

$$\mathbf{H}(\boldsymbol{\rho}) = \begin{cases} \mathbf{n}_{\varphi}\dfrac{I}{2\pi|\boldsymbol{\rho} - \boldsymbol{\rho}_0|} + \mathbf{n}_{\varphi'}\dfrac{I'}{2\pi|\boldsymbol{\rho} - \boldsymbol{\rho}'|}, & \text{for } y \geqslant 0, \\ \mathbf{n}_{\varphi}\dfrac{I''}{2\pi|\boldsymbol{\rho} - \boldsymbol{\rho}_0|}, & \text{for } y \leqslant 0, \end{cases} \qquad (***)$$

where $\boldsymbol{\rho}_0 = \{0, +d\}$, $\boldsymbol{\rho}' = \{0, -d\}$. At the interface, i.e. at $\boldsymbol{\rho} = \{x, 0\}$, this solution yields

$$H_x = \begin{cases} \dfrac{I}{2\pi}\dfrac{d}{x^2 + d^2} - \dfrac{I'}{2\pi}\dfrac{d}{x^2 + d^2}, & \text{for } y \geqslant 0, \\ \dfrac{I''}{2\pi}\dfrac{d}{x^2 + d^2}, & \text{for } y \leqslant 0, \end{cases}$$

$$B_y = \begin{cases} \dfrac{\mu_1 I}{2\pi}\dfrac{x}{x^2 + d^2} + \dfrac{\mu_1 I'}{2\pi}\dfrac{x}{x^2 + d^2}, & \text{for } y \geqslant 0, \\ \dfrac{\mu_2 I''}{2\pi}\dfrac{x}{x^2 + d^2}, & \text{for } y \leqslant 0. \end{cases}$$

These solutions evidently satisfy the boundary conditions (**) if I' and I'' are selected so that

$$I - I' = I'', \qquad \mu_1(I + I') = \mu_2 I''.$$

Solving this simple system of two equations, we obtain

$$I' = -\frac{\mu_1 - \mu_2}{\mu_1 + \mu_2}I, \qquad I'' = \frac{2\mu_1}{\mu_1 + \mu_2}I.$$

These relations, together with Eqs. (***) give the solution of the first task of the problem. Now we may use them to calculate the magnetic field at the position $\boldsymbol{\rho}_0$ of the actual wire (excluding its own, diverging contribution):

$$\mathbf{H}(\boldsymbol{\rho}_0) = \mathbf{n}_{\varphi'}\frac{I'}{2\pi|\boldsymbol{\rho}_0 - \boldsymbol{\rho}'|} = \mathbf{n}_{\varphi'}\frac{I'}{4\pi d} = -\mathbf{n}_{\varphi'}\frac{I}{4\pi d}\frac{\mu_1 - \mu_2}{\mu_1 + \mu_2}.$$

At this point, the unit vector $\mathbf{n}_{\varphi'}$ (see the figure above) is directed opposite to $\mathbf{n}_x$, so that the field is horizontal, with

$$H_x(\boldsymbol{\rho}_0) = \frac{I}{4\pi d}\frac{\mu_1 - \mu_2}{\mu_1 + \mu_2}, \qquad \text{i.e. } B_x(\boldsymbol{\rho}_0) = \frac{I}{4\pi d}\mu_1\frac{\mu_1 - \mu_2}{\mu_1 + \mu_2}.$$

According to the fundamental Eq. (5.15), the force exerted on the wire is vertical, with:

$$\frac{F_y}{l} = IB_x = \frac{I^2}{4\pi\ d}\mu_1\frac{\mu_1 - \mu_2}{\mu_1 + \mu_2}. .$$

In the particular case of a wire stretched in free space, parallel to a linear magnetic with the permeability μ, this formula yields

$$\frac{F_y}{l} = \frac{I^2}{4\pi\ d}\mu_0\frac{\mu_0 - \mu}{\mu_0 + \mu}.$$

Hence, the current is attracted by a paramagnetic material (with $\mu > \mu_0$) and repulsed by a diamagnetic one.

***Problem* 5.18.** Solve the magnetic shielding problem similar to that discussed in section 5.6 of the lecture notes, but for a *spherical*, rather than cylindrical shell, with the central cross-section shown in figure 5.16 (partly reproduced below). Compare the efficiency of those two shields, for the same permeability μ of the shell, and the same b/a ratio.

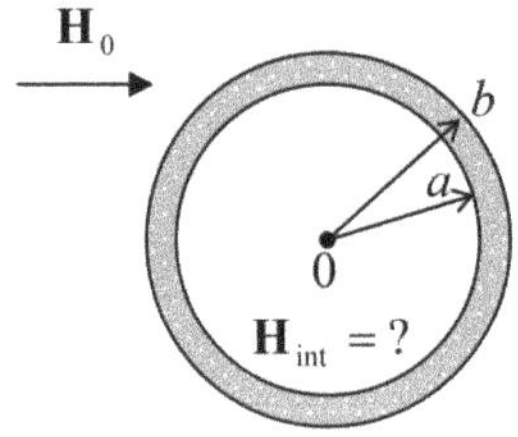

Solution: Guided by the general axially symmetric solution (2.172) of the Laplace equation in the spherical coordinates, we can look for the solution for the scalar potential of the magnetic field in the form

$$\phi_m = \begin{cases} \left(-H_0 r + b_1'/r^2\right)\cos\theta, & \text{for } b \leqslant r, \\ \left(a_1 r + b_1/r^2\right)\cos\theta, & \text{for } a \leqslant r \leqslant b, \\ -H_{\text{int}} r \cos\theta, & \text{for } r \leqslant a, \end{cases}$$

which differs from Eq. (5.127) of the lecture notes (for the cylindrical shield) only by the power of radius in the denominator. Plugging this solution into the boundary conditions, $\phi_{\text{m}} = \text{const}$ and $\mu\partial\phi_{\text{m}}/\partial r = \text{const}$, for both interfaces ($r = b$ and $r = a$), we obtain a system of four equations:

$$-H_0 b + b_1'/b^2 = a_1 b + b_1/b^2, \qquad \left(a_1 a + b_1/a^2\right) = -H_{\text{int}} a,$$

$$\mu_0\left(-H_0 - b_1'/2b^3\right)H_0 = \mu\left(a_1 - b_1/2b^3\right), \qquad \mu\left(a_1 - b_1/2a^3\right) = -\mu_0 H_{\text{int}},$$

for four unknown coefficients a_1, b_1, $b_{1'}$, and H_{int}. Solving the system, we obtain, in particular:

$$\frac{H_{\text{int}}}{H_0} = \frac{\alpha_s - 1}{\alpha_s - (a/b)^3}, \qquad \text{where } \alpha_s \equiv \frac{(\mu + 2\mu_0)(\mu + \mu_0/2)}{(\mu - \mu_0)^2}. \qquad (*)$$

Comparing the magnetic material factors α for the spherical (α_s) and cylindrical (α_c) geometries (see Eq. (5.129) of the lecture notes),

$$\frac{\alpha_s}{\alpha_c} = \frac{(\xi + 2)(\xi + 1/2)}{(\xi + 1)^2} \equiv \frac{\xi^2 + (5/2)\xi + 1}{\xi^2 + (4/2)\xi + 1}, \qquad \text{where } \xi \equiv \frac{\mu}{\mu_0},$$

we see that $\alpha_s > \alpha_c$ for any paramagnetic shell material (with $\mu > \mu_0$), but for the most important case of soft ferromagnets, with $\mu \gg \mu_0$, this difference is minor. More importantly, since $a < b$, the factor $(a/b)^3$ in the denominator of Eq. (*) is smaller than the factor $(a/b)^2$ in the denominator of Eq. (5.129). Hence, the spherical shield is more effective. This is very natural, because the long-cylinder geometry lacks the 'lids', which would help the shielding.

***Problem* 5.19.** Calculate the magnetic field distribution around a spherical permanent magnet with a uniform magnetization $\mathbf{M}_0 = \text{const}$.

Solution: Due to the absence of macroscopic ('stand-alone') currents in the region of our interest, the field $\mathbf{H}$ in the problem may be represented by Eq. (5.120) of the lecture notes, $\mathbf{H} = -\nabla\phi_{\text{m}}$, where the scalar potential ϕ_{m} satisfies the Laplace equation both inside the sphere and outside it. Due to the axial symmetry of the problem (about the direction of the constant magnetization $\mathbf{M}_0$, which may be conveniently taken for the polar axis), the general solution of the Laplace equation may be represented in a form similar to Eq. (2.172) of the lecture notes. Setting the arbitrary constant a_0 to zero, and taking into account that due to zero divergence of the vector $\mathbf{H}$, the constant b_0 is zero as well, this solution is reduced to

$$\phi_{\text{m}} = \sum_{l=1}^{\infty} \mathcal{P}_l(\cos\theta) \times \begin{cases} a_l r^l + b_l/r^{l+1}, & \text{for } r \leqslant R, \\ a_l' r^l + b_l'/r^{l+1}, & \text{for } R \leqslant r. \end{cases}$$

where R is sphere's radius, and $\mathcal{P}_l(\xi)$ are the Legendre polynomials (2.169). Next, since the field is produced by the sphere itself, it should vanish at large distances from it, so that all coefficients a_l' must equal zero. Similarly, the potential inside the cylinder has to be finite, so that all coefficients b_l have to vanish as well. As a result, the above expression is reduced to

$$\phi_{\text{m}} = \sum_{l=1}^{\infty} \mathcal{P}_l(\cos\theta) \times \begin{cases} a_l r^l, & \text{for } r \leqslant R, \\ b_l'/r^{l+1}, & \text{for } R \leqslant r. \end{cases}$$

In order to find the coefficients a_l and b_l', we have to write boundary conditions at the sphere's surface ($r = R$). From Eq. (5.117) for the tangential component of the field **H**, we obtain

$$\sum_{l=1}^{\infty} a_l R^l \mathcal{P}_l(\cos\theta) = \sum_{n=1}^{\infty} \frac{b_l'}{R^{l+1}} \mathcal{P}_l(\cos\theta) \qquad (*)$$

For the normal component of the field, we have to use the general equation (5.118), B_n = const (rather than Eq. (5.119), which is only valid for linear magnetics), because **B** is a linear function of **H** (equal to $\mu_0\mathbf{H}$) only outside the sphere, at $r \geqslant R$. At $r \leqslant R$ we should use the general relation (5.108), $\mathbf{B} = \mu_0(\mathbf{H} + \mathbf{M})$, for our geometry giving $B_n = \mu_0(H_n + M_n) = \mu_0(-\partial\phi_\mathrm{m}/\partial r + M_n)$. For the fixed, uniform polarization $\mathbf{M}_0$, $M_n = M_0 \cos\theta \equiv M_0\mathcal{P}_1(\cos\theta)$, so that the second boundary condition takes the form

$$\mu_0\left[-\sum_{l=1}^{\infty} l a_l R^{l-1}\mathcal{P}_l(\cos\theta) + M_0\mathcal{P}_1(\cos\theta)\right] = \mu_0\sum_{l=1}^{\infty}(l+1)\frac{b_l'}{R^{l+2}}\mathcal{P}_l(\cos\theta) \qquad (**)$$

Due to the linear independence of functions $\mathcal{P}_l(\cos\theta)$, Eqs. (*) and (**) fall apart into a set of independent couples of linear equations for each pair $\{a_l, b_l'\}$, but only for $l = 1$ such a pair of equations is inhomogeneous, i.e. gives a non-vanishing solution for coefficients a_1, b_1'. (All other coefficients *may* equal zero, and due to the uniqueness of the Laplace equation's solution, they *have to* equal zero.) Solving the system of two linear equations for $l = 1$, we obtain

$$a_1 = \frac{M_0}{3}, \qquad b_1' = \frac{M_0 R^3}{3}.$$

From here, the scalar potential of the magnetic field is

$$\phi_\mathrm{m}|_{r\leqslant R} = \frac{M_0}{3} r\cos\theta, \qquad \phi_\mathrm{m}|_{r\geqslant R} = \frac{M_0 R^3}{3r^2}\cos\theta.$$

The first of these expressions describes a uniform magnetic field with $\mathbf{H} = -\mathbf{M}/3$ and $\mathbf{B} = \mu_0(\mathbf{H} + \mathbf{M}_0) = (2/3)\mu_0\mathbf{M}_0$. (Note that the fields **H** and **B** have different directions!) The formula for the external potential shows that at $r > R$ the field exactly coincides with that of a magnetic dipole with the moment

$$\mathbf{m} = \frac{4\pi}{3}R^3\mathbf{M}_0 = V\mathbf{M}_0,$$

i.e. the sum of all elementary dipoles within the sphere—the result that could be conjectured even without the formal solution of the problem.

Note that this problem and its solution are very similar to problem 3.11 on a sphere with fixed electric polarization. Note also that the result may be also obtained by the method discussed in the next problem—though in this particular case this would involve a bulkier calculation.

***Problem* 5.20.** A limited volume V is filled with a magnetic material with a fixed (field-independent) magnetization $\mathbf{M}(\mathbf{r})$. Write explicit expressions for the magnetic field, induced by the magnetization, and its potential, and recast these expressions into the forms more convenient when $\mathbf{M}(\mathbf{r}) = \mathbf{M}_0 = \text{const}$ inside the volume V.

Solution: As was discussed in section 5.6 of the lecture notes, Eq. (5.132) for the magnetic field,

$$\nabla \cdot \mathbf{H} = -\nabla \cdot \mathbf{M},$$

is mathematically identical to the Maxwell equation (1.27) for the electric field:

$$\nabla \cdot \mathbf{E} = \frac{\rho}{\varepsilon_0}.$$

As was discussed in section 1.2 of the lecture notes, the last expression is equivalent to Eq. (1.9):

$$\mathbf{E}(\mathbf{r}) = \frac{1}{4\pi}\int_V \frac{\rho(\mathbf{r}')}{\varepsilon_0}\frac{\mathbf{r} - \mathbf{r}'}{|\mathbf{r} - \mathbf{r}'|^3}d^3r', \qquad (*)$$

for the field created by the electric charge contained inside volume V. Hence, if the only source of field $\mathbf{H}$ in a system is the magnetization of volume V, we may use this analogy to write

$$\mathbf{H}(\mathbf{r}) = \frac{1}{4\pi}\int_V \rho_\text{m}(\mathbf{r}')\frac{\mathbf{r} - \mathbf{r}'}{|\mathbf{r} - \mathbf{r}'|^3}d^3r', \qquad \text{where } \rho_\text{m}(\mathbf{r}') \equiv -\nabla \cdot \mathbf{M}(\mathbf{r}'). \qquad (**)$$

This means that in the free space outside of volume V,

$$\mathbf{B}(\mathbf{r}) = \mu_0\mathbf{H}(\mathbf{r}) = \frac{\mu_0}{4\pi}\int_V \rho_\text{m}(\mathbf{r}')\frac{\mathbf{r} - \mathbf{r}'}{|\mathbf{r} - \mathbf{r}'|^3}d^3r',$$

so that for this region the (minus) divergence of magnetization plays the role of the effective magnetic charge density. Note, however, that according to Eq. (5.108), inside volume V, the field $\mathbf{B}$ is different:

$$\mathbf{B}(\mathbf{r}) = \mu_0[\mathbf{M}(\mathbf{r}) + \mathbf{H}(\mathbf{r})] = \mu_0\mathbf{M}(\mathbf{r}) + \frac{\mu_0}{4\pi}\int_V [-\nabla \cdot \mathbf{M}(\mathbf{r}')]\frac{\mathbf{r} - \mathbf{r}'}{|\mathbf{r} - \mathbf{r}'|^3}d^3r'.$$

Also, as we know from chapter 1.3 of the lecture notes, Eq. (*) is equivalent to Eq. (1.38),

$$\phi(\mathbf{r}) = \frac{1}{4\pi}\int_V \frac{\rho(\mathbf{r}')}{\varepsilon_0}\frac{1}{|\mathbf{r} - \mathbf{r}'|}d^3r',$$

for the electrostatic potential defined by Eq. (1.33):

$$\mathbf{E} = -\nabla\phi.$$

This means that the magnetic field potential, defined by the similar Eq. (5.120)[30],

$$\mathbf{H} = -\nabla\phi_{\rm m},$$

may be calculated as

$$\phi_{\rm m}(\mathbf{r}) = \frac{1}{4\pi}\int_V [-\nabla \cdot \mathbf{M}(\mathbf{r}')]\frac{1}{|\mathbf{r} - \mathbf{r}'|}d^3r' \equiv \frac{1}{4\pi}\int_V \rho_{\rm m}(\mathbf{r}')\frac{1}{|\mathbf{r} - \mathbf{r}'|}d^3r'. \qquad (***)$$

As was also discussed in section 5.6, if $\mathbf{M}(\mathbf{r}) = \mathbf{M}_0$ = constant inside volume V, and $\mathbf{M}(\mathbf{r}) = 0$ outside it, the effective magnetic charge density $\rho_{\rm m} = -\nabla \cdot \mathbf{M}$ is different from zero (and formally infinite) only on the surface S of the volume. Hence it may be characterized by the effective surface charge density $\sigma_{\rm m}$, which may be calculated by the integration of $\rho_{\rm m}$ along the local coordinate n normal to the surface, from just inside the surface to just outside it (i.e. out of volume V):

$$\sigma_{\rm m} \equiv \int_{n_s-0}^{n_s+0} \rho_{\rm m}\, dn = -\int_{n_s-0}^{n_s+0} \nabla \cdot \mathbf{M}\, dn = -\int_{n_s-0}^{n_s+0} \frac{\partial M_n}{\partial n}\, dn = M_n\Big|_{n_s-0} = M_0 \cos\theta,$$

where θ is the angle between the direction of vector $\mathbf{M}_0$ and the outer normal $\mathbf{n}$ to the surface. (This just a more detailed derivation of Eq. (5.134) of the lecture notes.) This means that the effective surface charge is positive at the points where vector $\mathbf{M}_0$ is directed out of the magnetic material, negative in the opposite case, and equals zero at the points where $\mathbf{M}_0$ is parallel to the surface—see the figure below.

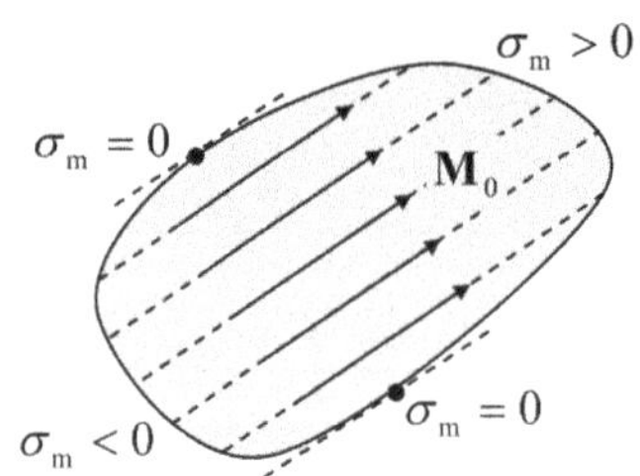

In this case, Eq. (**) takes the form

$$\mathbf{H}(\mathbf{r}) = \frac{1}{4\pi}\int_S \sigma_{\rm m}(\mathbf{r}')\frac{\mathbf{r} - \mathbf{r}'}{|\mathbf{r} - \mathbf{r}'|^3}d^2r', \qquad (****)$$

and Eq. (***) the form

$$\phi_{\rm m}(\mathbf{r}) = \frac{1}{4\pi}\int_S \sigma_{\rm m}(\mathbf{r}')\frac{1}{|\mathbf{r} - \mathbf{r}'|}d^3r'.$$

In the ultimate case of thin, long needle magnets (see figure 5.19 of the lecture notes), these expressions, for the free space outside the magnets, are reduced to the analog of Coulomb law for point charges—see Eq. (5.138) of the lecture notes.

[30] As a (I hope, unnecessary) reminder, this expression is only valid in the absence of vortex-like field induced by stand-alone currents—the condition satisfied in our case of the field induced by magnetization $\mathbf{M}(\mathbf{r})$.

***Problem* 5.21.** Use the results of the previous problem to calculate the distribution of the magnetic field **H** along the axis of a straight permanent magnet of length $2l$, with round cross-section of radius R, and a uniform magnetization $\mathbf{M}_0$ parallel to the axis—see the figure below.

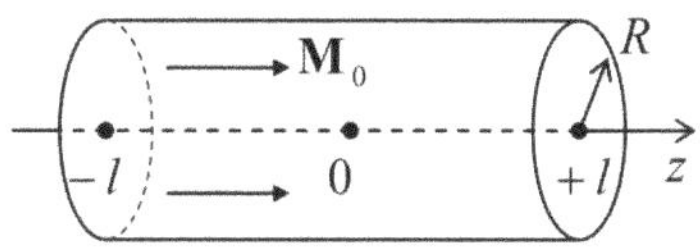

Solution: Due to the axial symmetry of the problem, the magnetic field on the axis has to be directed along it: $\mathbf{H} = \mathbf{n}_z H_z$. This is why it is sufficient to use the z-component of Eq. (****) of the model solution of the previous problem:

$$H_z(z) = \frac{1}{4\pi}\int_S \sigma_\text{m}(\mathbf{r}')\frac{z - z'}{|\mathbf{r} - \mathbf{r}'|^3}d^2r'.$$

Since the vector $\mathbf{M}_0$ is directed along the magnet's axis, the effective surface density σ_m of the effective magnetic charge is different from zero only on the magnet's ends, where it equals $\pm M_0$, so that the integration may be extended only over the ends' areas. Taking the magnet's center for $z = 0$ (so that the positions of its ends are $z = \pm l$, as shown in the figure above), we obtain

$$H_z(z) = \frac{M_0}{4\pi}\left\{(z - l)\int_0^R \frac{2\pi\rho d\rho}{\left[(z - l)^2 + \rho^2\right]^{3/2}} - (z + l)\int_0^R \frac{2\pi\rho d\rho}{\left[(z + l)^2 + \rho^2\right]^{3/2}}\right\},$$

where ρ is the distance of the elementary source point $\mathbf{r}'$ from the symmetry axis. These two integrals may be readily worked out by similar substitutions: respectively, $\xi \equiv (z \pm l)^2 + \rho^2$, so that in both cases $2\pi\rho d\rho = \pi d\xi$,

$$H_z(z) = \frac{M_0}{4\pi}\left[(z - l)\int_{(z-l)^2}^{(z-l)^2+R^2} \frac{\pi d\xi}{\xi^{3/2}} - (z + l)\pi\int_{(z+l)^2}^{(z+l)^2+R^2} \frac{\pi d\xi}{\xi^{3/2}}\right]$$
$$= \frac{M_0}{2}\left[-\frac{(z - l)}{\left[(z - l)^2 + R^2\right]^{1/2}} + \text{sgn}\,(z - l) + \frac{(z + l)}{\left[(z + l)^2 + R^2\right]^{1/2}} - \text{sgn}\,(z + l)\right].$$

This function is plotted in the figure below for three representative values of the R/l ratio. Please note, first of all, the reversal of the direction of field **H** at each end of the magnet—a result which should not be too surprising after looking at Eq. (5.134) of the lecture notes and the solution of the similar problem 5.19 for a sphere. Note that the magnitude of **H** is always below M_0 at all points, so that the field $\mathbf{B} = \mu_0(\mathbf{H} + \mathbf{M})$ is still directed along the magnetization vector: $B_z > 0$ for all z.

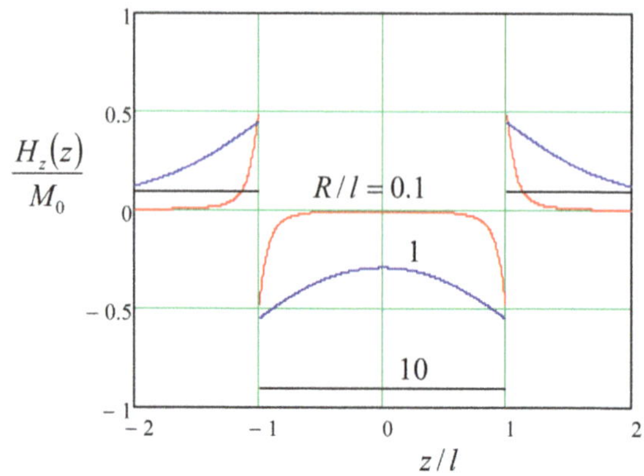

At $R \ll l$, **H** is just a sum of Coulomb-like fields of two equal and opposite point 'magnetic charges' $q_m = \pm M_0 A = M_0 \pi R^2$. In the opposite limit, the field inside the magnet is nearly constant, $H_z \approx -M_0$, while outside it is very small, and drops only at larger distances $|z| \sim R \gg l$. In one more limit, $|z| \gg R, l$, the result is reduced to

$$H_z = \frac{M_0 R^2 l}{2z^3}, \qquad \text{i.e. to } B_z = \mu_0 \frac{M_0 R^2 l}{2z^3} \equiv \frac{\mu_0}{4\pi} \frac{2(M_0 \pi R^2 l)}{z^3}.$$

This is just the field (5.99) of a magnetic dipole, with the dipole moment $\mathbf{m} = \mathbf{M}_0 V$, where $V = \pi R^2 l$ is magnet's volume, on its axis.

***Problem* 5.22.** A very broad film of thickness $2t$ is permanently magnetized normally to its plane, with a periodic checkerboard pattern, with the square of area $a \times a$:

$$\mathbf{M}\Big|_{|z|<t} = \mathbf{n}_z M(x, y), \qquad \text{with } M(x, y) = M_0 \times \operatorname{sgn}\left(\cos\frac{\pi x}{a} \cos\frac{\pi y}{a}\right).$$

Calculate the magnetic field's distribution in space[31].

Solution: Due to the absence of stand-alone currents in the system, we may describe the magnetic field in it using the scalar potential ϕ_m, defined by Eq. (5.120) of the lecture notes:

$$\mathbf{H} = -\nabla \phi_\mathrm{m}. \qquad (*)$$

According to Eq. (5.121), in each of the three uniform z-regions ($z < -t$, $-t < z < +t$, and $t < z$) the potential has to satisfy the Laplace equation. Taking into account the symmetry of the problem (which requires in particular that $\mathbf{H}(-z) = \mathbf{H}(z)$, so that if $\phi_\mathrm{m}(0)$ is set to zero, then $\phi_\mathrm{m}(-z) = \phi_\mathrm{m}(z)$), and its $a \times a$-periodicity is in the x–y

[31] This problem is of evident relevance for the *perpendicular magnetic recording* (PMR) technology, which currently dominates the field of high-density digital magnetic recording.

plane, it is natural to look for the solution in the following variable-separated series form (very similar to that in Eq. (2.95)):

$$\phi_{\rm m}(x, y, z) = \sum_{n,m=1}^{\infty} c_{nm} \cos\frac{\pi n x}{a} \cos\frac{\pi m y}{a} \times \begin{cases} \operatorname{sgn}(z) \exp\{-\gamma_{nm}|z|\}/\exp\{-\gamma_{nm}t\}, & \text{for } |z| \geqslant t, \\ \sinh\gamma_{nm} z/\sinh\gamma_{nm}t, & \text{for } |z| \leqslant t. \end{cases} \qquad (**)$$

(The denominators in these expressions are set up to have the first boundary condition at film's surfaces, the potential's continuity, satisfied for any choice of coefficients c_{nm}.) Indeed, the substitution of any term of the series to the Laplace equation shows that it is satisfied if the separation constant γ_{nm} obeys a relation similar to Eq. (2.93):

$$\gamma_{nm} = \frac{\pi}{a}\left(n^2 + m^2\right)^{1/2}. \qquad (***)$$

According to Eq. (*), the distribution (**) corresponds to the following normal component of the magnetic field on the film's top surface[32]:

$$H_z(x, y, t \pm 0) = -\left.\frac{\partial \phi_{\rm m}}{\partial z}\right|_{z=t\pm 0} = \sum_{n,m=1}^{\infty} c_{nm}\gamma_{nm} \cos\frac{\pi n x}{a} \cos\frac{\pi m y}{a} \times \begin{cases} 1, & \text{for } z = t + 0, \\ (-\cosh\gamma_{nm}t/\sinh\gamma_{nm}t), & \text{for } z = t - 0. \end{cases}$$

On the other hand, Eq. (5.134) of the lecture notes, applied to the film's surface at $z = t$, gives us the following second boundary condition:

$$H_z(x, y, t + 0) - H_z(x, y, t - 0) = M(x, y),$$

so that, for our particular magnetization pattern $M(x, y)$, the coefficients c_{nm} have to be found from the following system of equations:

$$\sum_{n,m=1}^{\infty} c_{nm}\gamma_{nm} \cos\frac{\pi n x}{a} \cos\frac{\pi m y}{a}\left(1 + \frac{\cosh\gamma_{nm}t}{\sinh\gamma_{nm}t}\right) = M_0 \times \operatorname{sgn}\left(\cos\frac{\pi x}{a}\cos\frac{\pi y}{a}\right).$$

As usual, let us multiply both parts of this equation by a function of the same orthogonal set, in this case $\cos(\pi n' x/a)\cos(\pi m' x/a)$, with arbitrary integer n' and m', and then integrate them over the x–y period of the structure, say over the square $0 \leqslant x,y \leqslant a$. In the left-hand side, only terms with $n = n'$ and $m = m'$ survive, and we obtain[33]

---

[32] The z-asymmetry of Eq. (**) automatically takes care of the boundary condition at $z = -t$.

[33] The integration on the right-hand side is easy using the fact that within our integration area, our sgn function equals (+1) if both x and y are either between 0 and $a/2$, or between $a/2$ and a, and (–1) in the complementary regions.

$$c_{nm}\gamma_{nm}\left(1+\frac{\cosh \gamma_{nm}t}{\sinh \gamma_{nm}t}\right)\frac{a^2}{4}=M_0\int_0^a dx\int_0^a dy \cos\frac{\pi nx}{a}\cos\frac{\pi my}{a}$$

$$\times \operatorname{sgn}\left(\cos\frac{\pi x}{a}\cos\frac{\pi y}{a}\right)$$

$$\equiv M_0\frac{4a^2}{\pi^2 nm}\sin\frac{\pi n}{2}\sin\frac{\pi m}{2},$$

so that the coefficients c_{nm} depend on a only via γ_{nm}:

$$c_{nm}=\frac{16}{\pi^2 nm}M_0\frac{\sinh \gamma_{nm}t}{\gamma_{nm}(\sinh \gamma_{nm}t+\cosh \gamma_{nm}t)}\sin\frac{\pi n}{2}\sin\frac{\pi m}{2}. \qquad (****)$$

(Note that the last two factors make c_{nm} non-vanishing only if both integers n and m are odd, and also provide its sign alternation for even n and m, speeding up the convergence of the series (**).)

Formulas (*)–(****) give a complete solution of the field distribution problem. For example, the left panel in the figure below shows the resulting H_z as a function of z at the center of one of the checkerboard squares (say, $x = y = 0$), for several values of the film thickness. The jumps of the field by $\pm M_0$ at the film surfaces, and the resulting reversal of its sign, qualitatively similar to that in the previous problem, are clearly visible. From the point of view of digital magnetic recording applications, these plots imply that in order to avoid a significant drop of the magnetic-field 'readout signal', the film's half-thickness t should be larger than at least $\sim 0.3a$, where a^2 is the recorded bit area.

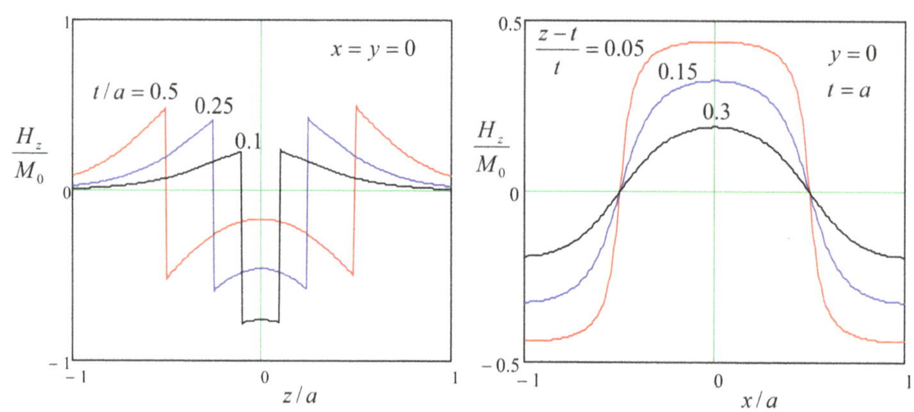

The right panel of the figure above shows H_z as a function of x at $y = 0$, for several values of z, for a sufficiently thick film ($t = a$). Both panels show that in order to avoid a substantial signal loss, the magnetic field sensing device (in modern technology, a GMR sensor) has to be rather close to the film's surface. Indeed, the distance $(z - t) \approx 0.3t$ already causes a more than two-fold drop of the field swing from its maximum value M_0.

Problem* 5.23. Based on the discussion of the quadrupole electrostatic lens in section 2.4 of the lecture notes, suggest permanent-magnet systems that may similarly focus particles moving close to the system's axis, and carrying:

(i) an electric charge,
(ii) no net electric charge, but a spontaneous magnetic dipole moment **m**.

Solutions:

(i) As was discussed in section 2.4, a quadrupole lens is a cylindrical system exerting a force **F** with Cartesian components

$$F_x = -cx, \qquad F_y = +cy \tag{*}$$

(where c is a constant), on a particle moving near and along axis z. The electrostatic system of four hyperbolically shaped electrodes, having this property in *all space* between them, is shown in figure 2.9 of the lecture notes. However, it is clear that any cylindrical system of four electrodes with a similar symmetry of their voltage biases, for example the one shown in the figure below, has the property (*) *near its center* (i.e. at $x, y \to 0$).

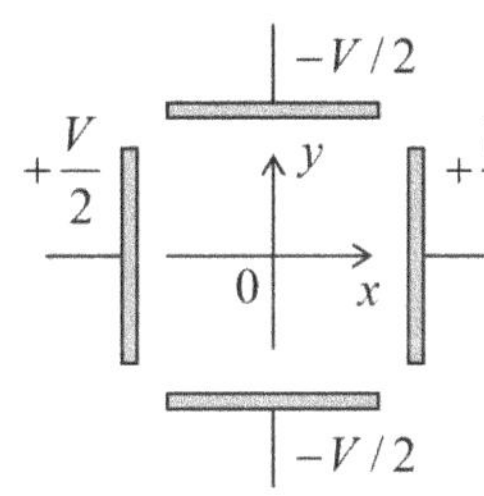

Indeed, due to the mirror symmetry of the system relative to planes $x = 0$ and $y = 0$, the electrostatic potential ϕ at the central line $x = y = 0$ should equal zero and have vanishing partial derivatives $\partial\phi/\partial x$ and $\partial\phi/\partial y$ (and actually all odd derivatives as well). Hence the Taylor expansion of the potential at that point has to start from quadratic terms, with equal but opposite coefficients for x and y:

$$\phi(x, y) = \kappa(x^2 - y^2) + O(\rho^4), \qquad \text{at } \rho \equiv (x^2 + y^2)^{1/2} \to 0,$$

with $\kappa \sim V/a^2$, where a is the spatial scale of the system's cross-section. (As a sanity check, the leading term of this expansion[34] satisfies the Laplace equation $\nabla^2\phi = 0$.) Now calculating the force $\mathbf{F} = q\mathbf{E} = -q\nabla\phi$ acting on a particle with an electric charge q, we get Eqs. (*) with $c = 2q\kappa$.

Now the analogy between the electrostatic field induced by surface electric charges, and the magnetic field induced by permanent magnets with the magnetization normal to their surface (see, e.g. the discussion following Eq. (5.132) of the

[34] Note that this term coincides with the exact potential distribution (2.75) in the hyperbolic-electrode system.

lecture notes and the model solutions of the three previous problems) implies that a similar distribution of the magnetic potential,

$$\phi_m(x, y) = \kappa_m(x^2 - y^2), \qquad \text{at } \rho \to 0, \qquad (**)$$

with $\kappa_m \propto M_0$, may be obtained, for example, in the system shown in the figure below. The magnetic field in the free space between the magnets, corresponding to this potential, equals $\mathbf{B} = -\mu_0\nabla\phi_m$, so that its Cartesian components are

$$B_x = -2\mu_0\kappa_m x, \quad B_y = 2\mu_0\kappa_m y.$$

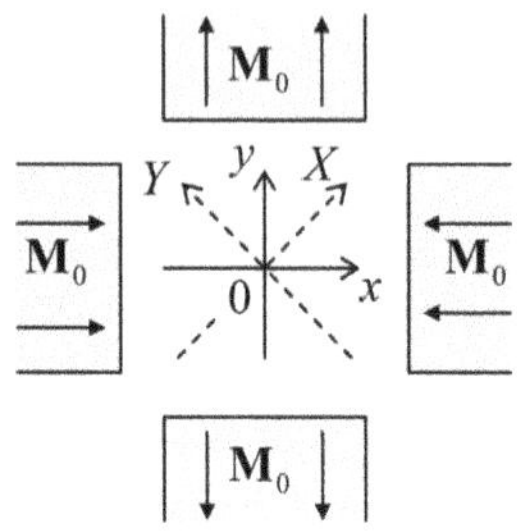

As a result, the Cartesian components of the magnetic Lorentz force $\mathbf{F} = q\mathbf{v} \times \mathbf{B}$ exerted on a charged particle, moving along the axis z, are

$$F_x = -qv_zB_y = -2\mu_0 qv_z\kappa_m y, \qquad F_y = qv_zB_x = -2\mu_0 qv_z\kappa_m x.$$

This result is different from Eqs. (*) not only by the signs, but also by field components' proportionality to different coordinates. However, relative to the Cartesian coordinates $\{X, Y\}$ turned by $\pi/4$ relative to $\{x, y\}$, the force has the required properties. Indeed, writing the relations between the two pairs of coordinates, which are evident from the figure above,

$$X = \frac{x + y}{\sqrt{2}}, \quad Y = \frac{-x + y}{\sqrt{2}}; \qquad \text{so that } x = \frac{X - Y}{\sqrt{2}}, \quad y = \frac{X + Y}{\sqrt{2}},$$

we may use them to recast Eq. (**) as

$$\phi_m = \kappa_m(x^2 - y^2) = \frac{\kappa_m}{2}[(X - Y)^2 - (X + Y)^2] = -2\kappa_m XY,$$

so that

$$B_X = 2\mu_0\kappa_m Y, \quad B_Y = 2\mu_0\kappa_m X, \qquad (***)$$

and the expressions for Lorentz force components,

$$F_X = -qv_zB_Y = -2\mu_0 qv_z\kappa_m X, \qquad F_Y = qv_zB_X = 2\mu_0 qv_z\kappa_m Y,$$

have the same functional form as in Eq. (*), with the coefficient c replaced with $2\mu_0 qv_z\kappa_m$.

(ii) According to Eq. (5.102) of the lecture notes, the force acting on magnetic dipole **m** in the external magnetic field **B** is

$$\mathbf{F} = \nabla(\mathbf{m} \cdot \mathbf{B}). \qquad (****)$$

For the quadrupole magnetic field (***), this force is coordinate-independent, i.e. does not have the required focusing properties. A clue in the search for a suitable system may be obtained by rewriting Eq. (**) in the complex form:

$$\phi_\text{m} = \kappa_\text{m}\,\text{Re}(z^2), \qquad \text{with } z \equiv x + iy \equiv \rho e^{i\varphi},$$

where φ is the polar angle on the $[x, y]$ plane. Now let us consider a different magnetic potential

$$\phi_\text{m} = \text{Re}(\lambda z^3) \equiv |\lambda|\rho^3 \cos(3\varphi + \theta),$$

where λ is some complex constant, and θ is its argument. Such function provides the extra power of z, necessary to make the force (****) proportional to particle's deviation from system's axis. Indeed, calculating the radial component of the force (****)[35], we obtain

$$F_\rho = -6\mu_0|\lambda|\rho\left[m_\rho \cos(3\varphi + \theta) + m_\theta \sin(3\varphi + \theta)\right].$$

This expression shows that the component is indeed proportional to the distance ρ of particle's trajectory from the system's axis, so that it indeed provides beam focusing, with the focal distance depending both on the angle φ and the dipole moment's orientation.

In addition, according to the discussion in section 2.4 of the lecture notes, any analytical complex function $f(z)$, in particular z^3, does satisfy the Laplace equation—and so does it real part. Hence, the above distribution $\phi_\text{m}(\rho, \varphi)$ may be obtained using some permanent magnet system. The form of this distribution implies that the system should be symmetric with respect to the rotation by $\Delta\varphi = 2\pi/3$ (rather than by $\pi \equiv 2\pi/2$ as the quadrupole system shown in the figure above), i.e. needs 3 rather than 2 pairs of magnets of alternating polarity—for example as shown in the figure below.

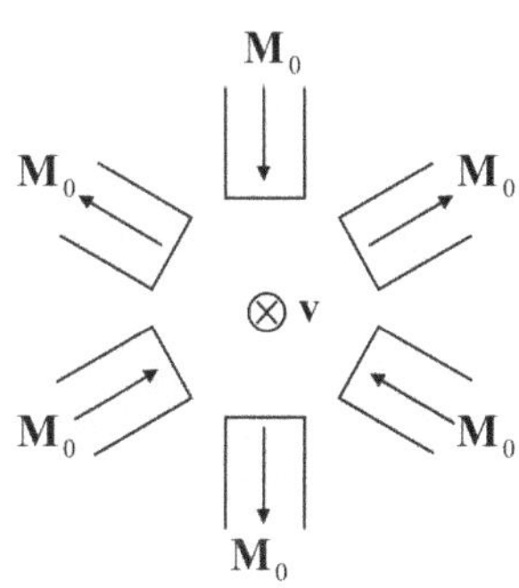

[35] See, e.g. Eq. (A.60).

Such *sextupole* (also called 'hexapole') magnet systems are actually used in experiment, in particular, for focusing of beams on neutrons—neutral particles with non-zero spins and hence non-zero dipole moments.

Reference

[1] Peierls R 1955 *Quantum Theory of Solids* (Oxford: Oxford University Press) pp 144–9

Chapter 6

Electromagnetism

***Problem* 6.1.** Prove that the electromagnetic induction e.m.f. $\mathscr{V}_{\text{ind}}$ in a conducting loop may be measured as shown on two panels of figure 6.1 of the lecture notes:

(i) by measuring the current $I = \mathscr{V}_{\text{ind}}/R$ induced in the closed loop with Ohmic resistance R, or
(ii) using a voltmeter inserted into the loop.

Solutions:

(i) The issue is not quite trivial, because initially the resistance was defined (see section 4.2) as the ratio V/I, where V is a *charge-induced* voltage, in its turn defined as the difference of electrostatic potentials ϕ at the resistor's ends—see Eq. (2.25) of the lecture notes. On the other hand, for the vortex field $\mathbf{E}_{\text{ind}}$ of the electromagnetic induction, such a representation is generally impossible. Moreover, the Faraday law does not prescribe the unique spatial distribution of the field—rather it only specifies its contour integral (6.2)—or, equivalently, the curl at each point—see Eq. (6.5). The exact particular distribution of $\mathbf{E}_{\text{ind}}$ depends not only on the inducing magnetic field $\mathbf{B}(\mathbf{r}, t)$, but also on the geometry and properties of the system.

Note, however, that for the 'usual' (charge-induced, potential) field $\mathbf{E}$, we may represent the voltage drop at a resistor as the integral,

$$V = \int_{\text{A}}^{\text{B}} \mathbf{E} \cdot d\mathbf{r},$$

along a path connecting the resistor's terminals A and B. Now using the differential form (4.8) of the Ohm law, we may write

$$\int_{\text{A}}^{\text{B}} \mathbf{E} \cdot d\mathbf{r} = \int_{\text{A}}^{\text{B}} \frac{\mathbf{j}(\mathbf{r}) \cdot d\mathbf{r}}{\sigma}.$$

The results of section 4.3 show that the normalized law,

$$\mathbf{f}(\mathbf{r}) \equiv \frac{\mathbf{j}(\mathbf{r})}{I}, \qquad (*)$$

of the current distribution in a conductor is uniquely defined by Eq. (4.6), equivalent to $\nabla \cdot \mathbf{f} = 0$, with the boundary conditions $f_n = 0$ on the conductor's surface, and $f_\tau = 0$ on its interfaces with the external electrodes, so that we may use Eq. (*) to express the above integral as

$$\int_A^B \mathbf{E} \cdot d\mathbf{r} = IR, \quad \text{with } R \equiv \int_A^B \frac{\mathbf{f}(\mathbf{r}) \cdot d\mathbf{r}}{\sigma}, \qquad (**)$$

the last equality giving just another definition of sample's resistance.

Now moving on to the electromagnetic induction, we may argue that since the current distribution law $\mathbf{f}(\mathbf{r})$, and hence the resistance R, are governed by relations including $\mathbf{j}$ alone, they do not depend on the nature of the applied electric field. Hence in the particular case of a conductor, it dictates the current distribution inside it:

$$\mathbf{E}_{\text{ind}} = \frac{\mathbf{j}(\mathbf{r})}{\sigma} = \frac{I}{\sigma}\mathbf{f}(\mathbf{r}), \qquad (***)$$

leaving Eq. (6.2) to govern only the magnitude of the current. (This internal distribution, in turn, dictates that of the field $\mathbf{E}_{\text{ind}}(\mathbf{r})$ outside the conductor, following the conductivity hierarchy that was discussed in section 4.3.)

Now let us consider a closed resistive loop, which may be represented as the result of a merger of the terminals A and B. Although such a closed loop does not have external electrodes, the condition $j_\tau = 0$ discussed above is still valid on an arbitrary cross-section surface perpendicular, in each point, to the vector $\mathbf{j}$. (This cross-section may be understood as a galvanic connection of two ultimately thin external electrodes.) So, all the above relations are valid for such a closed loop as well, so that integrating Eq. (***) along it, and then using Eq. (**), we obtain

$$\mathscr{V}_{\text{ind}} \equiv \oint_C \mathbf{E}_{\text{ind}} \cdot d\mathbf{r} = \frac{I}{\sigma}\oint_C \mathbf{f}(\mathbf{r}) \cdot d\mathbf{r} \equiv IR,$$

thus giving the required proof of the relation $I = \mathscr{V}_{\text{ind}}/R$.

(ii) The prevailing *electrodynamic* species of voltmeters[1] are essentially sensitive galvanometers that measure the weak current I induced inside them by the voltage under measurement, recalibrated into the voltmeter readout as

$$V = R_{\text{int}} I,$$

where R_{int} is the voltmeter's internal resistance—which is typically much higher that the characteristic resistance of the circuit under measurement. As it follows from the solution of task (i), for our system (voltmeter inserted into the loop) we may write

$$I = \frac{\mathscr{V}_{\text{ind}}}{R_{\text{int}} + R_{\text{ext}}},$$

[1] The alternative, *electrostatic voltmeter* species, will be briefly discussed in *Part SM* section 6.5. Those voltmeters are typically slow, and are not used for measurement of such transient effects as the electromagnetic induction.

where R_{ext} is the resistance of the loop itself (external to the voltmeter). Combining these two formulas, we obtain

$$V = \frac{R_{int}}{R_{int} + R_{ext}} \mathscr{V}_{ind}.$$

This result shows that if the internal resistance of the voltmeter is appropriately high, $R_{int} \gg R_{ext}$, it indeed measures the induced e.m.f. faithfully. The situation, however, changes if a voltmeter does not interrupt the induction loop, but is rather connected in parallel to its part—see, for example, the next problem.

***Problem* 6.2.** The flux Φ of the magnetic field that pierces a resistive ring is being changed in time, while the magnetic field outside of the ring is negligibly low. A voltmeter is connected to a part of the ring, as shown in the figure below. What would the voltmeter show?

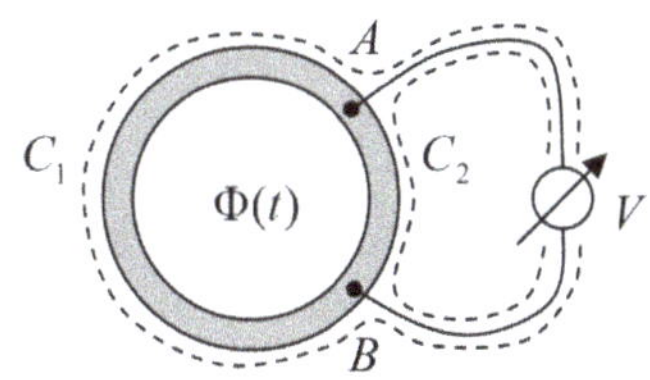

Solution: A naive application of the Faraday induction law (6.2) to the contour C_1 (see the external dashed line in the figure above) would give

$$V = \oint_{C_1} \mathbf{E}_{ind} \cdot d\mathbf{r} = -\dot{\Phi}. \qquad \text{(WRONG!)}$$

The fallacy of this argument becomes evident if we notice that it might be equally applied to the contour C_2, giving a totally different answer:

$$V = \oint_{C_2} \mathbf{E}_{ind} \cdot d\mathbf{r} = 0. \qquad \text{(WRONG!)}$$

The reason for these errors is the lack of account for the Ohmic voltage drops on the current-carrying parts of the ring.

The simplest way to obtain the correct answer is to use the *equivalent lumped circuit* language, which is broadly used in electrical engineering, and was briefly discussed in sections 4.1 and 6.6 of the lecture notes. First, note that in the absence of the voltmeter the system is similar to that analyzed in the previous problem, so that according to its solution the current I_r in the ring may be expressed as

$$I_r = \frac{\mathscr{V}_{ind}}{R} = \frac{\mathscr{V}_{ind}}{R_1 + R_2}, \quad \text{with } \mathscr{V}_{ind} \equiv \oint_{ring} \mathbf{E}_{ind} \cdot d\mathbf{r},$$

where R_1 and R_2 are the resistances of the two parts of the ring, separated by the contact points A and B, i.e. running along the contours C_1 and C_2, correspondingly. This result may be obtained from the equivalent circuit shown in panel (a) of the figure below, where each circle means a perfect voltage source, i.e. the imaginary two-terminal device that maintains the voltage equal to the specified e.m.f. (in this case, $\mathscr{V}_{\text{ind}}$) between its terminals, regardless of the external circuit—see the model solution of problem 4.2.

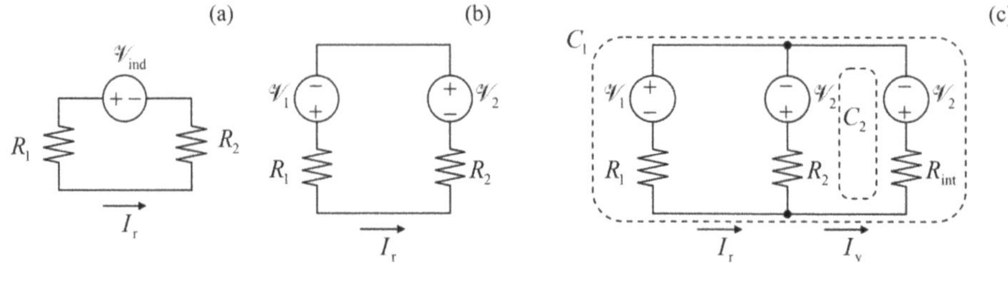

Since the integral defining $\mathscr{V}_{\text{ind}}$ may be always broken into two parts, corresponding to the two fragments of the ring,

$$\mathscr{V}_{\text{ind}} = \mathscr{V}_1 + \mathscr{V}_2, \quad \text{where } \mathscr{V}_1 \equiv \int_{\substack{A \\ \text{(along } C_1)}}^{B} \mathbf{E}_{\text{ind}} \cdot d\mathbf{r}, \quad \mathscr{V}_2 \equiv \int_{\substack{B \\ \text{(along } C_2)}}^{A} \mathbf{E}_{\text{ind}} \cdot d\mathbf{r},$$

the equivalent circuit (a) may be transformed into another one, shown in panel (b). The latter circuit gives the same current as the former one (hence the term 'equivalent'), but has the advantage that it may be readily generalized to the case of the voltmeter connected between the points A and B—see panel (c). The only non-trivial component of the last circuit is the additional source of the same e.m.f. $\mathscr{V}_2$ in the voltmeter branch, which is necessary to satisfy the Faraday induction law for the contour C_2:

$$\oint_{C_2} \mathbf{E}_{\text{ind}} \cdot d\mathbf{r} = \mathscr{V}_2 - \mathscr{V}_2 = 0.$$

Note that there is nothing unphysical in that additional e.m.f. source, because the induced electric field $\mathbf{E}_{\text{ind}}$ is *not* localized in the magnetic field region (only its curl is—see Eq. (6.5) of the lecture notes), and may extend to the whole space. For example, if the ring and the magnetic field distribution in our current problem are axially symmetric, so should be the field $\mathbf{E}_{\text{ind}}$,[2] and Eq. (6.2) is satisfied with the following solution

$$\mathbf{E}_{\text{ind}} = \mp\frac{|\dot{\Phi}|}{2\pi\rho}\mathbf{n}_{\varphi},$$

where ρ is the distance from system's axis, and the sign depends on the magnetic field's direction. This field extends to infinity and induces the e.m.f. not only in the ring, but also in the voltmeter's wires, so that in the contour C_2, they exactly compensate each other.

Now writing the usual Kirchhoff laws for the circuit (c) (see the discussion in sections 4.1 and 6.5) and solving the resulting system of equations for the voltmeter

[2] Strictly speaking, this is true in the absence of field perturbation by voltmeter's wires; this perturbation may be minimized by running wire fragments either exactly along, or exactly perpendicular to the concentric circles.

current I_v, we obtain a somewhat bulky result, which in the most important limit $R_{int} \gg R_{1,2}$ (only in this case the galvanometer may be indeed called a voltmeter) is reduced to a simple form[3]

$$I_v = \frac{\mathscr{V}_{ind}}{R_{int}} \frac{R_2}{R_1 + R_2}, \quad \text{for } R_i \gg R,$$

so that the voltmeter readout is

$$V \equiv I_v R_{int} = \frac{R_2}{R_1 + R_2} \mathscr{V}_{ind} \leqslant \mathscr{V}_{ind}.$$

Note that if the voltmeter was connected to the same points A and B, but physically located on the other side of the ring (see the figure below), the measurement result would be different:

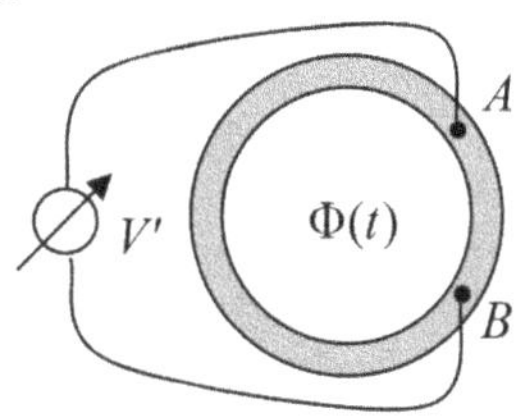

$$V' = -\frac{R_1}{R_1 + R_2} \mathscr{V}_{ind},$$

the minus sign being valid if the positive and negative terminals of the voltmeter are still connected to the same points of the ring. This sign difference enables popular and spectacular lecture demonstrations, in which two voltmeters, placed on opposite sides of the ring and connected simultaneously to the same points, with the same polarity, show readouts of opposite signs.

***Problem* 6.3.** A weak, uniform magnetic field **B** is applied to a small, axially-symmetric permanent magnet, with the dipole magnetic moment **m** directed along the symmetry axis, rapidly rotating about the same axis, with an angular momentum **L**. Calculate the electric field resulting from the field's application, and formulate the conditions of your result's validity.

Solution: According to Eq. (5.101) of the lecture notes, the magnetic field application creates the torque

$$\boldsymbol{\tau} = \mathbf{m} \times \mathbf{B}$$

[3] Actually the calculation may be much simplified by the replacement of the connection in series in each branch with the parallel connection of the same resistor and a perfect current generator—see the model solution of problem 4.2.

exerted on the dipole. In accordance with the basic law of rotational dynamics, $\dot{\mathbf{L}} = \boldsymbol{\tau}$.[4] Since, according to problem's conditions, the vectors **L** and **m** maintain the same direction (along the symmetry axis of the magnet), we may set $\mathbf{m} = \gamma\mathbf{L}$.[5] As a result, the dipole moment changes in time as

$$\dot{\mathbf{m}} = \gamma\dot{\mathbf{L}} = \gamma\boldsymbol{\tau} = \gamma\mathbf{m} \times \mathbf{B}. \tag{*}$$

According to the basic classical kinematics[6], this equation describes an additional, relatively slow rotation (the so-called *torque-induced precession*) of the vector **m**, and hence the magnet's symmetry axis, with the angular velocity

$$\Omega = \gamma B,$$

about the direction of the magnetic field's vector **B**, maintaining the initial angle θ between these two vectors. Directing the axis z along the magnetic field, we may describe this rotation as

$$\begin{aligned} &\mathbf{m} = \mathbf{m}_z + \mathbf{m}_\rho(t), \quad \text{with} \quad \mathbf{m}_z = \mathbf{n}_z m \cos\theta, \\ &\mathbf{m}_\rho(t) = \mathrm{Re}\left[(\mathbf{n}_x - i\mathbf{n}_y)\sin\theta\, e^{i\varphi}\right], \quad \text{and} \quad \varphi = \Omega t + \text{const}. \end{aligned} \tag{**}$$

Due to Faraday's induction, this rotation induces in the space around the magnet an electric field **E**, changing in time periodically with the same frequency Ω. In order to calculate the field, we may use the first of Eqs. (6.7) of the lecture notes, which, in the absence of an electrostatic potential ϕ, is reduced to

$$\mathbf{E} = -\frac{\partial \mathbf{A}}{\partial t}.$$

According to Eq. (5.90), the vector-potential of the magnetic dipole's field is

$$\mathbf{A} = \frac{\mu_0}{4\pi}\frac{\mathbf{m} \times \mathbf{r}}{r^3}, \tag{***}$$

so that the electric field at a fixed point **r** is

$$\mathbf{E} = -\frac{\mu_0}{4\pi}\frac{\dot{\mathbf{m}} \times \mathbf{r}}{r^3}.$$

Now using Eqs. (*)–(***), we finally obtain

$$\begin{aligned} \mathbf{E} &= \frac{\mu_0}{4\pi}\gamma\frac{(\mathbf{B} \times \mathbf{m}) \times \mathbf{r}}{r^3} = \frac{\mu_0}{4\pi}\Omega\frac{\left[\mathbf{n}_z \times \mathbf{m}_\rho(t)\right] \times \mathbf{r}}{r^3} \\ &= \frac{\mu_0}{4\pi}\Omega\frac{\sin\theta}{r^3}\mathrm{Re}\left[(\mathbf{n}_y - i\mathbf{n}_x) \times \mathbf{r} \exp\{i(\Omega t + \text{const})\}\right] \end{aligned}$$

---

[4] See, e.g. *Part CM*, Eq. (1.33).

[5] In atomic and nuclear physics, the coefficient γ is frequently a fundamental constant (see, e.g. Eq. (5.95) of the lecture notes), called the *gyromagnetic ratio*. It will be repeatedly discussed in *Part QM* of the lecture notes.

[6] See, e.g. *Part CM* section 4.1, in particular Eq. (4.8).

So, the oscillation of any Cartesian component of the field in time is exactly sinusoidal. Note also that the amplitude of these oscillations drops with the distance as $1/r^2$. This result is only valid in the quasi-static approximation; as will be extensively discussed in chapter 8 of the lecture notes, at larger distances, $r \gg c/\Omega$, Eq. (***) needs to be modified, leading to electromagnetic radiation, with the wave's electric and magnetic fields proportional to $1/r$.[7]

Another condition of validity of our result follows from the theory of torque-induced precession of such 'symmetric tops'[8]: the applied field **B** should be sufficiently low to keep the precession-induced addition $I\Omega$ to the basic angular momentum **L** negligibly small:

$$I\gamma B \ll L,$$

where I is the moment of inertia of the magnet's rotation about any axis normal to its symmetry axis.

***Problem* 6.4.** The similarity of Eq. (5.53), obtained in section 5.3 without any use of the Faraday induction law, and Eq. (5.54), proved in section 6.2 using the law, implies that the law may be derived from magnetostatics. Prove that this is indeed true for a particular case of a current loop, being slowly deformed in a fixed magnetic field **B**.

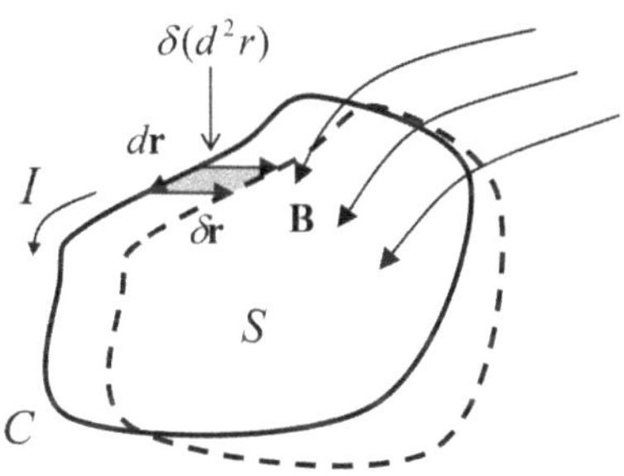

Solution: Consider a thin wire loop with current I, placed in a magnetic field **B**—see the figure above. According to Eq. (5.21) of the lecture notes, the magnetic force exerted by the field upon a small fragment dr of the wire is

$$d\mathbf{F} = I(d\mathbf{r} \times \mathbf{B}) \equiv -I(\mathbf{B} \times d\mathbf{r}),$$

where $d\mathbf{r}$ is a vector of length dr, tangential to the loop's contour and directed along the current I. Now let the wire be slightly (and slowly) deformed, so that this particular fragment is displaced by a small distance $\delta\mathbf{r}$. (Let me hope that the figure above makes the difference between the elementary vectors $d\mathbf{r}$ and $\delta\mathbf{r}$ absolutely clear.)

[7] See, in particular, Eqs (8.124)–(8.126). On the other hand, in quantum mechanics the torque-induced precession is quantized, and does not, as such, lead to radiation; only quantum transitions between the quantized precession states (crudely, between quantized values of $m_z = m\cos\theta$) lead to the emission of photons of frequency Ω.

[8] See, e.g. *Part CM* section 4.5, and in particular Eqs (4.73) and (4.85) and their discussion.

In order to keep the wire's acceleration negligibly small and thus neglect the kinetic energy of the system, some *external* (non-magnetic) forces should balance the force $d\mathbf{F}$, i.e. provide an equal and opposite force. With the account of this opposite direction, the work of these external forces at the displacement $\delta\mathbf{r}$, and hence the change of the magnetic potential energy U of the system, is

$$\delta(dU) = -d\mathbf{F} \cdot \delta\mathbf{r} = I\delta\mathbf{r} \cdot (\mathbf{B} \times d\mathbf{r}).$$

Let us apply to this mixed product the operand rotation rule of the vector algebra[9], so that the vector $\mathbf{B}$ comes out of the vector product:

$$\delta(dU) = I\mathbf{B} \cdot (d\mathbf{r} \times \delta\mathbf{r}). \qquad (*)$$

But the magnitude of the vector product in the parentheses is nothing more than the area $\delta(d^2r) \equiv \delta(dS)$ swept by the wire's fragment at the deformation (see the figure above again), while its direction is perpendicular to this elementary area dS, along the 'proper' normal vector $\mathbf{n} = (d\mathbf{r}/dr) \times (\delta\mathbf{r}/\delta r)$. The scalar multiplication of $\mathbf{B}$ by this vector is equivalent to taking its normal component. Hence, integrating Eq. (*) over all the wire length, we obtain the following result for the total variation of the magnetic energy of the system:

$$\delta U = I\oint_C B_n \delta(d^2r).$$

If $\mathbf{B}$ is fixed, i.e. does not change at the loop's deformation, the variation sign may be moved out from the integral, and we obtain

$$\delta U = I\delta\Phi, \qquad (**)$$

where Φ is the magnetic flux through the loop.

Now let the work $\delta\mathscr{W}$, necessary for this energy change, come from a generator of an external voltage V_{ext}, inserted somewhere in the loop. In order for the system to stay in quasi-equilibrium during the slow deformation of the loop, this voltage should counter-balance the electromagnetic induction's e.m.f.: $V_{ext} = -\mathscr{V}_{ind}$. The work of the voltage at the transfer of charge $\delta Q = I\delta t$, during a small time interval δt, is

$$\delta\mathscr{W} = V_{ext}\delta Q = -\mathscr{V}_{ind}\delta Q = -\mathscr{V}_{ind} I\delta t.$$

Requiring this work to be equal to the potential energy change (**) it causes, we arrive at the Faraday induction law (6.2)—for this particular case only. As was discussed in section 6.1 of the lecture notes, the law is actually much broader.

***Problem* 6.5.** Could problem 5.1 (i.e. the analysis of the mechanical stability of the system shown in the figure below) be solved using potential energy arguments?

---

[9] See, e.g. Eq. (A.48).

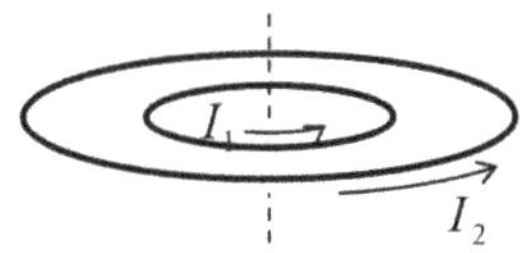

Solution: Since the currents are fixed, the stable equilibrium of the system corresponds to the minimum of the Gibbs potential energy given by Eqs. (6.17) and (6.20) of the lecture notes. In our non-magnetic case ($\mu = \mu_0$), it reads

$$U_G = -\frac{1}{2\mu_0}\int B^2 d^3r. \qquad (*)$$

Using the linear superposition principle, we may represent the vector **B** at each point as the sum $\mathbf{B}_1(\mathbf{r}) + \mathbf{B}_2(\mathbf{r})$, where each field is proportional to the corresponding current. Plugging this expression for **B** into Eq. (*), we may rewrite it as

$$U_G = U_1 + U_2 + U_{\text{int}},$$

where $U_k \propto I_k^2$, and only the last term,

$$U_{\text{int}} = -\frac{1}{\mu_0}\int \mathbf{B}_1(\mathbf{r}) \cdot \mathbf{B}_2(\mathbf{r}) d^3r, \qquad (**)$$

describing the energy of magnetic interaction of the loops, depends on their mutual orientation.

This energy is evidently smallest (and hence corresponds to system's equilibrium) if the directions of the fields $\mathbf{B}_1$ and $\mathbf{B}_2$ coincide in the regions where they are strongest, namely near the common center of the loops[10]. In particular, any turn of the inner loop by angle θ from the common axis decreases the magnitude of U_{int} by $\sim\cos\theta$, and hence increases the energy, so that the stable equilibrium corresponds to $\theta = 0$. (In the limit of a relatively small radius of the internal loop, when it may be well approximated by its dipole moment **m**, given by Eq. (5.97), whose direction coincides with that of the field $\mathbf{B}_1$ inside the ring, this conclusion is directly confirmed by Eq. (5.100) of the lecture notes.)

Similarly, if the inner loop is displaced from the plane of the outer loop in either direction along the common axis, the 'field overlap' integral in Eq. (**) decreases, so that the stable equilibrium corresponds to its coplanar position.

However, the analysis of the system's stability (or rather its instability :-) with respect to the lateral mutual displacement of the rings using the energy arguments is much less apparent than that using the interaction force arguments—see the model solution of problem 5.1.

[10] Strictly speaking, the individual fields $\mathbf{B}_1$ and $\mathbf{B}_2$ are the largest in close vicinity to the corresponding wires, but since here the field lines are closed circles, i.e. the regions with equal and opposite values of the scalar product $\mathbf{B}_1 \cdot \mathbf{B}_2$ are virtually equal in volume, they give a negligible net contribution to U_{int}.

***Problem* 6.6.** Use energy arguments to calculate the pressure exerted by the magnetic field **B** inside a long uniform solenoid of length l, and a cross-section of area $A \ll l^2$, with $N \gg l/A^{1/2} \gg 1$ turns, on its 'walls' (windings), and the force exerted by the field on the solenoid's ends, for two cases:

(i) the current through the solenoid is fixed by an external source, and
(ii) after the initial current setting, the ends of the solenoid's wire, with a negligible resistance, are connected so that it continues to carry a non-zero current.

Compare the results, and give a physical interpretation of the direction of these forces.

Solutions:

(i) If the current in the solenoid is maintained by an external current source, its equilibrium corresponds to the minimum of the total potential energy of the system (solenoid + source), i.e. of the Gibbs energy U_G of the solenoid. In the long solenoid (of length $l \gg A^{1/2}$), with dense winding ($N \gg l/A^{1/2} \gg 1$), the field **H** inside the solenoid is uniform, and its magnitude is given by Eq. (5.141):

$$H = \frac{NI}{l}.$$

Hence **H** is fixed at all points (it is virtually zero outside the solenoid) at the fixed current I, we may use Eqs. (6.17) and (6.20) to obtain

$$U_G = -\frac{B^2}{2\mu}V = -\frac{\mu H^2}{2}Al \equiv -\frac{\mu N^2 I^2 A}{2l}. \qquad (*)$$

This expression shows that in order to minimize U_G (i.e. to maximize its magnitude), the field 'tries' to *increase* the area A of the solenoid, and *decrease* its length l. These effects may be quantified by the calculation, respectively, of the pressure on the walls (windings)[11],

$$\mathcal{P} \equiv -\left.\frac{\partial U_G}{\partial V}\right|_{l=\text{const}} = -\frac{1}{l}\frac{\partial U_G}{\partial A} = \frac{\mu N^2 I^2 A}{2l^2} \equiv \frac{B^2}{2\mu}, \qquad (**)$$

and the effective force applied to its ends:

$$F \equiv -\frac{\partial U_G}{\partial l} = -\frac{\mu N^2 I^2 A}{2l^2} \equiv -\frac{B^2}{2\mu}A. \qquad (***)$$

The directions of these forces may be readily interpreted using the basic Eq. (5.1). Indeed, the lateral force on a current-carrying 'wall' of the solenoid is dominated by the oppositely directed current in the opposite wall, corresponding to their *repulsion*. (This is especially apparent if the cross-section has an elongated shape—see the figure below.) On the other hand, the force in the direction along the solenoid's axis,

[11] As will be shown in section 9.8 of the lecture notes, the last expression is very general.

exerted on each wire turn, is dominated by its adjacent turns, with the same current direction, corresponding to the *attraction* of the turns.

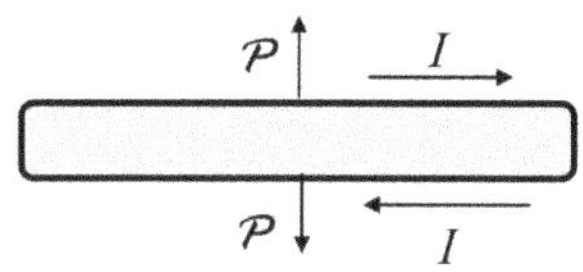

(ii) If the solenoid is disconnected from the current source, its equilibrium corresponds to the minimum of its own potential energy, U, which may be calculated using Eq. (6.15):

$$U = \frac{B^2}{2\mu}V \equiv \frac{\mu N^2 I^2 A}{2l}.$$

This expression differs from Eq. (*) only by its sign, and it may now seem that the results (**) and (***) would be opposite. However, this is not so. Indeed, if the wire's resistance is negligible, it cannot sustain any voltage, in particular the Faraday's e.m.f. (6.2) in the closed loop formed by the connection of the wire's ends, and hence the magnetic flux Φ in the loop, rather than the current I, remains constant at the solenoid's shape variations. (As was discussed in sections 6.4 and 6.5, if the wiring is superconducting, this flux conservation is perfect, but even in realistic large, high-μ magnets with 'normal' wires, the flux relaxation time $\tau = L/R$ may be in hours, much longer than needed for force measurements.)

Now using Eq. (5.143) of the lecture notes for the self-inductance L of the solenoid,

$$L \equiv \frac{\Phi}{I} = \mu n^2 A l \equiv \frac{\mu N^2 A}{l},$$

we may eliminate the current from the above expression for U:

$$U = \frac{LI^2}{2} \equiv \frac{\Phi^2}{2L} \equiv \frac{\Phi^2 l}{2\mu N^2 A}.$$

From here, we may readily calculate the pressure on the walls as

$$\mathcal{P} \equiv -\left.\frac{\partial U}{\partial V}\right|_{l=\text{const}} = -\frac{1}{l}\frac{\partial U}{\partial A} = \frac{\Phi^2}{2\mu N^2 A^2} \equiv \frac{B^2}{2\mu},$$

and the longitudinal force as

$$F \equiv -\frac{\partial U}{\partial l} = -\frac{\Phi^2}{2\mu N^2 A} \equiv -\frac{B^2}{2\mu}A,$$

obtaining exactly the same results as given by Eqs. (**) and (***).

Note that this problem, for a system with a uniform magnetic field (a long solenoid) is a virtually exact analog of problem 3.24 for a system with a uniform electric field (a plane capacitor), with the similar general conclusion: the (correctly calculated :-) forces do not depend on whether the system is connected to an external source, despite using different potential energies. Let me hope the solutions of these two problems make the notion of the Gibbs potential energy and the conditions of its application even more clear.

***Problem* 6.7.** The *electromagnetic railgun* is a projectile launch system consisting of two long, parallel conducting rails and a sliding conducting projectile, shorting the current I fed into the system by a powerful source—see panel (a) in the figure below. Calculate the force exerted on the projectile using two approaches:

(i) by a direct calculation, assuming that the cross-section of the system has the simple shape shown in panel (b) of the figure below, with $t \ll w$, l, and
(ii) using the energy balance (for simplicity, neglecting the Ohmic resistances in the system),

and compare the results.

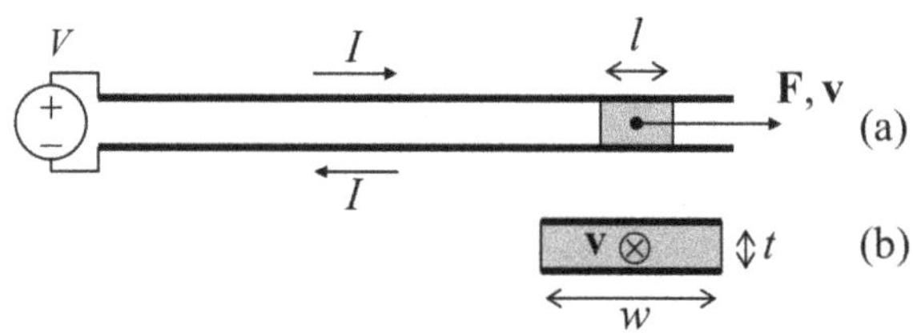

Solutions:

(i) Due to the condition $t \ll l$, w, the magnetic field induced by the shorting currents in the projective itself are negligible in comparison with those induced by the current in the rails. As was calculated in the solution of problem 5.3, the latter magnetic field is directed normally to the rails and its magnitude, in a particular cross-section of the projectile, is

$$B(x) = \mu \frac{I(x)}{w}.$$

Here x is the distance (along the rail) from one end of the projectile to its other end, so that $I(0) = I$, and $I(l) = 0$. According to Eq. (5.8) of the lecture notes, the force $d\mathbf{F}$ exerted by this field on a small fragment dx of the projectile which carries the shorting current $dI = J(x)dx$ (where J is its linear density) is directed along the rails, always away from the current source (see the figure above), and has the magnitude

$$dF = B(x)tdI = J(x)B(x)tdx = \frac{\mu t}{w} J(x)I(x)dx,$$

so that the total force exerted on the projectile is

$$F = \int_{x=0}^{x=l} dF = \frac{\mu t}{w} \int_0^l J(x)I(x)dx.$$

Since the current's conservation requires $dI(x)/dx = -J(x)$, we obtain the following simple result,

$$F = -\frac{\mu t}{w}\int_0^l \frac{dI(x)}{dx}I(x)dx \equiv -\frac{\mu t}{w}\int_{x=0}^{x=l} I(x)dI(x) \equiv -\frac{\mu t}{w}\left[\frac{I^2(x)}{2}\right]_{x=0}^{x=l} = \frac{\mu t}{w}\frac{I^2}{2}, \quad (*)$$

valid regardless of the distribution of the shorting currents inside the projective (which may be due to, for example, its non-uniform conductivity).

Note that according to another result of problem 5.3 (in an evident way generalized to the case $\mu \neq \mu_0$), the first fraction in the last form of Eq. (*) is just the self-inductance L_0 of the rail system (or if you like, the mutual inductance between the rails), per unit length, so that our result may be represented in a simpler form

$$F = \frac{L_0 I^2}{2}. \quad (**)$$

As we will see in a minute, this simplicity is not occasional.

(ii) At negligible resistance of all components of the system, the only voltage V at the current source is the Faraday induction e.m.f. $\mathscr{V}_{\text{ind}} = -d\Phi/dt$. So, the elementary work done by the source *against* the e.m.f. during an elementary time interval dt is $d\mathscr{W} = -\mathscr{V}_{\text{ind}}dQ = -\mathscr{V}_{\text{ind}}Idt = Id\Phi$. If the projectile is at a distance $X \gg l$ from the current source, the flux $\Phi = LI$ may be well approximated as L_0XI, so that we obtain

$$d\mathscr{W} = Id\Phi = Id(L_0XI) \equiv L_0XIdI + L_0I^2dX. \quad (***)$$

On the other hand, the change of the magnetic energy (5.71) of the loop during the same time interval is

$$dU = d\left(\frac{LI^2}{2}\right) = d\left(\frac{L_0XI^2}{2}\right) \equiv \frac{1}{2}L_0I^2dX + L_0XIdI,$$

so that Eq. (***) may be rewritten as

$$d\mathscr{W} = dU + \frac{1}{2}L_0I^2dX.$$

Due to the conservation of the total energy of the system, the last term has to describe the mechanical work $d\mathscr{W}_{\text{m}} = FdX$ of the magnetic field on the projectile, so that for the force F we again obtain Eq. (**), showing that this expression is independent of the particular cross-section of the system.

In the modern (so far, experimental) military railgun systems, the (pulsed) driving current, typically supplied by a capacitor battery, may be as high as $\sim 5 \times 10^6$ A, so that as Eq. (**), with $L_0 \sim \mu_0 \sim 10^{-6}$ H m^{-1}, shows that the force may be above 10^7 N, providing a projectile's acceleration approaching $\sim 10^6$ m s^{-2} (i.e. $\sim 10^5$ g!). Even much higher accelerations may be achieved in railguns with plasma projectiles, which are being explored as possible reaction triggers in inertial-confinement nuclear fusion systems.

***Problem* 6.8.** A uniform, static magnetic field **B** is applied along the axis of a long round pipe of radius R, and a very small wall thickness τ, made of a material with Ohmic conductivity σ. A sphere of mass M and radius $R' < R$, made of a linear magnetic with permeability $\mu \gg \mu_0$, is launched, with an initial velocity v_0 to fly ballistically along the pipe's axis—see the figure below. Use the quasi-static approximation to calculate the distance the sphere would pass before it stops. Formulate the conditions of validity of your result.

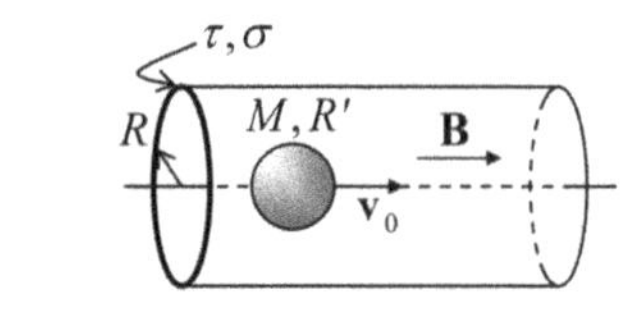

Solution: According to Eq. (5.125) of the lecture notes, taken in the limit $\mu \gg \mu_0$, the magnetic field distortion created by the such a sphere outside it coincides with the field of a point magnetic dipole with the moment[12]

$$\mathbf{m} = \frac{4\pi R'^3}{\mu_0}\mathbf{B}, \qquad (*)$$

whose position coincides with the sphere's center. In the quasi-static approximation, the vector-potential created by the dipole at a fixed point **r** may be found from Eq. (5.90) with the duly shifted origin:

$$\mathbf{A}(\mathbf{r}, t) = \frac{\mu_0}{4\pi}\frac{\mathbf{m} \times [\mathbf{r} - \mathbf{r}'(t)]}{|\mathbf{r} - \mathbf{r}'(t)|^3},$$

where $\mathbf{r}'$ is the radius-vector of the effective dipole, i.e. of the sphere's center. According to this formula, for our axially symmetric geometry (see the figure below), the vector-potential is purely azimuthal, $\mathbf{A} = A\mathbf{n}_\varphi$, and its magnitude at the pipe's wall is

$$A|_{\text{wall}} = \frac{\mu_0}{4\pi}\frac{mR}{\left\{[z - z'(t)]^2 + R^2\right\}^{3/2}}.$$

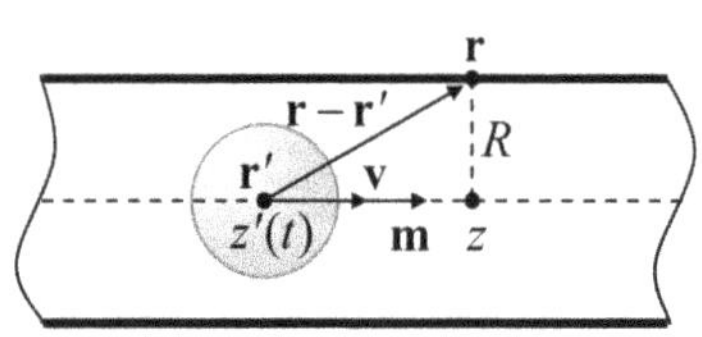

Now we may use Eq. (5.65), so that the magnetic flux through pipe's cross-section:

$$\Phi(z, t) \equiv \oint_{R'} \mathbf{A}|_{\text{wall}} \cdot d\mathbf{r} = 2\pi R A|_{\text{wall}} = \frac{\mu_0 m R^2}{2}\frac{1}{\left\{[z - z'(t)]^2 + R^2\right\}^{3/2}}.$$

[12] Note that the moment is automatically aligned with the applied magnetic field, so that all following results are robust with respect to the sphere's rotation—unless it is so fast that a finite re-magnetization process speed does not allow **m** to follow its quasi-stationary value (*).

According to the Faraday induction law (6.2), the change of this flux in time,

$$\frac{\partial\Phi(z,t)}{\partial t} = \frac{\partial}{\partial t}\left(\frac{\mu_0 m R^2}{2}\frac{1}{\{[z-z'(t)]^2+R^2\}^{3/2}}\right)$$
$$= \frac{\mu_0 m R^2}{2}\frac{\partial}{\partial z'}\left(\frac{1}{\{[z-z'(t)]^2+R^2\}^{3/2}}\right)\frac{dz'}{dt} \qquad (**)$$
$$= \frac{\mu_0 m R^2}{2}\left(\frac{3[z-z'(t)]}{\{[z-z'(t)]^2+R^2\}^{5/2}}\right)v$$

(where $v \equiv dz'/dt$ is the velocity of the sphere at moment t), causes an equal but opposite e.m.f. $\mathscr{V}_{\text{ind}}$. Due to the axial symmetry of our system, the induced electric field's magnitude does not depend on the azimuthal angle:

$$\mathbf{E}_{\text{ind}}(\mathbf{r},t) = \mathbf{n}_\varphi E_{\text{ind}}(z,t), \quad \text{with } E_{\text{ind}}(z,t) = \frac{\mathscr{V}_{\text{ind}}(z,t)}{2\pi R} = -\frac{\partial\Phi(z,t)}{\partial t}\frac{1}{2\pi R}.$$

In the thin wall of the pipe, this field creates loop-like eddy currents with the linear density

$$\mathbf{J}(z,t) = \mathbf{n}_z J(z,t), \qquad \text{with} \quad J(z,t) = j(z,t)\tau = \sigma E_{\text{ind}}(z,t)\tau = -\frac{\sigma\tau}{2\pi R}\frac{\partial\Phi(z,t)}{\partial t}.$$

These loop currents, in turn, induce an additional axially symmetric magnetic field $\mathbf{B}_J$. For a point z'' at the system's axis (generally, different from the sphere center's location z'), this field is directed along the axis, and its magnitude may be calculated by integration, using Eq. (5.23), with the proper replacement $z \to z'' - z$:

$$B_J(z'',t) = \frac{\mu_0 R^2}{2}\int_{-\infty}^{+\infty}\frac{J(z,t)dz}{[(z''-z)^2+R^2]^{3/2}}.$$

According to Eq. (5.102) of the lecture notes, this axially oriented field exerts the following *drag force* exerted on the effective magnetic dipole $\mathbf{m} = \mathbf{n}_z m$ (i.e. on the sphere), located at point $z'(t)$:

$$\mathbf{F} = \mathbf{n}_z F, \quad \text{with} \quad F = m\left.\frac{\partial B_J(z'',t)}{\partial z''}\right|_{z''=z'(t)}$$
$$= m\frac{\mu_0 R^2}{2}\int_{-\infty}^{+\infty}\left.\frac{\partial}{\partial z''}\frac{1}{[(z''-z)^2+R^2]^{3/2}}\right|_{z''=z'(t)} J(z,t)dz$$
$$= m\frac{\mu_0 R^2}{2}\int_{-\infty}^{+\infty}\frac{3[z-z'(t)]}{\{[z-z'(t)]^2+R^2\}^{5/2}}J(z,t)dz$$
$$= -m\frac{\mu_0 R}{4\pi}\sigma\tau\int_{-\infty}^{+\infty}\frac{3[z-z'(t)]}{\{[z-z'(t)]^2+R^2\}^{5/2}}\frac{\partial\Phi(z,t)}{\partial t}dz.$$

Plugging in Eq. (**), we obtain[13]

$$F = -\frac{9}{8\pi}(\mu_0 m)^2 \sigma\tau v\, R^3 \int_{-\infty}^{+\infty} \frac{[z - z'(t)]^2}{\{[z - z'(t)]^2 + R^2\}^5} dz$$
$$\equiv -\frac{9}{8\pi}(\mu_0 m)^2 \sigma\tau\, v \frac{1}{R^7} I, \quad \text{with } I \equiv \int_{-\infty}^{+\infty} \frac{\xi^2}{(\xi^2 + 1)^5} d\xi,$$

where $\xi \equiv [z - z'(t)]/R$. The dimensionless integral I may be worked out using the following variable replacement: $\xi = \tan \alpha$, so that $d\xi = d\alpha/\cos^2 \alpha$, and $1/(\xi^2 + 1) = \cos^2 \alpha$, giving:

$$I = \int_{-\pi/2}^{+\pi/2} \sin^2 \alpha \cos^6 \alpha d\alpha \equiv \int_{-\pi/2}^{+\pi/2} (1 - \cos^2 \alpha) \cos^6 \alpha d\alpha$$
$$\equiv \int_{-\pi/2}^{+\pi/2} (\cos^6 \alpha - \cos^8 \alpha) d\alpha.$$

Now the function under the integral may be represented as a sum of expressions proportional to power-free trigonometric functions, using a repeated application of well-known trigonometric identities[14]

$$\cos^2 a = \frac{1}{2} + \frac{1}{2} \cos 2a, \quad \text{and } \cos^3 a = \frac{3}{4} \cos a + \frac{1}{4} \cos 3a.$$

The final result of this (a bit tedious but elementary) calculation is

$$\cos^6 \alpha - \cos^8 \alpha = \frac{5}{128} + \frac{1}{32} \cos 2\alpha - \frac{1}{32} \cos 4\alpha - \frac{1}{32} \cos 6\alpha - \frac{1}{128} \cos 8\alpha.$$

At the integration over our interval, of length π, all terms of this expression but the first one give zeros, so that the integral is simply

$$I = \frac{5\pi}{128},$$

and taking into account Eq. (*), the drag force F may be rewritten as[15]

$$F = -\eta v, \quad \text{with } \eta \equiv \frac{9}{8\pi}(\mu_0 m)^2 \frac{\sigma\tau}{R^7} \frac{5\pi}{128} \equiv \frac{45}{1024}(4\pi R'^3 B)^2 \frac{\sigma\tau}{R^7}.$$

Since the force is proportional to the sphere's velocity, the equation of its motion along pipe's axis,

$$M\dot{v} = -\eta v,$$

[13] An alternative way to obtain this expression for the drag force $\mathbf{F}$ is to use the above result for $\mathbf{E}_{\text{ind}}(z, t)$ and Eq. (4.39) for the Joule power density, $\not{p} = \sigma E_{\text{ind}}^2$, to calculate the total power loss $\mathscr{P}$ as an integral of $\not{p}$ over pipe wall's volume. After that, we may argue that $\mathscr{P}$ should be equal to sphere's energy loss rate, $-\mathbf{F} \cdot \mathbf{v}$. The drawback of this approach is that it does not reveal the physical origin of the drag force as clearly as the calculation made above.

[14] See, e.g. Eqs. (A.18d) and (A.19).

[15] η is a usual notation for such a *drag coefficient*—see, e.g. *Part CM* section 5.1.

is linear, so that it may be readily solved to give an exponential law, $v(t) = v_0 \exp\{-t/\mathcal{T}\}$, with the time constant $\mathcal{T} = M/\eta$. From here, the distance passed by the sphere tends to

$$d = \int_0^\infty v(t)dt = v_0 \int_0^\infty \exp\left\{-\frac{t}{\mathcal{T}}\right\}dt = v_0\mathcal{T} = \frac{Mv_0}{\eta} = \frac{64}{45\pi^2}\frac{Mv_0R^7}{\sigma\tau R'^6B^2}.$$

There are two major conditions of validity of this result. First, the solution implied that the induced current j is uniformly distributed over pipe wall's thickness τ. This assumption is valid only if τ is much smaller than the radius R and the skin depth, $\delta_s = (2/\mu\sigma\omega)^{1/2}$, at the main frequency components, $\omega \sim v/R \leqslant v_0/R$, of the transient process of the current's rise and fall, i.e. if

$$\tau \ll R, \left(\frac{R}{\mu\sigma v_0}\right)^{1/2}.$$

Second, for a dipole moving with velocity v we have used the same Eq. (5.90) equation for **A** as for a static dipole. The discussion in section 6.7 implies that this is only valid if $v \sim v_0 \ll c$.

***Problem* 6.9.** AC current of frequency ω is passed through a long uniform wire with a round cross-section of a radius R comparable to the skin depth δ_s. In the quasi-static approximation, find the current's distribution across the cross-section, and analyze it in the limits $R \ll \delta_s$ and $\delta_s \ll R$. Calculate the effective ac resistance of the wire (per unit length) in these two limits.

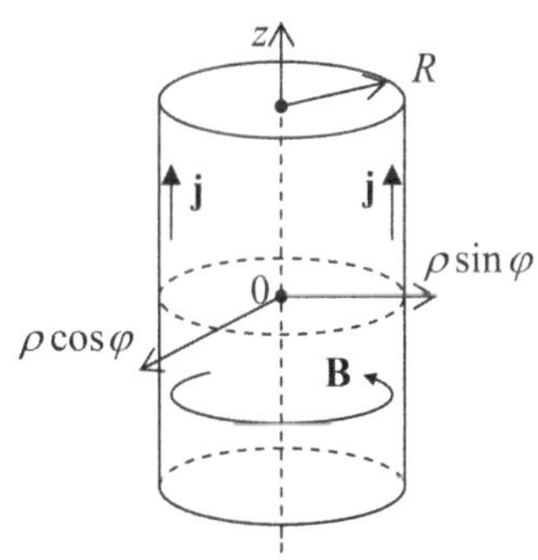

Solution: The current distribution at the skin effect obeys an equation similar to Eq. (6.23) for the magnetic field **B**. Indeed, let us stop at half-way of the derivation of that equation—see the second form of Eq. (6.23),

$$\frac{\partial \mathbf{B}}{\partial t} = -\frac{1}{\sigma}\nabla \times \mathbf{j},$$

and take the curl of both sides. According to the Maxwell equation $\mathbf{j} = \nabla \times \mathbf{H} = \nabla \times (\mathbf{B}/\mu)$, the left-hand side of the relation is equal to $\mu\partial\mathbf{j}/\partial t$, and according to vector algebra the right-hand one yields $\nabla^2\mathbf{j}/\sigma$ (since in the quasi-static limit, $\nabla \cdot \mathbf{j} = 0$), so that we obtain a similar differential equation:

$$\frac{\partial \mathbf{j}}{\partial t} = \frac{1}{\sigma\mu}\nabla^2\mathbf{j}.$$

Since for our axial symmetry $\mathbf{j} = \mathbf{n}_z j(\rho, t)$, we obtain a similar equation for the scalar function j, and spelling out the Laplace operator in the cylindrical coordinates[16], for the Fourier amplitude of the current density we obtain an ordinary differential equation:

$$-i\omega j_\omega = \frac{1}{\sigma\mu}\frac{1}{\rho}\frac{d}{d\rho}\left(\rho\frac{dj_\omega}{d\rho}\right).$$

Using the dimensionless coordinate $\xi \equiv \kappa\rho$, with $\kappa^2 \equiv i\omega\sigma\mu = -2i/\delta_s^2$, the equation may be recast as

$$\frac{d^2 j_\omega}{d\xi^2} + \frac{1}{\xi}\frac{dj_\omega}{d\xi} + j_\omega = 0.$$

This is the Bessel equation (2.130) with $\nu = 0$, and hence its general solution is

$$j_\omega(\rho) = c_1 J_0(\xi) + c_2 Y_0(\xi) \equiv c_1 J_0(\kappa\rho) + c_2 Y_0(\kappa\rho). \qquad (*)$$

The argument $\xi = \kappa\rho$ is now a complex number, but as was mentioned just after Eq. (2.158) of the lecture notes, this does not affect the general properties of the Bessel functions discussed in section 2.7. In particular, Eqs. (2.132), (2.135), (2.142), (2.143), and (2.152) all remain valid for arbitrary complex arguments. Hence, according to Eq. (2.152), the function $Y_0(\xi)$ diverges at $|\xi| \to 0$, so that the constant c_2 in Eq. (*) should equal 0. The other constant, c_1, may be calculated by requiring that the total current through the wire's cross-section has a certain complex amplitude I_ω:

$$I_\omega = 2\pi\int_0^R j_\omega \rho d\rho = 2\pi c_1\int_0^R J_0(\kappa\rho)\rho d\rho = \frac{2\pi c_1}{\kappa^2}\int_0^{\kappa R} J_0(\xi)\xi d\xi.$$

This integral may be readily calculated using Eq. (2.143) for $n = 1$:

$$I_\omega = \frac{2\pi c_1}{\kappa^2}[\xi J_1(\xi)]_0^{\kappa R} = \frac{2\pi c_1 R}{\kappa}J_1(\kappa R),$$

so that, finally,

$$j_\omega(\rho) = \frac{\kappa I_\omega J_0(\kappa\rho)}{2\pi R J_1(\kappa R)}.$$

The plots in the figure below show the radial distribution of $|j_\omega|$, i.e. the real amplitude of the current density, for several values of the ratio δ_s/R. In the

---

[16] See, e.g. Eq. (A.61) with $\partial/\partial z = \partial/\partial\varphi = 0$.

limit $R \ll \delta_s$, i.e. $|\kappa|R$, $|\kappa|\rho \ll 1$, we can use the Taylor series (2.132) to expand the functions $J_0(\xi)$ and $J_1(\xi)$ near the origin. In the lowest non-vanishing approximation, $J_0(\xi) \approx 1$ and $J_1(\xi) \approx \xi/2$, so that

$$j_\omega(\rho) \approx \frac{\kappa I_\omega}{2\pi R(\kappa R/2)} = \frac{I_\omega}{\pi R^2} = \text{const},$$

i.e. the current is uniformly distributed over the wire's cross-section. (As the figure below shows, the inequality $R \ll \delta_s$ should not actually be strong to ensure a virtually uniform current distribution.)

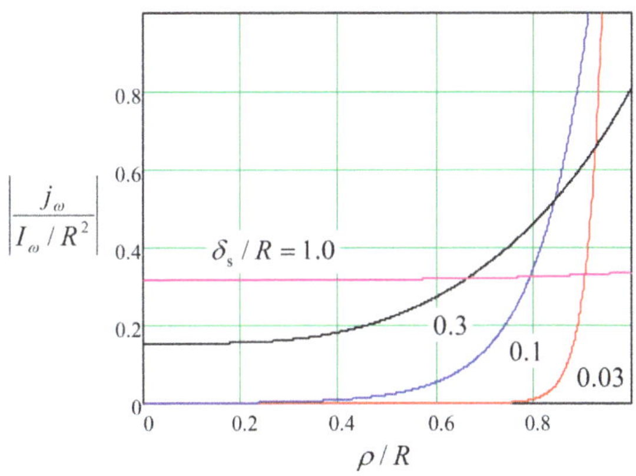

In the opposite limit, $\delta_s \ll R$, i.e. $1 \ll |\kappa|R$, $|\kappa|\rho$, we may use the asymptotic expansion (2.135) to obtain

$$J_0(\kappa\rho) \approx \left(\frac{2}{\pi\kappa\rho}\right)^{1/2} \cos\left(\kappa\rho - \frac{\pi}{4}\right) = \left(\frac{2}{\pi\kappa\rho}\right)^{1/2} \cos\left[(1+i)\frac{\rho}{\delta_s} - \frac{\pi}{4}\right],$$

$$J_1(\kappa R) \approx \left(\frac{2}{\pi\kappa R}\right)^{1/2} \cos\left(\kappa R - \frac{\pi}{4} - \frac{\pi}{2}\right) = \left(\frac{2}{\pi\kappa R}\right)^{1/2} \sin\left[(1+i)\frac{R}{\delta_s} - \frac{\pi}{4}\right].$$

Now using the well-known relations Eqs. (A.20) (which actually may serve as the definition of the hyperbolic functions), and the asymptotic relations $\sinh b \approx \cosh b \approx (1/2)\exp\{b\}$ at $b \gg 1$, we obtain

$$J_0(\kappa\rho) \approx \left(\frac{1}{2\pi\kappa\rho}\right)^{1/2} \exp\left\{\frac{\rho}{\delta_s}\right\} \exp\left\{-i\left(\frac{\rho}{\delta_s} - \frac{\pi}{4}\right)\right\},$$

$$J_1(\kappa R) \approx \left(\frac{1}{2\pi\kappa R}\right)^{1/2} \exp\left\{\frac{R}{\delta_s}\right\} i\exp\left\{-i\left(\frac{R}{\delta_s} - \frac{\pi}{4}\right)\right\},$$

so that, finally,

$$j_\omega(\rho) \approx -I_\omega \frac{i\kappa}{2\pi(R\rho)^{1/2}} \exp\left\{-\frac{R-\rho}{\delta_s}\right\} \exp\left\{-i\frac{R-\rho}{\delta_s}\right\}.$$

At $(R - \rho) \sim \delta_s \ll R$, this solution coincides with that following from Eqs. (6.32)–(6.34), with the replacement $x \to R - \rho$. This is natural, because in this limit, on the current decay scale $\delta_s \ll R$, the surface's curvature is negligible.

The simplest way to define the effective ac resistance $\mathcal{R}(\omega)$ of the wire is by writing the same relation for the average Joule power of energy loss,

$$\overline{\mathscr{P}} = \frac{\mathcal{R}(\omega)\, |I_\omega^2|}{2},$$

as follows from Eq. (4.40) of the lecture notes in the dc limit. (The factor ½ is due to the averaging over the period of sinusoidal oscillations of the current)[17]. For this power, per unit length of the wire, the time averaging of Eq. (4.39) gives the following expression:

$$\frac{\overline{\mathscr{P}}}{l} = \frac{1}{2\sigma}\int_A |j_\omega(\rho)|^2\, d^2\rho$$

so that we can calculate the effective resistance per unit length as

$$\frac{\mathcal{R}(\omega)}{l} = \frac{1}{\sigma}\int_A \left|\frac{j_\omega(\rho)}{I_\omega}\right|^2 d^2\rho \equiv \frac{1}{\sigma}2\pi\int_0^R \left|\frac{j_\omega(\rho)}{I_\omega}\right|^2 \rho d\rho.$$

Plugging in the current distributions calculated above, we obtain

$$\frac{\mathcal{R}(\omega)}{l} \approx \frac{1}{\pi\sigma R} \times \begin{cases} 1/R, & \text{at } R \ll \delta_s, \\ 1/2\delta_s, & \text{at } \delta_s \ll R. \end{cases}$$

In the first of these expressions we may readily recognize Eq. (4.21) for the dc resistance of the whole wire, with the cross-section area $A = \pi R^2$. The second of these results is the same as the dc resistance of a thin hollow tube, with radius R, thickness δ_s, and hence the cross-section area $A' = 2\pi R\delta_s$. Both results fit very nicely the physical picture of the current distribution—see the figure above.

Note that according to Eq. (6.33) of the lecture notes, $\delta_s \propto \omega^{-1/2}$, so that in the high-frequency limit ($\delta_s/R \to 0$), the ac resistance of metallic wires grows as $\omega^{1/2}$, providing serious problems for telecommunication technologies—see the discussion in section 7.9.

***Problem* 6.10.** A very long, round cylinder of radius R, made of a uniform conductor with Ohmic conductivity σ and magnetic permittivity μ, has been placed into a uniform ac magnetic field $\mathbf{H}_{ext} = \mathbf{H}_0 \cos \omega t$, directed along its symmetry axis. Calculate the spatial distribution of the magnetic field's amplitude, and in particular its value on the cylinder's axis. Spell out the last result in the limits of relatively small and large R.

[17] Another (perhaps less obvious) definition, giving the similar result, is $\mathcal{R}(\omega)/l \equiv \text{Re}[E_\omega(R)/I_\omega]$, where $E_\omega(R)$ is the complex amplitude of the ac electric field on the wire's surface.

Solution: This is just an axially symmetric version of the skin-effect problem solved in section 6.3 of the lecture notes. Due to the symmetry, the magnetic fields (both **B** and **H**) at any point are directed along cylinder's axis,

$$\mathbf{H}(\mathbf{r},\, t) = \mathbf{n}_z \operatorname{Re}[H_\omega(\rho)e^{-i\omega t}]\,,$$

where ρ is the distance for the axis (while the induced electric field and hence the eddy current density have only azimuthal components). As a result, the Laplace operator of the only component of the magnetic field takes a simple form[18]:

$$\nabla^2\mathbf{H} = \mathbf{n}_z \frac{1}{\rho}\frac{d}{d\rho}\left(\rho\frac{dH}{d\rho}\right),$$

and Eq. (6.23) yields the following equation for its complex amplitude inside the cylinder:

$$-i\omega H_\omega = \frac{1}{\sigma\mu}\frac{1}{\rho}\frac{d}{d\rho}\left(\rho\frac{dH_\omega}{d\rho}\right), \qquad \text{for } \rho \leqslant R.$$

This is exactly the same equation as was obtained for current's amplitude j_ω in the model solution of the previous problem, and we may borrow its solution (also giving a finite value at $\rho = 0$):

$$H_\omega(\rho) = cJ_0(\kappa\rho),$$

where J_0 is the zero-order Bessel function of the first kind, and

$$\kappa \equiv (-i\omega\mu\sigma)^{1/2} \equiv -\frac{1-i}{\delta_s}.$$

The numerical plots of $|H_\omega|$ as a function of ρ are also similar to those for $|H_\omega|$ in the model solution of the previous problem; however, their vertical scaling is different, because in our current case the constant c should be calculated from the boundary condition $H_\omega(R) = H_0$, giving

$$H_\omega(\rho) = H_0\frac{J_0(\kappa\rho)}{J_0(\kappa R)},$$

so that, using the fact that $J_0(0) = 1$ (see, e.g. either Eq. (2.132) or figure 2.18, with $n = 0$), for the field on cylinder's axis we obtain

$$H_\omega(0) = H_0\frac{J_0(0)}{J_0(\kappa R)} \equiv H_0\frac{1}{J_0(\kappa R)}.$$

A plot (in the appropriate semi-log scale) of the modulus of this value as a function of R, normalized by the skin depth δ_s, is shown in the figure below. For

[18] See, e.g. Eqs. (A.61) and (A.64) with $\mathbf{f} = f(\rho)\mathbf{n}_z$.

small values of the R/δ_s ratio, the result may be well approximated by using the leading two terms of the Taylor expansion (2.132):

$$\frac{H_\omega(0)}{H_0} = \frac{1}{J_0(\kappa R)} \approx \frac{1}{1-(\kappa R/2)^2} \equiv \frac{1}{1+iR^2/2\delta_s^2},$$

so that at $R/\delta_s \to 0$,

$$\frac{|H_\omega(0)|}{H_0} \approx \frac{1}{[1+(R^2/2\delta_s^2)^2]^{1/2}} \approx 1 - \frac{1}{8}\left(\frac{R}{\delta_s}\right)^4 \to 1.$$

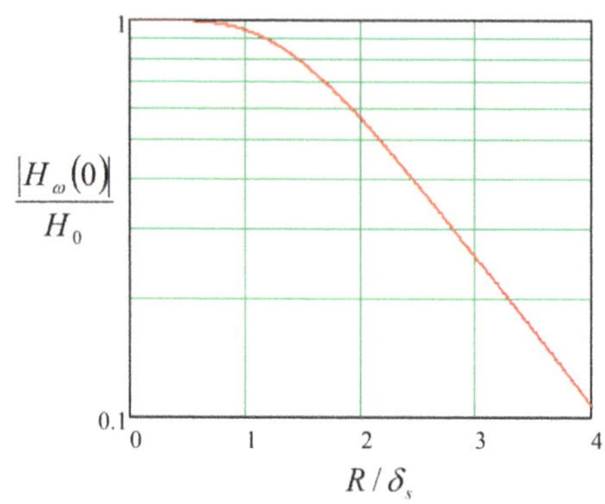

In this approximation the applied field is virtually uniform everywhere in the cylinder, because the diamagnetic effect of the Faraday-induced eddy currents is small.

In the opposite limit, we may reuse the large-argument asymptote of function J_0, which was employed in the model solution of the previous problem, to obtain

$$\frac{|H_\omega(0)|}{H_0} \approx |2\pi\kappa R|^{1/2} \exp\left\{-\frac{R}{\delta_s}\right\} = 2\left(\pi\frac{R}{\delta_s}\right)^{1/2} \exp\left\{-\frac{R}{\delta_s}\right\} \to 0, \quad \text{for } \frac{R}{\delta_s} \to \infty.$$

In this limit, the eddy currents shield most of the cylinder from the applied external magnetic field, which penetrates only into a relatively thin skin-layer, exactly as into a conducting semi-space—see section 6.3 of the lecture notes.

Problem* 6.11. Define and calculate an appropriate spatial-temporal Green's function for Eq. (6.25) of the lecture notes, and then use this function to analyze the dynamics of propagation of the external magnetic field, suddenly turned on at $t = 0$ and then left constant:

$$H(x<0,\,t) = \begin{cases} 0, & \text{at } t<0, \\ H_0, & \text{at } t>0, \end{cases}$$

into an Ohmic conductor occupying the half-space $x > 0$—see figure 6.2.

Hint: Try to use a function proportional to $\exp\{-(x-x')^2/2(\delta x)^2\}$, with a suitable time dependence of the parameter δx, and a properly selected pre-exponential factor.

Solution: As was discussed in section 2.10 of the lecture notes, the Green's function approach uses the linearity of a differential equation, giving an explicit expression of the linear superposition principle. For a *time-independent* but *inhomogeneous* equation, such as the Poisson equation (1.41), the superposition is applied to the addition of elementary right-hand sides, distributed over the space—see Eq. (2.203). In this case, the Green's function is purely spatial[19]. On the other hand, for a *time-dependent* but *homogeneous* equation, such as Eq. (6.25):

$$\frac{\partial H}{\partial t} = \frac{1}{\mu\sigma}\frac{\partial^2 H}{\partial x^2}, \qquad \text{for } x \geqslant 0,$$

the superposition principle may be applied to the spatial distribution of the initial conditions:

$$H(x, t) = \int_0^\infty G(x, t; x', t')H(x', t')dx'. \qquad (*)$$

With this definition, the Green's function G, considered as a function of x and t, is just the particular solution of Eq. (6.25) corresponding to delta-functional initial conditions:

$$G(x, t; x', t') = H(x, t), \quad \text{if } H(x, t') = \delta(x - x'), \quad \text{with } x' > 0. \qquad (**)$$

Let us first calculate the Green's function for Eq. (6.25), that would be valid in infinite space ($-\infty < x < +\infty$), using the provided hint. Selecting the pre-exponential coefficient to keep the function properly normalized[20],

$$\begin{aligned}\int_{-\infty}^{+\infty} G(x, t; x', t')dx &= \int_{-\infty}^{+\infty} G(x, t'; x', t')dx \\ &= \int_{-\infty}^{+\infty} H(x, t')dx = \int_{-\infty}^{+\infty} \delta(x - x')dx = 1,\end{aligned}$$

we obtain the well-known expression[21]

$$G(x, t; x', t') = \frac{1}{(2\pi)^{1/2}\delta x(t, t')}\exp\left\{-\frac{(x - x')^2}{2[\delta x(t, t')]^2}\right\}.$$

[19] Similarly, the Green's functions of an ordinary, inhomogeneous differential equation, describing the time dynamics alone, are purely temporal—see, e.g. *Part CM* section 5.1.

[20] This requirement follows from the fact that integrating both parts of Eq. (6.25) from $x = -\infty$ to $x = +\infty$, and assuming that field's gradient has a final spatial extent, i.e. that $\partial H/\partial x \to 0$ at $x \to \pm\infty$, we obtain

$$\frac{d}{dt}\int_{-\infty}^{+\infty} H(x, t)dx = 0.$$

[21] It describes, for example, the *Gaussian* (or 'normal') *probability distribution* with variance $(\delta x)^2$.

Now plugging this formula into Eq. (6.25), we see that it is indeed the solution of the equation, provided that $\delta x(t, t')$ satisfies the following (essentially, ordinary) differential equation:

$$\frac{\partial}{\partial t}\delta x = \frac{1}{\mu\sigma\delta x}, \qquad \text{i.e.} \quad \frac{\partial}{\partial t}[(\delta x)^2] = \frac{2}{\mu\sigma} = \text{const.}$$

With the requirement $\delta x \to 0$ at $t \to t'$, following from the initial condition (**), the solution of this equation is

$$(\delta x)^2 = \frac{2(t - t')}{\mu\sigma}, \qquad \text{so that } \delta x(t, t') = \left[\frac{2(t - t')}{\mu\sigma}\right]^{1/2}.$$

Since the Green's function depends on the difference $(t - t')$ only, in all formulas below I will take $t' = 0$, and use shorthand notations $G(x, t; x') \equiv G(x, t; x', 0)$ and $\delta x(t) \equiv \delta x(t, 0)$. As Eq. (**) says, the physical sense of $\delta x(t)$ is the gradually expanding width of the magnetic field distribution, induced by a delta-functional pulse of the external field at $t = 0$.

Now we may use the equation's linearity, and the mirror-image themes discussed in chapters 2–5 of the lecture notes, to construct the Green's function for the half-space $x > 0$, which obeys the boundary condition $G(0, t; x') = 0$:

$$G(x, t; x', t') = \frac{1}{(2\pi)^{1/2}\delta x(t)}\left[\exp\left\{-\frac{(x - x')^2}{2[\delta x(t)]^2}\right\} - \exp\left\{-\frac{(x + x')^2}{2[\delta x(t)]^2}\right\}\right], \quad \text{with } \delta x(t) = \left(\frac{2t}{\mu\sigma}\right)^{1/2}, \qquad (***)$$

and hence automatically generates solutions (*) which satisfy this solution.

(ii) Due to the eventual decay of the eddy currents in time, the constant applied magnetic field cannot be prevented from its eventual penetration into the conductor: $H(x, t) \to H_0$. Since the Green's function (***) tends to zero at $t \to 0$, we better reduce the problem to finding a function $H'(x, t)$ with the same trend. Evidently, this may be done by representing the field as the difference

$$H(x, t) \equiv H_0 - H'(x, t),$$

with the initial value $H'(x, 0) = H_0$. For this auxiliary field, Eqs. (*) and (***) yield

$$H'(x, t) = H_0\int_0^{+\infty} G(x, t; x')dx' = \frac{H_0}{(2\pi)^{1/2}\delta x(t)}\int_0^{\infty}\left[\exp\left\{-\frac{(x - x')^2}{2[\delta x(t)]^2}\right\} - \exp\left\{-\frac{(x + x')^2}{2[\delta x(t)]^2}\right\}\right]dx'. \qquad (****)$$

Note that the integration limits automatically exclude the non-physical, mirror image delta-function at $-x' < 0$, used to construct the Green's function (***).

This integral may be readily expressed via the so-called *error function*

$$\mathrm{erf}(\zeta) \equiv \frac{2}{\pi^{1/2}} \int_0^{\zeta} \exp\{-\xi^2\} d\xi;$$

however, for all practical purposes (such as plotting, asymptote analysis, etc) the explicit integral form (****) is preferable. Note that introducing normalized variables: field $h \equiv H/H_0$ and coordinate $\xi \equiv x/\delta x(t) = x/(2Dt)^{1/2}$, we may rewrite our result in the form of a universal function

$$h(\xi) = 1 - \frac{1}{(2\pi)^{1/2}} \int_0^{\infty} \left[\exp\left\{-\frac{(\xi - \xi')^2}{2}\right\} - \exp\left\{-\frac{(\xi + \xi')^2}{2}\right\} \right] d\xi',$$

plotted in the figure below, so that the field's profile in the actual space just stretches with time, as $\delta x(t) \propto t^{1/2}$.

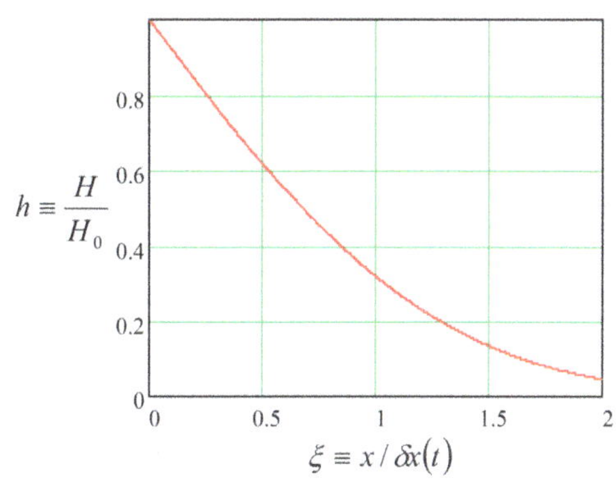

***Problem* 6.12.** Solve the previous problem using the variable separation method, and compare the results.

Solution: Let us look for the solution of the same Eq. (6.25),

$$\mu\sigma \frac{\partial H}{\partial t} = \frac{\partial^2 H}{\partial x^2}, \qquad \text{for } x \geqslant 0,$$

in the usual variable-separated form, but (just as in the solution of the previous problem), singling out the eventual uniform distribution of the field:

$$H(x, t) = H_0 - \sum_k T_k(t) X_k(x), \tag{*}$$

so that the second term of Eq. (*) equals H_0 at $t = 0$, and has to vanish at $t \to \infty$. Plugging a particular term of this series into the equation, and dividing both parts by $T_k X_k$, we obtain

$$\mu\sigma \frac{1}{T_k}\frac{dT_k}{dt} = \frac{1}{X_k}\frac{d^2X_k}{dx^2} = \text{const} = -k^2.$$

Solving the two resulting ordinary differential equations for T_k and X_k, we obtain[22]

$$T_k = a_k \exp\left\{-\frac{k^2}{\mu\sigma}t\right\}, \qquad X_k = b_k \sin kx,$$

with (at this stage) arbitrary a_k and b_k. Since the length of the segment at which our equation is valid ($0 \leqslant x \leqslant \infty$) is infinite, the spectrum of the eigenvalues k is continuous, so that plugging these solutions into Eq. (*), we need to replace the summation over k by integration:

$$H(x, t) = H_0 - \int_{-\infty}^{+\infty} c_k \sin kx \exp\left\{-\frac{k^2}{\mu\sigma}t\right\}dk, \qquad \text{where } c_k \equiv a_k b_k. \qquad (**)$$

What remains is to find the function c_k from the initial condition:

$$\int_{-\infty}^{+\infty} c_k \sin kx\, dk = H_0, \qquad \text{for } x > 0. \qquad (***)$$

As usual at the reciprocal Fourier transform, we may calculate c_k from this equation by multiplying both sides by a similar eigenfunction, but an arbitrary eigenvalue, in our case $\sin k'x$, and integrating them over x. Changing the integration order on the left-hand side, we obtain

$$\int_{-\infty}^{+\infty} dk\, c_k \int_0^{\infty} dx \sin kx \sin k'x = H_0 \int_0^{\infty} \sin k'x\, dx.$$

Since the expression under the inner integral on the left-hand side is an even function of x, the integral may be transformed as

$$\begin{aligned}\int_0^{\infty} \sin kx \sin k'x\, dx &= \frac{1}{2}\int_{-\infty}^{\infty} \sin kx \sin k'x\, dx \\ &= \frac{1}{4}\int_{-\infty}^{\infty} [\cos (k - k')x - \cos (k + k')x]\, dx \\ &= \frac{1}{4}\operatorname{Re}\left[\int_{-\infty}^{\infty} e^{i(k-k')x}dx - \int_{-\infty}^{\infty} e^{i(k+k')x}dx\right]\end{aligned}$$

According to Eq. (A.88), these integrals equal, respectively, $2\pi\delta(k \mp k')$, so that Eq. (***) yields

$$\frac{\pi}{2}(c_{k'} - c_{-k'}) = H_0 \int_0^{\infty} \sin k'x\, dx \equiv -H_0 \frac{\cos k'x}{k'}\Bigg|_{x=0}^{x\to\infty}.$$

[22] The other possible eigenfunctions, proportional to $\cos kx$, do not satisfy our boundary condition $H(0, t) = H_0$.

As Eq. (**) shows, c_k has to be an odd function of k, so that $(c_{k'} - c_{-k'}) = 2c_{k'}$. Also, the upper limit on the right-hand side may be ignored, because the solution should not be sensitive to a very slow quenching of the right-hand side of Eq. (***) at $x \to \infty$. As a result, changing the index notation from k' to k, we obtain

$$c_k = \frac{H_0}{\pi k},$$

so that Eq. (**) yields

$$H(x, t) = H_0\left[1 - \frac{1}{\pi}\int_{-\infty}^{+\infty} \frac{\sin kx}{k} \exp\left\{-\frac{k^2}{\mu\sigma}t\right\}dk\right]$$
$$\equiv H_0\left[1 - \frac{2}{\pi}\int_{0}^{+\infty} \frac{\sin kx}{k} \exp\left\{-\frac{k^2}{\mu\sigma}t\right\}dk\right].$$

Introducing dimensionless variables $\xi \equiv x/(2t/\mu\sigma)^{1/2}$ and $\kappa \equiv k(2t/\mu\sigma)^{1/2}$ (so that $kx = \kappa\xi$), and $h \equiv H/H_0$, we may rewrite this result in a parameter-free form

$$h = h(\xi) \equiv 1 - \frac{2}{\pi}\int_{0}^{+\infty} \frac{\sin \kappa\xi}{\kappa} \exp\left\{-\frac{\kappa^2}{2}\right\}d\kappa, \qquad (****)$$

which shows that the field distribution profile in the conductor is universal, and all its time evolution is reduced to the increase of its spatial scale $(2t/\mu\sigma)^{1/2}$. Indeed, the numerical plot of this result exactly coincides with that shown in the model solution of the previous problem, despite the rather different form of the analytical formulas. This fact is not quite surprising, because it is straightforward (but still recommended for the reader :-) to show that Eq. (****) is just as the Fourier-integral expansion of the result derived in the previous problem.

***Problem* 6.13.** A small, planar wire loop, carrying a fixed current I, is located far from a plane surface of a superconductor. Within the 'coarse-grain' (ideal-diamagnet) description of the Meissner–Ochsenfeld effect, calculate:

(i) the energy of the loop-superconductor interaction,
(ii) the force and torque acting on the loop, and
(iii) the distribution of supercurrents on the superconductor surface.

Solution: As was discussed in section 6.4 of the lecture notes, in the ideal-diamagnet approximation, the field outside the superconductor has to satisfy the boundary condition (6.59):

$$B_n = 0.$$

The field of a wire loop in an unlimited free space, at distances much larger than the loop size, may be described in the magnetic dipole approximation. An elementary generalization of Eq. (5.99) of the lecture notes to the arbitrary position $\mathbf{r}'$ of the loop's center yields

$$\mathbf{B}'(\mathbf{r}) = \frac{\mu_0}{4\pi}\frac{3(\mathbf{r}-\mathbf{r}')[(\mathbf{r}-\mathbf{r}')\cdot\mathbf{m}'] - \mathbf{m}'|\mathbf{r}-\mathbf{r}'|^2}{|\mathbf{r}-\mathbf{r}'|^5},$$

where $\mathbf{m}'$ is its magnetic moment (with $m' = IA$, where A is the loop area). Note that a simultaneous flipping of signs of a certain Cartesian component of the vectors $\mathbf{m}$ and $(\mathbf{r} - \mathbf{r}')$ reverses the sign of the same component of the field, but not its other components. From this symmetry, it is clear that the boundary condition (*) may be satisfied by representing the total magnetic field outside the superconductor as a sum of this field $\mathbf{B}'$, and of a similar field $\mathbf{B}''$ of a mirror-image magnetic dipole $\mathbf{m}''$ that is located and oriented as shown in the figure below. Selecting the coordinate axes as shown in this figure (with the axis x in the same vertical plane as the vector $\mathbf{m}'$), we may write

$$x'' = x' = 0, \quad y'' = y' = 0, \quad z'' = -z' = -d,$$

$$m_x'' = m_x' = m\sin\theta, \quad m_y'' = m_y' = 0, \quad m_z'' = -m_z' = -m\cos\theta.$$

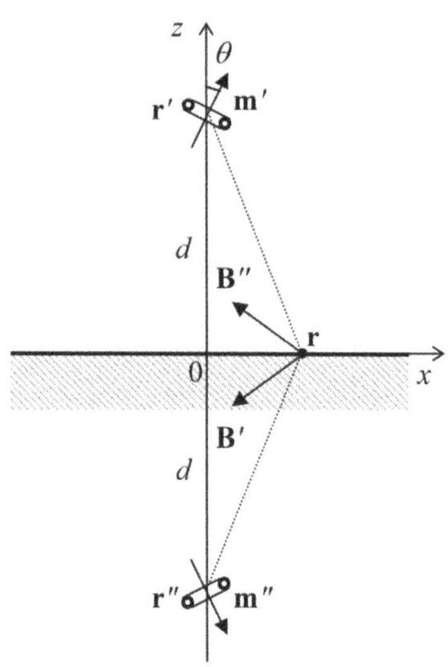

Note that the image dipole orientation is exactly opposite to that in problem 3.6, due to the different boundary conditions. Now we may reuse the solution of that problem, just minding the signs and the fundamental constants, to obtain the following results:

(i) The interaction energy,

$$U_{\text{int}} = -\frac{1}{2}\mathbf{m}'\cdot\mathbf{B}''(\mathbf{r}') = -\frac{1}{2}\left(m_x'B_x'' + m_z'B_z''\right)_{x=y=0,z=d} = \frac{\mu_0}{8\pi}\frac{m^2}{(2d)^3}(1+\cos^2\theta),$$

corresponds to the *repulsion* of the dipole from the superconductor, at any dipole's orientation. This is very natural, because a superconductor pushes out the external field lines, and hence repels their source. Such repulsion is the basis of all *magnetic levitation* projects (including the 'maglev' trains) using superconductivity.

(ii) The force and the torque may be readily calculated from U_{int}:

$$F = F_z = -\frac{\partial U_{\text{int}}}{\partial d} = \frac{\mu_0}{4\pi}\frac{3m^2}{16d^4}(1 + \cos^2\theta), \quad \tau = \tau_y = -\frac{\partial U_{\text{int}}}{\partial \theta} = \frac{\mu_0}{4\pi}\frac{m^2}{16d^3}\sin 2\theta.$$

Note that the system's stable equilibrium with respect to rotation is at $\theta = \pm\pi/2$, with the magnetic moment parallel to the surface.

(iii) The surface supercurrent density may be calculated from its universal relation (6.38) with the magnetic field just outside the surface:

$$\mathbf{J} = \mathbf{n}_z \times \mathbf{H}_\tau \equiv \mathbf{n}_z \times \mathbf{B}_\tau/\mu_0,$$

where $\mathbf{B}_\tau$ is the tangential component of magnetic field (in our coordinates, $\mathbf{B}_\tau = B_x\mathbf{n}_x + B_y\mathbf{n}_y$) near the surface. Calculating the net field of the two dipoles, $\mathbf{B} = \mathbf{B}' + \mathbf{B}''$, at an arbitrary point $\mathbf{r} = x\mathbf{n}_x + y\mathbf{n}_y$ near the surface ($z = 0$), we obtain

$$\mathbf{B}_\tau = \mathbf{B}'_\tau + \mathbf{B}''_\tau = 2\mathbf{B}'_\tau$$
$$= \frac{\mu_0 m}{2\pi}\frac{\mathbf{n}_x[(2x^2 - y^2 - z^2)\sin\theta - 3dx\cos\theta] + \mathbf{n}_y 3y(x\sin\theta - d\cos\theta)}{(x^2 + y^2 + d^2)^{5/2}},$$

so that, finally,

$$\mathbf{J} = \frac{m}{2\pi}\frac{\mathbf{n}_x 3y(-x\sin\theta + d\cos\theta) + \mathbf{n}_y[(2x^2 - y^2 - z^2)\sin\theta - 3dx\cos\theta]}{(x^2 + y^2 + d^2)^{5/2}}.$$

In the particular case when the current loop plane is parallel to the superconductor surface ($\sin\theta = 0$, $\cos\theta = \pm 1$), the current lines are concentric circles, with $\rho \equiv (x^2 + y^2)^{1/2} = \text{const}$, and its magnitude,

$$J = \frac{m}{2\pi}\frac{3d\rho}{(\rho^2 + d^2)^{5/2}},$$

does not depend on the azimuthal angle. The direction of this vortex current is opposite to that of that in the loop, thus fulfilling its function of shielding the superconductor's interior from the penetration of the loop's field.

***Problem* 6.14.** A straight, uniform magnet of length $2l$, cross-section area $A \ll l^2$, and mass m, with a permanent longitudinal magnetization M_0, is placed over a horizontal surface of a superconductor—see the figure below. Within the ideal-diamagnet description of the Meissner–Ochsenfeld effect, find the stable equilibrium position of the magnet.

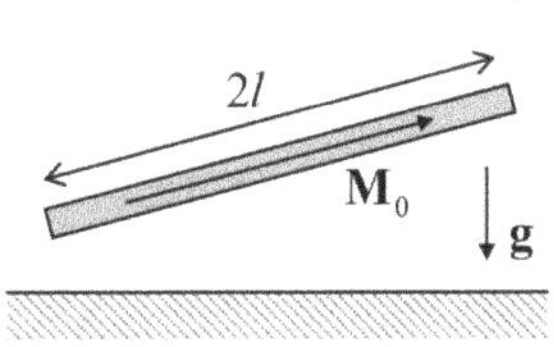

Solution: As was discussed in section 5.6 of the lecture notes, the magnetic field induced by a long and thin ($A \ll l^2$) permanent magnet, with the magnetization $\mathbf{M}_0$ parallel to its length, is equivalent to that of two point magnetic charges $q_\mathrm{m} = \pm AM_0$ at its ends, each producing a Coulomb-like, radial magnetic field

$$\mathbf{B}(\mathbf{r}) = \frac{\mu_0}{4\pi} q_m \frac{\mathbf{r} - \mathbf{r}'}{|\mathbf{r} - \mathbf{r}'|^3},$$

where $\mathbf{r}'$ is the charge's (i.e. the magnet end's) position. In the coarse-grain (ideal-diamagnet) picture of the Meissner–Ochsenfeld effect, the magnetic field cannot penetrate into the superconductor. Thus the boundary problem for a magnetic point charge near superconductor's surface is similar to that of an electric point charge near conductor's surface, and may be solved exactly as was discussed in section 2.9, i.e. by introduction of its mirror image. However, the image charge sign should be similar (rather than opposite) to the initial charge, because it should compensate the normal (rather than tangential) component of the field at the surface—see Eq. (6.59).

Hence, in our case we obtain the system of two real and two image charges shown in the figure below. Introducing the vertical coordinate y of the magnet's center and the angle φ of its slope, as shown in the figure, we may readily express distances $d_\pm$ and d between the charges as functions of these two generalized coordinates[23]:

$$d_\pm = 2(y \pm l \sin \varphi),$$
$$d^2 = (2y)^2 + (2l \cos \varphi)^2 \equiv 4(y^2 + l^2 \cos^2 \varphi).$$

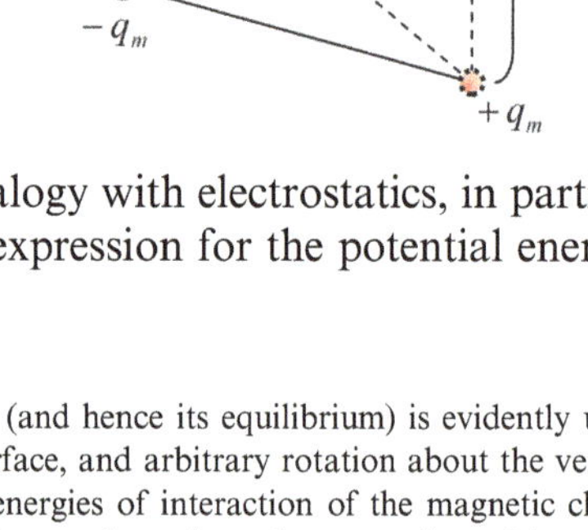

Continuing to use the analogy with electrostatics, in particular with Eq. (2.192), we may write the following expression for the potential energy of the system as[24]

[23] The magnet's potential energy (and hence its equilibrium) is evidently unaffected by its arbitrary translational displacement along the surface, and arbitrary rotation about the vertical axis.

[24] We may ignore the potential energies of interaction of the magnetic charge pairs belonging to the same magnet, because these energies do not depend on the magnet's position, i.e. on coordinates y and φ. (The longitudinal forces resulting from these interactions are compensated by the mechanical stress forces in the rigid magnets.)

$$U = \frac{\mu_0}{4\pi}\frac{q_\mathrm{m}^2}{2}\left(\frac{1}{d_+} + \frac{1}{d_-} - \frac{2}{d}\right) + mgy$$

$$= \frac{\mu_0}{4\pi}\frac{q_\mathrm{m}^2}{2}\left[\frac{1}{2(y + l\sin\varphi)} + \frac{1}{2(y - l\sin\varphi)} - \frac{1}{(y^2 + l^2\cos^2\varphi)^{1/2}}\right] + mgy$$

$$\equiv \frac{\mu_0}{4\pi}\frac{q_\mathrm{m}^2}{2}\left\{\left[\frac{y}{y^2 - l^2\sin^2\varphi} - \frac{1}{(y^2 + l^2\cos^2\varphi)^{1/2}}\right] + \frac{y}{a^2}\right\}, \quad \text{with } a^2 \equiv \frac{\mu_0}{4\pi}\frac{q_\mathrm{m}^2}{2}\bigg/ mg.$$

For any fixed y, the function $U(\varphi)$ has two similar minima at $\varphi = 0$ and π, i.e. at $\sin^2\varphi = 0$ and $\cos^2\varphi = 1$. Perhaps the easiest way to see that is to expand the expression in the square brackets (which alone depends on the angle) in the Taylor series at small φ, keeping only two leading terms:

$$\frac{y}{y^2 - l^2\sin^2\varphi} - \frac{1}{(y^2 + l^2\cos^2\varphi)^{1/2}} \equiv \frac{1}{y(1 - l^2\sin^2\varphi/y^2)} - \frac{1}{(y^2 + l^2)^{1/2}[1 - l^2\sin^2\varphi/(y^2 + l^2)]^{1/2}}$$

$$\approx \frac{1}{y(1 - l^2\varphi^2/y^2)} - \frac{1}{(y^2 + l^2)^{1/2}[1 - l^2\varphi^2/(y^2 + l^2)]^{1/2}}$$

$$\approx \frac{1}{y}\left(1 + \frac{l^2\varphi^2}{y^2}\right) - \frac{1}{(y^2 + l^2)^{1/2}}\left[1 + \frac{l^2\varphi^2}{2(y^2 + l^2)}\right]$$

$$\equiv \left[\frac{1}{y} - \frac{1}{(y^2 + l^2)^{1/2}}\right] + l^2\varphi^2\left[\frac{1}{y^3} - \frac{1}{2(y^2 + l^2)^{3/2}}\right].$$

Since the second term in the last square brackets is always smaller than the first one, the potential energy always grows with φ^2.

Hence, the stable stationary position of the magnet is horizontal[25]. Its stationary height y over the superconducting surface should be calculated from the requirement $dU/dy = 0$ (at $\varphi = 0$ or π). Formally, for arbitrary values of the only dimensionless parameter of the problem, a/l, the equation for y, resulting from this condition,

$$\frac{1}{y^2} - \frac{y}{(y^2 + l^2)^{3/2}} = \frac{1}{a^2}, \qquad (*)$$

may be solved only numerically. However, this equation may be readily used for plotting the reciprocal function, $a(y)$, and then transposing the plot—see the solid

[25] Note that this conclusion is in agreement with the solution of task (ii) of the previous problem. Indeed, in the particular case $y \gg l$ (see the balance of this solution), the magnet may be approximated as a magnetic dipole—just as a small current loop.

curve in the figure below. Moreover, Eq. (*) may be easily solved analytically in two limits:

$$y \to \begin{cases} a, & \text{for } a \ll l, \\ (2/3)^{1/4}(la)^{1/2}, & \text{for } l \ll a, \end{cases} \qquad (**)$$

see the dashed lines in the figure below. As the plots show, the exact solution describes a simple, smooth crossover between these asymptotes, taking place at $a/l \approx 1.3$.

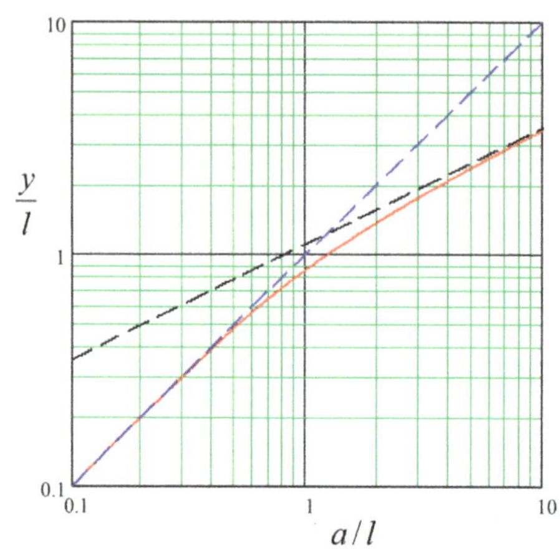

Physically, the top line of Eq. (**) corresponds to the limit of relatively weak magnetic interaction, when the gravity presses the magnet very close to the superconductor surface, and only the closest magnetic poles ('magnetic charges') interact substantially. On the other hand, if the magnetic interaction is strong enough, the magnet's repulsion by the superconducting surface brings it up so high ($y \gg l$) that its field near the surface is close to that of a dipole with moment $m = VM_0 = 2lAM_0 = 2lq_m$. As a result, the bottom line of Eq. (**) could be calculated from Eqs. (5.99) and (5.100), with this value of m, and the inclusion of factor ½ due to the induced nature of the dipole image.

***Problem* 6.15.** A plane superconducting wire loop, of area A and inductance L, may rotate, without friction, about a horizontal axis 0 (in the figure below, perpendicular to the plane of drawing) passing through its center of mass. Initially the loop was horizontal (with $\theta = 0$), and carried supercurrent I_0 in such a direction that its magnetic dipole vector was directed down. Then a uniform magnetic field **B**, directed vertically up, was applied. Using the ideal-diamagnet description of the Meissner–Ochsenfeld effect, find all possible equilibrium positions of the loop, analyze their stability, and give a physical interpretation of the results.

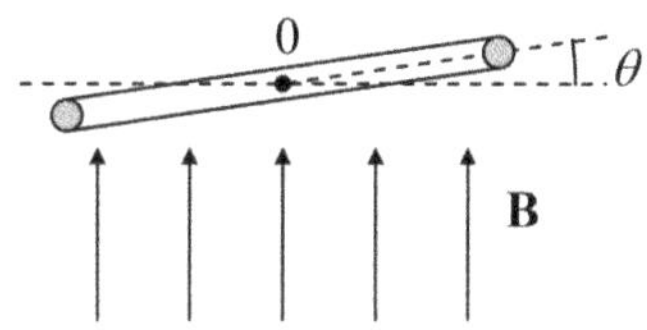

Solution: Due to the magnetic flux quantization (see Eq. (6.62) of the lecture notes and its discussion), the total flux (6.78) through the loop,

$$\Phi = \Phi_{\text{ext}} - LI = AB\cos\theta - LI,$$

has to be constant and equal to its value,

$$\Phi = -LI_0,$$

before the external field's application. Combining these two relations, we obtain

$$I = I_0 + I_B, \quad \text{with } I_B = \frac{\Phi_{\text{ext}}}{L} = \frac{AB\cos\theta}{L}.$$

This means that the vertical component m of the magnetic moment (5.97) of the loop may be represented as the sum of two parts: $m_0 = -AI_0\cos\theta$ created by the pre-existing current, and $m_B = -AI_B\cos\theta$ created by the current I_B that is induced by the external field. Such decomposition of m is important, because while the former part's contribution to the potential energy U of the loop may be calculated using Eq. (5.100):

$$U_0 = -m_0 B = ABI_0\cos\theta,$$

the latter part's contribution should be multiplied by the usual factor ½:

$$U_B = -\frac{1}{2}m_B B = \frac{1}{2}ABI_B\cos\theta = \frac{A^2B^2\cos^2\theta}{2L}.$$

As a result, the total potential energy of the loop is[26]

$$U = U_0 + U_B = ABI_0\cos\theta + \frac{A^2B^2\cos^2\theta}{2L}.$$

Since this is a system with just one degree of freedom (θ), its equilibrium positions correspond to the extrema of the function $U(\theta)$, in which $dU/d\theta = 0$, while their stability is determined by the sign of the second derivative of this function: if $d^2U/d\theta^2 \geqslant 0$, the point is stable. For our function, the equilibrium requirement gives the equation

$$\frac{dU}{d\theta} \equiv -ABI_0\sin\theta - \frac{A^2B^2\sin\theta\cos\theta}{L} \equiv -AB\sin\theta\left(I_0 + \frac{AB\cos\theta}{L}\right) = 0. \quad (*)$$

In relatively weak fields, when

$$AB < LI_0, \quad (**)$$

[26] Due to the rotation axis passage through the center of mass, gravity does not give any contribution to U.

the expression in the parentheses of the right-hand side of Eq. (*) is positive for all values of θ, so this equation has only two physically distinguishable roots, both with $\sin\theta = 0$:

$$\theta_0 = 0, \qquad \text{and } \theta_1 = \pi,$$

i.e. both corresponding to horizontal position of the loop. Calculating the second derivative,

$$\frac{d^2U}{d^2\theta} = \frac{d}{d\theta}\left(-ABI_0\sin\theta - \frac{A^2B^2\sin\theta\cos\theta}{L}\right) = \frac{d}{d\theta}\left(-ABI_0\sin\theta - \frac{A^2B^2\sin 2\theta}{2L}\right)$$
$$= -ABI_0\cos\theta - \frac{A^2B^2\cos 2\theta}{L} = -AB\left(I_0\cos\theta + \frac{AB\cos 2\theta}{L}\right),$$

we see that if the low-field condition (**) is satisfied, the second term in the last parentheses cannot affect the sign of the whole expression, so that it is positive for θ_0 and negative for θ_1. This means that the initial horizontal position of the loop, with $\theta = \theta_0 = 0$, is unstable, while the flipped state, with $\theta = \theta_1 = \pi$, is its only stable position.

However, in higher fields, when the condition (**) is not valid, the second term in the parentheses (at $\theta = \theta_1 = \pi$, equal to AB/L), overrules the first term (at $\theta = \theta_1 = \pi$, equal to $-I_0$), so that the flipped horizontal state becomes unstable as well. Instead, at such fields, Eq. (*) acquires two other roots,

$$\theta_{3,4} = \pm\cos^{-1}\left(-\frac{LI_0}{AB}\right);$$

since

$$\left.\frac{d^2U}{d^2\theta}\right|_{\theta=\theta_{3,4}} = -AB\left(I_0\cos\theta_{3,4} + \frac{AB\cos 2\theta_{3,4}}{L}\right)$$
$$= -AB\left(I_0\cos\theta_{3,4} + \frac{AB(2\cos^2\theta_{3,4} - 1)}{L}\right)$$
$$= AB\left(I_0\frac{LI_0}{AB} - \frac{AB(2[LI_0/AB]^2 - 1)}{L}\right) = \frac{1}{L}(A^2B^2 - L^2I_0^2),$$

both states are stable as soon as they exist (at $AB > LI_0$).

The physics of this curious behavior is follows. In weak fields, the role of the induced current I_B is small, and the applied field just tries to flip the loop, so that its magnetic moment vector $\mathbf{m} \approx \mathbf{m}_0$ would be aligned with the vector $\mathbf{B}$. However, in high magnetic fields, the effects related to I_B, dominate, and in the limit $B \gg LI_0/A$ the field tries to turn loop's *plane* parallel to vector $\mathbf{B}$: $\theta \to \pm\pi/2$ ($\cos\theta \to 0$). This fact may be interpreted using the discussion in the end of section 6.2 of the lecture notes, even though that discussion is quantitatively valid only for a cylindrical geometry. As was shown there, the external magnetic field tends to fill all available space. In the case of superconductor ring with $I_0 \to 0$, the Meissner–Ochsenfeld effect pushes the field out of a volume of the order of $A^{3/2}\cos\theta$, so that the system 'tries' to minimize this volume by making $\cos\theta \to 0$.

***Problem* 6.16.** Use the London equation to analyze the penetration of a uniform external magnetic field into a thin ($t \sim \delta_L$), planar superconducting film, whose plane is parallel to the field.

Solution: According to the definition $\mathbf{B} = \nabla \times \mathbf{A}$ of the vector potential, in this, essentially 1D situation the field's magnitude B is just a derivative of that (A) of the vector potential, and hence obeys an equation similar to the 1D version of Eq. (6.56) of the lecture notes:

$$\frac{d^2B}{dx^2} = \frac{1}{\delta_L^2}B,$$

where axis x is perpendicular to the film. In our current problem, the magnetic field's distribution should be symmetric relative to the plane passing through the middle of film's thickness. This is why, selecting the position of this plane for $x = 0$, we obtain

$$B(x) = \text{const} \times \cosh\frac{x}{\delta_L}.$$

The constant may be found from the usual (not coarse-grain!) boundary condition (5.117) applied to film surfaces: $H(\pm t/2) = H_0 = B_0/\mu_0$, i.e. $B(\pm t/2) = (\mu/\mu_0)B_0$. (For most superconductors, the difference between the ratio (μ/μ_0) and 1 is negligible, but I still keep it spelled out for clarity.) As a result, we obtain

$$B(x) = B_0\frac{\mu}{\mu_0}\frac{\cosh(x/\delta_L)}{\cosh(t/2\delta_L)}.$$

This result shows that if the film is thin ($t \ll \delta_L$), the field is almost uniformly distributed over its thickness. In the opposite limit, the field only penetrates into two δ_L-scale layers at the film's surfaces, and is shielded out of the film's interior by the supercurrent in these surface layers.

***Problem* 6.17.** Use the London equation to calculate the distribution of the supercurrent density j inside a long, straight superconducting wire, with a circular cross-section of radius $R \sim \delta_L$, carrying dc current I.

Solution: For this axially symmetric problem, the solution of the London equation (see Eq. (6.56) of the lecture notes) may be looked for in the form $\mathbf{A}(\mathbf{r}) = A(\rho)\mathbf{n}_z$, where axis z coincides with that of the wire, and ρ is the distance of the observation point $\mathbf{r}$ from that axis. With the well-known expression for the Laplace operator in cylindrical coordinates[27], the equation becomes[28]

[27] See, e.g. Eq. (A.61) with $\partial/\partial\varphi = \partial/\partial z = 0$.

[28] Note that this equation, and the balance of the solution, are very similar to those in problem 6.9, with the 'only' exception that the complex parameter κ, given Eq. (6.30), is now replaced with the real parameter $1/\delta_L$.

$$\frac{1}{\rho}\frac{d}{d\rho}\left(\rho\frac{dA}{d\rho}\right) = \frac{1}{\delta_{\rm L}^2}A.$$

Introducing the natural dimensionless variable $\xi \equiv \rho/\delta_{\rm L}$, this equation may be brought to the form,

$$\frac{d^2A}{d\xi^2} + \frac{1}{\xi}\frac{dA}{d\xi} - A = 0,$$

which is the modified Bessel equation (2.155) with $\nu = 0$. Its general solution is

$$A = a_1 I_0(\xi) + a_2 K_0(\xi),$$

so that the supercurrent density j (which, within the London gauge we are using, is proportional to A—see Eq. (6.55) of the lecture notes):

$$j(\rho) = c_1 I_0\left(\frac{\rho}{\delta_{\rm L}}\right) + c_2 K_0\left(\frac{\rho}{\delta_{\rm L}}\right).$$

Since the modified Bessel function of the second kind, $K_0(\xi)$ diverges at $\xi \to 0$ (see, e.g. Eq. (2.157), or the right-hand panel of figure 2.22), while j has to stay finite at all points, the coefficient c_2 has to equal 0. The remaining coefficient c_1 may be found from the given total current:

$$\int_A j(\rho)d^2\rho = 2\pi\int_0^R j(\rho)\rho d\rho = 2\pi\ c_1\int_0^R I_0\left(\frac{\rho}{\delta_{\rm L}}\right)\rho d\rho = 2\pi\ \delta_{\rm L}^2 c_1\int_0^{R/\delta_L} I_0(\xi)\xi d\xi = I.$$

The dimensionless integral in this relation may be calculated using the recurrence relation (2.143) (valid for any Bessel functions), in our current case with $n = 1$. As a result, we obtain

$$c_1 = \frac{I}{2\pi\delta_{\rm L}^2}\left[\int_0^{R/\delta_L} I_0(\xi)\xi d\xi\right]^{-1} = \frac{I}{2\pi\delta_{\rm L}^2[\xi I_1(\xi)]_0^{R/\delta_{\rm L}}} = \frac{I}{2\pi\delta_{\rm L}RI_1(R/\delta_{\rm L})},$$

so that, finally,

$$j = \frac{I}{2\pi\delta_{\rm L}R}\frac{I_0(\rho/\delta_{\rm L})}{I_1(R/\delta_{\rm L})}.$$

If the wire is very thin ($R \ll \delta_{\rm L}$), then we may use the first of Eqs. (2.157) to write $I_0(\rho/\delta_{\rm L}) \approx 1$, $I_1(R/\delta_{\rm L}) \approx R/2\delta_{\rm L}$, so that $j \approx I/\pi R^2$ = const, i.e. the supercurrent is uniformly distributed over the wire's cross-section. In the opposite limit $\delta_{\rm L} \ll R$, we may use the first of asymptotic formulas (2.158) to reduce our result to

$$j \approx \frac{I}{2\pi\delta_L R}\left(\frac{R}{\rho}\right)^{1/2}\frac{\exp\{\rho/\delta_L\}}{\exp\{R/\delta_L\}} = \frac{I}{2\pi\delta_L(R/\rho)^{1/2}}\exp\left\{-\frac{R-\rho}{\delta_L}\right\}.$$

This expression shows that appreciable supercurrent only flows in a δ_L-thin sheet at the wire's surface—exactly as in the plane-surface problem, see Eq. (6.57).

***Problem* 6.18.** Use the London equation to calculate the inductance (per unit length) of a long, uniform superconducting strip placed close to the surface of a similar superconductor—see the figure below, which shows the structure's cross-section.

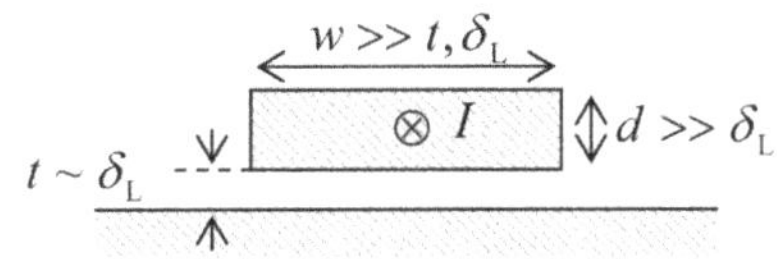

Solution: DC supercurrent does not require any electric field to sustain it, so that it is free to distribute over the strip's cross-section to minimize the total energy of the system. In particular, in our limit t, $\delta_L \ll w$, all current in the strip has to be localized at its surface facing the bulk superconductor (in electrical engineering, called the *ground plane*), to minimize the magnetic field outside the gap t, because any substantial field there, decaying only on distances $\sim w$, would give very high contributions, with a weight $d^2r \sim w^2$, into the magnetic energy (per unit length),

$$\frac{U}{l} = \int \frac{B^2(\mathbf{r})}{2\mu(\mathbf{r})} d^2r. \qquad (*)$$

Next, for a gap of constant thickness t, the linear density J of the supercurrent should be constant across the gap's width w, because otherwise the magnetic flux (per unit length),

$$\frac{\Phi}{l} = \int B(x)dx \propto J, \qquad (**)$$

of field $B(x)$ it creates in the gap would change across the width. (Here x is the Cartesian coordinate across the gap.) This is impossible, because the thick superconducting films prevent the magnetic field lines' from escaping the gap. So, $J = I/w = \text{const}$.

The last (but not least) fact we need is that in order to prevent the magnetic field in the gap from spreading into the ground plane, its surface has to carry an equal and opposite supercurrent J. (This fact is the basis of *dc current transformers*, which are an important component of superconductor electronics.)

Now the problem may be readily solved, using Eq. (6.57) of the lecture notes. Indeed, the integral in Eq. (**) is the sum of three integrals: one over the gap (giving the contribution $Bt = \mu_0 Ht = \mu_0 Ht = \mu_0 Ht = \mu_0 Jt = \mu_0 It/w$), and two similar integrals through the London penetration layers of the top electrode and the ground plane:

$$\int_0^\infty B(x)dx = B(0)\int_0^\infty \exp\left\{-\frac{x}{\delta_L}\right\}dx = \mu H(0)\delta_L = \mu J\delta_L = \frac{\mu I\delta_L}{w}.$$

As a result, for the inductance L per unit length we obtain

$$\frac{L}{l} \equiv \frac{\Phi}{lI} = \frac{1}{I}\int B(x)dx = \frac{1}{I}\left(\frac{\mu_0 It}{w} + 2\frac{\mu I\delta_L}{w}\right) = \mu_0\frac{t_{ef}}{w},$$
$$\text{with } t_{ef} \equiv t + 2\frac{\mu}{\mu_0}\delta_L. \qquad (***)$$

This expression is similar to the solution of problem 5.3(iv), apart from the effective increase of the gap between the strips due to the magnetic field penetration into the superconductors. In real superconductor electronic circuits, the insulating gap thicknesses t may be from a few nanometers to ~100 nm, while δ_L is of the order of 100 nm, so that this effect is quite noticeable (and may be used in particular for the London depth measurement).

Note also that in contrast to the coarse-grain, ideal-diamagnet description of superconductivity (by taking $\mu = 0$ inside it), the London equation takes supercurrents into account explicitly (as 'stand-alone' currents), so that the magnetic permeability participating in this result describes only the effects of *localized* atomic currents. In typical superconductors, these effects are relatively weak ($\mu \approx \mu_0$),[29] so that the effective gap width t_{ef} is very close to just $(t + 2\delta_L)$.

It might be tempting to derive Eq. (***) by calculating the magnetic field energy (*) and requiring it to equal $LI^2/2$. However, such calculation gives a wrong result for t_{ef}, without the factor of 2 before the second term. The reason for this discrepancy is pretty subtle: since the Cooper pairs move in superconductors without dissipation, we cannot neglect their *kinetic energy*. Indeed, according to the London equation (6.55), the effective velocity of the Cooper pair condensate is

$$\mathbf{v} \equiv \frac{\mathbf{j}}{qn_p} = -\frac{q}{m}\mathbf{A}.$$

For the Meissner–Ochsenfeld effect in a 1D geometry, described by Eq. (6.57), the vector-potential definition $\nabla \times \mathbf{A} \equiv \mathbf{B}$ yields $A^2 = \delta_L^2 B^2 = [m/\mu q^2 n_p]B^2$, so that the kinetic energy of Cooper pairs per unit volume,

$$\frac{T}{V} = n_p\frac{mv^2}{2} = n_p\frac{m}{2}\left(\frac{q}{m}\right)^2 A^2 = n_p\frac{m}{2}\left(\frac{q}{m}\right)^2 \frac{m}{\mu q^2 n_p}B^2 \equiv \frac{B^2}{2\mu},$$

is exactly equal to that of the magnetic field. Naturally, its account properly doubles the contribution to energy from the London penetration layers.

[29] In particular because ferromagnetism suppresses superconductivity, so that the two effects rarely coexist in the same material.

***Problem* 6.19.** Calculate the inductance (per unit length) of a superconducting cable with a round cross-section, shown in the figure below, in the following limits:

(i) $\delta_L \ll a, b, c - b$, and
(ii) $a \ll \delta_L \ll b, c - b$.

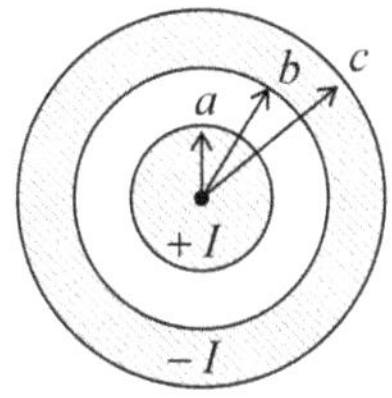

Solutions: Due to problem's symmetry, the direction of the magnetic field is perpendicular to the direction $\boldsymbol{\rho}$ from the system's axis, and its magnitude is a function of ρ alone. Hence the *magnetic* energy (per unit length of the cable) may be calculated just as

$$\frac{U}{l} = \frac{1}{2\mu_0}\int B^2(\rho)d^2\rho = \frac{\pi}{\mu_0}\int B^2(\rho)\rho d\rho,$$

where the integral should be extended to all distances at which B is substantial. Note that the net current in the cable (i.e. the algebraic sum of currents in its conductors) is zero, so that according to the Ampère law (5.37), the field vanishes at $\rho > c$.

(i) In the limit $\delta_L \ll a, b, c - b$, we can neglect the magnetic field's penetration into both conductors of the cable[30], and use Eq. (5.20) for the field between them:

$$B(\rho) = \frac{\mu_0 I}{2\pi\rho}.$$

Now an elementary integration gives

$$\frac{U}{l} = \frac{\pi}{\mu_0}\int_a^b B^2(\rho)\rho d\rho = \frac{\pi}{\mu_0}\left(\frac{\mu_0 I}{2\pi}\right)^2\int_a^b \frac{d\rho}{\rho} = \frac{\mu_0 I^2}{4\pi}\ln\frac{b}{a},$$

so that, according to Eq. (5.71), the self-inductance per unit length is

$$\frac{L}{l} \equiv \frac{U/(I^2/2)}{l} = \frac{\mu_0}{2\pi}\ln\frac{b}{a}.$$

Comparing this result with Eq. (5.79) of the lecture notes, we see that this inductance is somewhat lower than that of a normal-conductor cable (at low frequencies), due to the negligible magnetic field penetration into the conductors. Note, however, that if the frequency of current is so high that the skin-depth δ_s (6.33) is much smaller than all transverse dimensions of the cable, the above expression is valid for the normal-conductor cables as well.

[30] This essentially means that in this limit we are using the coarse-grain (ideal-diamagnetic) description of the Meissner–Ochsenfeld effect.

(ii) As Eq. (6.56) shows, in the limit $a \ll \delta_L$, the supercurrent is uniformly distributed across the inner conductor, with density $j = I/\pi a^2$, just as in a uniform normal conductor. Thus we may repeat the calculations of section 5.3, that led to Eq. (5.79), with the integration from 0 to b only, and obtain

$$\frac{L_m}{l} \equiv \frac{U/(I^2/2)}{l} = \frac{1}{2\pi}\left(\mu_0 \ln \frac{b}{a} + \frac{\mu}{4}\right),$$

i.e. the same answer as for a normal-metal cable with $(c - b) \to 0$—see Eq. (5.80). However, this is not the end of the story. As was discussed in the solution of the previous problem, in superconductors, the magnetic energy has to be summed up with the kinetic energy T of the Cooper-pair condensate. Moreover, in our current case of a very thin inner conductor ($a \ll \delta_L$, i.e. $A_a \equiv \pi a^2 \ll \delta_L^2$), T is much larger than U. Indeed, for the uniformly distributed supercurrent, the kinetic energy per unit length is

$$\frac{T}{l} = A_a n_p \frac{mv^2}{2} = A_a n_p \frac{m}{2}\left(\frac{j}{qn_p}\right)^2 = A_a n_p \frac{m}{2}\left(\frac{I}{qn_p A_a}\right)^2 = \frac{m}{q^2 n_p A_a}\frac{I^2}{2},$$

so that the *kinetic inductance* associated with this effect,

$$\frac{L_k}{l} \equiv \frac{T/(I^2/2)}{l} = \frac{m}{n_p(T)q^2 A_a} = \mu\frac{\delta_L^2}{A_a} = \mu\frac{\delta_L^2}{\pi a^2} \gg \mu,$$

is much larger than the magnetic inductance L_m.

***Problem* 6.20.** Use the London equation to analyze the magnetic field shielding by a superconducting thin film of thickness $t \ll \delta_L$, by calculating the penetration of the field induced by current I flowing in a thin wire which runs parallel to a wide, planar thin film, at distance $d \gg t$ from it, into the half-space behind the film.

Solution: This problem is conceptually very close to that of problem 2.35 on electrostatic screening, so the reader is advised to review its solution first[31]. The only substantial novelty of the magnetic field is the vector character of its potential **A**. However, as was discussed in section 5.2 of the lecture notes[32], the potential created by a current in a straight wire has only one Cartesian component, directed along the wire, and it is clear that a superconducting film parallel to it cannot change this alignment. (The proof of this statement is simple: if **A** had obtained a transverse component, would it be directed clockwise or counter-clockwise?) So, in our geometry the London equation (6.56) is valid for the magnitude A of the only (say, z-) component of $\mathbf{A} = A\mathbf{n}_z$.

Hence, we may return to the same reasoning as was used in the model solution of problem 2.35. Namely, due to the imposed condition $t \ll \delta_L$, the gradient of the

[31] The current problem is more realistic, because (as was discussed in section 6.4) the London penetration depth δ_L is typically of the order of 10^{-7} – 10^{-6} m, i.e. much larger than the Thomas–Fermi screening length ($\sim 10^{-10}$ m in metals, see section 2.1), so that superconductor films of thickness $t \ll \delta_L$ are quite practicable.

[32] See, e.g. the discussion leading to Eq. (5.49).

vector-potential A in the film is dominated by its component along the axis (say, y) normal to the film's plane, so that the London equation (6.56) is well approximated by its 1D version

$$\frac{\partial^2 A}{\partial y^2} = \frac{1}{\delta_L^2} A,$$

even if A is relatively slowly (on distances of the order of $d \gg t$) changing in the plane of the film. Integrating both sides of this equation over the film's thickness t, we obtain

$$\frac{\partial A_+}{\partial y} - \frac{\partial A_-}{\partial y} = \frac{1}{\delta_L^2} \int_{-t/2}^{+t/2} A dy, \qquad (*)$$

where the indices ± refer to two film's surfaces. This relation shows that if $t \ll \delta_L$, the film can create only a small difference (of the order of At/δ_L^2) of the derivative $\partial A/\partial y$ through its thickness, and hence just a negligible variation (of the order of At^2/δ_L^2) of the vector-potential inside it. As a result, Eq. (*) yields two boundary conditions for the potential's distribution outside the film:

$$A_+ = A_- \equiv A, \qquad \frac{\partial A_+}{\partial y} - \frac{\partial A_-}{\partial y} = \frac{t}{\delta_L^2} A. \qquad (**)$$

Now, inspired by the success of numerous charge-image analyses in chapters 2–5, we may try to describe the magnetic field (and hence the potential A) at the half-space where the actual current resides (say, $y > 0$) as a superposition of the fields induced by that current I, and by its image I' running in parallel, at the same distance d, but on the opposite side of the film—see the figure below.

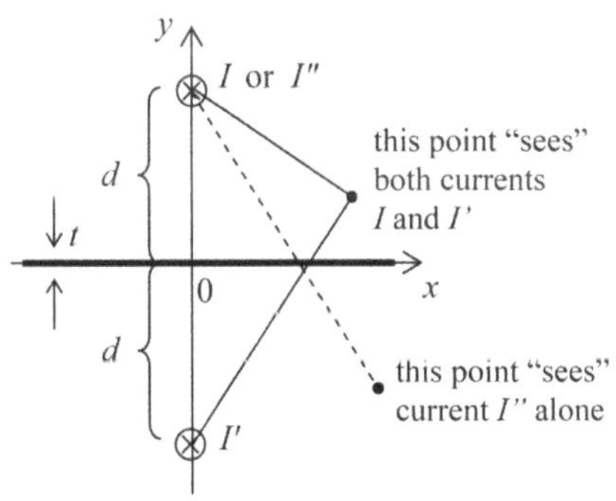

In addition, in order to describe some field's remnants at the other side of the film (at $z < 0$), we may use the effective current I'' co-located with the real current. As a result, using Eq. (5.51) of the lecture notes for each current's potential, we may look for the potential distribution in the form[33]

[33] As was discussed in section 3.3 of the lecture notes, absolutely the same solution (although with different constants q' and q'') describes the electric field penetration into a half-space filled with a linear dielectric.

$$A(x, y) = -\frac{\mu_0}{2\pi} \times \begin{cases} I \ln [x^2 + (y - d)^2]^{1/2} + I' \ln [x^2 + (y + d)^2]^{1/2}, & \text{for } y > 0, \\ I'' \ln [x^2 + (y - d)^2]^{1/2}, & \text{for } y < 0. \end{cases} \qquad (***)$$

Plugging this solution into the boundary conditions (6.37), we see that they are indeed satisfied (so that Eq. (***) is indeed the unique solution of our boundary problem), if the effective currents I' and I'' obey the following relations:

$$I + I' = I'', \qquad (I - I') - I'' = \frac{td}{\delta_L^2} I'',$$

similar to those (for image charges) in the similar electrostatic screening problem. As a result, the solution of this simple system of equations for I' and I'' is also similar. In particular, for the ratio I''/I, which is the adequate measure of the field's penetration behind the film, we obtain

$$\frac{I''}{I} = \frac{1}{1 + td/2\delta_L^2}.$$

As in electrostatics, the most interesting feature of this result is that the ratio starts to drop as soon as t is increased to $\sim\delta_L^2/d \ll \delta_L$, i.e. the field may be well (though not exponentially well) screened even by a film much thinner than the screening length, if only d is large enough. Another formulation of the same fact, which may be more revealing in some problems, is that a thin ($t \ll \delta_L$) superconductor film has a certain characteristic *lateral* size scale

$$\delta_\perp \equiv \frac{2\delta_L^2}{t},$$

usually called the *perpendicular penetration length*, which characterizes its diamagnetic properties. In particular, it is easy to use Eqs. (5.28) and (6.55) to show (a task highly recommended to the reader) that the supercurrent, run through a strip of width w, made of such a thin film, is uniformly distributed over its width only if $w \ll \delta_\perp$. This is one of the reasons why in applications, superconducting strips are typically run very close to a superconducting ground plane, which enforces the uniform current distribution even in wide thin films—see, e.g. problem 6.18.

***Problem* 6.21.** Use the Ginzburg–Landau equations (6.54) and (6.63) to calculate the largest ('critical') value of the supercurrent in a uniform, long superconducting wire of a small cross-section $A_w \ll \delta_L{}^2$.

Solution: As was discussed in the solution of problem 6.17, in a wire with a cross-section so small, the current density, and hence the gauge-invariant combination $(\hbar\nabla - iq\mathbf{A})\psi$, participating in the Ginzburg–Landau equations, are uniformly distributed along the cross-section, so that Eqs. (6.54), with both sides multiplied by A_w, yields

$$I = j_z A_\text{w} = \frac{\hbar q n_\text{p} A_\text{w}}{m} |\psi|^2 \left(\frac{\partial \varphi}{\partial z} - \frac{q}{\hbar} A \right), \tag{*}$$

where A is the (only) Cartesian component of the vector-potential, directed along the wire's length. As was discussed at the derivation of Eq. (6.55) of the lecture notes, in this geometry we may (just for notation simplicity) select the gauge $\varphi = \text{const}$, so that Eq. (*) yields

$$A = -\frac{m}{q^2 n_\text{p} |\psi|^2} \frac{I}{A_\text{w}} \equiv -\frac{\mu \delta_\text{L}^2}{|\psi|^2} \frac{I}{A_\text{w}}.$$

Plugging this expression into the second form of Eq. (6.63), taken in the same gauge, we obtain the following equation:

$$\xi^2 \left(\frac{\mu q \delta_\text{L}^2}{\hbar} \frac{I}{A_\text{w}} \right)^2 = |\psi|^4 (1 - |\psi|^2). \tag{**}$$

At $I \to 0$, its stable solution gives the unperturbed value (equal to 1) of the modulus of the wavefunction ψ of the Cooper pair condensate, but an increase of the current suppresses it—see the figure below, which shows the right-hand side of Eq. (**) as a function of $|\psi|^2$. As an elementary differentiation shows, the function reaches its maximum at

$$|\psi|^2 = \frac{2}{3}, \quad \text{with } [|\psi|^4(1 - |\psi|^2)]_\text{max} = \left(\frac{2}{3}\right)^2 \left(1 - \frac{2}{3}\right) \equiv \frac{4}{27} \approx 0.148.$$

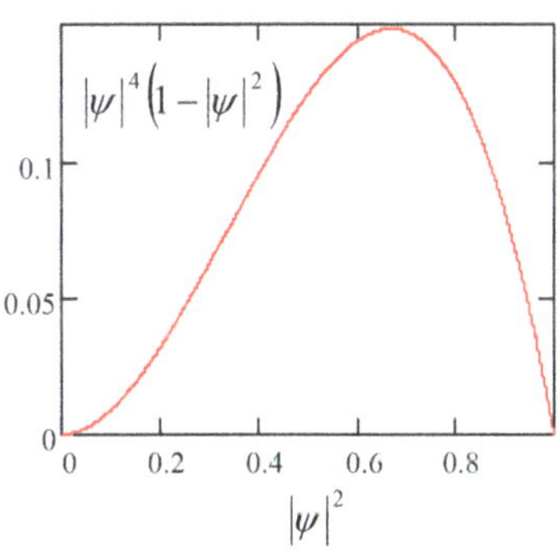

From here, the maximum ('critical') value of the supercurrent density is

$$j_\text{c} \equiv \frac{I_\text{c}}{A_\text{w}} = \left(\frac{4}{27}\right)^{1/2} \frac{\hbar}{\mu |q| \delta_\text{L}^2 \xi}. \tag{***}$$

Increasing j above this value destroys the superconductivity, driving the wire into its 'normal' state, with $\psi = 0$. Note that the external magnetic field **H**, applied to a bulk superconductor, has a qualitatively similar effect. Indeed, as Eqs. (6.38) and (6.57) of the lecture notes, it induces a surface supercurrent with a linear density

$J \sim H$,[34] distributed within a layer of thickness δ_L, i.e. with the areal density $j \sim H/\delta_L$. When the density approaches the critical value (***), i.e. at

$$H \sim H_c \equiv \frac{\hbar}{\sqrt{2}\mu|q|\delta_L\xi}, \quad \text{i.e. } B \sim B_c \equiv \mu H \equiv \frac{\hbar}{\sqrt{2}|q|\delta_L\xi} \equiv \frac{\Phi_0}{2\sqrt{2}\pi\delta_L\xi}. \qquad (****)$$

the superconductivity in the surface layer starts to be suppressed. As the detailed solutions of the Ginzburg–Landau equations show[35], the behavior in higher fields is rather complicated and depends on the relation between the superconductor's characteristic lengths δ_L and ξ. In the so-called *type-I* superconductors, with $\delta_L < \xi/\sqrt{2}$, the superconductivity (at least in the cylindrical geometry) is suppressed, at $H = H_c$, simultaneously in the whole sample. However, in the *type-II superconductors*, which have $\xi/\sqrt{2} < \delta_L$, the magnetic field may penetrate into the superconductor in the form of separate Abrikosov vortices—see the discussion in section 6.5 of the lecture notes[36]. Moreover, in the case $\xi/\sqrt{2} \ll \delta_L$, such vortices may penetrate the superconductor even at much lower fields—see Eq. (6.72) of the lecture notes, and the next problem.

***Problem* 6.22.** Use the discussion of a long, straight Abrikosov vortex, in the approximation $\xi \ll \delta_L$, in section 6.5 to prove Eqs. (6.71) and (6.72) for its energy per unit length and the first critical field.

Solution: As follows from the model solutions of problems 6.18 and 6.19, the total energy of the vortex is the sum of its magnetic field energy,

$$U = \frac{1}{2\mu}\int B^2 d^3r,$$

and the kinetic energy of the Cooper-pair condensate,

$$T = \frac{m}{2}\int n_p v^2 d^3r \equiv \frac{m}{2}\int n_p\left(\frac{j}{qn_p}\right)^2 d^3r \equiv \frac{m}{2q^2 n_p}\int j^2 d^3r \equiv \frac{\mu\delta_L^2}{2}\int j^2 d^3r,$$

where, for the last step, Eq. (6.46) of the lecture notes was used. Plugging, into these expressions, Eqs. (6.68) and (6.70), respectively, we obtain the following energies per unit length of the vortex:

[34] As Eq. (6.38) shows, in the cylindrical geometry (such as shown in figure 5.15 or figure 6.2) the relation $J = H_{ext}$ is exact; however, in the general case there is an additional geometric factor in their relation—see, e.g. Eqs. (6.58).

[35] See, e.g. [1].

[36] For such superconductors, the above result for the critical current is valid for a wire with the cross-section's area smaller than not only δ_L^2, but also than ξ^2.

$$\frac{U}{l} = \frac{1}{2\mu}\int B^2 d^2r = \frac{1}{2\mu}2\pi\int_{\sim\xi}^{\infty} B^2(\rho)\rho d\rho$$
$$= \frac{\pi}{\mu}\left(\frac{\Phi_0}{2\pi\delta_L^2}\right)^2\int_{\sim\xi}^{\infty} K_0^2\left(\frac{\rho}{\delta_L}\right)\rho d\rho \equiv \frac{\Phi_0^2}{4\pi\mu\delta_L^2}\int_{\varepsilon}^{\infty} K_0^2(\zeta)\zeta d\zeta,$$
$$\frac{T}{l} = \frac{\mu\delta_L^2}{2}\int j^2 d^2r = \frac{\mu\delta_L^2}{2}2\pi\int_{\sim\xi}^{\infty} j^2(\rho)\rho d\rho$$
$$= \pi\mu\delta_L^2\left(\frac{\Phi_0}{2\pi\mu\delta_L^2}\right)^2\int_{\sim\xi}^{\infty} K_1^2\left(\frac{\rho}{\delta_L}\right)\rho d\rho \equiv \frac{\Phi_0^2}{4\pi\mu\delta_L^2}\int_{\varepsilon}^{\infty} K_1^2(\zeta)\zeta d\zeta,$$

so that the total energy of the vortex per unit length, i.e. its tension, is

$$\mathscr{T} \equiv \frac{U+T}{l} = \frac{\Phi_0^2}{4\pi\mu\delta_L^2}\int_{\varepsilon}^{\infty} [K_0^2(\zeta) + K_1^2(\zeta)]\,\zeta d\zeta,$$

where $\varepsilon \sim \xi/\delta_L \ll 1$. (This lower cutoff is due to the fact, that Eqs. (6.68) and (6.70) are only valid at distances much larger than the coherence distance ξ—see the discussion in section 6.5 of the lecture notes.) Such integrals are well known[37]:

$$\int K_n^2(\zeta)\zeta d\zeta = \frac{\zeta^2}{2}[K_n^2(\zeta) - K_{n-1}(\zeta)K_{n+1}(\zeta)];$$

taking into account that, according to Eq. (2.150), $K_{-1}(\zeta) = -K_{-1}(\zeta)$, for our particular case they give

$$\int_{\varepsilon}^{\infty}\left[K_0^2(\zeta) + K_1^2(\zeta)\right]\zeta d\zeta = \left\{\frac{\zeta^2}{2}\left[K_0^2(\zeta) + 2K_1^2(\zeta) - K_0(\zeta)K_2(\zeta)\right]\right\}_{\varepsilon}^{\infty}.$$

As Eq. (2.158) (and figure 2.22) of the lecture notes shows, on the upper limit all these functions vanish. Now using the second of Eqs. (2.157) for their approximate values on the lower limit, we obtain

$$\int_{\varepsilon}^{\infty}\left[K_0^2(\zeta) + K_1^2(\zeta)\right]\zeta d\zeta \approx \frac{\varepsilon^2}{2}K_0(\varepsilon)K_2(\varepsilon) \approx \ln\frac{1}{\varepsilon} \approx \ln\frac{\delta_L}{\xi},$$

thus proving Eq. (6.71):

$$\mathscr{T} \approx \frac{\Phi_0^2}{4\pi\mu\delta_L^2}\ln\frac{\delta_L}{\xi}.$$

A more accurate theory[38], involving an explicit solution of the Ginzburg–Landau equation (6.53) in the vortex's core (i.e. at distances at $\rho \sim \xi \ll \delta_L$ from its axis), confirms this result, offering just a small correction: $\ln(\delta_L/\xi) \to \ln(\delta_L/\xi) + 0.50$.

[37] This expression, valid for any Bessel functions, follows from the recurrence relations (2.142).

[38] See [2].

In order to calculate the first critical field H_{c1}, let us use Eq. (6.17) of the lecture notes for the Gibbs potential energy, corrected for the kinetic energy T of the Cooper-pair condensate, by the replacement $U \to U + T \equiv \mathscr{T}l$, for a single vortex inside a superconductor:

$$\frac{U_G}{l} = \mathscr{T} - \int \mathbf{H}_{\text{ext}} \cdot \mathbf{B} d^2r,$$

where the integral may be limited by the cross-section of the vortex. In the cylindrical geometry we are considering, the field $\mathbf{B}$ of the vortex is directed along the external field $\mathbf{H}_{\text{ext}}$, which is constant. Hence we may take $\mathbf{H}_{\text{ext}}$ out of the integral, and the remaining integral of $\mathbf{B}$ is just the quantized magnetic flux Φ_0 of the vortex; hence

$$\frac{U_G}{l} = \mathscr{T} - H_{\text{ext}}\Phi_0.$$

It is energy-favorable for a vortex to enter the superconductor if this value is lower than U_G in the absence of the vortex (with our choice of the arbitrary constant, this value is zero), i.e. if $H_{\text{ext}} > H_{c1}$, where

$$H_{c1} = \frac{\mathscr{T}}{\Phi_0} \approx \frac{\Phi_0}{4\pi\mu\delta_L^2} \ln \frac{\delta_L}{\xi},$$

thus proving Eq. (6.72) of the lecture notes.

Note that the condition $H_{\text{ext}} > H_{c1}$ makes the vortex entrance possible, but does not guarantee it, because it may be prevented by a substantial surface potential barrier, which is fully suppressed only when the field reaches the (much higher) value H_c, calculated (or rather estimated) in the previous problem:

$$H_c \equiv \frac{\hbar}{\sqrt{2}\mu|q|\delta_L\xi} \equiv \frac{\Phi_0}{2\pi\sqrt{2}\mu\delta_L\xi} \sim \frac{\delta_L}{\xi}H_{c1} \gg H_{c1}.$$

However, certain experimental tricks (such as using slightly non-cylindrical geometries or surface plane roughening) may enable reproducible vortex entry at fields very close to H_{c1}.

Problem* 6.23. Use the Ginzburg–Landau equations (6.54) and (6.63) to prove the Josephson's relation (6.76) for a small superconducting weak link, and express its critical current I_c of via the Ohmic resistance R_n of the same weak link in its normal state.

Solution[39]: Let weak link size scale a be much smaller than both the Cooper pair size ξ and the London penetration depth δ_L. The first of these relations ($a \ll \xi$) makes the left-hand side of the second form of Eq. (6.63), which scales as ξ^2/a^2, much larger than 1, i.e. much larger that its right-hand side. On the other hand, according to the

[39] This simple and elegant calculation belongs to L Aslamazov and A Larkin [3].

discussion in section 6.4 of the lecture notes, the latter relation ($a \ll \delta_L$) allows us to neglect magnetic field effects, and hence drop the term $(-q\mathbf{A})$ from the parenthesis in Eq. (6.63), reducing it to just our familiar Laplace equation, but now for the complex wavefunction:

$$\nabla^2 \psi = 0.$$

Since the weak coupling cannot change $|\psi|$ in the superconducting electrodes' bulk, this differential equation may be solved with the following simple boundary conditions:

$$\psi(\mathbf{r}) \to \begin{cases} |\psi| e^{i\varphi_1}, & \text{for } \mathbf{r} \to \mathbf{r}_1, \\ |\psi| e^{i\varphi_2}, & \text{for } \mathbf{r} \to \mathbf{r}_2, \end{cases}$$

where $\mathbf{r}_1$ and $\mathbf{r}_2$ are some points well inside the corresponding superconductors, i.e. at distances much larger than a from the weak link. It is straightforward to verify that the (unique) solution of this boundary problem may be expressed as follows,

$$\psi(\mathbf{r}) = |\psi| e^{i\varphi_1} f(\mathbf{r}) + |\psi| e^{i\varphi_2} [1 - f(\mathbf{r})],$$

via the *real* function $f(\mathbf{r})$ that satisfies the Laplace equation and the following boundary conditions:

$$f(\mathbf{r}) \to \begin{cases} 1, & \text{for } \mathbf{r} \to \mathbf{r}_1, \\ 0, & \text{for } \mathbf{r} \to \mathbf{r}_2. \end{cases}$$

The function $f(\mathbf{r})$ depends on the weak link geometry and may be rather complicated, but we do not need to know it to obtain the most important result. Indeed, plugging this solution into Eq. (6.54), with the term $-q\mathbf{A}$ again ignored as being negligibly small, we obtain

$$\mathbf{j} = -\frac{\hbar q n_p}{m} \nabla f \sin \varphi.$$

Integrating this relation over any cross-section S of the weak link, we arrive at the Josephson's result (6.76), with the following critical current:

$$I_c = -\frac{\hbar q n_p(T)}{m} \int_S (\nabla f)_n d^2 r.$$

This expression may be readily evaluated via the resistance of the same weak link in the 'normal' (non-superconducting) state, say at $T > T_c$. Indeed, as we know from section 4.3, the distribution of the electrostatic potential ϕ at normal conduction also obeys the Laplace equation, with boundary conditions that may be taken in the form

$$\phi(\mathbf{r}) \to \begin{cases} V, & \text{for } \mathbf{r} \to \mathbf{r}_1, \\ 0, & \text{for } \mathbf{r} \to \mathbf{r}_2. \end{cases}$$

Comparing the boundary problem for the function $\phi(\mathbf{r})$ with that for the function $f(\mathbf{r})$, we obtain $\phi = Vf$. This means that the gradient ∇f, which participates in the expression for the critical current, is just $(-\mathbf{E}/V) = (-\mathbf{j}/\sigma V)$. Hence the integral in that formula is just $-I/\sigma V = -1/\sigma R_n$, where R_n is the resistance of the Josephson junction in its normal state. As a result, we get the expression

$$I_c = \frac{\hbar q n_p}{m\sigma}\frac{1}{R_n},$$

showing that the I_cR_n product is independent of the junction geometry, though it does depend on temperature, vanishing, together with n_p, at $T \to T_c$. (The microscopic theory of superconductivity, which I do not have time to cover, shows that well below the critical temperature, I_cR_n of sufficiently short weak links is of the order of $\Delta(0)/e$, i.e. typically a few mV.)

***Problem* 6.24.** Use Eqs. (6.76) and (6.79) of the lecture notes to calculate the coupling energy of a Josephson junction, and the full potential energy of the SQUID shown in figure 6.4c.

Solution: The elementary work of an external voltage V, applied to a lumped, two-terminal device that carries current $I = dQ/dt$ may be calculated as $d\mathscr{W} = VdQ = VIdt$. Combining this expression with Eqs. (6.76) and (6.79), we obtain

$$d\mathscr{W} = VIdt = \frac{\hbar}{2e}\frac{d\varphi}{dt}I_c \sin\varphi\, dt \equiv \frac{\hbar I_c}{2e}\sin\varphi\, d\varphi \equiv -\frac{\hbar I_c}{2e}d(\cos\varphi).$$

From here, the work to increase the junction's phase to a certain value φ, i.e. its coupling ('Josephson') energy U_J is[40]

$$U_J(\varphi) = \int^{\varphi} d\mathscr{W} = -E_J\cos\varphi + \text{const}, \qquad \text{where } E_J \equiv \frac{\hbar I_c}{2e}.$$

Note that the function $U_J(\varphi)$ is highly nonlinear, and quadratic only at $|\varphi| \ll 1$ (when Eq. (6.76) is reduced to $I \approx I_c\varphi$)[41]:

$$U_J(\varphi) \approx E_J\frac{\varphi^2}{2} + \text{const} = \frac{E_J}{I_c^2}\frac{I^2}{2} + \text{const}.$$

In a SQUID, the Josephson energy U_J should be added to the magnetic energy $U_m = LI^2/2$ of the superconducting loop, so that using Eqs. (6.78), the energy of the system may be represented as

$$U = \frac{LI^2}{2} - E_J\cos\varphi + \text{const} = \frac{(\Phi - \Phi_{\text{ext}})^2}{2L} - \frac{\hbar I_c}{2e}\cos\frac{2\pi\Phi}{\Phi_0} + \text{const}. \qquad (*)$$

[40] Physically, U_J is just a form of the kinetic energy of the Cooper-pair condensate, similar to that discussed in the solutions of problems 6.19, 6.20, and 6.22, but for the specific case of a Josephson junction, with its highly nonlinear dynamics.

[41] By the way, the last expression shows that the non-magnetic ('kinetic' or 'Josephson' *inductance* of the junction for small current is $L_J \equiv E_J/I_C^2 \equiv \hbar/2eI_c$. This formula is broadly used in superconductor electronics.

If the magnetic flux Φ is understood as the generalized coordinate of the system, this function $U = U(\Phi)$ represents its potential energy (despite the 'kinetic energy' terminology for its component U_J), so that its minima correspond to stable states of the system[42]. The figure below shows this potential energy profile, plotted for several values of the external flux, and two characteristic values of the normalized LI_c product,

$$\lambda \equiv \frac{2\pi L I_c}{\Phi_0},$$

one below 1, and the other above 1.

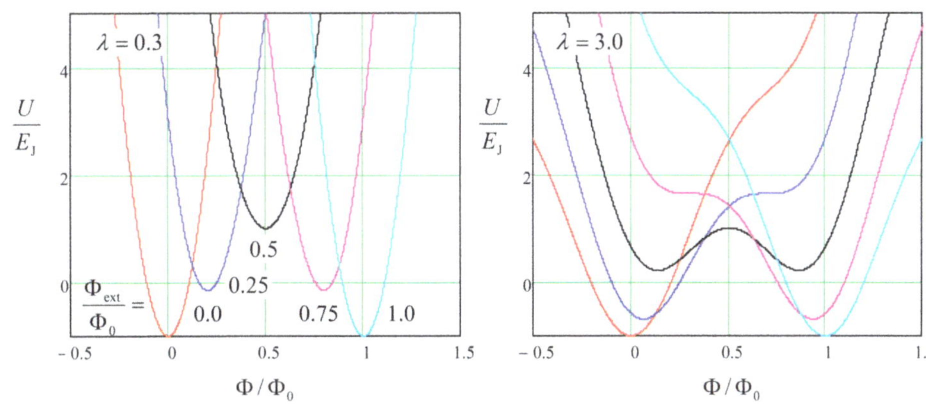

An elementary analysis of Eq. (*) shows that the fixed-point requirement $\partial U/\partial\Phi = 0$ yields the stationary relation $\Phi(\Phi_{ext})$ given by Eqs. (6.77) and (6.78) and shown in figure 6.6 of the lecture notes. The negative-slope branches of those plots correspond to the maxima of energy U (with $\partial^2 U/\partial\Phi^2 < 0$) and hence are unstable. As these plots show, a SQUID with $\lambda < 1$ always has just one stable state Φ, at $\lambda \ll 1$ closely following Φ_{ext}. In contrast, a SQUID with $\lambda > 1$ may have several (at $\lambda \gg 1$, many) stable states, corresponding to quasi-quantization of the flux.

The flexibility of $U(\Phi)$ profile to the change of parameters Φ_{ext} and λ is broadly used in superconductor electronics; for example, it was used for the first experimental observation of the quantum interference of two macroscopic flux states[43].

Finally, let me note that for a full analysis of the system's dynamics, the potential energy (*) has to be augmented by the electrostatic energy of the Josephson junction $CV^2/2$. According to Eq. (6.79), this energy equals

$$\frac{C}{2}\left(\frac{\hbar}{2e}\right)^2\left(\frac{d\Phi}{dt}\right)^2,$$

[42] Since U is the potential energy of the system in a fixed magnetic external field, which plays the role of the generalized external force, its fairer name is the Gibbs potential energy—see, e.g. section 6.2.

[43] See [4].

and hence may be considered the kinetic energy T of the SQUID. The sum $(T + U)$ determines the full energy of the SQUID, which is conserved in the absence of dissipation. In many applications, however, such dissipation, mostly due to the Ohmic current of 'normal' (unpaired) electrons through the Josephson junction, flowing at $V \neq 0$, is essential for a fair description of the system's dynamics.

***Problem* 6.25.** Analyze the possibility of wave propagation in a long, uniform chain of lumped inductances and capacitances—see the figure below.

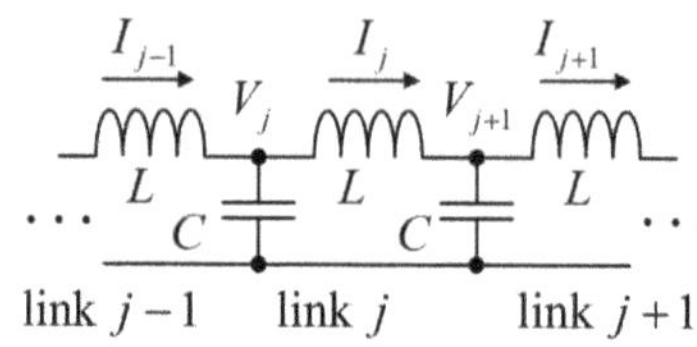

Hint: Readers without prior experience with electromagnetic wave analysis may like to use a substantial analogy between this effect and mechanical waves in a 1D chain of elastically coupled particles[44].

Solution: Applying the first Kirchhoff law (the 'node rule', see Eq. (6.88a) of the lecture notes) to the jth link of the chain, numbered as the figure above shows, we obtain

$$I_{j-1} - I_j - C\dot{V}_j = 0, \qquad (*)$$

where the last term is the time derivative of the electric charge $Q_j = CV_j$ of the jth capacitor. On the other hand, the second Kirchhoff law (the 'loop rule', see Eq. (6.88b) of the lecture notes), applied to the contour containing the jth inductance and two adjacent capacitors, yields

$$V_j - V_{j+1} - L\dot{I}_j = 0, \qquad (**)$$

where Eq. (6.85) was used for the voltage drop on the inductance L carrying current $I_j(t)$.

Generally, the system of differential equations (*) and (**) for a long chain, i.e. for many functions $V_j(t)$ and $I_j(t)$, may have rather complex solutions, but for our task it is sufficient to look for the variables as sinusoidal functions of time, with a constant phase shift (say, α) between adjacent links[45]:

$$I_j = \text{Re}\,[I_\omega \exp\{i(\alpha j - \omega t)\}], \qquad V_j = \text{Re}\,[V_\omega \exp\{i(\alpha j - \omega t)\}]. \qquad (***)$$

Plugging this solution into Eqs. (*) and (**), and then cancelling the non-zero common factor $\exp\{i(\alpha j - \omega t)\}$, we obtain the following system of two linear, homogeneous algebraic equations for two complex amplitudes I_ω and V_ω,

$$I_\omega(e^{-i\alpha} - 1) + i\omega C V_\omega = 0,$$
$$V_\omega(1 - e^{i\alpha}) + i\omega L I_\omega = 0.$$

[44] See, e.g. *Part CM* section 6.3.

[45] See, e.g. *Part CM*, Eq. (6.26) and its discussion.

The equations are compatible if the system's determinant equals zero:

$$\begin{vmatrix} e^{-i\alpha} - 1 & i\omega C \\ i\omega L & 1 - e^{i\alpha} \end{vmatrix} = 0.$$

This condition yields the following *dispersion relation* between ω and α:[46]

$$\omega^2 = \omega_{\max}^2 \sin^2 \frac{\alpha}{2}, \qquad \text{where } \omega_{\max} \equiv \frac{2}{(LC)^{1/2}}.$$

So, the LC-chain indeed supports the solutions (***) with a real parameter α, i.e. sinusoidal traveling waves[47], if their frequency is below $\omega_{\max}$. (If the chain is excited at a higher frequency, the solution (***) is still valid, but with purely imaginary values of α, and describes an exponential decay of the excitation magnitude with the link number j.)

***Problem* 6.26.** A sinusoidal e.m.f. of amplitude V_0 and frequency ω is applied to an end of a long chain of similar, lumped resistors and capacitors, shown in the figure below. Calculate the law of decay of the ac oscillation's amplitude in the chain.

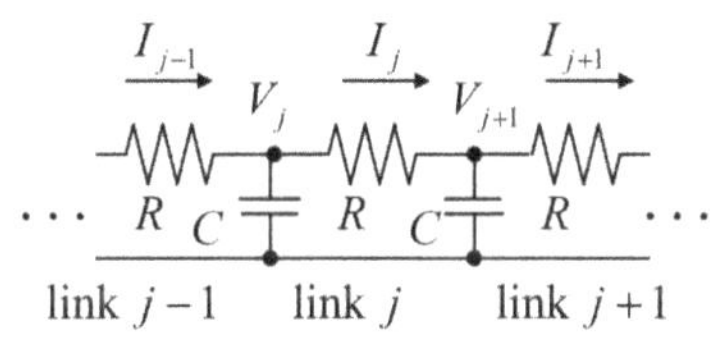

Solution: Applying the first Kirchhoff law (the 'node rule', see Eq. (6.88a) of the lecture notes) to jth link of the system, numbered as the figure above shows, we obtain

$$I_{j-1} - I_j - C\dot{V}_j = 0, \qquad (*)$$

where the last term is the time derivative of the electric charge $Q_j = CV_j$ of jth capacitor. On the other hand, the second Kirchhoff law (the 'loop rule', Eq. (6.88b)), applied to the contour containing the jth resistor and two adjacent capacitors, yields

$$V_j - V_{j+1} - RI_j = 0. \qquad (**)$$

Taking into account that the chain is driven by a sinusoidal e.m.f. of frequency ω, we may look for solution of the system of Eqs. (*) and (**) in the following natural form:

$$I_j = \text{Re}\,[I_0 \exp\{-\kappa j - i\omega t\}], \qquad V_j = \text{Re}\,[V_0 \exp\{-\kappa j - i\omega t\}]. \qquad (***)$$

[46] A similar dispersion relation for mechanical elastic waves plays an important role in the theory of solids—see, e.g. *Part CM* section 6.3 and *Part SM* section 2.5.

[47] For a chain with a spatial period a along axis z, the exponents in Eq. (***) may be rewritten as $\exp\{i(kz - \omega t)\}$, with $k = \alpha/d$, so that α is just the normalized (dimensionless) wave number: $\alpha = kd$.

Plugging this solution into Eqs. (*) and (**), and then cancelling the common factor $\exp\{-\kappa j - i\omega t\}$, we obtain the following system of two linear, homogeneous algebraic equations for two complex amplitudes I_ω and V_ω,

$$I_0(e^{\kappa} - 1) + i\omega C V_0 = 0,$$
$$V_0(1 - e^{-\kappa}) + RI_0 = 0$$

The equations are compatible if the system's determinant equals zero:

$$\begin{vmatrix} e^{\kappa} - 1 & i\omega C \\ R & 1 - e^{-\kappa} \end{vmatrix} = 0.$$

This condition yields the following expression for the propagation constant κ:

$$\kappa = \cosh^{-1}(1 + 2i\omega\tau) \equiv \cosh^{-1}[(1 + \omega^2\tau^2)^{1/2} + \omega\tau] + i\cos^{-1}\frac{1}{(1 + \omega^2\tau^2)^{1/2} + \omega\tau}.$$

where $\tau \equiv RC/4$. If we are not interested in the phase shift, described by the second term of this expression, we may write the following law of decay for the real amplitude V_j of the ac oscillations in the jth link of a semi-infinite chain:

$$V_j = V_0 \exp\{-(\text{Re}\,\kappa)j\}$$
$$= V_0 \exp\{-\cosh^{-1}[(1 + \omega^2\tau^2)^{1/2} + \omega\tau]j\}.$$

The dependence of the decay factor Reκ on frequency (or rather on the dimensional combination $\omega\tau$) is shown in the figure below. At relatively low frequencies ($\omega\tau \ll 1$), Re$\kappa \approx (2\omega\tau)^{1/2} \ll 1$, i.e. the characteristic depth $J \equiv 1/\text{Re}\kappa$ of the oscillations' penetration (expressed in the number of links), is much larger than 1. However, as ω is increased beyond $\sim 1/2\tau$, i.e. beyond $\sim 2/RC$, we obtain Re$\kappa \gtrsim 1$, i.e. the ac excitation essentially does not go beyond the first link of the chain. Due to this property, electrical engineers call such a circuit, used most typically with just a few links, the *low-pass RC filter*.

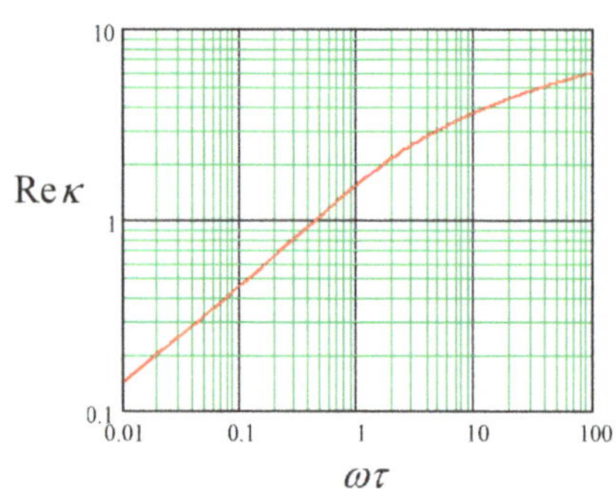

***Problem* 6.27.** As was discussed in section 6.7 of the lecture notes, the displacement current concept allows one to generalize the Ampère law to time-dependent processes as

$$\oint_C \mathbf{H} \cdot d\mathbf{r} = I_S + \frac{\partial}{\partial t}\int_S D_n d^2r.$$

We also have seen that such generalization makes $\int \mathbf{H} \cdot d\mathbf{r}$ over an external contour, such as the one shown in figure 6.10, independent of the choice of the surface S limited by the contour. However, it may look like the situation is different for a contour drawn inside the capacitor—see the figure below. Indeed, if the contour's size is much larger than the capacitor's thickness, the magnetic field $\mathbf{H}$ created by the linear current I on the contour's line is virtually the same as that of a continuous wire, and hence the integral $\int \mathbf{H} \cdot d\mathbf{r}$ along the contour apparently does not depend on its area, while the magnetic flux $\int D_n d^2r$ does, so that the equation displayed above seems invalid. (The current I_S piercing this contour evidently equals zero.) Resolve the paradox, for simplicity considering an axially symmetric system.

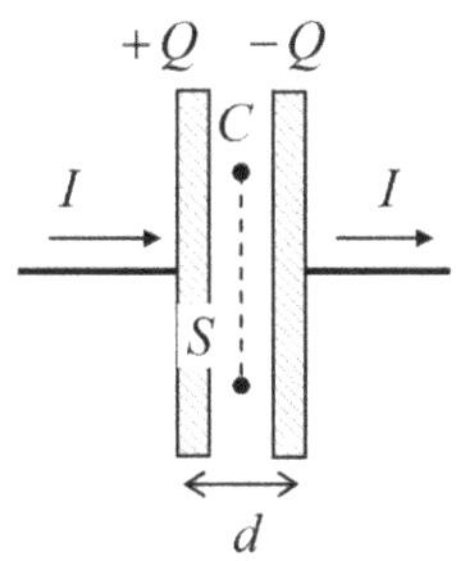

Solution: Let us consider a circular contour C, of radius $\rho < R$, where R is the radius of circular plates of the capacitor. If $\rho \gg d$, we can indeed use Eq. (5.20) of the lecture notes for the magnetic field $\mathbf{H}_I$ induced by the current I in the wire, so that $\int \mathbf{H}_I d\mathbf{r}$ along the contour C is indeed equal to I. However, at $dI/dt \neq 0$ the system also has currents flowing in the capacitor plates in the radial direction, necessary to change the surface density $\sigma = Q/\pi R^2$ of the full charges $\pm Q$ of the plates in time. The left-plate's charge sitting outside a circle of radius ρ equals $\pi(R^2 - \rho^2)\sigma$, so that the linear density $J(\rho)$ of the surface currents in the plate may be found from the charge conservation requirement:

$$2\pi\rho J(\rho) = \frac{d}{dt}[\pi(R^2 - \rho^2)\,\sigma] \equiv \pi(R^2 - \rho^2)\frac{d\sigma}{dt}.$$

The magnetic field between the plates obeys Eq. (6.38), so it is also tangential to the circular contour C, opposite in direction to that of straight wire, and has the magnitude

$$H_J(\rho) = J(\rho) = \frac{(R^2 - \rho^2)}{2\rho}\frac{d\sigma}{dt} = \frac{(R^2 - \rho^2)}{2\pi\rho R^2}\frac{dQ}{dt} = \frac{(R^2 - \rho^2)}{2\pi\rho R^2}I,$$

As a result, the net integral of the field is

$$\oint_C \mathbf{H} \cdot d\mathbf{r} = \oint_C \mathbf{H}_I \cdot d\mathbf{r} + \oint_C \mathbf{H}_J \cdot d\mathbf{r} = I - \oint_C H_J dr = I - 2\pi\rho\frac{(R^2 - \rho^2)}{2\pi\rho R^2}I \equiv I\frac{\rho^2}{R^2},$$

so that it exactly matches the integral of the displacement current over the surface area $\pi\rho^2$:

$$\frac{\partial}{\partial t}\int_S D_n d^2r = \frac{\partial}{\partial t}\int_S \sigma d^2r = \frac{d\sigma}{dt}\int_S d^2r = \frac{d\sigma}{dt}\pi\rho^2 = \frac{1}{\pi R^2}\frac{dQ}{dt}\pi\rho^2 = \frac{\rho^2}{R^2}I,$$

thus resolving the paradox.

***Problem* 6.28.** A straight, uniform, long wire with circular cross-section of radius R, is made of an Ohmic conductor with conductivity σ, and carries dc current I. Calculate the flux of the Poynting vector through its surface, and compare it with the Joule energy losses.

Solution: As was discussed in section 4.3 of the lecture notes, in such a uniform, long wire, the dc current is distributed uniformly, with a constant density

$$j = \frac{I}{\pi R^2}.$$

According to the Ohm law, in this case the electric field's distribution is also uniform,

$$E = \frac{j}{\sigma} = \frac{I}{\pi\sigma R^2},$$

and is naturally directed, as the current density, along the wire's length.

The magnetic field distribution in such a wire was (easily) calculated in section 5.2; for the wire's surface, the second of Eqs. (5.38) yields

$$H = \frac{I}{2\pi R}.$$

Since the direction of the vector **H** is tangential to wire's cross-section (see figure 5.5 of the lecture notes), i.e. perpendicular to the vector **E**, the Poynting vector (6.114) on the wire's surface is directed perpendicular to it, into the wire's bulk, and has the magnitude

$$S = EH = \frac{I}{\pi\sigma R^2}\frac{I}{2\pi R} \equiv \frac{I^2}{2\pi^2\sigma R^3},$$

so that the power flow into a wire's section of length l is

$$\mathscr{P}_{\text{Poynting}} = 2\pi R S l = \frac{I^2}{\pi\sigma R^2}l.$$

On the other hand, according to the Joule law (4.39), the total power of energy dissipation in this section is

$$\mathscr{P}_{\text{Joule}} = \not{p}V = \frac{j^2}{\sigma}\pi R^2 l = \left(\frac{I}{\pi R^2}\right)^2\frac{1}{\sigma}\pi R^2 l \equiv \frac{I^2}{\pi\sigma R^2}l.$$

So, we have arrived at a simple result, $\mathscr{P}_{\text{Joule}} = \mathscr{P}_{\text{Poynting}}$, thus confirming the natural picture of the persistent power flow in the system—from the current source to the external fields, then through the wire's surface into its bulk, and then to heat.

References

[1] Tinkham M 1996 *Introduction to Superconductivity* 2nd edn (New York: McGraw-Hill) chs. 1 and 5
[2] Hu C 1972 *Phys. Rev.* B **6** 1956
[3] Aslamazov L and Larkin A 1969 *JETP Lett.* **9** 87
[4] Freedman J *et al* 2000 *Nature* **406** 43

Chapter 7

Electromagnetic wave propagation

***Problem* 7.1.** Find the temporal Green's function of a medium whose complex dielectric constant obeys the Lorentz oscillator model, given by Eq. (7.32) of the lecture notes, using:

(i) the Fourier transform, and
(ii) the direct solution of Eq. (7.30).

Hint: For the Fourier-transform approach, you may like to use the Cauchy integral[1].

Solutions:

(i) The general relation between $\varepsilon(\omega)$ and the Green's function $G(\theta)$ is given by Eq. (7.26b) of the lecture notes:

$$\varepsilon(\omega) - \varepsilon_0 = \int_0^\infty G(\theta) e^{i\omega\theta} d\theta. \qquad (*)$$

Although the actual ('physical') Green's function $G(\theta)$ does not make sense at $\theta < 0$, mathematically nothing prevents us from extending the integral (*) to the whole axis $-\infty < \theta < +\infty$, taking $G(\theta) = 0$ at $\theta < 0$. Now we can use the Fourier theorem to write the reciprocal transform:

$$G(\theta) = \frac{1}{2\pi} \int_{-\infty}^{+\infty} [\varepsilon(\omega) - \varepsilon_0] e^{-i\omega\theta} d\omega.$$

For the particular form of function $\varepsilon(\omega)$, given by Eq. (7.32), this equality becomes

$$G(\theta) = \frac{nq^2}{2\pi\, m} \int_{-\infty}^{+\infty} \frac{1}{(\omega_0^2 - \omega^2) - 2i\omega\delta} e^{-i\omega\theta} d\omega. \qquad (**)$$

In order to calculate this integral, let us first represent the denominator as the product $-(\omega - \boldsymbol{\omega}_+)(\omega - \boldsymbol{\omega}_-)$, where $\boldsymbol{\omega}_\pm$ are the values of the complex variable $\boldsymbol{\omega}$ at

[1] See, e.g. Eq. (A.92).

doi:10.1088/978-0-7503-1407-7ch7

which the denominator vanishes, i.e. the roots of the corresponding quadratic equation. An elementary calculation yields

$$\boldsymbol{\omega}_\pm = \pm\omega_0' - i\delta, \quad \text{where } \omega_0' \equiv (\omega_0^2 - \delta^2)^{1/2}.$$

Now we may rewrite the whole fraction under the integral in Eq. (**) as a sum of two simple singularities (*poles*):

$$\frac{1}{(\omega_0^2 - \omega^2) - 2i\omega\delta} = -\frac{1}{(\omega - \boldsymbol{\omega}_+)(\omega - \boldsymbol{\omega}_-)} = \frac{a}{\omega - \boldsymbol{\omega}_+} + \frac{b}{\omega - \boldsymbol{\omega}_-}.$$

The constants a and b may be calculated from the requirement that the two last expressions are identical; this gives

$$a = -b = -\frac{1}{2\omega_0'}.$$

As a result, Eq. (**) turns into

$$G(\theta) = \frac{q^2}{4\pi\, m\omega_0'}\left[-\int_{-\infty}^{+\infty} e^{-i\omega\theta}\frac{d\omega}{\omega - \boldsymbol{\omega}_+} + \int_{-\infty}^{+\infty} e^{-i\omega\theta}\frac{d\omega}{\omega - \boldsymbol{\omega}_-}\right].$$

Since at $\theta > 0$ the factor $e^{-i\omega\theta}$, considered as a function of the complex argument $\boldsymbol{\omega} \equiv \omega + i\omega''$, tends to zero at $|\boldsymbol{\omega}| \to \infty$ and $\omega'' < 0$, each integral in the last displayed formula may be replaced with one over the closed contour C shown in the figure below on the plane of complex values of argument ω. This integral may be worked out using the Cauchy integral formula:

$$\oint_C f(\boldsymbol{\omega})\frac{d\boldsymbol{\omega}}{\boldsymbol{\omega} - \boldsymbol{\Omega}} = 2\pi i f(\boldsymbol{\Omega}).$$

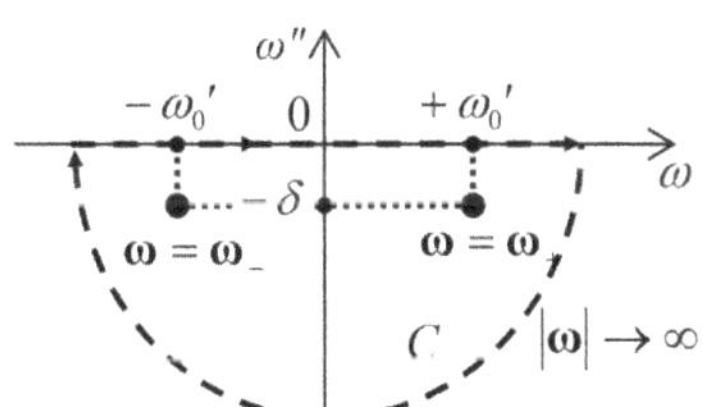

The formula is valid if the function $f(\omega)$ is analytical (as our $e^{-i\omega\theta}$ is), and the pole $\boldsymbol{\Omega}$ is located within the area encircled by contour C. This is correct in our case, because both poles $\boldsymbol{\omega}_\pm$ are in the negative semi-plane—see the figure above. However, the Cauchy theorem implies the positive (counter-clockwise) integration's direction along a closed contour, while our direction is negative, so we have to change the signs before the integrals. As a result, we obtain

$$G(\theta) = \frac{nq^2}{4\pi\, m\omega_0'}2\pi i[e^{-i\boldsymbol{\omega}_+\theta} - e^{-i\boldsymbol{\omega}_-\theta}] \equiv \frac{nq^2}{m\omega_0'}e^{-\delta\theta}\sin\omega_0'\theta.$$

(ii) In section 7.2 of the lecture notes, Eq. (7.32) was derived from the Lorentz oscillator model: a set of independent, similar classical oscillators obeying the differential equation (7.30):

$$\frac{d^2x}{dt^2} + 2\delta\frac{dx}{dt} + \omega_0^2 x = \frac{q}{m}E(t).$$

Its temporal Green's function $g(\theta)$ is just the solution of this differential equation for $t = \theta$ and $E = \delta(\theta)$, while the Green's function for the polarization of the media with n independent similar oscillators per unit volume is $G(\theta) = nqg(\theta)$, so that we can write

$$\frac{d^2G}{d\theta^2} + 2\delta\frac{dG}{d\theta} + \omega_0^2 G = \frac{nq^2}{m}\delta(\theta).$$

At $\theta > 0$, this equation is homogeneous,

$$\frac{d^2G}{d\theta^2} + 2\delta\frac{dG}{d\theta} + \omega_0^2 G = 0, \qquad (***)$$

and has a simple, well-known solution[2]

$$G(\theta) = \left[c_c \cos\omega_0'\theta + c_s \sin\omega_0'\theta\right]e^{-\delta\theta}, \quad \text{where } \omega_0' \equiv \left(\omega_0^2 - \delta^2\right)^{1/2}.$$

In order to find the constants $c_{c,s}$, we need to formulate the appropriate initial conditions for Eq. (***). For that, let us integrate both parts of the initial (inhomogeneous) equation for $G(\theta)$ over an infinitesimal time interval of width $2\Delta \to 0$ centered to point $\theta = 0$:

$$\left(\left.\frac{dG}{d\theta}\right|_{\theta=+\Delta} - \left.\frac{dG}{d\theta}\right|_{\theta=-\Delta}\right) + 2\delta\left(G|_{\theta=+\Delta} - G|_{\theta=-\Delta}\right) + \omega_0^2\int_{-\Delta}^{+\Delta} G d\theta = \frac{nq^2}{m}.$$

Due to the causality principle, both the Green's function and its derivative should equal zero for all moments $t < t'$, i.e. $\theta < 0$. Moreover, $G(\theta)$ should be a continuous function, because is proportional to the oscillator's coordinate, which cannot change instantly even if the acting force is infinite at some moment. On the other hand, under such infinite force, the derivative $dG/d\theta$ may experience a finite jump[3], so that $dG/d\theta|_{\theta=+\Delta}$ may be different from zero. All other terms on the left-hand side vanish, and as a result, we obtain the following initial conditions for the homogeneous equation (***)

$$G(0) = 0, \quad \frac{dG}{d\theta}(0) = \frac{nq^2}{m}.$$

[2] If this is not evident for the reader, he or she may consult *Part CM* section 5.1.

[3] Recall any of the standard collision problems in classical mechanics, solved assuming a finite momentum transfer between two bodies during an infinitesimal time interval.

Applying them to the above general solution of this equation, we finally obtain

$$G(\theta) = \frac{nq^2}{m\omega_0'} e^{-\delta\theta} \sin \omega_0'\theta \,,$$

i.e. (luckily :-) the same result as using the Fourier transform.

***Problem* 7.2.** The electric polarization of a material responds in the following way to an electric field step[4]:

$$P(t) = \varepsilon_1 E_0 (1 - e^{-t/\tau}) \,, \qquad \text{if } E(t) = E_0 \times \begin{cases} 0, & \text{for } t < 0, \\ 1, & \text{for } 0 < t, \end{cases}$$

where τ is a positive constant. Calculate the complex permittivity $\varepsilon(\omega)$ of this material and discuss a possible simple physical model giving such dielectric response.

Solution: Let us calculate the dielectric response of an arbitrary system, described by Eq. (7.23) of the lecture notes, to such an electric field step:

$$P(t) = \int_0^\infty E(t - \theta) G(\theta) d\theta = E_0 \int_0^\infty \begin{Bmatrix} 0, & \text{for } t < \theta \\ 1, & \text{for } \theta < t \end{Bmatrix} G(\theta) d\theta = E_0 \int_0^t G(\theta) d\theta.$$

Comparing this response with that given in the problem's assignment, we see that the temporal Green's function of this system obeys the following relation:

$$\int_0^t G(\theta) d\theta = \varepsilon_1 (1 - e^{-t/\tau}), \qquad \text{for } t > 0.$$

Differentiating both sides of this relation over t, we obtain

$$G(t) = \varepsilon_1 \frac{d}{dt} (1 - e^{-t/\tau}) = \frac{\varepsilon_1}{\tau} e^{-t/\tau}, \qquad \text{for } t > 0.$$

Now it is straightforward to use Eq. (7.26b) to calculate

$$\varepsilon(\omega) = \varepsilon_0 + \int_0^\infty G(\theta) e^{i\omega\theta} d\theta = \varepsilon_0 + \frac{\varepsilon_1}{\tau} \int_0^\infty e^{-\theta/\tau} e^{i\omega\theta} d\theta = \varepsilon_0 + \frac{\varepsilon_1}{1 - i\omega\tau}.$$

Comparing this expression with Eq. (7.32) of the lecture notes, which follows from the Lorentz oscillator model (7.30), we see that they coincide in the low-frequency limit $\omega \ll \omega_0$, if we take

$$\varepsilon_1 = \frac{nq^2}{\kappa} \quad \text{and } \tau = \frac{\eta}{\kappa}, \quad \text{with } \kappa \equiv m\omega_0^2 \quad \text{and } \eta \equiv 2m\delta_0.$$

[4] This function $E(t)$ is of course proportional to the well-known step function $\theta(t)$ —see, e.g., Eq. (A.87). I am not using this notion just to avoid possible confusion between the two different uses of the Greek letter θ.

In other words, such a dielectric response is pertinent, for example, to a medium whose electric dipoles $p = qx$ are due to the heavily overdamped charge displacements $x(t)$,[5] obeying the following equation of motion:

$$\eta\dot{x} + \kappa x = qE(t),$$

i.e. having negligible inertia.

***Problem* 7.3.** Calculate the complex dielectric constant $\varepsilon(\omega)$ for a material whose temporal Green's function, defined by (7.23) of the lecture notes, is

$$G(\theta) = G_0(1 - e^{-\theta/\tau}),$$

with some positive constants G_0 and τ. What is the difference between this dielectric response and the apparently similar one considered in the previous problem?

Solution: With the given $G(\theta)$, Eq. (7.26) of the lecture notes yields

$$\varepsilon(\omega) = \varepsilon_0 + G_0\int_0^\infty (1 - e^{-\theta/\tau})\ e^{i\omega\theta}d\theta = \varepsilon_0 + G_0\left[\left(\frac{1}{i\omega} - \frac{e^{-\theta/\tau}}{i\omega - 1/\tau}\right)e^{i\omega\theta}\right]_{\theta=0}^{\theta=\infty}.$$

We see that the exponential factor is uncertain at the upper limit. This technical problem may be treated in the following standard way: adding to ω (before the integration) a small imaginary part $i\mu$, and then (after the integration) taking the limit $\mu \to 0$:

$$\varepsilon(\omega) = \varepsilon_0 + G_0 \lim_{\mu\to 0}\int_0^\infty (1 - e^{-\theta/\tau})\ e^{i(\omega+i\mu)\theta}d\theta = \varepsilon_0 - \frac{G_0}{i\omega\,(1 - i\omega\tau)}.$$

The temporal Green's function is, by definition, the response of a system to a delta-functional 'excitation' (in our case, the electric field), rather than to a step-like excitation specified in the previous problem, so that the Green's functions and hence all dielectric properties of the systems considered in this and in the previous problem are rather different.

***Problem* 7.4.** Use the Lorentz oscillator model of an atom, given by Eq. (7.30) of the lecture notes, to calculate the average potential energy of the atom in a uniform, sinusoidal ac electric field, and use the result to calculate the potential profile created for the atom by a standing electromagnetic wave with the electric field amplitude $E_\omega(\mathbf{r})$.

Solution: For any linear, isotropic model of the charged particle's motion, its oscillations, induced by a sinusoidal electric field $\mathbf{E}(t) = E_x(t)\mathbf{n}_x$, with

$$E_x(t) = \text{Re}\,(E_\omega e^{-i\omega t}) = \frac{1}{2}(E_\omega e^{-i\omega t} + \text{c.c.}),$$

[5] See, e.g. *Part CM* section 5.1.

are also sinusoidal and directed along the electric field: $\mathbf{r}(t) = x(t)\mathbf{n}_x$, with

$$x(t) = \text{Re}\,(x_\omega e^{-i\omega t}) = \frac{1}{2}(x_\omega e^{-i\omega t} + \text{c.c.})$$

This motion creates the oscillating dipole moment $\mathbf{p}(t) = q\mathbf{r}(t)$, also directed along the applied field, and hence the potential energy of its interaction with the applied field, may be calculated using Eq. (3.15b) of the lecture notes[6]:

$$U(t) = -\frac{1}{2}\mathbf{p}(t)\cdot\mathbf{E}(t) = -\frac{q}{8}(E_\omega e^{-i\omega t} + \text{c.c.})(x_\omega e^{-i\omega t} + \text{c.c.})$$
$$= -\frac{q}{8}(E_\omega x_\omega e^{-2i\omega t} + E_\omega x_\omega^* + \text{c.c.}).$$

Averaging the energy over the period of the applied field kills the terms proportional to $\exp\{\pm 2i\omega t\}$, giving

$$\overline{U} = -\frac{q}{8}(E_\omega x_\omega^* + \text{c.c.}) = -\frac{q}{4}\,\text{Re}\,(E_\omega x_\omega^*).$$

In particular, in the simple oscillator model (7.30), the complex amplitude x_ω is given by Eq. (7.31):

$$x_\omega = \frac{q}{m}\frac{E_\omega}{(\omega_0^2 - \omega^2) - 2i\omega\delta_0},$$

so that

$$\overline{U} = -\frac{q^2}{4m}E_\omega E_\omega^* \text{Re}\left[\frac{1}{(\omega_0^2 - \omega^2) - 2i\omega\delta_0}\right] \equiv \frac{q^2}{4m}E_\omega E_\omega^* \frac{\omega^2 - \omega_0^2}{(\omega^2 - \omega_0^2)^2 + (2\omega\delta_0)^2}. \quad (*)$$

This expression shows that the potential energy is negative at $\omega < \omega_0$ and positive in the opposite case. In a more general model (7.33) corresponding to several oscillators with different resonance frequencies ω_j, such behavior repeats near each of them—see, e.g. the red line in figure 7.5 of the lecture notes[7].

Proceeding to the second part of the problem, we should take into account that according to Eqs. (5.10) and Eqs. (7.6)–(7.8)[8], the interaction of a non-relativistic charged particle (moving with velocity $v \ll c$) with any electromagnetic wave is dominated by the wave's electric field. Hence, for a standing wave we may use Eq. (*) with E_ω replaced with $E_\omega(\mathbf{r})$, where $\mathbf{r}$ is the atom's position[9]. This expression

---

[6] Note that a non-zero net charge also has another, polarization-independent contribution (1.31), $U_0(\mathbf{r}, t) = q\phi(\mathbf{r}, t)$, to its potential energy in the applied field, but for a sinusoidal field, the electric potential ϕ is also a sinusoidal function of time, so that the time average of this part of the interaction vanishes.

[7] Moreover, quantum mechanics confirms the same behavior, near each quantum transition frequency, for a charged particle confined in a potential well of arbitrary shape—not only of the quadratic shape corresponding to a harmonic oscillator—see, e.g. Part *QM*, chapter 6.

[8] The last formulas have been derived for plane waves, but according to the analyses in sections 7.6–7.9, by the order of magnitude they are valid for virtually any traveling or standing wave.

[9] This conclusion hinges on the atom size a being much smaller than the wavelength λ, but for typical atoms and molecules, with $a \sim 10^{-10}$ m, this assumption is valid up to extremely high frequencies corresponding to x-rays.

shows that in a standing wave with $\omega < \omega_0$, the particle is repulsed from electric field maxima and hence is pushed to an adjacent field's minimum ('node'), while in the opposite case ($\omega > \omega_0$) it is pushed to the adjacent maximum of the standing wave.

As Eq. (*) shows, the effect is particularly large if the oscillator's damping is low ($\delta \ll \omega_0$), as is typical for atoms in gases, and the laser's frequency is close to one of the atom's eigenfrequencies ω_0, but still outside the narrow interval $\omega_0 \pm \delta_0$. Note, however, that the effect exists even for a free particle, which may be described by the Lorentz oscillator model with $\omega_0 = \delta_0 = 0$, when Eq. (*) is reduced to

$$\overline{U} = \frac{q^2}{4m\omega^2} E_\omega E_\omega^*, \qquad (**)$$

showing that the particle is pushed toward the field's minimum. (The physics of this force is the subject of the next problem.)

This effect is widely used in experimental practice for *optical* (practically, laser) *trapping* of atoms. Moreover, the effect exists for any (even electrically neutral) particle with $\varepsilon(\omega) \neq \varepsilon_0$, and may be used for trapping of macroscopic (in particular, biological) samples at the focus of a paraxial (e.g. Gaussian) laser beam—the so-called *optical tweezers*[10].

***Problem* 7.5.** The solution of the previous problem shows that a standing plane wave exerts a time-averaged force on a non-relativistic charged particle. Reveal the physics of this force by writing and solving the equations of motion of a free, charged particle in:

(i) a linearly polarized, monochromatic, plane traveling wave, and
(ii) a similar but standing wave.

Solutions:

(i) Directing axis z along the wave's propagation, and axis x along its electric field, we may use Eqs. (7.6)–(7.8) and (7.11) of the lecture notes to express the wave's fields as

$$\mathbf{E}(\mathbf{r}, t) = \mathbf{n}_x \operatorname{Re}[E_\omega e^{i(kz-\omega t)}],$$
$$\mathbf{B}(\mathbf{r}, t) = \mu_0 \mathbf{H}(\mathbf{r}, t) = \mu_0 \frac{\mathbf{n}_z \times \mathbf{n}_x}{Z_0} \operatorname{Re}[E_\omega e^{i(kz-\omega t)}] \equiv \frac{\mathbf{n}_y}{c} \operatorname{Re}[E_\omega e^{i(kz-\omega t)}],$$

so that the Lorentz force (5.10) exerted by the wave on a particle with charge q, located at $\mathbf{r} = \{x, y, z\}$,

$$\mathbf{F}(\mathbf{r}, t) = q[\mathbf{E} + \mathbf{v} \times \mathbf{B}] = q\mathbf{n}_x \operatorname{Re}[E_\omega e^{i(kz-\omega t)}] + q\frac{\mathbf{v} \times \mathbf{n}_y}{c} \operatorname{Re}[E_\omega e^{i(kz-\omega t)}].$$

The electric force, expressed by the first term on the right-hand side, clearly has no time average, while the second term does. In the non-relativistic limit, when

[10] See, e.g. [1].

$v/c \ll 1$, the magnetic force, expressed by the second term, is much smaller than the electric one, so that we may first calculate **v** from Newton's second law in which the magnetic force is ignored:

$$m\dot{\mathbf{v}} = q\mathbf{n}_x \operatorname{Re}[E_\omega e^{i(kz-\omega t)}].$$

Solving this simple (linear) equation, we obtain

$$\mathbf{v} = \mathbf{n}_x \frac{q}{m\omega} \operatorname{Re}[iE_\omega e^{i(kz-\omega t)}],$$

where we have neglected that the possible free motion of the particle by inertia, unrelated to the wave effect. (Indeed, in any realistic system, such motion eventually stops, because of the energy loss due to some inevitable interaction of the particle with its environment.)

Now plugging this result into the full expression (5.10) for the Lorentz force, we may calculate the magnetic force:

$$\begin{aligned}\mathbf{F}(\mathbf{r}, t) &= \frac{q^2}{m\omega c}\mathbf{n}_x \times \mathbf{n}_y \operatorname{Re}[iE_\omega e^{i(kz-\omega t)}] \operatorname{Re}[E_\omega e^{i(kz-\omega t)}] \\ &= \frac{q^2}{m\omega c}\mathbf{n}_z \frac{1}{2}[iE_\omega e^{i(kz-\omega t)} + \text{c.c.}]\frac{1}{2}[E_\omega e^{i(kz-\omega t)} + \text{c.c.}].\end{aligned}$$

The product of the square brackets gives four terms, but only two of them, both proportional to $E_\omega E_\omega{}^*$, have the time dependence cancelled, thus possibly contributing to the average force:

$$\overline{\mathbf{F}}(\mathbf{r}) = \frac{q^2}{4m\omega c}\mathbf{n}_z E_\omega E_\omega^*(ie^{ikz}e^{-ikz} + \text{c.c.}) = \frac{q^2}{4m\omega c}\mathbf{n}_z E_\omega E_\omega^*(i + \text{c.c.}) = 0.$$

Hence, in this case the Lorentz force has no average. This result is in accordance with the solution of the previous problem, in which the average interaction energy is constant (and hence has no gradient) if $E_\omega E_\omega{}^*$ is constant as it is in a traveling plane wave.

(ii) The situation changes if the wave is standing. Taking its fields, for example, in the form given by Eqs. (7.61)–(7.62) of the lecture notes, with the same linear polarization as above,

$$\mathbf{E}(\mathbf{r}, t) = 2\mathbf{n}_x \operatorname{Re}(iE_\omega e^{-i\omega t})\sin kz, \quad \mathbf{B}(\mathbf{r}, t) = \mu_0\mathbf{H}(\mathbf{r}, t) = 2\frac{\mathbf{n}_y}{c}\operatorname{Re}(E_\omega e^{-i\omega t})\cos kz,$$

we see that the average electric component of the Lorentz force equals zero again. Using the same approach as in the previous task to find the magnetic component of the force, we obtain

$$m\dot{\mathbf{v}} = 2q\mathbf{n}_x \operatorname{Re}(iE_\omega e^{-i\omega t})\sin kz, \qquad \text{giving } \mathbf{v} = -\mathbf{n}_x \frac{2q}{m\omega}\operatorname{Re}(E_\omega e^{-i\omega t})\sin kz,$$

so that

$$\mathbf{F}(\mathbf{r}, t) = q\mathbf{v} \times \mathbf{B} = -\frac{4q^2}{m\omega c}\mathbf{n}_z \operatorname{Re}(E_\omega e^{-i\omega t}) \sin kz \operatorname{Re}(E_\omega e^{-i\omega t}) \cos kz$$

$$\equiv -\frac{4q^2}{m\omega c}\mathbf{n}_z \frac{1}{2}(E_\omega e^{-i\omega t} + \text{c.c.})\frac{1}{2}(E_\omega e^{-i\omega t} + \text{c.c}) \sin kz \cos kz.$$

Here again, only two terms of four, resulting from the multiplication of the parentheses, have their time dependences cancelled, and hence contribute to the average force, but now they add up rather than cancel:

$$\overline{\mathbf{F}}(\mathbf{r}) = -\frac{q^2}{m\omega c}\mathbf{n}_z(E_\omega E_\omega^* + E_\omega^* E_\omega) \sin kz \cos kz \equiv -\frac{q^2}{m\omega c}\mathbf{n}_z E_\omega E_\omega^* \sin 2kz. \qquad (*)$$

This force vanishes only in the points where $2kz = n\pi$ (with $n = 0, -1, -2, \ldots$, because our result is valid only for $z \leqslant 0$—see figure 7.8), but only at the points with even n, in which the electric field vanishes, it leads to a stable equilibrium[11].

Let us compare this result with that following from Eq. (**) of the model solution of the previous problem:

$$\overline{U} = \frac{q^2}{4m\omega^2}E_\omega E_\omega^*.$$

Now we should recall that obtaining this result, we took $\mathbf{E}$ in the form $\mathbf{n}_x\operatorname{Re}(E_\omega e^{-i\omega t})$, so that in order to comply with our current assumptions, we have to make the following replacement: $E_\omega \to 2iE_\omega \sin kz$, obtaining

$$\overline{U} \to \overline{U}(\mathbf{r}) = \frac{q^2}{4m\omega^2}(2iE_\omega \sin kz)(2iE_\omega \sin kz)^* = \frac{q^2}{m\omega^2}E_\omega E_\omega^* \sin^2 kz.$$

This average potential energy, depending only on z, yields the average force with just one Cartesian component:

$$\overline{\mathbf{F}}(\mathbf{r}) \equiv -\nabla \overline{U}(\mathbf{r}) = -\mathbf{n}_z \frac{q^2}{m\omega^2}E_\omega E_\omega^* \frac{d}{dz}(\sin^2 kz) = -\mathbf{n}_z \frac{q^2}{m\omega c}E_\omega E_\omega^* \sin 2kz,$$

i.e. exactly the same result as Eq. (*) of this solution.

Hence, we may conclude that the average force acting on a particle is essentially the average magnetic component of the Lorentz force, but it would not arise without the particle's velocity induced by the (much larger) electric component of the force.

[11] The last statement may be readily verified by exploring the equation of motion along axis z.

***Problem* 7.6.** Calculate, sketch, and discuss the dispersion relation for electromagnetic waves propagating in a medium described by the Lorentz oscillator model (7.32), for the case of negligible damping.

Solution: The above assumption, $\delta_0 = 0$, reduces Eq. (7.32) to

$$\varepsilon(\omega) = \varepsilon_0 + n\frac{q^2}{m}\frac{1}{(\omega_0^2 - \omega^2)} \equiv \varepsilon_0\left(1 + \frac{\omega_p^2}{\omega_0^2 - \omega^2}\right), \qquad (*)$$

where ω_0 is the resonance frequency of the oscillators, and ω_p is the plasma frequency defined by Eq. (7.36) of the lecture notes. (For similar oscillators, that expression is reduced to

$$\omega_p^2 = \frac{nq^2}{\varepsilon_0 m};$$

note that the parameters ω_p and ω_0 are completely independent, because ω_p is proportional to the oscillator density n, while ω_0 is a property of a single oscillator.) Plugging Eq. (*) into the general dispersion relation (7.28) for a uniform, linear medium, assuming that our medium is not magnetically active, $\mu(\omega) \equiv \mu_0$, and recalling that $\varepsilon_0\mu_0 \equiv 1/c^2$, we obtain

$$k(\omega) = \frac{\omega}{c}\left(1 + \frac{\omega_p^2}{\omega_0^2 - \omega^2}\right)^{1/2}.$$

The figure below shows a typical plot of this function (for a particular ratio ω_p/ω_0). It reminds the standard level-anticrossing diagram[12], but coincides with it only in the limit $\omega_p \ll \omega_0$, i.e. for very low density of the oscillators. For a non-zero ratio ω_p/ω_0, there is a substantial frequency gap[13],

$$\omega_0 < \omega < \omega_0', \quad \text{with } \omega_0' \equiv \left(\omega_0^2 + \omega_p^2\right)^{1/2},$$

in which $\varepsilon(\omega) < 0$, and hence the electromagnetic wave propagation is impossible. At frequencies far from (both below and above) the gap, the dispersion relation is close to that of waves in free space, $k \approx \omega/c$, testifying that here the coupling between the wave and the medium oscillators is weak. However, at both approaches to the forbidden frequency gap, the dispersion relation plots bend strongly, indicating a substantial energy sharing between the wave and the oscillators. In quantum mechanics, the elementary excitations of such strongly coupled fields are called *polaritons*, because of the electric-polarization origin of Eq. (7.32). (Note that the

[12] See, e.g. *Part CM* section 6.1, and *Part QM* sections 2.7 and 6.1.

[13] Note that according to Eq. (*), the ratio ω_0'/ω_0 has a very simple physical sense: $(\omega_0'/\omega_0)^2 = \varepsilon(0)/\varepsilon_0 = \varepsilon(0)/\varepsilon(\infty)$.

term 'polaritons' is currently used in a much broader context than the Lorentz oscillator model—for virtually any excitations resulting from the interaction of traveling electromagnetic waves with a set of linear resonant dipoles.)

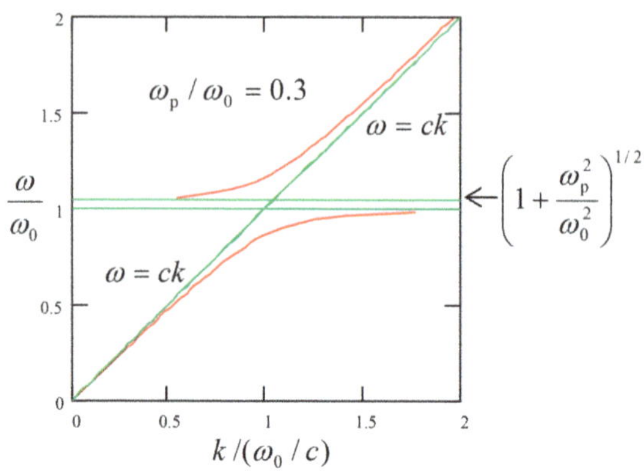

***Problem* 7.7.** As was briefly discussed in section 7.2 of the lecture notes[14], a wave pulse of a finite but relatively large spatial extension $\Delta z \gg \lambda \equiv 2\pi/k$ may be represented with a *wave packet*—a sum of sinusoidal waves with wave vectors **k** within a relatively narrow interval. Consider an electromagnetic plane wave packet of this type, with the electric field distribution

$$\mathbf{E}(\mathbf{r}, t) = \text{Re}\int_{-\infty}^{+\infty} \mathbf{E}_k e^{i(kz-\omega_k t)}dk, \qquad \text{with } \omega_k[\varepsilon(\omega_k)\mu(\omega_k)]^{1/2} \equiv |k|,$$

propagating along axis z in an isotropic, linear, and loss-free (but not necessarily dispersion-free) medium. Express the full energy of the packet (per unit area of wave's front) via the complex amplitudes $\mathbf{E}_k$, and discuss its dependence on time.

Solution: For a linear and isotropic medium we may use the first of Eqs. (6.113)[15]:

$$u = \frac{\mathbf{E}\cdot\mathbf{D}}{2} + \frac{\mathbf{H}\cdot\mathbf{B}}{2},$$

so that the total energy of the pulse (per unit area of wave's front) is

$$\frac{U}{A} = \int_{-\infty}^{+\infty} u dz = \frac{1}{2}\int_{-\infty}^{+\infty} \mathbf{E}\cdot\mathbf{D}dz + \frac{1}{2}\int_{-\infty}^{+\infty} \mathbf{H}\cdot\mathbf{B}dz. \qquad (*)$$

Just as it was done in section 7.2 of the lecture notes for the derivation of Eq. (7.42), let us use the Fourier expansions of all these fields over real k, at arbitrary time t:

[14] And in more detail in *Part CM* section 6.3, and in *Part QM* section 2.2.

[15] Note that the second of Eqs. (6.113) is valid only for dispersion-free media.

$$\mathbf{E}(\mathbf{r}, t) = \text{Re} \int_{-\infty}^{+\infty} \mathbf{E}_k e^{i\psi_k} dk$$
$$\equiv \frac{1}{2}\left[\int_{-\infty}^{+\infty} \mathbf{E}_k e^{i\psi_k} dk + \text{c.c.}\right], \qquad \text{with total phases } \psi_k \equiv kz - \omega_k t,$$
$$\mathbf{D}(\mathbf{r}, t) = \text{Re} \int_{-\infty}^{+\infty} \mathbf{D}_{k'} e^{i\psi_{k'}} dk$$
$$\equiv \frac{1}{2}\left[\int_{-\infty}^{+\infty} \mathbf{D}_{k'} e^{i\psi_{k'}} dk' + \text{c.c.}\right] \equiv \frac{1}{2}\left[\int_{-\infty}^{+\infty} \varepsilon(\omega_{k'}) \mathbf{E}_{k'} e^{i\psi_{k'}} dk' + \text{c.c.}\right],$$

and absolutely similarly for **H** and **B**. Plugging these expressions into the first integral of Eq. (*), and changing the integration order, we obtain

$$\int_{-\infty}^{+\infty} \mathbf{E} \cdot \mathbf{D} dz = \frac{1}{4} \int_{-\infty}^{+\infty} dk \int_{-\infty}^{+\infty} dk' \Big[\varepsilon(\omega_{k'}) \mathbf{E}_k \cdot \mathbf{E}_{k'} \int_{-\infty}^{+\infty} e^{i(\psi_k + \psi_{k'})} dz$$
$$+ \varepsilon^*(\omega_{k'}) \mathbf{E}_k \cdot \mathbf{E}_{k'}^* \int_{-\infty}^{+\infty} e^{i(\psi_k - \psi_{k'})} dz + \text{c.c.} \Big]. \qquad (**)$$

Using Eq. (A.88), we may readily work out both internal integrals, which do not depend on $\mathbf{E}_k$:

$$\int_{-\infty}^{+\infty} e^{i(\psi_k \pm \psi_{k'})} dz \equiv \exp\{-i(\omega_k \pm \omega_{k'})t\} \int_{-\infty}^{+\infty} e^{i(k \pm k')z} dz$$
$$= 2\pi \exp\{-i(\omega_k \pm \omega_{k'})t\} \delta(k \pm k')$$
$$= 2\pi \times \begin{cases} \exp\{-2i\omega_k t\}\ \delta(k + k'), & \text{for the upper sign,} \\ \delta(k - k'), & \text{for the lower sign,} \end{cases}$$

so that the integral over k' in Eq. (**) is also elementary, and that equality is reduced to

$$\int_{-\infty}^{+\infty} \mathbf{E} \cdot \mathbf{D} dz - \frac{\pi}{2} \int_{-\infty}^{+\infty} [\varepsilon(\omega_k) \mathbf{E}_k \cdot \mathbf{E}_{-k} \exp\{-2i\omega_k t\} + \varepsilon^*(\omega_k) E_k E_k^* + \text{c.c.}]\, dk,$$

because due to our definition of the partial wave frequencies, $\omega_{-k} = \omega_k$, and hence $\varepsilon(\omega_{-k}) = \varepsilon(\omega_k)$. According to the discussion in section 7.2 of the lecture notes, in a lossless medium ω_k is real, and $\varepsilon(\omega_k)$ is also a real, even function of frequency—and hence of k, so that $\varepsilon^*(\omega_k) = \varepsilon(\omega_k)$. As a result, spelling out the complex conjugate term, we obtain

$$\int_{-\infty}^{+\infty} \mathbf{E} \cdot \mathbf{D} dz = \frac{\pi}{2} \int_{-\infty}^{+\infty} \varepsilon(\omega_k)$$
$$\times [\mathbf{E}_k \cdot \mathbf{E}_{-k} \exp\{-2i\omega_k t\} + \mathbf{E}_k^* \cdot \mathbf{E}_{-k}^* \exp\{2i\omega_k t\} + 2E_k E_k^*]\, dk.$$

So far, the calculations have been valid for any linear superposition of sinusoidal waves. If the packet is narrow as discussed in the assignment, the first two terms in the square brackets are rapidly oscillating functions of time, with a zero average, for all essential values of k. Neglecting these oscillations, we obtain

$$\int_{-\infty}^{+\infty} \mathbf{E} \cdot \mathbf{D} dz = \pi \int_{-\infty}^{+\infty} \varepsilon(\omega_k) E_k E_k^* \, dk.$$

Carrying out an absolutely similar calculation for the magnetic energy, we obtain

$$\int_{-\infty}^{+\infty} \mathbf{H} \cdot \mathbf{B} dz = \pi \int_{-\infty}^{+\infty} \mu(\omega_k) H_k H_k^* \, dk.$$

Now taking into account that Eqs. (7.6) and (7.7), generalized in accordance with Eq. (7.28), in our current notation give

$$\mathbf{H}_k = \frac{\mathbf{n}_z \times \mathbf{E}_k}{Z(\omega_k)} \operatorname{sgn}(k), \quad \text{with } Z(\omega_k) = \left[\frac{\mu(\omega_k)}{\varepsilon(\omega_k)}\right]^{1/2},$$

we obtain

$$\mu(\omega_k) H_k H_k^* = \mu(\omega_k) \frac{E_k E_k^*}{|Z(\omega_k)|^2} = \varepsilon(\omega_k) E_k E_k^*,$$

and both contributions to the energy are equal, so that Eq. (*) yields

$$\frac{U}{A} = 2\pi \int_0^{+\infty} \varepsilon(\omega_k) E_k E_k^* dk.$$

This expression shows that the packet's energy does not depend on time, even though its shape may be deformed (most typically, spreads out in time) very significantly by the frequency dispersion of the medium[16]. This result certainly corresponds to the intuitive notion of the wave packet motion in the absence of attenuation.

Problem* 7.8. Analyze the effect of a constant, uniform magnetic field $\mathbf{B}_0$, parallel to the direction $\mathbf{n}$ of electromagnetic wave propagation, on the wave's dispersion in plasma, within the same simple model that was used in section 7.2 of the lecture notes for the derivation of Eq. (7.38). (Limit your analysis to relatively weak waves, whose magnetic field is negligible in comparison with $\mathbf{B}_0$.)

Hint: You may like to represent the incident wave as a linear superposition of two circularly polarized waves, with opposite polarization directions.

Solution: For clarity, let us first repeat the derivation of Eq. (7.38), valid for $\mathbf{B}_0 = 0$, for a plasma of free, non-interacting particles, spelling out the spatial orientation of the field and of the field-induced particle motion. For a free particle ($\omega_0 = 0$, $\delta_0 = 0$), Eq. (7.30) may be rewritten as

$$m\ddot{\mathbf{r}} = q\mathbf{E}(t), \qquad (*)$$

[16] Again, see *Part QM* section 2.1 for a detailed discussion and examples of this effect.

so that for a monochromatic, linearly polarized wave $\mathbf{E}(t) = \mathbf{n}_x\text{Re}[E_\omega\exp\{-i\omega t\}]$ (where $\mathbf{n}_x$ is the polarization direction), we may look for the solution in the similar form, $\mathbf{r}(t) = \mathbf{n}_x\text{Re}[r_\omega\exp\{-i\omega t\}]$. Plugging these expressions into Eq. (*), we obtain the following equation for the complex amplitude of particle's forced oscillations:

$$-m\omega^2 r_\omega = qE_\omega, \qquad \text{giving } r_\omega = -\frac{q}{m\omega^2}E_\omega.$$

Now calculating the complex amplitude,

$$P_\omega = np_\omega = nqr_\omega = -\frac{nq^2}{m\omega^2}E_\omega,$$

of the resulting electric polarization of the plasma, $\mathbf{P}(t) = \mathbf{n}_x\text{Re}[P_\omega\exp\{-i\omega t\}]$, we may use Eq. (3.33) to calculate its frequency-dependent permittivity,

$$\varepsilon(\omega) \equiv \frac{D_\omega}{E_\omega} \equiv \varepsilon_0 + \frac{P_\omega}{E_\omega} = \varepsilon_0\left(1 - \frac{\omega_p^2}{\omega^2}\right),$$

with the plasma frequency ω_p expressed similarly to Eq. (7.37):

$$\omega_p^2 \equiv \frac{nq^2}{\varepsilon_0 m},$$

i.e. reproduce Eq. (7.36) of the lecture notes. As was discussed in section 7.2 of the lecture notes, such permittivity immediately leads to the plasma dispersion law (7.38),

$$k(\omega) = \omega[\varepsilon(\omega)\mu_0]^{1/2} = \frac{\omega}{c}\left(1 - \frac{\omega_p^2}{\omega^2}\right)^{1/2},$$

which in particular forbids the propagation of waves with frequencies $\omega < \omega_p$ in the plasma.

In the case of non-vanishing dc field $\mathbf{B}_0$, we have to generalize Eq. (*), taking into account the full Lorentz force (5.10) acting on the particle:

$$m\ddot{\mathbf{r}} = q[\mathbf{E}(t) + \dot{\mathbf{r}} \times \mathbf{B}_0]. \qquad (**)$$

This equation is still linear in $\mathbf{r}(t)$, and hence the plasma response still may be characterized by a (frequency-dependent) permittivity. However, Eq. (**) shows that if the field $\mathbf{B}_0$ is directed along the wave's propagation (say, axis z), i.e. $\mathbf{B}_0 = \mathbf{n}_z B_0$, the particle's motion in the direction x of wave's electric field causes a magnetic force in the direction y, and vice versa, so that the electric polarization acquires not only the x-, but also the y-component. This means that the plasma's response becomes anisotropic, and its permittivity for linearly polarized waves has to be described with a 2 × 2 tensor rather than a scalar.

However, the situation may be simplified using the given hint, i.e. representing the incident wave, with an arbitrary polarization, as a sum of two circularly polarized ones—see the discussion in the end of section 7.1 of the lecture notes. For example, a wave polarized in the direction x,

$$\begin{aligned}\mathbf{E}(t) &= \mathbf{n}_x \operatorname{Re}\left[E_\omega \exp\{-i\omega t\}\right] \\ &= \sum_\pm \operatorname{Re}\left[\mathbf{n}_x \frac{E_\omega}{2} \exp\{-i\omega t\} \pm i\mathbf{n}_y \frac{E_\omega}{2} \exp\{-i\omega t\}\right] \\ &= \sum_\pm \operatorname{Re}\left[(\mathbf{n}_x \pm i\mathbf{n}_y) E_\omega^\pm \exp\{-i\omega t\}\right], \quad \text{with } E_\omega^\pm = \frac{E_\omega}{2}.\end{aligned} \qquad (***)$$

Indeed, looking for the solution of Eq. (**) in a similar form, i.e. as a sum of two simultaneous circular motions:

$$\mathbf{r}(t) = \sum_\pm \operatorname{Re}\left[(\mathbf{n}_x \pm i\mathbf{n}_y)\, r_\omega^\pm \exp\{-i\omega t\}\right],$$

we see that since

$$\begin{aligned}(\mathbf{n}_x \pm i\mathbf{n}_y) \times \mathbf{B}_0 = B_0(\mathbf{n}_x \pm i\mathbf{n}_y) \times \mathbf{n}_z &\equiv B_0(\mathbf{n}_x \times \mathbf{n}_z \pm i\mathbf{n}_y \times \mathbf{n}_z) \\ &\equiv B_0(-\mathbf{n}_y \pm i\mathbf{n}_x) \equiv \pm iB_0(\mathbf{n}_x \pm i\mathbf{n}_y),\end{aligned}$$

Eq. (**) separates into two independent relations for their complex amplitudes:

$$-m\omega^2 r_\omega^\pm = q[E_\omega^\pm - i\omega r_\omega^\pm(\pm iB_0)],$$

giving

$$r_\omega^\pm = -\frac{q}{m\omega^2 \pm q\omega B} E_\omega^\pm = -\frac{q}{m\omega(\omega \pm \omega_c)} E_\omega^\pm,$$

where $\omega_c \equiv qB_0/m$ is the well-known *cyclotron frequency* of particle's motion in the magnetic field[17]. This result is very natural: it shows that the magnetic field speeds up the particle's rotation in one direction and slows its rotation in another direction, but does not entangle these two motion modes.

Now calculating the resulting electric permittivity of the plasma as was done above, we see that it may be still characterized by scalar functions of frequency, but these functions are different for waves with different circular polarizations:

$$\varepsilon^\pm(\omega) = \varepsilon_0 + \frac{P_\omega^\pm}{E_\omega^\pm} = \varepsilon_0\left[1 - \frac{\omega_p^2}{\omega\,(\omega \pm \omega_c)}\right],$$

and so are their dispersion relations:

$$k^\pm(\omega) = \omega[\varepsilon^\pm(\omega)\mu_0]^{1/2} = \frac{\omega}{c}\left[1 - \frac{\omega_p^2}{\omega\,(\omega \pm \omega_c)}\right]^{1/2}. \qquad (****)$$

[17] For a detailed discussion of this motion (in both the non-relativistic and relativistic cases), see section 9.6 of the lecture notes.

This relation shows, in particular, that the magnetic field increases the cut-off frequency $\omega_{\rm min}$ for wave propagation at which the wave vector vanishes for one circular polarization (depending on the particle charge sign) and decreases it for the counterpart polarization—although never suppresses it:

$$\omega_{\rm min}^{\pm} = \left(\omega_{\rm p}^2 + \frac{\omega_{\rm c}^2}{4}\right)^{1/2} \pm \frac{|\omega_{\rm c}|}{2} \rightarrow \begin{cases} |\omega_{\rm c}|, \\ 2\omega_{\rm p}^2/|\omega_{\rm c}|, \end{cases} \quad \text{for } |\omega_{\rm c}| \gg \omega_{\rm p}.$$

Moreover, even for frequencies above both values of $\omega_{\rm min}$, the phase and group velocity for each of the two modes is different, so that an incident wave of any polarization (even a linear one), upon entering the plasma, splits into two independently propagating circularly polarized waves. (For example, for the Earth's ionosphere in the magnetic field of the planet itself, $\omega_{\rm p}$ is typically below 10^8 s^{-1}, while $|\omega_{\rm c}| \approx 10^7$ s^{-1}, so that the wave splitting effects may be rather noticeable.) In the case of small splitting, when

$$\Delta k \equiv k^+ - k^- \approx \frac{\omega_c \omega_{\rm p}^2}{c\omega^2} \ll k^{\pm}, \qquad \text{for } |\omega_{\rm c}| \ll \omega,\ \omega_{\rm p},$$

its effect may be conveniently represented as a slow rotation of the linear polarization plane angle θ. Indeed, after propagation by distance z, the initially x-polarized wave (***) becomes

$$\mathbf{E}(z, t) = \sum_{\pm} \text{Re}\left[(\mathbf{n}_x \pm i\mathbf{n}_y)\frac{E_\omega}{2} \exp\{i(k^{\pm}z - \omega t)\}\right]$$
$$\equiv \sum_{\pm} \text{Re}\left[(\mathbf{n}_x \pm i\mathbf{n}_y)\frac{E_\omega}{2} e^{\pm i\theta} \exp\{i(kz - \omega t)\}\right],$$

where k is the wave vector in the absence of the magnetic field, and $\theta \equiv (\Delta k)z/2$. Representing it as the sum of two Cartesian components,

$$\mathbf{E}(z, t) = \text{Re}\left\{\frac{E_\omega}{2}\left[\mathbf{n}_x(e^{i\theta} + e^{-i\theta}) + i\mathbf{n}_y(e^{i\theta} - e^{-i\theta})\right] \exp\{i(kz - \omega t)\}\right\},$$

we see that the ratio of the complex amplitudes of the components,

$$\frac{E_{\omega y}}{E_{\omega x}} = \frac{i(e^{i\theta} - e^{-i\theta})}{e^{i\theta} + e^{-i\theta}} \equiv \tan\theta,$$

remains real. According to Eq. (7.17) of the lecture notes, this means that the wave remains linearly polarized, but with the polarization plane slowly rotated by angle θ, which is proportional to the traveled distance z, and to $\omega_{\rm c}$, i.e. the applied magnetic field. This is one of manifestations of the *Faraday effect* (or 'Faraday rotation')—just one of several *magneto-optical phenomena*.

It is also an example of the so-called *non-reciprocal effects*, which violate the Lorentz reciprocity relation (6.121). Indeed, as was discussed in the end of section 6.8 of the lecture notes, that relation predicts, in particular, that the electromagnetic waves propagate through a medium similarly in both directions. In the case of the Faraday rotation effect, however, at the wave's direction reversal (say, at its reflection from a mirror), the direction of the magnetic field 'as seen by the wave' changes. As a result, the sign of ω_c in Eq. (****) changes, so that the direction of the wave's quasi-linear polarization's rotation reverses from the wave's 'point of view', and hence remains the same as observed from the lab frame. In other words, in the lab frame, the rotation goes in the same direction regardless of the wave's direction, accumulating as the wave's roundtrip, so that with the appropriate choice of the plasma layer's thickness, the returning wave may have the alternative (y-)linear polarization, and may be readily separated from the incident one, using a polarization filter.

This effect is used for the implementation of very important non-reciprocal microwave and optical devices—*circulators*, operating as shown in the figure below. (Typically, they use dc-magnetic-field-biased ferromagnetic materials rather than a gaseous plasma.) The circulators allow, for example, to stabilize sensitive *reflection amplifiers* based on the effective negative resistance of nonlinear active electron devices.

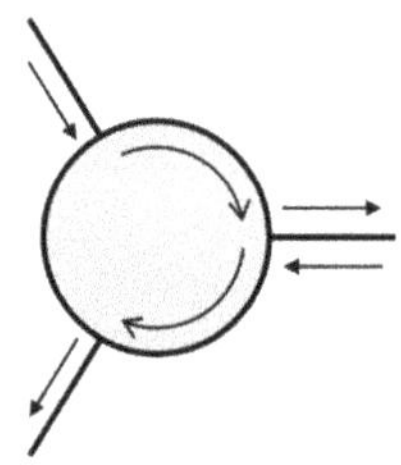

***Problem* 7.9.** A monochromatic, plane electromagnetic wave is normally incident, from the free space, on a uniform slab of a material with electric permittivity ε and magnetic permeability μ, with the slab thickness d comparable with the wavelength.

(i) Calculate the power transmission coefficient $\mathcal{T}$, i.e. the fraction of the incident power that is transmitted through the slab.
(ii) Assuming that ε and μ are frequency-independent and positive, analyze in detail the frequency dependence of $\mathcal{T}$. In particular, how does the function $\mathcal{T}(\omega)$ depend on the slab's thickness d and the wave impedance $Z = (\mu/\varepsilon)^{1/2}$ of its material?

Solutions:

(i) Since the wave may be partly reflected not only from the first, but also from the second interface (see the figure below), we should look for the solution in the form of five traveling waves:

$$E = E_\omega e^{-i\omega t} \times \begin{cases} e^{ik_0 z} + R_0\, e^{-ik_0 z}, & \text{for } z \leqslant 0, \\ Te^{ikz} + R\, e^{-ikz}, & \text{for } 0 \leqslant z \leqslant d, \\ T_0 e^{ik_0 z}, & \text{for } d \leqslant z. \end{cases}$$

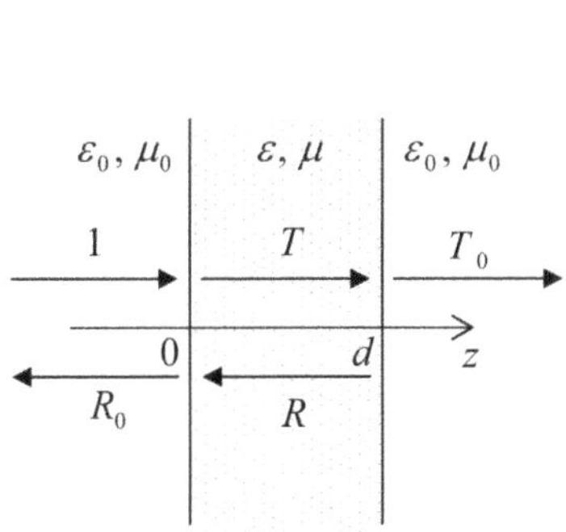

The four unknown (complex) coefficients R, R_0, T, and T_0, participating in these expressions, may be found by plugging them into four boundary conditions that express the continuity of the electric field E and the magnetic field $H = \pm E/Z$ at both surfaces ($z = 0$ and d). These conditions give us the following system of four linear equations:

$$1 + R_0 = T + R, \qquad \frac{1 - R_0}{Z_0} = \frac{T - R}{Z},$$

$$Te^{ikd} + Re^{-ikd} = T_0 e^{ik_0 d}, \qquad \frac{Te^{ikd} - Re^{-ikd}}{Z} = \frac{T_0 e^{ik_0 d}}{Z_0}.$$

Solving this system, we obtain, in particular,

$$T_0 = \frac{4ZZ_0}{(Z + Z_0)^2 \exp\{-ikd\} - (Z - Z_0)^2 \exp\{ikd\}} \exp\{-ik_0 d\}, \qquad (*)$$

and since the wave impedances in the regions in front of the slab and behind it are equal, the power transmission coefficient may be now calculated just as $\mathcal{T} = |T_0|^2$.

(ii) The simplest way to analyze the dependence of T_0 and $\mathcal{T}$ on the system's parameters (at constant and positive ε and μ, and hence constant and positive Z and Z_0), is to consider the diagram in the figure below, which shows the denominator of Eq. (*), a complex-variable function, as the difference of two 2D vectors of lengths $(Z + Z_0)^2$ and $(Z - Z_0)^2 < (Z + Z_0)^2$ with angle $2kd$ between them.

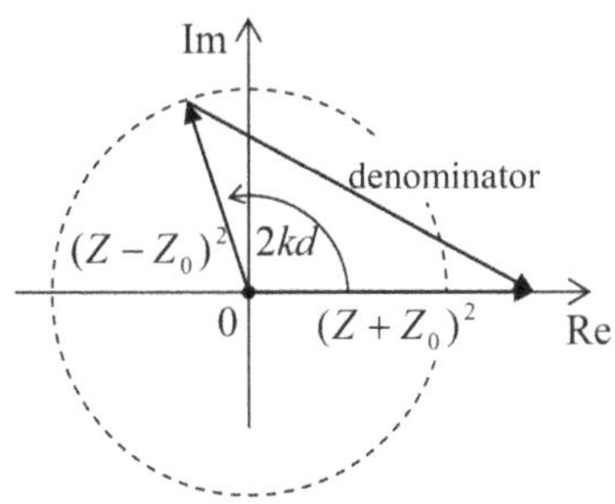

It is clear, first of all, that the denominator, and hence $|T_0|$ and $\mathcal{T}$, oscillate as functions of both the wave frequency ω and the slab's thickness d, with the period $\Delta(kd) = \pi$ (see the figure below), with the transmitted power having equal maxima at[18]

$$kd = \pi m, \quad m = 1, 2, \ldots.$$

This is the well-known *constructive interference* condition, taking place when the slab thickness d equals an integer number of π/k, i.e. of half-wavelengths $\lambda/2$.[19]

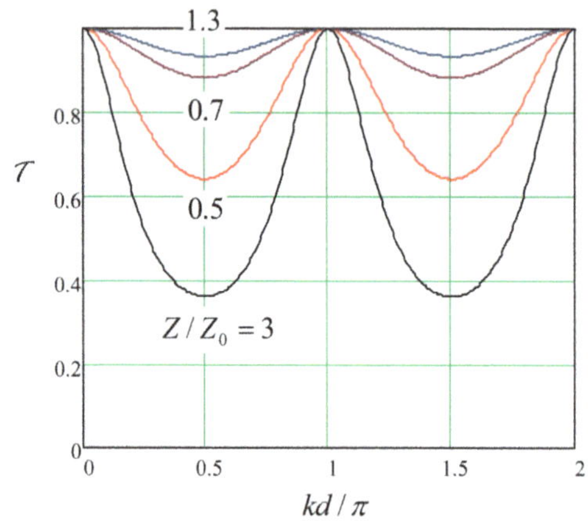

A much less trivial fact is that in each maximum, the transmission is perfect:

$$|T_0|_{\max} = |T_0|_{kd=\pi m} = \frac{4ZZ_0}{(Z + Z_0)^2 - (Z - Z_0)^2} = 1,$$

regardless of the Z/Z_0 ratio. This ratio, however, does affect the sharpness of these resonances and the depth of the transmission minima: the farther is Z/Z_0 from 1, the lower is the transmission minimum at *destructive interference*,

$$|T_0|_{\min} = |T_0|_{kd=\pi(m+1/2)} = \frac{4ZZ_0}{(Z + Z_0)^2 + (Z - Z_0)^2} = \frac{2(Z/Z_0)}{(Z/Z_0)^2 + 1},$$

$$\mathcal{T}_{\min} = \frac{4(Z/Z_0)^2}{[(Z/Z_0)^2 + 1]^2} \leqslant 1.$$

***Problem* 7.10.** A monochromatic, plane electromagnetic wave, with free-space wave number k_0, is normally incident on a planar, conducting film of thickness $d \sim \delta_s \ll 1/k_0$.

[18] Formally, the value $m = 0$ should be in this list as well, but it corresponds to the trivial cases of either $d = 0$ (no slab at all), or $k = 0$, i.e. $\omega = 0$.

[19] A similar effect in quantum mechanics is called *resonant tunneling*—see, e.g. *Part QM* section 2.5.

Calculate the power transmission coefficient of the system, i.e. the fraction of the incident wave's power propagating beyond the film. Analyze the result in the limits of small and large ratio d/δ_s.

Solution: This problem may be solved by re-deriving (or just using) the result of the previous problem:

$$\mathcal{T} = |T_0|^2, \quad \text{where}$$
$$T_0 = \frac{4ZZ_0}{(Z + Z_0)^2 \exp\{-ikd\} - (Z - Z_0)^2 \exp\{ikd\}} \exp\{-ik_0 d\}. \qquad (*)$$

Reviewing the calculations that have led to this result, we may see that it is valid even if the wave impedance Z and the wave number k inside the film are complex numbers, evaluated at the wave's frequency ω—see Eqs. (7.28) of the lecture notes:

$$Z = Z(\omega) \equiv \left[\frac{\mu(\omega)}{\varepsilon(\omega)}\right]^{1/2}, \qquad k = k(\omega) \equiv \omega\, [\varepsilon(\omega)\mu(\omega)]^{1/2}.$$

The problem's assignment implies that the film is non-magnetic, so that $\mu(\omega) = \mu_0$, and that its conductance is Ohmic and non-dispersive, so that we may account for it by using Eq. (7.46) with real $\sigma(\omega) = \sigma = \text{const}$:

$$\varepsilon(\omega) = \varepsilon_{\text{opt}}(\omega) + i\frac{\sigma}{\omega}.$$

Since, according to Eq. (6.33), the skin depth δ_s equals $(2/\mu_0\sigma\omega)^{1/2}$, the given condition $\delta_s \ll 1/k_0 \equiv 1/\omega(\varepsilon_0\mu_0)^{1/2}$ means that $\sigma/\omega \gg \varepsilon_0$, so that unless we are in a very close vicinity of some optical resonance (which should be specified in the assignment), $\sigma/\omega \gg \varepsilon_{\text{opt}}(\omega)$ as well, and we may neglect the first contribution to the complex permittivity, taking[20,21]

$$\varepsilon(\omega) = i\frac{\sigma}{\omega}, \qquad \text{so that } Z = \left(\frac{\mu_0\omega}{i\sigma}\right)^{1/2}, \text{ and } k = \left(i\sigma\omega\mu_0\right)^{1/2} \equiv \frac{1+i}{\delta_s}. \qquad (**)$$

Plugging these expressions into Eq. (*) and taking into account that the condition $k_0\delta_s \ll 1$ means, in particular, that

$$\frac{Z}{Z_0} = \frac{(\mu_0\omega/i\sigma)^{1/2}}{(\mu_0/\varepsilon_0)^{1/2}} = \frac{1}{(2i)^{1/2}}\left[\omega(\varepsilon_0\mu_0)^{1/2}\right]\left(\frac{2}{\mu_0\sigma\omega}\right)^{1/2} = \frac{k_0\delta_s}{1+i},$$

[20] Actually this means that for the thin film description we are using the quasi-static approximation (see section 6.3 of the lecture notes), i.e. are neglecting the displacement currents in the film in comparison with the actual (Ohmic) currents in it. Still, using Eq. (*) for $\mathcal{T}$, which follows from the full system of Maxwell equations, means that we are still keeping the due account of the displacement currents in the space around the film.

[21] You may notice that this expression for k differs from Eq. (6.30) for $\kappa_\pm$ only by the multiplication by the imaginary unity. The difference is due to a different definition of $\kappa_\pm$—see Eq. (6.29).

so that $|Z| \ll Z_0$, we obtain

$$\begin{aligned} T_0 &= \frac{4ZZ_0}{(Z_0^2 + 2ZZ_0)\exp\{-ikd\} - (Z_0^2 - 2ZZ_0)\exp\{ikd\}} \exp\{-ik_0 d\} \\ &= \left\{\cos kd + \frac{1-i}{2k_0\delta_s}\sin kd\right\}^{-1} \exp\{-ik_0 d\}. \end{aligned}$$

Calculating the power transparency $\mathcal{T}$ from the last, innocently looking expression, we have to be careful, because according to the last of Eqs. (**), the argument kd of the involved trigonometric functions is complex. After the separation of their real and imaginary parts[22] we get

$$\begin{aligned} T_0 &= \left\{\cos\left[(1+i)\frac{d}{\delta_s}\right] + \left[\frac{1-i}{2k_0\delta_s}\right]\sin\left[(1+i)\frac{d}{\delta_s}\right]\right\}^{-1} \exp\{-ik_0 d\} \\ &= \left\{\left[\cosh\frac{d}{\delta_s}\left(\cos\frac{d}{\delta_s} + \frac{1}{2k_0\delta_s}\sin\frac{d}{\delta_s}\right) + \frac{1}{2k_0\delta_s}\sinh\frac{d}{\delta_s}\cos\frac{d}{\delta_s}\right]\right. \\ &\quad \left. - i\left[\sinh\frac{d}{\delta_s}\left(\sin\frac{d}{\delta_s} - \frac{1}{2k_0\delta_s}\cos\frac{d}{\delta_s}\right) + \frac{1}{2k_0\delta_s}\cosh\frac{d}{\delta_s}\sin\frac{d}{\delta_s}\right]\right\}^{-1} \exp\{-ik_0 d\}, \end{aligned}$$

so that the final result is given by a rather bulky formula

$$\begin{aligned} \mathcal{T} = |T_0|^2 &= \left\{\left[\cosh\frac{d}{\delta_s}\left(\cos\frac{d}{\delta_s} + \frac{1}{2k_0\delta_s}\sin\frac{d}{\delta_s}\right) + \frac{1}{2k_0\delta_s}\sinh\frac{d}{\delta_s}\cos\frac{d}{\delta_s}\right]^2\right. \\ &\quad \left. + \left[\sinh\frac{d}{\delta_s}\left(\sin\frac{d}{\delta_s} - \frac{1}{2k_0\delta_s}\cos\frac{d}{\delta_s}\right) + \frac{1}{2k_0\delta_s}\cosh\frac{d}{\delta_s}\sin\frac{d}{\delta_s}\right]^2\right\}^{-1}. \end{aligned}$$

The formula may be readily simplified in the limits of large and small film thickness:

$$\mathcal{T} \to \begin{cases} \dfrac{(k_0\delta_s)^2}{2}\exp\left\{-\dfrac{2d}{\delta_s}\right\} \to 0, & \text{for } \dfrac{d}{\delta_s} \gg 2\ln\dfrac{1}{k_0\delta_s}, \\ \left(1 + \dfrac{d}{k_0\delta_s^2}\right)^{-2}, & \text{for } \dfrac{d}{\delta_s} \ll 1 . \end{cases} \qquad (***)$$

Plots in the figure below show $\mathcal{T}$ as a function of ratio d/δ_s for several values of the parameter $k_0\delta_s$, typical in particular for the penetration of microwave radiation through metallic films. The plots clearly show the crossover between the asymptotic

[22] See, e.g. Eq. (A.20).

behaviors (***), taking place at $d \sim \delta_s$. They also show that it makes sense to spell out the second of the asymptotes, for $d/\delta_s \ll 1$, into two sub-limits:

$$\mathcal{T} \to \begin{cases} \left(\dfrac{k_0\delta_s^2}{d}\right)^2 \ll 1, & \text{for } k_0\delta_s \ll \dfrac{d}{\delta_s} \ll 1, \\ 1 - \dfrac{2d}{k_0\delta_s^2} \to 1, & \text{for } \dfrac{d}{\delta_s} \ll k_0\delta_s \ll 1. \end{cases}$$

The first of these functions describes the quasi-linear part of the log–log plots in the figure below; the smaller is the parameter $k_0\delta_s$, the broader is this region.

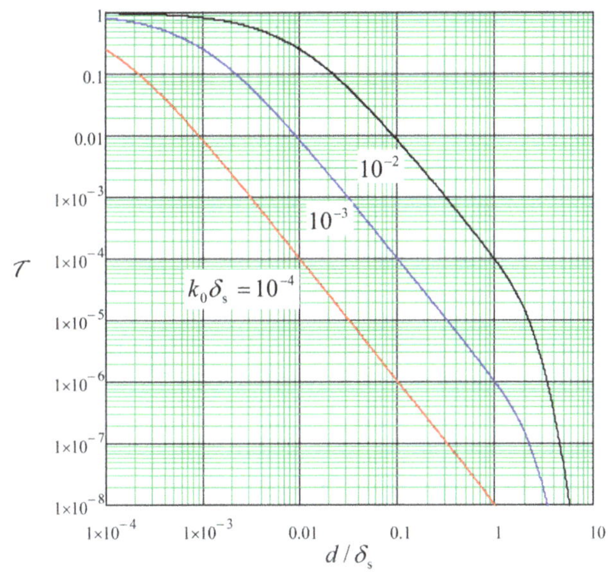

The second of these results may be rewritten in a different form, using Eq. (6.33) for the skin depth, and Eq. (7.8) for the free-space wave impedance:

$$\mathcal{T} = 1 - \frac{\sigma d\mu_0\omega}{k_0} \equiv 1 - \frac{Z_0}{R_s}, \qquad \text{for } Z_0/R_s \ll 1, \qquad (****)$$

(where $R_s \equiv 1/\sigma d$)[23], and derived in a much simpler way. Indeed, if $R_s \gg Z_0$, the film almost does not disturb the incident wave (so that in the crudest approximation, $\mathcal{T} \to 1$), and the electric field **E** applied to the film is just that of the incident wave—at normal incidence, parallel to the film. According to the first form of Eq. (4.39), the field leads to the following Joule power loss per unit area of the film:

$$\frac{d\mathcal{P}}{dA} = p d = \sigma E^2 d = \frac{E^2}{R_s}.$$

[23] As was already mentioned in the model solution of problem 4.7, R_s so defined is commonly called the *sheet resistance* (or sometimes the 'resistance per square') of the thin film.

On the other hand, according to Eq. (7.9), the power of the wave itself (also per unit area) is $S = E^2/Z_0$, so that the lost fraction of power is $(d\mathscr{P}/dA)/S = Z_0/R_s$, immediately leading to Eq. (****) for the transmitted fraction of the power.

Note that the second of Eqs. (***) may be also rewritten in terms of the Z_0/R_s ratio (in this more general case, not limited to small values of this parameter):

$$\mathcal{T} = \frac{1}{(1 + Z_0/2R_s)^2}, \qquad \text{for } d \ll \delta_s.$$

A good additional exercise: derive this expression from scratch, utilizing the smallness of d from the very beginning.

***Problem* 7.11.** A plane wave of frequency ω is normally incident, from free space, on a planar surface of a material with real values of the electric permittivity ε' and magnetic permeability μ'. To minimize the wave reflection from the surface, you may cover it with a layer of thickness d of another transparent material—see the figure below. Calculate the optimal values of ε, μ, and d.

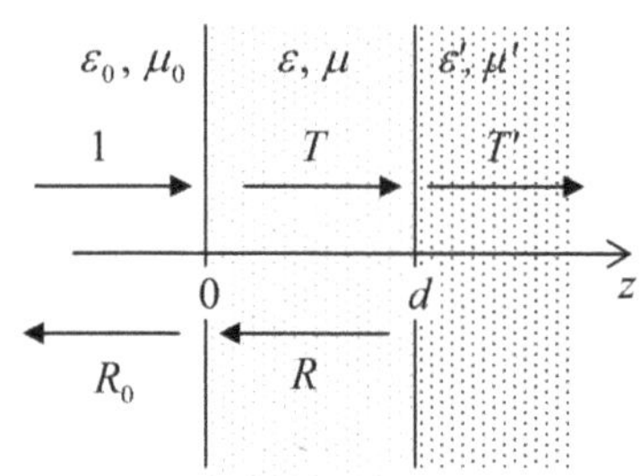

Solution: This problem is an evident generalization of problem 7.9 and may be solved very similarly, in the form of five plane waves with the complex amplitudes shown in the figure above. The system of linear equations relating these amplitudes, resulting from the boundary conditions of two interfaces, is also a straightforward generalization of that derived and solved in problem 7.9:

$$1 + R_0 = T + R, \qquad \frac{1 - R_0}{Z_0} = \frac{T - R}{Z},$$

$$Te^{ikd} + Re^{-ikd} = T'e^{ik'd}, \qquad \frac{Te^{ikd} - Re^{-ikd}}{Z} = \frac{T'e^{ik'd}}{Z'},$$

where Z_0, Z, and Z' are the wave impedances (7.7) of the corresponding materials, and $k = \omega(\varepsilon\mu)^{1/2}$. However, now the equations have more parameters, so that their solutions are bulkier, and may be harder to analyze. The simplest way toward such an analysis is to solve the system for the reflected wave's amplitude R_0, first eliminating variables T and R, and then the product $T'e^{ik'd}$ (treated as one variable)[24]. The result is

$$R_0 = \frac{(Z' - Z_0)Z \cos kd + i(Z_0 Z' - Z^2) \sin kd}{(Z' + Z_0)Z \cos kd + i(Z_0 Z' + Z^2) \sin kd}. \qquad (*)$$

As a sanity check, for the constructive interference case, i.e. $kd = \pi m$ with $m = 0, 1, 2,\ldots$ (in particular for $d = 0$, i.e. for the surface with no coating at all), we obtain $\sin kd = 0$, and the result reduces to Eq. (7.68), with the appropriate notation change:

$$R_0|_{kd=2\pi m} = \frac{Z' - Z_0}{Z' + Z_0}.$$

The result we are seeking for, $R_0 = 0$, may be obtained in the middle between those resonance values of d, i.e. at $kd = \pi(m + ½)$, when $\cos kd = 0$, so that

$$R_0|_{kd=2\pi(m+1/2)} = \frac{Z_0 Z' - Z^2}{Z_0 Z' + Z^2}.$$

Indeed, by selecting

$$Z = Z_{\text{opt}} \equiv (Z_0 Z')^{1/2}, \qquad (**)$$

we obtain $R_0 = 0$, i.e. the system does not reflect—at least at the frequency providing the above condition of kd. It is easy to use Eq. (*) to verify that the frequency bandwidth around such a value, within which the reflection is relatively low[25], is largest for $m = 0$, i.e. at

$$kd = \frac{\pi}{2}, \qquad \text{i.e. } d = \frac{\pi}{2k} \equiv \frac{\lambda}{4},$$

where $\lambda = 2\pi/k$ is the wavelength inside the 'coating' layer.

Nowadays, such a quarter-wavelength *antireflective coating* is broadly used in optics (from billion-dollar telescopes all the way to consumer glasses), so it is a bit surprising that it had been invented as recently as in 1935 and was a guarded military secret until the end of the World War II. Since most optical materials are not magnetic, $\mu \approx \mu' \approx \mu_0$, so that $Z \propto \varepsilon^{1/2} \propto n$, the condition (**) is more frequently represented as[26]

$$n = n_{\text{opt}} = (n')^{1/2},$$

---

[24] Another popular approach is the *transfer matrix* method. For the solution of this simple problem, its introduction and use would take more space than the given solution, but for more complex cases, such as multilayer systems, the method is invaluable—not only in electromagnetism, but also in quantum mechanics and statistical physics; see, e.g. *Part QM* section 2.5, and *Part SM* section 4.5.

[25] Even larger bandwidth may be reached using multilayer coatings.

[26] For a typical optical glass, $n' \approx 1.5$, so the optimal coating material would have $n_{\text{opt}} = (n')^{1/2} \approx 1.22$. However, there are no convenient solid materials with such a refractive index, so that the most popular practical choice is magnesium fluoride (Mg_2F), having $n \approx 1.38$, and obtaining $|R_0|^2 \approx 0.12$. A better suppression of the reflection may be achieved using multilayer coatings, which are becoming more and more popular as their fabrication technology improves.

where n is the index of refraction—see Eq. (7.84). However, this form of Eq. (**) should not obscure the important physical fact that it relates the wave *impedances* of the materials, and has nothing to do with their wave propagation *speed*, which is most frequently associated with the refractive index in undergraduate (and some graduate :-) textbooks.

Problem **7.12.** A monochromatic, plane wave is incident from inside a medium with $\varepsilon\mu > \varepsilon_0\mu_0$ onto its plane surface, at an angle of incidence θ larger than the critical angle $\theta_c = \sin^{-1}(\varepsilon_0\mu_0/\varepsilon\mu)^{1/2}$. Calculate the depth δ of the evanescent wave penetration into the free space, and analyze its dependence on θ. Does the result depend on the wave's polarization?

Solution: Selecting axes x and z in the plane of incidence (see the figure below), we can claim that in our problem $\partial/\partial y = 0$, so that the 3D wave equation, which describes the electric and magnetic fields at $z > 0$ (see, e.g. Eq. (7.3) of the lecture notes), is reduced to the 2D form

$$\left(\frac{\partial^2}{\partial x^2} + \frac{\partial^2}{\partial z^2} - \frac{1}{c^2}\frac{\partial^2}{\partial t^2}\right)f = 0,$$

where f is any field component, and $c = 1/(\varepsilon_0\mu_0)^{1/2}$ is the speed of light in a vacuum. Let us look for the solution of this differential equation in the natural 2D wave form

$$f = f_\omega \exp\{i(k_x x + k_z z - \omega t)\}.$$

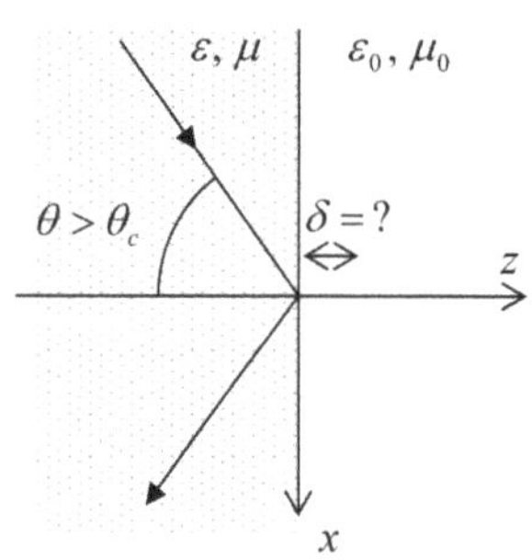

In order to satisfy the boundary conditions at $z = 0$, the horizontal component k_x of the wave vector has to match those of the incident and reflected waves:

$$k_x = k \sin\theta = \omega(\varepsilon\mu)^{1/2}\sin\theta, \qquad (*)$$

c.f. Eq. (7.80) of the lecture notes. On the other hand, in the free-space region, the wave vector's magnitude should be equal $k_0 \equiv \omega/c$, so that

$$k_x^2 + k_z^2 = k_0^2. \qquad (**)$$

At $\theta > \theta_c \equiv \sin^{-1}(k_0/k)$, we have $k_x > k_0$, so that Eq. (**) may be only satisfied by purely imaginary $k_z = i/\delta$. Plugging this expression into the above expression for the wave, we obtain

$$f = f_\omega \exp\{i(k_x x - \omega t)\}\exp\{-z/\delta\},$$

so that δ is exactly the penetration depth we are seeking for. As a result, Eqs. (*) and (**) yield

$$\delta = \frac{1}{(k_x^2 - k_0^2)^{1/2}} = \frac{c \sin\theta_c}{\omega(\sin^2\theta - \sin^2\theta_c)^{1/2}} = \frac{\lambda_0}{2\pi}\frac{\sin\theta_c}{(\sin^2\theta - \sin^2\theta_c)^{1/2}}.$$

This result shows that δ is smallest at the virtually parallel ('grazing-angle') incidence, $\theta \to 0$:

$$\delta \to \frac{\lambda_0}{2\pi}\frac{\sin\theta_c}{(1 - \sin^2\theta_c)^{1/2}} = \frac{\lambda_0}{2\pi}\tan\theta_c \sim \frac{\lambda_0}{2\pi},$$

but diverges as θ approaches θ_c. Finally note that the result for δ has been derived by a 'kinematics' analysis, and hence (just like the reflection and Snell laws) does not depend on the incident wave's polarization.

***Problem* 7.13.** Analyze the possibility of propagation of surface electromagnetic waves along a plane boundary between a plasma and the free space. In particular, calculate and analyze the dispersion relation of the waves.

Hint: Assume that the magnetic field of the wave is parallel to the boundary and perpendicular to the wave's propagation direction. (After solving the problem, justify this choice.)

Solution: With the suggested assumption, the problem may be readily solved for the interface between two linear, isotropic media with arbitrary frequency dispersions of both $\varepsilon_\pm(\omega)$ and $\mu_\pm(\omega)$.[27] With the coordinate axis selection shown in the figure below, where the axis x is directed along the wave propagation, the fields of a monochromatic surface wave may be represented in the form

$$\begin{aligned}\mathbf{E}^\pm(\mathbf{r}, t) &= \text{Re}\left[\left(E_x^\pm\mathbf{n}_x + E_y^\pm\mathbf{n}_y + E_z^\pm\mathbf{n}_z\right)\exp\{i(kx - \omega t) \mp \kappa_\pm z\}\right],\\ \mathbf{H}^\pm(\mathbf{r}, t) &= \text{Re}\left[H^\pm\mathbf{n}_y \exp\{i(kx - \omega t) \mp \kappa_\pm z\}\right].\end{aligned} \qquad (*)$$

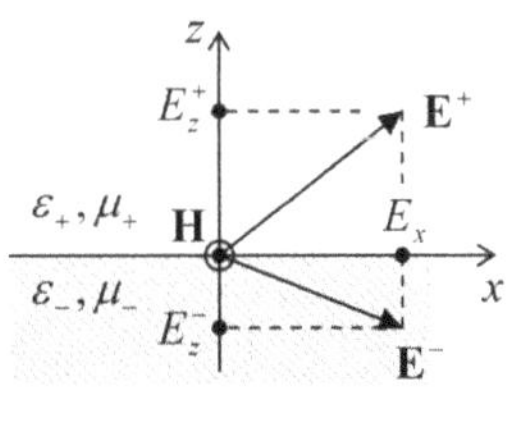

Here the x-component k of the wave vector is taken the same for both media to satisfy the boundary conditions (see below) simultaneously at all points of the interface, while its z-components are assumed imaginary to ensure that the wave

[27] As was discussed in section 7.2 of the lecture notes, any non-zero Ohmic conductivity $\sigma(\omega)$ of a medium may be incorporated into its complex electric permittivity $\varepsilon(\omega)$—see Eq. (7.46).

fields decay in the media bulks[28]. (Actually, in dissipative media, the function $\kappa(\omega)$ may have an imaginary part, but to describe an interface wave of frequency ω, its real part should be positive at that frequency.)

Plugging Eqs. (*) into the source-free (homogeneous) macroscopic Maxwell equations (7.2), we see that they indeed represent valid solutions, provided that the following relations are satisfied:

$$\kappa_{\pm}^{2}(\omega) = k^2 - \omega^2 \varepsilon_{\pm}(\omega)\mu_{\pm}(\omega), \qquad (**)$$

$$E_x^{\pm} = \mp \frac{\kappa^{\pm}}{i\omega\varepsilon_{\pm}(\omega)} H^{\pm}, \qquad E_y^{\pm} = 0, \qquad E_z^{\pm} = -\frac{k}{\omega\varepsilon_{\pm}(\omega)} H^{\pm}. \qquad (***)$$

These relations show, in particular, that if the both media are lossless (i.e. if $\varepsilon_{\pm}$ and $\mu_{\pm}$ are both real), the exponents $\kappa_{\pm}$ are real as well, and the phases of oscillations of the Cartesian components E_x and E_z are shifted by $\pi/2$. The last fact means that the vector $\mathbf{E} = \{E_x, 0, E_z\}$ in each medium rotates, its end following an ellipse with semi-axes proportional, respectively, to the coefficients κ and k.

In order to find these coefficients, Eq. (**) alone is insufficient. An additional equation may be obtained from applying the boundary conditions (3.37), (3.56), and (5.117):

$$E_x^{+} = E_x^{-} \equiv E_x, \qquad \varepsilon_{+}(\omega)E_z^{+} = \varepsilon_{-}(\omega)E_z^{-}, \qquad H^{+} = H^{-}.$$

These linear, homogeneous equations are compatible with Eqs. (***) only if

$$\frac{\kappa_{+}}{\kappa_{-}} = -\frac{\varepsilon_{+}(\omega)}{\varepsilon_{-}(\omega)}. \qquad (****)$$

This is the key relation for the interface/surface waves. It shows that such waves cannot propagate on the interface between two 'usual' wave media, with real and positive dielectric constants. However, as was discussed in detail in section 7.2, dilute gases feature frequency ranges with negative electric permittivity immediately above resonance (more generally, any quantum-transition) frequencies—see, e.g. figure 7.5. Moreover, according to Eq. (7.36), in a collision-free plasma, the function $\varepsilon(\omega)$ is real and negative at all frequencies below the plasma frequency ω_p given by Eq. (7.37). For the waves on the surface of such a plasma (i.e. its 'interface' with the free space), Eqs. (**), (****), and (7.36) yield[29]

$$k = \frac{\omega}{c}\left(\frac{\omega_p^2 - \omega^2}{\omega_p^2 - 2\omega^2}\right)^{1/2}, \qquad \text{for } \omega < \omega_p.$$

---

[28] This means that in contrast to the situation at the total internal reflection (see its discussion in section 7.4), the interface wave is evanescent in both media rather than in just one of them.

[29] Note that if the interface waves can propagate, i.e. if $\varepsilon_{+}(\omega)$ and $\varepsilon_{-}(\omega)$ have different signs, the signs of the z-components of the electric field in the two bordering media are opposite, as shown in the figure above. As a result, the electric field vectors in the media rotate in opposite directions—see, e.g. the nice animation available online at http://en.wikipedia.org/wiki/Surface_plasmon_polariton#Dispersion_relation.

This dispersion relation is shown with the red line in the figure below, while the blue line shows the relation (see figure 7.6 of the lecture notes) for the usual ('3D') waves in the same plasma.

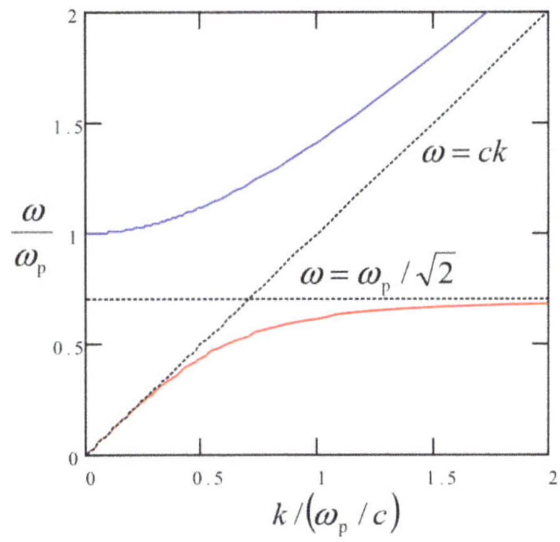

At low frequencies ($\omega \ll \omega_p$), the surface waves[30] have almost the same properties as the usual electromagnetic waves in free space: $k \approx \omega/c$, $v_{gr} \approx v_{ph} \approx c$. (This is natural, because in this case, according to Eq. (****) with $\varepsilon_+(\omega) = \varepsilon_0$, and $\varepsilon_-(\omega) \approx -\varepsilon_0\omega_p^2/\omega^2 \to -\infty$, $\kappa_+ \ll \kappa_-$, i.e. the wave propagates mostly is free space, only weakly 'guided' by the plasma surface.) As the frequency grows, the wave vector diverges at $\omega \to \omega_p/\sqrt{2}$, so that the surface plasmons cannot propagate at higher frequencies. In this limit, $\varepsilon_-(\omega) \approx -\varepsilon_0 = -\varepsilon_+(\omega)$, so that $\kappa_+ \approx \kappa_-$, i.e. the wave is equally shared between the plasma and free space.

Note that the conceptual importance of the plasmons extends well beyond plasma physics and engineering. For example, let us return to the simplest (and very common) situation shown in figure 2.26 of the lecture notes, and ask ourselves: if a charge q has been brought from to a metallic surface from afar very fast, how long would it take for the familiar picture of the charge image to become quantitatively valid? There is of course the relativistic limitation $\Delta t \gg d/c$, but the limitation imposed by surface plasmon propagation, $\Delta t \gg 1/\omega_p$, where is the plasma frequency of the conduction electrons in the metal, may be much more significant.

Finally, let us return to the initial assumption $H_z = 0$. Due to the symmetry of the homogeneous Maxwell equations (6.100) to the simultaneous swap $\mathbf{E} \leftrightarrow \mathbf{H}$, $\mathbf{D} \leftrightarrow -\mathbf{B}$, it is evident that similar surface waves with $H_z \neq 0$, but $E_z = 0$, are also possible, but only if $\mu(\omega) < 0$ in one of the media, which is not common.

Note that conceptually similar, but qualitatively different, mechanical surface waves may propagate at an interface between two elastic media[31].

[30] The quantized excitations of these waves are called either *surface plasmons* or (due to their similarity to the polaritons discussed in problem 7.6) *surface plasmon polaritons*.

[31] See, e.g. *Part CM* section 7.7, and references therein.

***Problem* 7.14.** Light from a very distant source arrives to an observer through a planar layer of a non-uniform medium with a certain refraction index distribution, $n(z)$, at angle θ_0—see the figure below. What is the actual direction θ_i to the source, if $n(z) \to 1$ at $z \to \infty$? (This problem is obviously important for high-precision astronomical measurements from the Earth's surface.)

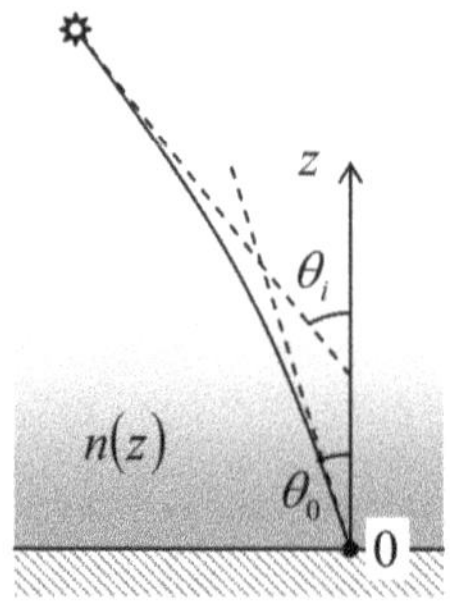

Solution: If the layer thickness' scale d is much larger than the light's wavelength λ, we may represent it locally (on a distance Δz such that $\lambda \ll \Delta z \ll d$) as a plane wave —see, e.g. Eq. (7.79) of the lecture notes. Hence, if n changes only in the direction z, we may repeat the arguments in the beginning of section 7.4 to conclude that the wave vector component in the direction perpendicular to this axis, equal to $k \sin \theta \propto n \sin \theta$, cannot change at the wave's refraction between any two elementary sub-layers. Hence, the product $n(z) \sin \theta(z)$ should stay constant through the whole layer. Applying this conclusion to the initial direction θ_i (at $z \gg d$, where $n(z) \to 1$) and observer's location ($z = 0$, where $n(z) = n(0) \equiv n_0$), we finally obtain a very simple answer:

$$\theta_i = \sin^{-1}(n_0 \sin \theta_0), \qquad (*)$$

i.e. just the Snell law, regardless of the details of the function $n(z)$. As table 3.1 of the lecture notes shows, for the air of Earth's surface, $n_0 \approx \kappa^{1/2} \approx 1.0003$, so that the angular corrections to the actual position of stars and planets (called *atmospheric refraction*) may be rather substantial—about one angular minute at $\theta_i \sim \pi/4$. Note that Eq. (*) should be applied to practical observations at angles $\theta_i \to \pi/2$ with care, because here the Earth surface's curvature, ignored in its derivation, becomes important. With the account of these effects, the atmospheric refraction at the horizon is as large as ~35 angular minutes—quite comparable to the visible diameters of the Sun and the Moon.

***Problem* 7.15.** Calculate the impedance Z_W of the long, straight TEM transmission lines formed by metallic electrodes with the cross-sections shown in the figure below:

(i) two round, parallel wires, separated by distance $d \gg R$,
(ii) a *microstrip line* of width $w \gg d$,
(iii) a *stripline* with $w \gg d_1 \sim d_2$,

in all cases using the macroscopic boundary conditions on metallic surfaces. Assume that the conductors are embedded into a linear dielectric with constant ε and μ.

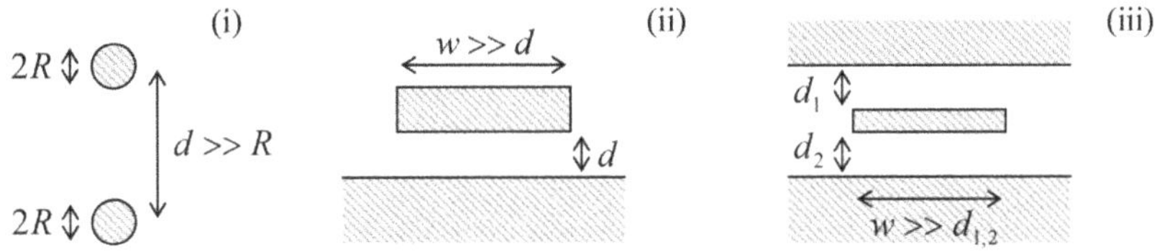

Solutions:

(i) From the solution of problem 2.7, with $R_1 = R_2 = R$, generalized to the dielectric filling by replacing of ε_0 with ε, the mutual capacitance per unit length is

$$C_0 = \frac{\pi\varepsilon}{\ln(d/R)}.$$

The inductance L_0 per unit length may be found either from magnetostatics, or (more easily) from the general Eq. (7.114) of the lecture notes:

$$L_0 = \frac{\varepsilon\mu}{C_0} = \frac{\mu}{\pi}\ln\frac{d}{R}.$$

Combining these expressions, we obtain

$$Z_W \equiv \left(\frac{L_0}{C_0}\right)^{1/2} = \frac{Z}{\pi}\ln\frac{d}{R}, \qquad \text{where } Z \equiv \left(\frac{\mu}{\varepsilon}\right)^{1/2}.$$

This formula is close in structure to that for the coaxial cable—see Eq. (7.120)—with an extra factor of 2 due to two (rather than one) thin conductors, and with the distance d between the wires playing a role similar to the diameter of the outer conductor of the cable.

As was mentioned in section 7.6 of the lecture notes, such wire pairs, lighter and cheaper than the coaxial cable, have been the basis of all earlier electric communication technologies (such as telegraphy) and are still used, in the form of twisted pairs, in land-line telephony at frequencies up to a few hundred kHz, and in relatively short Ethernet cables at frequencies up to a few hundred MHz.

(ii) Neglecting the fringe field effects (whose relative contribution scales as $d/w \ll 1$), we may use Eq. (3.55) for the line's capacitance per unit length:

$$C_0 = \frac{\varepsilon w}{d}.$$

The inductance per unit length may be found, again, either from magnetostatics[32] or from the general Eq. (7.114):

$$L_0 = \frac{\varepsilon\mu}{C_0} = \frac{\mu d}{w}.$$

Combining these expressions, we obtain

$$Z_W = Z\frac{d}{w} \ll Z.$$

This expression shows that the microstrip lines may have very low impedance. Such low-impedance lines are broadly used in electronics, e.g. on printed circuit boards. Due to the low fringe fields, they may ensure low mutual interference ('crosstalk') even if spaced rather closely—at distances $\sim w \gg d$.

(iii) The stripline geometry, which is also broadly used in multi-layer printed circuit boards and for on-chip interconnects in microelectronics, provides even lower crosstalk between adjacent transmission lines. Its parameters may be readily found by calculating C_0 as a parallel connection of two plane capacitors:

$$C_0 = \varepsilon w\left(\frac{1}{d_1} + \frac{1}{d_2}\right), \quad L_0 = \frac{\varepsilon\mu}{C_0} = \frac{\mu}{w}\left(\frac{1}{d_1} + \frac{1}{d_2}\right)^{-1},$$

$$Z_W = \left(\frac{L_0}{C_0}\right)^{1/2} = Z\frac{1}{w}\left(\frac{1}{d_1} + \frac{1}{d_2}\right)^{-1}.$$

If one of the gaps is much narrower than the other one (say, $d_1 \ll d_2$), the electromagnetic field is concentrated in it, and the result is reduced to that for the microstrip line, with the replacement $d \to d_1$.

***Problem* 7.16.** Modify the solution of task (ii) of the previous problem for a superconductor microstrip line, taking into account the magnetic field penetration into both the strip and the ground plane.

Solution: In this case, the expression for L_0 has to be replaced with the solution of problem 6.18:

$$L_0 = \frac{\mu}{w}(d + 2\delta_L),$$

while the capacitance C_0 per unit length remains the same as in the solution of the previous problem, because the electric field penetrates into conductors (including

[32] As was discussed in the solution of problem 6.18 in the context of superconducting electrodes, if the width d of the gap between the conductors and the depth of the field penetration into the line electrodes (say, the skin-effect depth δ_s) are much smaller than w, the field in the gap is uniform, $B = \mu I/w$. If, in addition, the field penetration depth is also much less than d, then the magnetic field is localized in the gap and the magnetic flux (per unit length) is $\Phi/l = Bd = \mu Id/w$, immediately giving the above formula for L_0.

superconductors) by a much smaller, typically negligible screening length—see section 2.1. As a result, the TEM wave propagation speed in such a line,

$$v' = \frac{1}{(L_0C_0)^{1/2}} = \frac{1}{(\varepsilon\mu)^{1/2}}\left(\frac{d}{d+2\delta_L}\right)^{1/2} = \frac{v}{(1+2\delta_L/d)^{1/2}},$$

is lower than that (v) of plane waves[33]. Note that in this case the 'general' relation (7.114) is not satisfied, because it is based on the assumption of similar boundary conditions for the magnetic and electric fields on the conductor surfaces.

Problem* 7.17. What lumped ac circuit would be equivalent to the TEM-line system shown in figure 7.19 of the lecture notes, with an incident wave's power $\mathscr{P}_i$? Assume that the wave reflected from the lumped load does not return to it.

Solution: It is intuitively clear that the circuit has to include:

(i) the lumped load with the (generally, complex) impedance $Z_L(\omega)$,
(ii) a lumped element representing passive properties of the transmission line, with the ac impedance equal to line's impedance $Z_W(\omega)$, and
(iii) some ac generator representing the incident wave.

Following these arguments, let us try to use the circuit shown in the figure below, where $\mathscr{V}(t)$ is an e.m.f. representing the incident wave. An elementary calculation of the complex amplitude of the current in the load yields $I_\omega = \mathscr{V}_\omega/[Z_L(\omega) + Z_W(\omega)]$, so that the voltage drop across the load is $V_\omega = I_\omega Z_L(\omega) = \mathscr{V}_\omega Z_L(\omega)/[Z_L(\omega) + Z_W(\omega)]$. Now the calculation similar to that carried out at the derivation of Eq. (7.42) yields the following expression for the average power absorbed in the load:

$$\mathscr{P}_L = \frac{1}{2}V_\omega I^*_\omega = \frac{1}{2}\frac{|\mathscr{V}_\omega|^2}{|Z_L(\omega) + Z_W(\omega)|^2}\,\mathrm{Re}\,Z_L(\omega). \qquad (*)$$

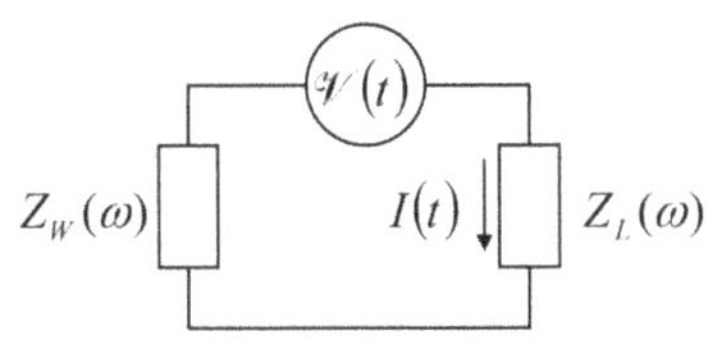

On the other hand, we can calculate the same power in the genuine circuit (figure 7.19) by subtracting, from the incident power $\mathscr{P}_i$, the reflected wave power $\mathscr{P}_r = \mathscr{P}_i|R|^2$, where R is given by Eq. (7.118):

[33] This difference is particularly large ($v'/v \sim 0.1$) in the so-called *long* (or 'distributed') *Josephson junctions* with $d \sim 1$ nm and $\delta_L \sim 100$ nm, enabling very interesting 'super-relativistic' effects of interactions between the slow TEM waves and magnetic-field-induced 'waves' of supercurrent—see, e.g. [2].

$$\mathscr{P}_L = \mathscr{P}_i(1 - |R|^2) = \mathscr{P}_i\left(1 - \frac{|Z_L(\omega) - Z_W(\omega)|^2}{|Z_L(\omega) + Z_W(\omega)|^2}\right)$$
$$\equiv \mathscr{P}_i\frac{|Z_L(\omega) + Z_W(\omega)|^2 - |Z_L(\omega) - Z_W(\omega)|^2}{|Z_L(\omega) + Z_W(\omega)|^2}.$$

In the most important case of a loss-free line, its impedance Z_W is real and this expression reduces to

$$\mathscr{P}_L = \mathscr{P}_i\frac{4Z_W(\omega)}{|Z_L(\omega) + Z_W(\omega)|^2}\,\mathrm{Re}\,Z_L(\omega). \qquad (**)$$

Comparing Eqs. (*) and (**), we see that they give similar results (and hence the lumped circuit is equivalent to the wave system shown in figure 7.19), if the effective ac e.m.f. has the amplitude

$$|\mathscr{V}_\omega| = [8\mathscr{P}_i Z_W(\omega)]^{1/2}.$$

Such representation of a transmission line, carrying an incident wave, by its impedance and an e.m.f. is valid even if the lumped load is nonlinear and/or time-dependent, and is broadly used for the analysis of not only linear, but also nonlinear and parametric microwave devices. Note, however, that this result is only valid under the assumption mentioned in the assignment, that the wave reflected from the load does not return to it. This condition (generally beneficial for applications, e.g. for amplifier stability) may be most simply satisfied using such non-reciprocal devices as circulators—see the discussion in the model solution of problem 7.8.

***Problem* 7.18.** Find the lumped ac circuit equivalent to a loss-free TEM transmission line of length $l \sim \lambda$, with a small cross-section area $A \ll \lambda^2$, as 'seen' (measured) from one end, if the line's conductors are galvanically connected ('shortened') at the other end—see the figure below. Discuss the result's dependence on the signal frequency.

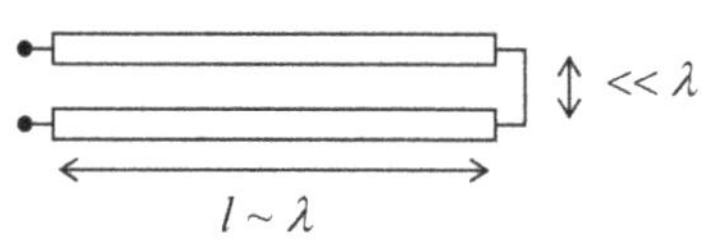

Solution: Neglecting losses in the transmission line, we can describe any 'global' variable, e.g. voltage $V(z, t)$ or current $I(z, t)$, by a sum of two sinusoidal waves: one traveling to the shortened end and another one reflected from it. Taking the shortened end's position for $z = 0$, for the complex amplitudes of these variables we can write

$$V_\omega(z \leqslant 0) = V_0(e^{ikz} + Re^{-ikz}), \qquad I_\omega(z \leqslant 0) = \frac{V_0}{Z_W}(e^{ikz} - Re^{-ikz}),$$

—cf. Eqs. (7.64) and (7.65) for plane waves. Now requiring the ratio V_ω/I_ω to vanish at the shorted end (at $z = 0$), we obtain $R = -1$, so that at the other end of the line (at $z = -l$)

$$Z(\omega) \equiv \left.\frac{V_\omega}{I_\omega}\right|_{z=-l} = Z_W \frac{e^{-ikl} - e^{ikl}}{e^{-ikl} + e^{ikl}} \equiv -iZ_W \tan kl. \qquad (*)$$

This result shows that in contrast to the TEM line impedance $Z_W = (L_0/C_0)^{1/2}$, which does not depend on frequency, the effective ac impedance $Z(\omega)$ of the shorted line segment is a strong function of ω. In particular, at low frequencies $\omega \ll v/l$ (i.e. $kl \ll 1$) this expression reduces to

$$Z(\omega) \approx -iZ_W kl = -i\left(\frac{L_0}{C_0}\right)^{1/2} \omega(L_0C_0)^{1/2} l = -i\omega L_0 l = -i\omega L,$$

where $L = L_0 l$ is the segment's inductance. This is a well-known expression for the complex impedance of a lumped inductance. (The negative sign is due to the fact that we are using the 'physics' convention $\exp\{-i\omega t\}$ for the time dependence, while in electrical engineering the positive sign is common.)

However, as kl is increased, the impedance given by Eq. (*) grows faster than that of a lumped inductance, and diverges at $kl = \pi/2$, i.e. at wavelength $l = \pi/2k \equiv \lambda/4$. This divergence means that in the absence of power losses, a non-vanishing voltage V_ω may be sustained by vanishing current. Such resonant behavior is similar to that of an ac circuit consisting of a lumped inductance and a lumped capacitance connected in parallel; it repeats at all frequencies where

$$k_n l = \frac{\pi}{2} + n\pi, \qquad (**)$$

i.e. $(n + 1/2)\lambda_n/2 = l$. These so-called *parallel resonances* are interleaved, at wave number values

$$k'_n l = n\pi, \qquad (***)$$

by resonances of a different kind, where $Z(\omega) = V_\omega/I_\omega$ vanishes, and hence a finite ac current may be sustained with little, if any, ac voltage. These are so-called *series resonances*, similar to those in the series lumped LC circuits.

Such *distributed resonators* are broadly used in electrical engineering at the high frequencies where the implementation of genuinely lumped circuits leads to unacceptably high energy losses.

***Problem* 7.19.** Represent the fundamental H_{10} wave in a rectangular waveguide (see figure 7.22 of the lecture notes) as a sum of two plane waves, and discuss the physics behind such a representation.

Solution: Let us use Eq. (7.138) of the lecture notes to write the following expression for the instant electric field of this wave:

$$\mathbf{E}(\mathbf{r},\,t) = \mathrm{Re}\left[i\frac{ka}{\pi}ZH_l\,\sin\frac{\pi x}{a}\,\exp\,\{i(k_z z - \omega t)\}\right]\mathbf{n}_y$$
$$\equiv \mathrm{Re}\left[E_\omega\,\exp\left\{i\left(\frac{\pi}{a}x + k_z z - \omega t\right)\right\} - E_\omega\,\exp\left\{i\left(-\frac{\pi}{a}x + k_z z - \omega t\right)\right\}\right]\mathbf{n}_y,$$

with $E_\omega \equiv (ka/2\pi)ZH_l$. The last of these expressions shows that the mode is nothing more than the sum of two plane waves, linearly polarized along the axis y, with equal amplitudes E_ω, and wave vectors $\mathbf{k}_\pm$ with the Cartesian components $k_x = \pm\pi/a$, $k_y = 0$, and the k_z defined by Eqs. (7.122) and (7.132) (the latter one, with $n = 1$, $m = 0$), so that

$$k_x^2 + k_y^2 + k_z^2 = k_\pm^2 = k^2 = \frac{\omega^2}{v^2},$$

see panel (a) of the figure below.

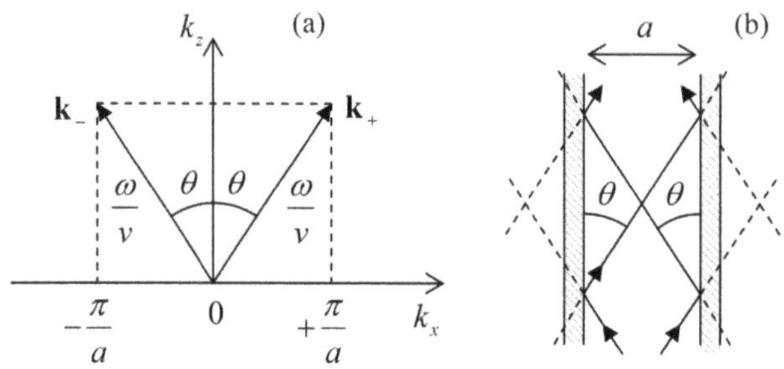

This figure shows that the plane wave propagation angle θ satisfies the following relation:

$$\sin\theta = \frac{|k_x|}{k} = \frac{\pi}{ka} = \frac{\lambda_0/2}{a},$$

where $\lambda_0 = 2\pi/k$ is the TEM wavelength. This angle asymptotically approaches zero at very high frequencies ($\omega \gg \omega_c$), but grows as the frequency is reduced toward the cutoff value ω_c, so that at $\omega \to \omega_c$, $\theta \to \pi/2$, i.e. $k_z \to 0$. Note that the cutoff frequency corresponds to the half-wavelength being exactly equal to the wide side of the waveguide, so that longer waves cannot propagate in it.

The fact that we are dealing with two usual plane waves may be confirmed by the electromagnetic field amplitude ratio calculation. Indeed, the similar decomposition of the H_{10} mode's magnetic field, given by Eqs. (7.131) and (7.137), yields two plane waves,

$$\mathbf{H}(\mathbf{r},\,t) = \frac{1}{2}H_l\,\mathrm{Re}\left[\left(-\frac{k_z a}{\pi}\mathbf{n}_x + \mathbf{n}_z\right)\exp\left\{i\left(\frac{\pi}{a}x + k_z z - \omega t\right)\right\}\right.$$
$$\left.+\left(\frac{k_z a}{\pi}\mathbf{n}_x + \mathbf{n}_z\right)\exp\left\{i\left(-\frac{\pi}{a}x + k_z z - \omega t\right)\right\}\right],$$

each with amplitude

$$|H_\omega| = \frac{H_l}{2}\left[\left(\frac{k_z a}{\pi}\right)^2 + 1\right]^{1/2} = H_l\frac{ka}{2\pi} = \frac{1}{Z}|E_\omega|,$$

whose ratio to E_ω is given by plane wave's impedance $Z = (\mu/\varepsilon)^{1/2}$, regardless of the ratio ω/ω_c.

Now, let us discuss why such decomposition of waves in waveguide into plane waves is possible. If these two plane waves propagated in unlimited isotropic media, we could notice, first of all, that an insertion of metallic walls normal to the axis y at *any* locations (say, at $y = 0$ and $y = b$, see figure 7.22) would not perturb the field between them. Indeed, as we know from chapter 2, the electric field perpendicular to the conductor surface is screened from penetration into the conductor's bulk by surface charges with density $\sigma(x,y) = \varepsilon E_z(x,y)$, without any perturbation of the applied field. Similarly, the magnetic fields of these two plane waves are tangential to the wall surfaces and are shielded from penetration inside the walls by the skin-effect currents, without any field perturbation outside the walls.

The situation with the other two walls (at $x = 0$ and $x = a$, see figure 7.22) is more involved, since at an arbitrary location of such a wall, the wave sum would have some tangential component of the electric field, and some normal component of the magnetic field, which cannot interact with the wall without incident wave's perturbation. However, as the above formulas for **E** and **H** show, at positions $k_x x_n = n\pi$, with any integer n, the component sums vanish and the wall insertion leaves the field between the walls intact.

These results may be also interpreted by saying that the fundamental H_{10} wave is formed by a TEM wave repeatedly reflected from the side walls of the waveguide—see panel (b) of the figure above.

Problem* 7.20. For the metallic coaxial cable with circular cross-section (see figure 7.20 of the lecture notes), find the lowest non-TEM mode and calculate its cutoff frequency.

Solution: The analysis may repeat that of the hollow circular waveguide (figure 7.23a) in section 7.7 of the lecture notes, up to the derivation of the Bessel equation (7.140) for the radial factor $\mathcal{R}(\rho)$ of the longitudinal field f—depending on the mode, either E_z or H_z. However, in the coaxial cable the axial points with $\rho = 0$ are inaccessible for the field, and hence instead of the simple solution given by Eq. (7.141), we have to look for its radial part in the form of a linear superposition of the Bessel functions of the fist and the second kind:

$$f_{nm} = [c_1 J_n(k_{nm}\rho) + c_2 Y_n(k_{nm}\rho)] \cos n(\varphi - \varphi_0), \quad \text{with } n = 0, 1, 2, \ldots,$$

where the eigenvalues k_{nm} of the transverse wave number k_t have to be chosen to satisfy the boundary conditions at $\rho = a$ and $\rho = b$, and (for a non-TEM field) give a non-uniform longitudinal field, $f_{nm} \neq \text{const}$.

For the E-modes, the boundary condition Eq. (7.124) yields the following system of two equations for the constants $c_{1,2}$:

$$c_1 J_n(k_{nm}a) + c_2 Y_n(k_{nm}a) = 0,$$
$$c_1 J_n(k_{nm}b) + c_2 Y_n(k_{nm}b) = 0 .$$

These two linear, homogeneous equations are compatible if the denominator of the system equals zero. This requirement yields

$$J_n(k_{nm}a)\, Y_n(k_{nm}b) = J_n(k_{nm}b)\, Y_n(k_{nm}a),$$

where the integer index m (taking values 1, 2, 3,...) numbers the roots of this characteristic equation—for each fixed angular index n (taking values 0, 1, 2,...). Introducing the dimensionless variable $\xi_{nm} \equiv k_{nm}a$, we may rewrite this characteristic equation as

$$J_n(\xi_{nm})\, Y_n\!\left(\xi_{nm}\frac{b}{a}\right) = J_n\!\left(\xi_{nm}\frac{b}{a}\right) Y_n(\xi_{nm}). \qquad (*)$$

The table below shows the results for the product $(b/a - 1)\xi_{nm} \equiv k_{nm}(b - a)$, obtained by numerical solution of this equation for a few values of the ratio b/a, and a few lowest numbers n and m.[34] It shows that the product $k_{nm}(b - a)$ changes very slowly with the ratio b/a, staying close to the asymptotic value $n\pi$ (reached at $b/a \to 1$) for all reasonable values of that ratio. This result may be readily understood physically: at the cutoff frequency $\omega_c = k_{nm}v$ (when $k_t = k = 2\pi/\lambda$), approximately m TEM half-wavelengths fit the distance between the external and internal conductors—almost independently of n, provided that the latter number is not too high. (Still, as could be expected, the axially symmetric distribution with $n = 0$ gives the lowest k_t, and hence the lowest cutoff frequency.)

	$n = 0$	$n = 1$	$n = 0$	$n = 1$
b/a	$m = 1$	$m = 1$	$m = 2$	$m = 2$
4	3.073	3.336	6.243	6.403
2	3.123	3.197	6.273	6.312
1	π	π	2π	π

For the H-modes, we need to use another boundary condition, Eq. (7.126), that gives, instead of Eq. (*), a different characteristic equation:

$$J_n'(\xi_{nm})\, Y_n'\!\left(\xi_{nm}\frac{b}{a}\right) = J_n'\!\left(\xi_{nm}\frac{b}{a}\right) Y_n'(\xi_{nm}), \qquad (**)$$

[34] See, for example, [3].

where the prime sign means the derivative of the Bessel function over its whole argument. The table below gives results for a different dimensionless combination, $(1 + b/a)\xi_{nm} \equiv k_{nm}(a + b)$, which is more relevant (virtually parameter-independent) in this case, for two lowest modes with $n = 1$.[35] It shows that for any realistic b/a ratio, this combination is close to $2m$. Physically, this means that at the cutoff frequency, m TEM wavelengths $\lambda = 2\pi/k_t$ fit the average circumference $p = \pi(a + b)$ of the cable's cross-section.

	$n = 1$	$n = 1$
b/a	$m = 1$	$m = 2$
4	2.055	3.760
2	2.031	4.023
$\to 1$	$\to 2$	$\to 4$

Now we can compare the approximate values of the lowest k_{nm} for the E- and H-modes:

$$\left.\frac{\pi}{b - a}\right|_{\text{for } E_{01}} \leftrightarrow \left.\frac{2}{b + a}\right|_{\text{for } H_{11}},$$

so that (just as in the circular waveguide, see section 7.7 of the lecture notes), the lowest non-TEM mode is H_{11}, with the cutoff frequency $\omega_c \approx 2v/(a + b)$, i.e. the TEM wavelength $\lambda_{max} \approx \pi(a + b)$. This result (which, at $a/b \to 0$, is close to Eq. (7.145) for the circular single-hole waveguide) is very important, because it imposes a practical limit $\lambda > \pi(a + b)$ for using coaxial cables as TEM transmission lines, in order to avoid unintentional excitation of the non-TEM modes on unavoidable small inhomogeneities.

Note, however, that in practical systems with long cables, the wavelength may be restricted even more severely by the wave attenuation effects—see section 7.9 of the lecture notes.

***Problem* 7.21.** Two coaxial cable sections are connected coaxially—see the figure below, which shows the cut along the system's symmetry axis. Relations (7.118) and (7.120) of the lecture notes seem to imply that if the ratios b/a of these sections are equal, their impedance matching is perfect, i.e. a TEM wave incident from one side on the connection would pass it without any reflection at all: $R = 0$. Is this statement correct?

[35] For $n = 0$, we can take advantage of the recurrence relations (2.143), which in particular yield $J_0' = -J_1$, $Y_0' = -Y_1$, to rewrite Eq. (**) in a form similar to Eq. (*) with $n = 1$:

$$J_1(\xi_{0m})Y_1\left(\xi_{0m}\frac{b}{a}\right) = J_1\left(\xi_{0m}\frac{b}{a}\right)Y_1(\xi_{0m}).$$

This comparison shows that values k_{0m} of the H-modes are equal to those of k_{1m} of the E-modes (for the same m), and hence are higher than the values k_{0m} of those modes—and hence, than the k_{1m} of the H modes.

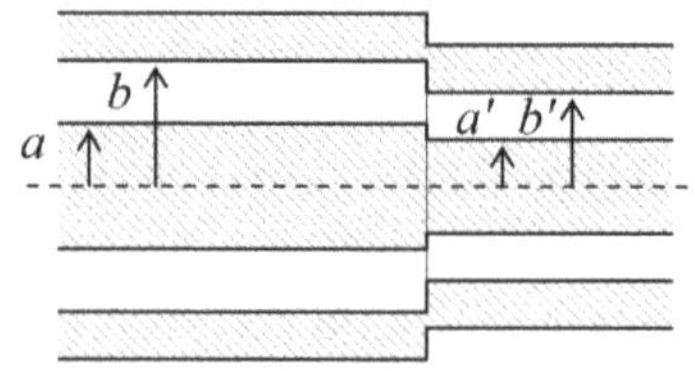

Solution: The derivations of both formulas mentioned in the assignment were based on the assumption of that the wave fields in both sections are strictly of the TEM type, with vectors **E** and **H** exactly normal to the wave vector **k**, i.e. to the system axis. This cannot be true close to the connection; for example, in the connection plane, this would lead to the vector **E** being parallel to the 'quasi-lid' surfaces of both conductors, thus contradicting the boundary condition (7.104). Thus, there is always some field perturbation, and as a result, some wave reflection from such a sharp jump of cable's parameters. Moreover, if the wavelength is smaller than $\lambda_c \approx \pi(a+b)$, according to the solution of the previous problem, waves of other type(s) may propagate at least in a wider section, and may be effectively excited by the TEM wave incident on the step[36].

However, if the wavelength λ is much longer than this λ_c (as it is in most practical applications of coaxial cables), such propagation is impossible, and the field perturbations may be only local, spreading at distances of the order of a, b from the connection plane. Since the telegrapher's equations (7.110) and (7.111), whose corollary Eq. (7.118) is, are derived by averaging the field effects at distances much larger than cross-section dimensions (in our case, a and b), the effect of these perturbations scales as $\lambda_c/\lambda \ll 1$. Thus if the matching condition $a'/b' = a/b$ is fulfilled, the reflection coefficient R is not exactly zero, but is small in proportion to the λ_c/λ ratio.

***Problem* 7.22.** Prove that TEM-like waves may propagate, in the radial direction, in the free space between two coaxial, round, metallic cones—see the figure below. Can this system be characterized by a certain transmission line impedance Z_W, as defined by Eq. (7.115) of the lecture notes?

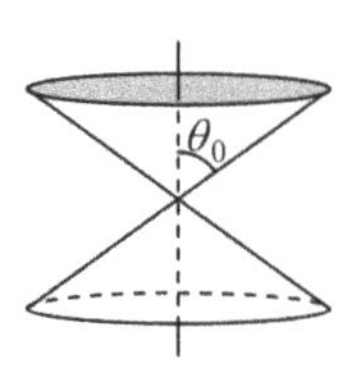

[36] Note that some (even very popular) textbooks describe over-simplified calculations of the higher mode excitation at waveguide connections by a simple expansion of the perturbed field over the sum of all possible *traveling-wave* modes, in the cross-section of the connection. Such calculations, neglecting the localized field perturbations (which are, in particular, responsible for the deviation of R from 1 in our problem), are not strictly valid, and their results may be considered as estimates at best.

Solution: The 'radial direction' means that the wave vector $\mathbf{k}$ has to be directed along the system's radius (from the point where the cones meet). The 'TEM-like' wave should have its vectors $\mathbf{E}$ and $\mathbf{H}$ directed normally to the vector $\mathbf{k}$, and to each other. Due to the axial symmetry of the system, this means that the complex amplitudes of the (monochromatic) wave fields, defined by Eq. (7.98), may have only one of the following two spatial structures: either

$$\mathbf{E}_\omega(\mathbf{r}) = \mathbf{n}_\theta E(r, \theta), \quad \text{and } \mathbf{H}_\omega(\mathbf{r}) = \mathbf{n}_\varphi H(r, \theta), \qquad (*)$$

or, vice versa,

$$\mathbf{E}_\omega(\mathbf{r}) = \mathbf{n}_\varphi E(r, \theta), \quad \text{and } \mathbf{H}_\omega(\mathbf{r}) = \mathbf{n}_\theta H(r, \theta), \qquad (**)$$

where r, θ, and φ are the usual spherical coordinates—see the figure below.

However, only the wave (*) may satisfy the macroscopic boundary conditions (7.104) at the cone surfaces. For its complex amplitudes, the four homogeneous Maxwell equations (7.2), written in the spherical coordinates[37], with $\varepsilon = \varepsilon_0$, and $\mu = \mu_0$, are reduced to

$$\frac{1}{r}\frac{\partial(rE)}{\partial r} - i\omega\mu_0 H = 0, \qquad -\frac{1}{r}\frac{\partial(rH)}{\partial r} + i\omega\varepsilon_0 E = 0, \qquad (***)$$

$$\frac{1}{r\sin\theta}\frac{\partial(E\sin\theta)}{\partial\theta} = 0, \qquad 0 = 0, \qquad (****)$$

the last, trivial relation meaning that the equation $\nabla\cdot\mathbf{H} = 0$ is automatically satisfied for any function $\mathbf{H}(\mathbf{r}) = \mathbf{n}_\varphi H(r, \theta)$. From the non-trivial equation (****) we immediately obtain that $e \equiv E\sin\theta$ is a function of r alone. Since Eqs. (***) do not involve the polar angle θ, they may be identically satisfied at all θ only if H follows the same dependence on the polar angle, i.e.

$$E(r, \theta) = \frac{e(r)}{\sin\theta}, \qquad H(r, \theta) = \frac{h(r)}{\sin\theta}.$$

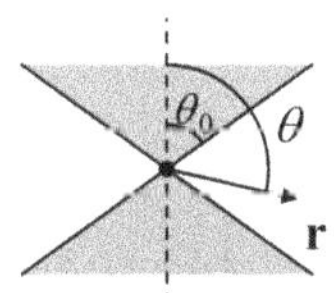

Now Eqs. (****), multiplied by r, may be rewritten as

$$\frac{d(re)}{dr} - i\omega\mu_0(rh) = 0, \qquad -\frac{d(rh)}{dr} + i\omega\varepsilon_0(re) = 0.$$

But this means that the products $re \equiv E(r, \theta)r\sin\theta$ and $rh \equiv H(r, \theta)r\sin\theta$ are related exactly as the complex wave amplitudes in the usual plane waves in free space,

[37] See, e.g. Eqs. (A.68) and (A.69).

i.e. have r-independent amplitudes, propagate along radius r (in any direction) with velocity c, and have an r- and θ-independent ratio (7.8):

$$\frac{re}{rh} \equiv \frac{E(r, \theta)}{H(r, \theta)} = Z_0 \equiv \left(\frac{\mu_0}{\varepsilon_0}\right)^{1/2} = \text{const.}$$

Let us check whether the wave impedance Z_W of such a 'transmission line', defined by the first of Eqs. (7.115), depends on r. The complex amplitude V_ω of the voltage between two points of the opposite cones, located as the same distance r from the center of the system, may be calculated as

$$\begin{aligned} V_\omega &= -\int_{\theta=\theta_0}^{\theta=\pi-\theta_0} \mathbf{E}_\omega \cdot d\mathbf{r} = \int_{\theta_0}^{\pi-\theta_0} Erd\theta = 2(re)\int_{\theta_0}^{\pi} \frac{d\theta}{\sin\theta} \\ &= -2(re)\int_{\theta=\theta_0}^{\theta=\pi} \frac{d(\cos\theta)}{1-\cos^2\theta} = -2(re)\int_{\theta=\theta_0}^{\theta=\pi} \frac{d\xi}{1-\xi^2} \\ &= -(re)\int_{\theta=\theta_0}^{\theta=\pi}\left(\frac{1}{1+\xi}+\frac{1}{1-\xi}\right)d\xi = (re)\ln\frac{1+\cos\theta_0}{1-\cos\theta_0}, \end{aligned}$$

i.e. is independent of r. Similarly, let us calculate the complex amplitude I_ω of the current flowing in each cone at distance r from the center, where the cone's cross-section radius is $\rho(r) = r\sin\theta_0$. Using the universal relation (6.38) between the linear density of the cone surface current (flowing in the radial direction) and the azimuthal magnetic field near the surface, we obtain

$$I_\omega = 2\pi\rho J_\omega = 2\pi r \sin\theta_0 H(r, \theta_0) = 2\pi(rh),$$

so that

$$Z_W \equiv \frac{V_\omega}{I_\omega} = \frac{1}{2\pi}\frac{(re)}{(rh)}\ln\frac{1+\cos\theta_0}{1-\cos\theta_0} \equiv \frac{Z_0}{2\pi}\ln\frac{1+\cos\theta_0}{1-\cos\theta_0}.$$

This result is close in structure to Eq. (7.120) of the lecture notes for the wave impedance of the usual coaxial cable, in particular featuring a similar weak (logarithmic) divergence at $\theta_0 \to 0$, i.e. at the vanishing cone's 'thickness' $\rho(r) \propto \sin\theta_0$.

Problem* 7.23. Use the recipe outlined in section 7.7 of the lecture notes to prove the characteristic equation (7.161) for the *HE* and *EH* modes in a round, step-index optical fiber.

Solution: Let us start with calculating the constant coefficient in Eq. (7.160), from the requirement that at the core-to-cladding interface ($\rho = R$), both E_z and H_z have to be continuous (as the components parallel to the interface), i.e. $f_+ = f_-$. This relation, combined with Eq. (7.156) for f_-, allows making Eq. (7.160) more specific:

$$f_+ = f_l \frac{J_n(k_t R)}{K_n(\kappa_t R)} K_n(\kappa_t \rho)\cos n(\varphi - \varphi_0).$$

Since we are now dealing with a linear superposition of two longitudinal fields, it is easier to operate with the complex-exponential form of this expression and Eq. (7.156):

$$f_+ = f_l \frac{J_n(k_t R)}{K_n(\kappa_t R)} K_n(\kappa_t\rho)e^{in\varphi}, \quad f_- = f_l J_n(\kappa_t\rho)e^{in\varphi}, \quad \text{where } f_l = \begin{cases} E_l, & \text{for } f_\pm = E_\pm, \\ H_l, & \text{for } f_\pm = H_\pm, \end{cases}$$

and the constant phases φ_0 (which may be different for E_z and H_z) are incorporated into the complex amplitudes f_l.[38] Now plugging these expressions into the first of Eqs. (7.121), and using the general vector-algebra expressions for the cylindrical coordinates[39],

$$(\nabla_t f)_\rho = \frac{\partial f}{\partial \rho}, \qquad (\nabla_t f)_\varphi = \frac{1}{\rho}\frac{\partial f}{\partial \varphi}, \qquad (\mathbf{n}_z \times \nabla_t f)_\rho = \frac{1}{\rho}\frac{\partial f}{\partial \varphi}, \qquad (\mathbf{n}_z \times \nabla_t f)_\varphi = -\frac{\partial f}{\partial \rho},$$

we obtain the following formulas for the transverse components of the electric field:

$$E_\rho^- = \frac{i}{k_t^2}\left[k_z k_t J_n'(k_t\rho)E_l - k_- Z_- \frac{in}{\rho} J_n(k_t\rho)H_l\right] e^{in\varphi},$$

$$E_\varphi^- = \frac{i}{k_t^2}\left[k_z \frac{in}{\rho} J_n(k_t\rho)E_l + k_- Z_- k_t J_n'(k_t\rho)H_l\right] e^{in\varphi},$$

$$E_\rho^+ = -\frac{i}{\kappa_t^2}\frac{J_n(k_t R)}{K_n(\kappa_t R)}\left[k_z \kappa_t K_n'(\kappa_t\rho)E_l - k_+ Z_+ \frac{in}{\rho} K_n(\kappa_t\rho)H_l\right] e^{in\varphi},$$

$$E_\varphi^+ = -\frac{i}{\kappa_t^2}\frac{J_n(k_t R)}{K_n(\kappa_t R)}\left[k_z \frac{in}{\rho} K_n(\kappa_t\rho)E_l + k_+ Z_+ \kappa_t K_n'(\kappa_t\rho)H_l\right] e^{in\varphi}.$$

where the prime signs denote the differentiation of each Bessel function over its whole argument.

Now we should write the boundary conditions for the transverse components of the electric field:

$$\varepsilon_- E_\rho^- = \varepsilon_+ E_\rho^+, \quad E_\varphi^- = E_\varphi^+, \qquad \text{at } \rho = R.$$

Requiring that these conditions are satisfied at all angles φ, we obtain two linear equations for the complex constants E_l and H_l. At $\mu_- = \mu_+ = \mu_0$, the boundary conditions for the transverse components of the magnetic field would give an equivalent system. The same relation ensures that the ratio $\varepsilon_+/\varepsilon_-$ may be

[38] Indeed, we are working with the complex amplitudes of the fields anyway—see Eqs. (7.98).

[39] See, e.g. Eq. (A.60) with $\partial/\partial z = 0$.

represented as k_+^2/k_-^2, and that $k_+Z_+ = k_-Z_- = \omega\mu_0$, so that the system of two equations reduces to

$$k_z\left(\frac{k_-^2}{k_t}\frac{J_n'}{J_n} + \frac{k_+^2}{\kappa_t}\frac{K_n'}{K_n}\right)E_l - \frac{in}{R}\omega\mu_0\left(\frac{k_-^2}{k_t^2} + \frac{k_+^2}{\kappa_t^2}\right)H_l = 0,$$

$$k_z\frac{in}{R}\left(\frac{1}{k_t^2} + \frac{1}{\kappa_t^2}\right)E_l + \omega\mu_0\left(\frac{1}{k_t}\frac{J_n'}{J_n} + \frac{1}{\kappa_t}\frac{K_n'}{K_n}\right)H_l = 0.$$

Now requiring the determinant of this system to be zero, so they would be consistent, we indeed arrive at Eq. (7.161):

$$\left(\frac{k_-^2}{k_t}\frac{J_n'}{J_n} + \frac{k_+^2}{\kappa_t}\frac{K_n'}{K_n}\right)\left(\frac{1}{k_t}\frac{J_n'}{J_n} + \frac{1}{\kappa_t}\frac{K_n'}{K_n}\right) = \frac{n^2}{R^2}\left(\frac{k_-^2}{k_t^2} + \frac{k_+^2}{\kappa_t^2}\right)\left(\frac{1}{k_t^2} + \frac{1}{\kappa_t^2}\right).$$

***Problem* 7.24.** Neglecting the skin-effect depth δ_s, find the lowest eigenfrequencies, and the corresponding field distributions, of the standing electromagnetic waves inside a round cylindrical resonant cavity—see the figure below.

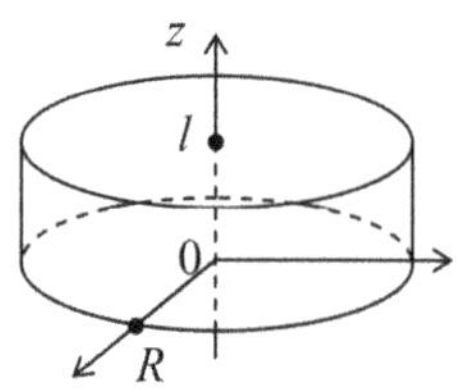

Solution: Due to the simple geometry of the system, and hence the simple structure of the fields in it, we can use the first approach to the eigenfrequency calculation, based on the analysis of circular metallic waveguides in section 7.6 of the lecture notes. In particular, for the H-modes we may use Eqs. (7.121) with $E_z = 0$, and an evident generalization of Eq. (7.144),

$$H_z = H_l J_n\left(\xi'_{nm}\frac{\rho}{R}\right)\cos n(\varphi - \varphi_0),$$

to calculate the transverse fields $\mathbf{E}_t$ and $\mathbf{H}_t$ in the waveguide. Without writing these formulas explicitly, we may note that according to Eq. (7.121), $\mathbf{E}_t$ has the same phase as H_z, which does not depend on the point's position on the cross-section z = const. Hence if two metallic walls, parallel to the axis z, are inserted into the waveguide at the distance $\Delta z = p(\lambda_z/2) = \pi p/k_z$, where $p = 1, 2, \ldots$, they do not disturb the field distribution—besides turning the traveling wave into the standing one.

Now requiring this Δz to be equal l, and using the general relation (7.122) in the form

$$k^2 \equiv \frac{\omega^2}{v^2} = k_t^2 + k_z^2,$$

and Eq. (7.143) for the eigenvalues of the transverse component k_t of the wave vector, we obtain the following eigenfrequency spectrum of the H-modes:

$$\omega_{nmp}^2 = v^2 k^2 = v^2\left[\left(\frac{\xi'_{nm}}{R}\right)^2 + \left(\frac{\pi p}{l}\right)^2\right], \qquad (*)$$

Since the sets of possible integers m and p start with 1, according to table 7.1, the lowest of these frequencies is

$$\omega_{011}^2 = v^2\left[\left(\frac{\xi'_{11}}{R}\right)^2 + \left(\frac{\pi}{l}\right)^2\right] \approx v^2\left[\left(\frac{1.841}{R}\right)^2 + \left(\frac{\pi}{l}\right)^2\right]. \qquad (**)$$

(As a reminder, v is plane wave's velocity in the dielectric filling the resonator: $v^2 = 1/\varepsilon\mu$.)

However, this is not necessarily the lowest frequency of the resonator. Indeed, using the absolutely similar analysis of the E-waves, whose longitudinal electric field is described by the natural modification of Eq. (7.144) of the lecture notes[40]:

$$E_z = E_l J_n\left(\xi_{nm}\frac{\rho}{R}\right)\cos n(\varphi - \varphi_0),$$

for the corresponding modes of the resonator we obtain an expression very similar to Eq. (*):

$$\omega_{nmp}^2 = v^2\left[\left(\frac{\xi_{nm}}{R}\right)^2 + \left(\frac{\pi p}{l}\right)^2\right].$$

Note, however, that for the E modes, the boundary conditions at the lid surfaces ($\mathbf{E}_t = 0$, $\partial H_z/\partial z = 0$) allow non-zero fields even if $p = 0$. (In this case neither component of the fields $\mathbf{E}$ or $\mathbf{H}$ depends on z.) Hence the lowest eigenfrequency of these modes is independent of l:

$$\omega_{010}^2 = v^2\left(\frac{\xi_{01}}{R}\right)^2 \approx v^2\left(\frac{2.405}{R}\right)^2. \qquad (***)$$

Note that the field distribution is this E_{010} mode is very similar to that in the fundamental mode of the rectangular resonator—see figure 7.30, where b should be replaced with l, and a with $2R$.

A direct comparison of Eqs. (**) and (***) shows that that the latter frequency is lower, and hence E_{010}, with the eigenfrequency (***), is the fundamental mode of the resonator, if $l < \pi R/(\xi^2_{01} - \xi'^2_{11})^{1/2} \approx 2.03R$. In the opposite case, H_{011} is the fundamental mode, with the eigenfrequency given by Eq. (**)[41].

[40] Note the change from the roots ξ'_{nm} to roots ξ_{nm}, due to the different relevant boundary conditions on the side wall of the resonator (at $\rho = R$): for the E waves, $E_z = 0$, instead of the $dH_z/d\rho = 0$ for the H waves.

[41] A similar crossover in the rectangular resonator, disguised by the similarity of field distribution along its sides, takes place at $l = \min[a,b]$, i.e. essentially at the same condition as in the circular resonator, if we identify its diameter $2R$ with the smallest lateral size of the rectangular resonator.

Just for the reader's reference, let me note that the E-modes with $p = 0$ (i.e. with the fields independent of the coordinate z directed along cylinder's axis), high indices $n \gg 1$, and lower indices $m \sim 1$ are called the *whispering gallery* modes. This name[42], which came from similar acoustic wave modes, is due to the fact that (as, for example, figure 2.18 shows) the fields of such modes are localized mostly at the cavity's walls (at $\rho \approx R$)—just as human whispers may be virtually localized at the walls of a round hall or gallery.

***Problem* 7.25.** A plane, monochromatic wave propagates through a medium with an Ohmic conductivity σ, and negligible electric and magnetic polarization effects. Calculate the wave's attenuation, and relate the result with a certain calculation carried out in chapter 6 of the lecture notes.

Solution: In a plane wave the only energy losses are those in the propagation medium, so that the attenuation coefficient is completely described by Eq. (7.218) of the lecture notes:

$$\alpha = 2\omega \mathrm{Im}[\varepsilon^{1/2}(\omega)\mu^{1/2}(\omega)]. \qquad (*)$$

For the medium described in the problem's assignment, Eq. (7.46) takes the form

$$\varepsilon(\omega) = \varepsilon_0 + i\frac{\sigma}{\omega},$$

so that Eq. (*), with $\mu(\omega) = \mu_0$, yields

$$\alpha = 2\omega \mathrm{Im}\left[\left(\varepsilon_0 + i\frac{\sigma}{\omega}\right)^{1/2}\mu_0^{1/2}\right] \equiv \frac{\omega}{c}\mathrm{Im}\left[\left(1 + \frac{i}{\omega\tau_r}\right)^{1/2}\right]$$
$$\equiv \frac{1}{\sqrt{2}}\frac{\omega}{c}\left[\left(1 + \frac{1}{\omega^2\tau_r^2}\right)^{1/2} - 1\right]^{1/2}, \qquad (**)$$

where $\tau_r \equiv \varepsilon_0/\sigma$ is the charge relaxation time defined by Eq. (4.10), for the particular case $\varepsilon_0 = 1$.

This result may be simplified in two limiting cases. At relatively high frequencies (or at relatively small conductivity) the formula is reduced to

$$\alpha \approx \frac{1}{\sqrt{2}}\frac{\omega}{c}\frac{1}{(2\omega^2\tau_r^2)^{1/2}} \equiv \frac{1}{2c\tau_r} \equiv \frac{\sigma}{2c\varepsilon_0} \equiv \frac{1}{2}\sigma Z_0 \ll k_0, \quad \text{for } \omega\tau_r \gg 1, \quad \text{i.e. for } \alpha \ll k_0,$$

where $Z_0 \equiv (\mu_0/\varepsilon_0)^{1/2}$ and $k_0 = \omega/c$ are, respectively, the wave impedance and the wave number in the free space. In this case the wave decays at a distance ($l_d \equiv 1/\alpha$) much larger than its wavelength $\lambda_0 \equiv 2\pi/k_0$.

In the opposite limit of low frequencies (or high conductivity), the result yields

$$\alpha \approx \frac{1}{\sqrt{2}}\frac{\omega}{c}\frac{1}{(\omega\tau_r)^{1/2}} \equiv \left(\frac{\mu_0\omega\sigma}{2}\right)^{1/2} \gg k_0, \quad \text{for } \omega\tau_r \ll 1, \quad \text{i.e. } \alpha \gg k_0.$$

---

[42] Due to Lord Rayleigh who first (in 1878) noticed this effect (for acoustic waves) in St Paul's Cathedral.

In its reciprocal $l_d \equiv 1/\alpha$, i.e. the wave decay length, we may readily recognize Eq. (6.33) for the skin depth δ_s, which was derived in section 6.3 without the account of the displacement currents. Indeed, in this limit the decay length l_d is much smaller than the free-space wavelength λ_0 at this frequency, so that the quasi-static approximation is fully justified—see section 6.8. Note that in this limit it is hardly possible to speak about the wave propagation: upon entering the medium, the wave immediately decays—physically, because the high Ohmic conductivity results in very high Joule losses (4.39).

***Problem* 7.26.** Generalize the telegrapher's equations (7.110) and (7.111) by accounting for small energy losses:

(i) in the transmission line's conductors, and
(ii) in the medium separating the conductors,

using their simplest (Ohmic) models. Formulate the conditions of validity of the resulting equations.

Solutions:

(i) Physically, Eq. (7.111) is just a balance between the Faraday's e.m.f. induced on a unit length of the transmission line (described by the right-hand side) and its drop on the inductance of that segment. If the line's conductors have the (total) resistance R_0 per unit length, it makes an additional contribution $R_0 I$ to the balance, so that the equation becomes

$$L_0\frac{\partial I}{\partial t} = -\frac{\partial V}{\partial z} - R_0 I. \qquad (*)$$

(ii) Similarly, Eq. (7.110) has a clear physical sense of the charge conservation law: its left-hand side is the rate of change of the charge of the capacitance of a unit length of the transmission line, while the right-hand side is the balance of currents flowing in and out of such a unit-length fragment. A non-vanishing Ohmic conductance G_0 (per unit length) of the media separating the wires evidently brings into this balance an additional leakage current $G_0 V$, so that the equation takes the following form:

$$C_0\frac{\partial V}{\partial t} = -\frac{\partial I}{\partial z} - G_0 V. \qquad (**)$$

The correctness of the signs of the additional terms in these generalized telegrapher's equations may be double-checked from the natural requirement that if R_0 or G_0 are positive (dissipative) they lead to wave's attenuation. In order to calculate the attenuation, we may plug into Eqs. (*) and (**) the standard monochromatic traveling-wave solution $I = \text{Re}[I_\omega e^{i(kz - \omega t)}]$, $V = \text{Re}[V_\omega e^{i(kz - \omega t)}]$, obtaining the following system of two linear equations for their complex amplitudes:

$$-i\omega L_0 I_\omega = -ikV_\omega - R_0 I_\omega, \qquad -i\omega C_0 V_\omega = -ikI_\omega - G_0 V_\omega,$$

cf. Eqs. (7.113). The condition of their consistency,

$$\begin{vmatrix} -i\omega L_0 + R_0 & ik \\ ik & -i\omega C_0 + G_0 \end{vmatrix} = 0,$$

immediately yields the dispersion relation,

$$k = [(\omega L_0 + iR_0)(\omega C_0 + iG_0)]^{1/2},$$

which is evidently a generalization of Eq. (7.114) to the case of non-vanishing Ohmic losses. In particular, in the most important case when the losses are relatively low ($\text{Re}\, k \equiv k' \gg k'' \equiv \text{Im}\, k$)[43], the relation reduces to

$$k = k' + ik'', \quad \text{with } k' = \omega(L_0C_0)^{1/2}, \quad k'' = \frac{1}{2}\left(\frac{R_0}{Z_W} + G_0Z_W\right),$$

where $Z_W \equiv (L_0/C_0)^{1/2}$ is the lossless transmission line's impedance (7.115). This result shows that, indeed, positive R_0 and G_0 do give positive contributions to the attenuation constant (7.216), $\alpha \equiv 2k''$.

Since the above derivation of Eqs. (*) and (**) was based on such global (or 'instant') characteristics of the transmission line as R_0 and G_0, the necessary condition of validity of these equations is the smallness of the linear scale a of transmission line's cross-section in comparison with wave's decay length:

$$a \ll l_\text{d} \equiv \frac{1}{2k''} \approx \min\left[\frac{Z_W}{R_0}, \frac{1}{G_0Z_W}\right].$$

***Problem* 7.27.** Calculate the skin-effect contribution to the attenuation coefficient α, defined by Eq. (7.214) of the lecture notes, for the fundamental (H_{10}) mode propagating in a metallic-wall waveguide with a rectangular cross-section—see figure 7.22. Use the results to evaluate the wave decay length $l_\text{d} \equiv 1/\alpha$ for a 10 GHz wave in the standard X-band waveguide WR-90 (with copper walls, a = 23 mm, b = 10 mm, and no dielectric filling), at room temperature. Compare the result with that, made in section 7.9 of the lecture notes, for the standard TV coaxial cable, at the same frequency.

Solution: As was discussed in section 7.6 of the lecture notes, in the H_{10} wave the electric field has just one Cartesian component, with the complex amplitude

$$E_y(x) = i\frac{ka}{\pi}ZH_l \sin\frac{\pi x}{a},$$

while the magnetic field has two components, with complex amplitudes

$$H_z(x) = H_l \cos\frac{\pi x}{a}, \quad H_x(x) = -i\frac{k_z a}{\pi}H_l \sin\frac{\pi x}{a}.$$

[43] If they are not, the wave decays so fast that we cannot really speak about its *propagation* in the line.

Of those two components, only H_x contributes to the longitudinal (z) component of the time-averaged Poynting vector

$$\overline{S_z} = \frac{E_x H_y^* - E_y H_x^*}{2} = \frac{kk_z a^2}{2\pi^2} Z\, |H_l|^2 \sin^2 \frac{\pi x}{a},$$

which gives the following total power flow along the waveguide:

$$\mathscr{P} = \int_0^a dx \int_0^b dy \overline{S_z} = \frac{kk_z a^3 b}{4\pi^2} Z\, |H_l|^2, \qquad (*)$$

giving us the denominator on the right-hand side of Eq. (7.214).

In order to calculate the nominator of that fraction, i.e. the power dissipation per unit length, we have to integrate the losses per unit area, given by Eq. (7.219), over the cross-section's perimeter:

$$\begin{aligned}
\frac{d\mathscr{P}_{\text{loss}}}{dz} &= \frac{\mu_0 \omega \delta_s}{4} \oint_C |H_0(x)|^2 dl \\
&= \frac{\mu_0 \omega \delta_s}{4}\left[2\int_0^a [|H_x(x)|^2 + |H_z(x)|^2] dx + \int_0^b |H_z(0)|^2 dy + \int_0^b |H_z(b)|^2 dy\right] \\
&= \frac{\mu_0 \omega \delta_s}{4} |H_l|^2 \left\{2\int_0^a \left[\left(\frac{k_z a}{\pi}\right)^2 \sin^2 \frac{\pi x}{a} + \cos^2 \frac{\pi x}{a}\right] dx + 2b\right\} \\
&= \frac{\mu_0 \omega \delta_s}{4} |H_l|^2 \left\{\left[\left(\frac{k_z a}{\pi}\right)^2 + 1\right] a + 2b\right\}.
\end{aligned}$$

According to the waveguide's dispersion relation, given by Eqs. (7.122) and (7.133),

$$k_z^2 + k_t^2 = k^2 = \omega^2 \varepsilon \mu, \quad \text{with } (k_t)_{10} = \frac{\pi}{a},$$

the expression in the last square brackets is just $(ka/\pi)^2$, so that, using Eq. (*) for $\mathscr{P}$, for the attenuation constant we finally obtain

$$\alpha \equiv \frac{1}{\mathscr{P}} \frac{d\mathscr{P}_{\text{loss}}}{dz} = \pi^2 \frac{\mu \omega \delta_s}{Zkk_z a^2 b}\left[\left(\frac{ka}{\pi}\right)^2 + \frac{2b}{a}\right] = \pi \frac{\delta_s}{(k_z a/\pi)ab}\left[\left(\frac{ka}{\pi}\right)^2 + \frac{2b}{a}\right].$$

Note that α scales approximately as δ_s/A, where $A \equiv ab$ is waveguide's cross-section area. More particularly, α diverges at $b \to 0$, because the transmitted power vanishes while the skin-effect loss in the broader walls of the waveguide remain constant (at a fixed field amplitude). Another interesting fact is that the dependence of the attenuation on frequency is non-monotonic: α diverges at $\omega \to \omega_c$, where $k_z \to 0$, but also grows (as $\omega^{1/2}$) at $\omega \to \infty$, so that the lowest attenuation is reached at a frequency ~30% above the threshold, where the higher modes (H_{11}, H_{20}, and E_{11}) still cannot propagate in the waveguide.

For a 10 GHz wave in the WR-90 waveguide we obtain $\lambda = \lambda_0 \approx 30$ mm, $ka/\pi = 2a/\lambda \approx 1.53$, $k_z a/\pi = [(ka/\pi)^2 - 1]^{1/2} \approx 1.16$, while for copper at room temperature and that frequency, $\delta_s \approx 6.5 \times 10^{-7}$ m; as a result, our final formula yields $\alpha \approx 0.025$ m^{-1} (i.e. ~0.1 dB m^{-1}). Hence $l_d \equiv 1/\alpha \approx 40$ m, so that the waveguide provides an attenuation well below that ($\alpha \approx 0.16$ m^{-1}, see section 7.9) of the standard TV coaxial cable RG-6/U at the same frequency. (Admittedly, the difference is mostly due to the waveguide's larger cross-section, rather than its different wave mode.)

***Problem* 7.28.** Calculate the skin-effect contribution to the attenuation coefficient α of

(i) the fundamental (H_{11}) wave, and
(ii) the H_{01} wave,

in a metallic-wall waveguide with the circular cross-section (see figure 7.23a of the lecture notes), and analyze the low-frequency ($\omega \to \omega_c$) and high-frequency ($\omega \gg \omega_c$) behaviors of α for each of these modes.

Solutions:

(i) For the H_{11} mode, the longitudinal component H_z of the magnetic field (or rather its complex amplitude) is described by Eq. (7.144) of the lecture notes:

$$H_z(\rho, \varphi) = H_l J_1\left(\xi'_{11}\frac{\rho}{R}\right)\cos\varphi, \quad \text{with } \xi'_{11} \approx 1.84,$$

where the arbitrary constant φ_0, which evidently does not affect such a global parameter as α, was taken for 0. Plugging this result into Eqs. (7.121) with $E_z = 0$, and using the general vector-algebra expressions[44]

$$(\nabla_t f)_\rho = \frac{\partial f}{\partial \rho}, \quad (\nabla_t f)_\varphi = \frac{1}{\rho}\frac{\partial f}{\partial \varphi}, \quad (\mathbf{n}_z \times \nabla_t f)_\rho = \frac{1}{\rho}\frac{\partial f}{\partial \varphi}, \quad (\mathbf{n}_z \times \nabla_t f)_\varphi = -\frac{\partial f}{\partial \rho},$$

for transverse components of the electric and magnetic fields, we obtain the following formulas:

$$E_\rho(\rho, \varphi) = \frac{ikZ}{k_t^2} H_l \frac{1}{\rho} J_1\left(\xi'_{11}\frac{\rho}{R}\right)\sin\varphi, \qquad E_\varphi(\rho, \varphi) = \frac{ikZ}{k_t} H_l J'_1\left(\xi'_{11}\frac{\rho}{R}\right)\cos\varphi,$$

$$H_\rho(\rho, \varphi) = i\frac{k_z}{k_t} H_l J'_1\left(\xi'_{11}\frac{\rho}{R}\right)\cos\varphi, \qquad H_\varphi(\rho, \varphi) = -\frac{ik_z}{k_t^2} H_l \frac{1}{\rho} J_1\left(\xi'_{11}\frac{\rho}{R}\right)\sin\varphi,$$

where $k \equiv \omega(\varepsilon\mu)^{1/2}$, $Z \equiv (\mu/\varepsilon)^{1/2}$, and $k_t = \xi'_{11}/R$, while k_z should be found from the general equation (7.102):

$$k_z = (k^2 - k_t^2)^{1/2}.$$

[44] See, e.g. Eq. (A.60) with $\partial/\partial z = 0$; see also the model solution of problem 7.23.

From here, the time-averaged areal density of the power loss due to the skin effect, given by Eq. (7.219), is

$$\frac{d\mathscr{P}_{\text{loss}}}{dA} = \frac{\mu\omega\delta_s}{4}\left[\left|H_\varphi(R,\varphi)\right|^2 + |H_z(R,\varphi)|^2\right]$$
$$= \frac{\mu\omega\delta_s}{4}|H_l|^2 J_1^2(\xi'_{11})\left(\frac{k_z^2}{k_t^4 R^2}\sin^2\varphi + \cos^2\varphi\right).$$

Integrating this expression over the cross-section's perimeter, we find the average power loss per unit length of the waveguide:

$$\frac{d\mathscr{P}_{\text{loss}}}{dz} = \frac{\pi}{4}J_1^2(\xi'_{11})\mu\omega\delta_s R\,|H_l|^2\left[\frac{1}{(k_t R)^2}\frac{k_z^2}{k_t^2} + 1\right].$$

By the definition of ξ'_{11}, $J_1(\xi'_{11})$ is just the first maximum of Bessel function $J_1(\xi'_{11})$, close to 0.58 (see, e.g. figure 2.18), while $(k_t R)^2 = (\xi'_{11})^2 \approx 3.39$.

The average longitudinal component of the Poynting vector is

$$\overline{S_z} = \frac{E_\rho H_\varphi^* - E_\varphi H_\rho^*}{2} = \frac{Z}{2}|H_l^2|\frac{kk_z}{k_t^4}\left[\frac{1}{\rho^2}J_1^2\left(\xi'_{11}\frac{\rho}{R}\right)\sin^2\varphi + \frac{\xi_{11}'^2}{R^2}J_1'^2\left(\xi'_{11}\frac{\rho}{R}\right)\cos^2\varphi\right].$$

Integrating this expression over the waveguide's cross-section, for the average propagating power we obtain

$$\mathscr{P} = \int_0^{2\pi} d\varphi \int_0^R \rho\, d\rho\ \overline{S_z} = \frac{\pi}{2}Z|H_l^2|\frac{kk_z}{k_t^2}\int_0^R\left[\frac{1}{k_t^2\rho^2}J_1^2\left(\xi'_{11}\frac{\rho}{R}\right) + J_1'^2\left(\xi'_{11}\frac{\rho}{R}\right)\right]\rho\, d\rho$$
$$= \frac{\pi I}{\xi_{11}'^2}Z|H_l^2|kk_z R^4,$$

where I is a dimensionless constant:

$$I \equiv \int_0^{\xi'_{11}}\left[\frac{J_1^2(\xi)}{\xi^2} + J_1'^2(\xi)\right]\xi\, d\xi \equiv \frac{1}{4}\int_0^{\xi'_{11}}[J_0^2(\xi) + J_2^2(\xi)]\,\xi\, d\xi \approx 0.291.$$

From here, the attenuation constant is

$$\alpha \equiv \frac{1}{\mathscr{P}}\frac{d\mathscr{P}_{\text{loss}}}{dz} = \frac{J_1^2(\xi'_{11})}{2I}\frac{\delta_s}{k_z R^3}\left(\frac{k_z^2}{k_t^2} + \xi_{11}'^2\right).$$

The attenuation's dependence on the wave frequency ω, following from this formula, is very similar to that for the rectangular waveguide (see the model solution of the previous problem): α diverges when ω is reduced to the cutoff frequency $\omega_c = k_t/(\varepsilon\mu)^{1/2}$, because at that point $k_z \to 0$, while it increases as $\delta_s k \propto \omega^{1/2}$ at high frequencies $\omega \gg \omega_c$ where $k_z \approx k \gg k_t$. (The minimum of α is reached in-between these two extremes, at $\omega \approx 3\omega_c$.)

(ii) For the H_{01} mode, the longitudinal field is described by Eqs. (7.141) and (7.143) with $n = 0$ and $m = 1$:

$$H_z(\rho, \varphi) = H_l J_0\left(\xi'_{01}\frac{\rho}{R}\right), \qquad \xi'_{01} \approx 3.83,$$

i.e. is independent of the azimuthal angle. Because of that independence, the field calculations, absolutely similar to those carried out for case (i), yield even simpler results:

$$E_\rho(\rho, \varphi) = 0, \qquad E_\varphi(\rho, \varphi) = -i\frac{k}{k_t^2}ZH_l\frac{\xi'_{01}}{R}J'_0\left(\xi'_{01}\frac{\rho}{R}\right) = i\frac{k}{k_t^2}ZH_l\frac{\xi'_{01}}{R}J_1\left(\xi'_{01}\frac{\rho}{R}\right),$$

$$H_\rho(\rho, \varphi) = i\frac{k_z}{k_t^2}H_l\frac{\xi'_{01}}{R}J'_0\left(\xi'_{01}\frac{\rho}{R}\right) = -i\frac{k_z}{k_t^2}H_l\frac{\xi'_{01}}{R}J_1\left(\xi'_{01}\frac{\rho}{R}\right), \qquad H_\varphi(\rho, \varphi) = 0.$$

As a result of the vanishing azimuthal magnetic field, the skin-effect losses are determined only by the longitudinal component of **H**:

$$\frac{d\mathscr{P}_{\text{loss}}}{dA} = \frac{\mu\omega\delta_s}{4}\,|H_z(R, \varphi)|^2 = \frac{\mu\omega\delta_s}{4}\,|H_l|^2 J_0^2(\xi'_{01}), \quad \text{so that}$$

$$\frac{d\mathscr{P}_{\text{loss}}}{dz} = R\int_0^{2\pi}\frac{\mathscr{P}_{\text{loss}}}{A}d\varphi = \frac{\pi}{2}J_0^2(\xi'_{01})\mu\omega\delta_s R\,|H_l|^2.$$

The longitudinal component of the Poynting vector is also contributed by only one component product:

$$\overline{S}_z = -\frac{E_\varphi H_\rho^*}{2} = \frac{Z}{2}|H_l^2|\frac{kk_z}{k_t^4 R^2}J_1^2\left(\xi'_{01}\frac{\rho}{R}\right),$$

so that the total average power flow is

$$\mathscr{P} = \int_0^{2\pi} d\varphi\int_0^R \rho\, d\rho\;\overline{S}_z = \pi Z|H_l^2|\frac{kk_z}{k_t^4}\int_0^R J_1^2\left(\xi'_{01}\frac{\rho}{R}\right)\rho\, d\rho$$

$$= \pi\, I' Z|H_l^2|\frac{kk_z}{k_t^4}R^2, \quad \text{with } I' \equiv \int_0^{\xi'_{01}} J_1^2(\xi)\xi\, d\xi \approx 0.632.$$

As a result, the expression for the attenuation constant,

$$\alpha \equiv \frac{1}{\mathscr{P}}\frac{\mathscr{P}_{\text{loss}}}{dz} = \frac{J_0^2(\xi'_{01})}{2I'}\frac{\delta_s k_t^4}{k_z R},$$

has a rather different frequency dependence: while still diverging at $\omega \to \omega_c$ (where $k_z \to 0$), it is proportional to $\delta_s/k \propto \omega^{-1/2} \to 0$ at $\omega/\omega_c \to \infty$, where $k_z \to k \gg k_t$.

This property, which makes H_{01} (as well as higher H_{0m} modes) so attractive for long-distance microwave energy transfer, is due to the absence of the azimuthal component of the magnetic field and the radial component of the electric field. At the wall surface ($\rho = R$), where the function $J_1(\xi'_{01}\rho/R) \equiv -J'_0\,(\xi'_{01}\rho/R)$ vanishes, two other transverse components of the fields also vanish. As a result, the skin-effect losses are due only to the longitudinal component of the magnetic field, which decreases (at a fixed wave power) as wave's frequency is increased and its structure becomes closer and closer to TEM.

***Problem* 7.29.** For a rectangular metallic-wall resonance cavity with dimensions $a \times b \times l\,(b \leqslant a, l)$, calculate the Q-factor in the fundamental oscillation mode, due to the skin-effect losses in the walls. Evaluate the factor for a $23 \times 23 \times 10$ mm^3 resonator with copper walls, at room temperature.

Solution: Selecting the Cartesian coordinates as in figure 7.29 of the lecture notes (partly reproduced below), the magnetic field distribution in the fundamental (H_{101}) mode, discussed in section 7.8, is as follows:

$$H_x = \frac{a}{l}H_l \sin\frac{\pi x}{a}\cos\frac{\pi z}{l}, \quad H_y = 0, \quad H_z = H_l \cos\frac{\pi x}{a}\sin\frac{\pi z}{l}.$$

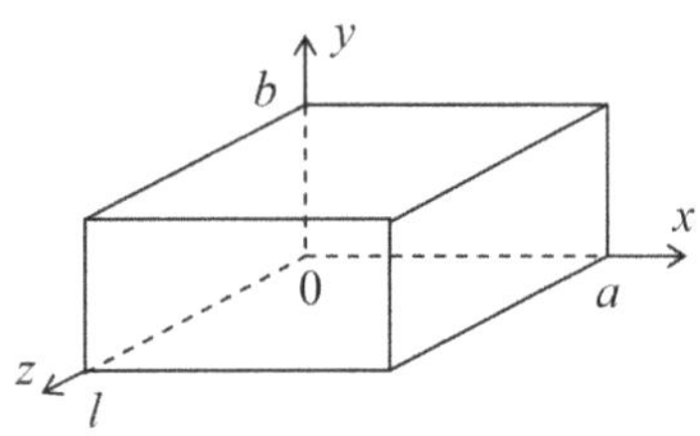

The electromagnetic field energy in the resonator may be calculated, for example, as the maximum energy of the magnetic field:

$$\begin{aligned} U &= \frac{\mu_0}{2}\int_0^a dx\int_0^b dy\int_0^l dz(H_x^2 + H_z^2) \\ &= \frac{\mu_0}{2}\,|H_l|^2 b\int_0^a dx\int_0^l dz\left[\frac{a^2}{l^2}\sin^2\frac{\pi x}{a}\cos^2\frac{\pi z}{l} + \cos^2\frac{\pi x}{a}\sin^2\frac{\pi z}{l}\right] \\ &= \frac{\mu_0}{8}\,|H_l|^2 abl\left(\frac{a^2}{l^2} + 1\right). \end{aligned}$$

The time-averaged power losses due to the skin-effect may be calculated just as for a waveguide (see the model solution of problem 7.27), by integrating Eq. (7.219) over all the walls:

$$\begin{aligned} \mathscr{P}_{\text{loss}} &= \frac{\mu_0\omega\delta_s}{4}\left\{2b\int_0^a |H_x(x,0)|^2\,dx + 2b\int_0^l |H_z(0,z)|^2\,dz \right. \\ &\quad \left. + 2\int_0^a dx\int_0^l dz[|H_x(x,z)|^2 + |H_z(x,z)|^2]\right\} \\ &= \frac{\mu_0\omega\delta_s}{4}\,|H_l|^2 l\left\{b\left(\frac{a^3}{l^3} + 1\right) + \frac{a}{2}\left(\frac{a^2}{l^2} + 1\right)\right\}. \end{aligned}$$

As a result, for the Q-factor, defined by Eq. (7.227), we obtain

$$Q \equiv \omega \frac{U}{\mathscr{P}_{\text{loss}}} = \frac{1}{2\delta_s}\left[ab\left(\frac{a^2}{l^2}+1\right)\right] \Big/ \left[b\left(\frac{a^3}{l^3}+1\right)+\frac{a}{2}\left(\frac{a^2}{l^2}+1\right)\right].$$

As could be expected from Eq. (7.78), Q scales as the ratio of some effective size of the resonator (of the order of the TEM wavelength λ), to the skin depth δ_s. According to Eq. (7.206) of the lecture notes, the resonance frequency of the resonator specified in the assignment is $\omega \approx 6 \times 10^{10}\ \text{s}^{-1}$ ($f \approx 10$ GHz), so that the skin depth δ_s for copper walls is close to 6.5×10^{-7} m, and the above formula yields $Q \approx 8.2 \times 10^3 \sim 10^4$. For a microwave metallic resonator, operating at its fundamental mode, this is almost as high quality as you can get at room temperature[45]. Note that for these numbers the condition $Q \gg 1$ is well satisfied, so that our approximate method of calculation of the Q-factor is indeed legitimate.

Problem* 7.30. Calculate the lowest eigenfrequency and the Q-factor (due to the skin effect) of the toroidal (axially symmetric) resonance cavity with metallic walls, and interior's cross-section shown in the figure below, in the case when $d \ll r, R$.

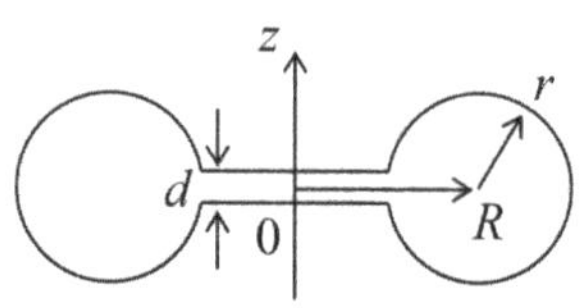

Solution: Looking at the field distribution in the fundamental H_{101} mode of a rectangular resonator (figure 7.30 of the lecture notes), and the virtually similar distribution in the E_{010} mode of the circular cylindrical resonator (see the model solution of problem 7.24), it is easy to understand that the lowest mode of the toroidal resonator has the electric field concentrated almost exclusively in its central part, with distance d between the parallel walls, and the magnetic field virtually limited to the toroidal ('doughnut') part of radii r and R.[46] Due to this field separation, the resonance frequency may be calculated just as for the lumped LC circuit,

$$\omega = \frac{1}{(LC)^{1/2}},$$

with the capacitance equal to that of the plane capacitor of thickness d and area $A = \pi(R - r)^2$,

$$C = \varepsilon \frac{\pi(R - r)^2}{d},$$

[45] Much higher Q-factors (up to $\sim 10^{12}$ at microwave frequencies) may be obtained using resonators with superconducting walls.

[46] Such resonators are broadly used in particle accelerators and also in vacuum devices for microwave generation (e.g. the so-called *klystrons*), where the electric field has to be concentrated in the region of charged particle passage—typically, along the symmetry axis (in the figure above, axis z), through a pair of small holes, which do not affect the field distribution substantially.

and the inductance calculated as in problem 5.16, but with $N = 1$,

$$L = \mu[R - (R^2 - r^2)^{1/2}],$$

so that, finally,

$$\omega^2 = \frac{d}{\pi\varepsilon\mu(R - r)^2[R - (R^2 - r^2)^{1/2}]}. \qquad (*)$$

This means, in particular, that the TEM wavelength corresponding to the frequency is

$$\lambda \equiv \frac{2\pi}{k} = \frac{2\pi}{(\varepsilon\mu)^{1/2}\omega} = 2\pi^{3/2}\frac{(R - r)}{d^{1/2}}[R - (R^2 - r^2)^{1/2}]^{1/2}.$$

This formula shows that at $r \sim R$, the wavelength scales as $(R^3/d)^{1/2} \gg R, r \gg d$, so that our quasi-stationary treatment of the electric and magnetic fields in indeed valid.

To calculate the Q factor, we need to evaluate both the nominator and denominator of the right-hand side of Eq. (7.227) of the lecture notes; assuming that $Q \gg 1$, this may be done neglecting the effect of losses on the field distribution. In this approximation, the average energy of oscillations may be found as the maximum energy of the electric field:

$$U = \frac{|Q_\omega|^2}{2C} = \frac{|Q_\omega|^2 d}{2\pi\varepsilon(R - r)^2}, \qquad (**)$$

where Q_ω is the complex amplitude of the total electric charge of each plane 'lid' of resonator.

In order to calculate the energy loss rate due to the skin effect, we can again use Eq. (7.219). With the account of the fundamental Eq. (6.38), it may be rewritten as

$$\frac{d\mathscr{P}_{\text{loss}}}{dA} = \frac{\mu\omega\delta_s}{4}|J_\omega(\rho)|^2, \qquad (***)$$

where $J_\omega(\rho)$ is the complex amplitude of the linear density of the surface current in a point of the resonator wall surface, separated from the axis z by distance ρ. Let us express this density via its value $J_0 \equiv J_\omega(R - r)$ in the resonator's 'neck', i.e. at the connection of its plane and toroidal parts, because J_0 may be readily related to Q_ω, using the charge conservation law, $dQ/dt = -2\pi(R - r)J_0$. With the account of the sinusoidal law of time evolution of all the variables ($\partial/\partial t \to -i\omega$), this relation yields

$$J_0 = i\omega\frac{1}{2\pi(R - r)}Q_\omega.$$

The current density in the toroidal part of the resonator, with negligible displacement currents, may be calculated from the conservation of the total surface current, $2\pi\rho J_\omega(\rho) = 2\pi(R - r)\, J_0$, giving

$$J_\omega(\rho) = J_0\frac{R - r}{\rho}, \qquad \text{for } \rho \geqslant R - r.$$

However, in the plane part (with $\rho < R - r$), the displacement currents are substantial, because of the smallness of d. The easiest way to calculate J_ω here is to write the charge conservation law similar for that written above for J_0, $dQ(\rho)/dt = -2\pi\rho J(\rho)$, where $Q(\rho) = Q[\rho/(R - r)]^2$ is the part of the full charge Q, located within a circle of radius ρ.[47] As a result, we obtain

$$J_\omega(\rho) = J_0\left(\frac{\rho}{R - r}\right)^2, \qquad \text{for } \rho \leqslant R - r.$$

As the sanity checks, $J_\omega(\rho) = J_0$, and $J_\omega(0) = 0$.

Now we have everything ready for the energy loss calculation using Eq. (***):

$$\begin{aligned}\mathscr{P}_{\text{loss}} &= \frac{\mu\omega\delta_s}{4}\int_{\substack{\text{over all}\\ \text{surfaces}}} |J_\omega(\rho)|^2 d^2r \\ &= \frac{\mu\omega\delta_s}{4}|J_0|^2\left[2\int_0^{R-r}\left(\frac{\rho}{R - r}\right)^4 2\pi\rho\, d\rho + 2\int_0^{\pi}\left(\frac{R - r}{\rho}\right)^2 2\pi\rho\, r d\theta\right],\end{aligned}$$

where θ is the angle pointing into the given point of the torus from the center of its cross-section, so that $\rho = R + r\cos\theta$. An easy integration yields

$$\begin{aligned}\mathscr{P}_{\text{loss}} &= \frac{\mu\omega\delta_s}{4}|J_0|^2\left\{\frac{2\pi}{3}(R - r)^2 + 4\pi^2 rR^2\left[\frac{R - r}{(R + r)^3}\right]^{1/2}\right\} \\ &= \frac{\mu\omega\delta_s}{4}\left[\frac{\omega|Q_\omega|}{2\pi(R - r)}\right]^2\left[\frac{2\pi}{3}(R - r)^2 + 4\pi^2 rR^2\left(\frac{R - r}{(R + r)^3}\right)^{1/2}\right].\end{aligned}$$

Finally, combining this expression with Eqs. (*) and (**), for the Q-factor we obtain

$$Q \equiv \omega\frac{U}{\mathscr{P}_{\text{loss}}} = \frac{R - (R^2 - r^2)^{1/2}}{\delta_s}\Big/\left[\frac{2}{3\pi} + \frac{8\, rR^2}{(R^2 - r^2)^{3/2}}\right].$$

This formula shows that in our quasi-static approximation, the Q-factor does not depend on d, and at $r \sim R$, it scales as $\sim R/\delta_s$, i.e. is lower than the value $\sim\lambda/\delta_s$ that we could obtain from the 'usual' resonance cavity, with all three linear dimensions comparable. This is the price we have to pay for the system's convenience for applications. This result explains why in such unique applications as particle accelerators, where refrigeration costs are not the primary concern, accelerating

[47] This particular calculation is similar to the one carried out in the model solution of problem 6.27.

RF cavities are fabricated from a superconducting material (usually, niobium with the critical temperature 9.2 K) and are kept at a 'helium' temperature of a few kelvins.

***Problem* 7.31.** Express the contribution to the damping coefficient (the reciprocal Q-factor) of a resonator from small energy losses in the dielectric that fills it, via the complex functions $\varepsilon(\omega)$ and $\mu(\omega)$ of the material.

Solution: For the dispersion-free, loss-free case, Eq. (7.202) of the lecture notes has partial variable-separated solutions (7.203) whose temporal factors obey the ordinary differential equation similar to the second of Eqs. (7.201):

$$\ddot{\mathcal{T}} + \omega^2\mathcal{T} = 0, \quad \text{where } \omega^2 = k^2v^2 \equiv \frac{k^2}{\varepsilon\mu}, \qquad (*)$$

and k is the corresponding eigenvalue of the Helmholtz equation (7.204). For a lossless medium, ε and μ are real, so that $\mathcal{T}$ describes sinusoidal oscillations of frequency ω:

$$\mathcal{T}(t) = \text{Re}\,[\mathcal{T}_\omega e^{-i\omega t}] = \text{Re}\,\mathcal{T}_\omega \cos\omega t - \text{Im}\,\mathcal{T}_\omega \sin\omega t.$$

Acting in analogy with the derivation of Eq. (7.28), we may generalize this relation for a dispersive medium as

$$\omega^2 \equiv \frac{k^2}{\varepsilon(\omega)\mu(\omega)}. \qquad (**)$$

If the dielectric is lossy, then $\varepsilon(\omega) = \varepsilon'(\omega) + i\varepsilon''(\omega)$, and $\mu(\omega) = \mu'(\omega) + i\mu''(\omega)$,[48] the solution of this equation is complex, $\omega = \omega' + i\omega''$. This means that the solution of Eq. (*) is not exactly sinusoidal, but decays with time:

$$\mathcal{T}(t) = \text{Re}\,[\mathcal{T}_\omega e^{-i\omega t}] = \text{Re}\,[\mathcal{T}_\omega e^{-i(\omega+i\omega'')t}] = e^{\omega'' t}\,\text{Re}[\mathcal{T}_\omega e^{-i\omega' t}],$$

so that the oscillations' energy, proportional to $\mathcal{T}^2$, decays as

$$U(t) = U(0)e^{2\omega'' t}.$$

Comparing this expression with Eq. (7.227), we see that the contribution of the complex ω to the damping coefficient $1/Q$ is

$$\left(\frac{1}{Q}\right)_{\text{dielectric}} = -\frac{2\omega''}{\omega'}.$$

In the most important case when the losses are small, $|\varepsilon''(\omega)| \ll \varepsilon'(\omega)$, and $|\mu''(\omega)| \ll \mu'(\omega)$, so that $|\omega''| \ll \omega'$, the characteristic equation (**), rewritten as

$$(\omega' + i\omega'')^2(\varepsilon' + i\varepsilon'')(\mu' + i\mu'') = k^2, \qquad (***)$$

[48] For most dielectrics, $\mu(\omega)$ is very close to μ_0, so that $\mu''(\omega)$ is negligible, and in this solution the component is kept mostly for the sake of generality.

may be solved by successive approximations. In the zeroth approximation, $\omega'' = 0$, while ω' is just the unperturbed, real eigenfrequency ω_0 of the resonator, that should be found self-consistently from the equation $\omega_0^2 = k^2/\varepsilon'(\omega_0)\mu'(\omega_0)$. In the first approximation, we may plug this result, $\omega' = \omega_0$, into Eq. (***) linearized with respect to small ω'', ε'', and μ'', to obtain

$$2\omega'\omega''\varepsilon'\mu' + \omega'^2(\varepsilon'\mu'' + \varepsilon''\mu') \approx 0,$$

so that

$$\omega'' \approx -\frac{\omega'(\varepsilon'\mu'' + \varepsilon''\mu')}{2\varepsilon'\mu'} \equiv -\frac{\omega'}{2}\left(\frac{\varepsilon''}{\varepsilon'} + \frac{\mu''}{\mu'}\right),$$

and, finally, we obtain a very simple and natural expression:

$$\left(\frac{1}{Q}\right)_{\text{dielectric}} \approx \frac{\varepsilon''}{\varepsilon'} + \frac{\mu''}{\mu'} \equiv \frac{\operatorname{Im}\varepsilon(\omega_0)}{\operatorname{Re}\varepsilon(\omega_0)} + \frac{\operatorname{Im}\mu(\omega_0)}{\operatorname{Re}\mu(\omega_0)}.$$

***Problem* 7.32.** For the dielectric Fabry–Pérot resonator (figure 7.31 of the lecture notes) with the normal wave incidence, calculate the Q-factor due to radiation losses, in the limit of a strong impedance mismatch ($Z \gg Z_0$), using two approaches:

(i) from the energy balance, using Eq. (7.227), and
(ii) from the frequency dependence of the power transmission coefficient, using Eq. (7.229).

Compare the results.

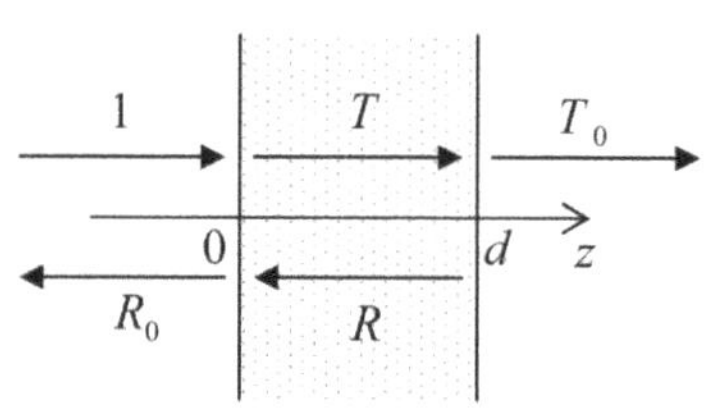

Solutions:

(i) The calculation of the field distribution in the resonator, with a unit-amplitude external wave incident from one side, was the subject of problem 7.9. The last two equations of the full system for the traveling wave amplitudes T, T_0, R, and R_0, normalized to the incident wave amplitude E_ω (for notation, see the figure above) may be readily solved to give

$$Te^{ikd} = \frac{Z_0 + Z}{2Z_0}T_0, \qquad Re^{-ikd} = \frac{Z_0 - Z}{2Z_0}T_0.$$

At the exact resonance ($kd = k_m d = \pi m$, where $m = 1, 2, \ldots$), both exponential factors are equal to $(-1)^m$, and from the solution of problem 7.9 we know that $T_0 = 1$, so that at $Z \gg Z_0$, $R \approx -T$, with

$$|T|^2 \approx |R|^2 \approx \left(\frac{Z}{2Z_0}\right)^2 \gg 1.$$

Thus the field inside the resonator ($0 \leqslant z \leqslant d$) may be approximated with a standing wave

$$E(z) = E_\omega(Te^{ikz} + R\, e^{-ikz}) \approx -E_\omega \frac{Z}{2Z_0}(e^{ikz} - e^{-ikz}) = -iE_\omega \frac{Z}{Z_0} \sin\frac{\pi mz}{d}.$$

Note that the wave amplitude is much higher than that (E_ω) of the incident wave. (This is due to the well-known energy accumulation at a resonance.). Also notable is the $\pi/2$ phase shift between the standing wave and the incident wave (represented by the coefficient $-i = \exp\{-i\pi/2\}$), which is also typical for the exact resonance[49].

Now we can use Eq. (3.79) to calculate the full (time-independent) oscillation energy as the maximum electric field energy in the resonator (per unit area):

$$\frac{U}{A} = \frac{\varepsilon}{2}\int_0^d \max_t E^2(z, t)dz = \frac{\varepsilon}{2}|E_\omega|^2\left(\frac{Z}{Z_0}\right)^2\int_0^d \sin^2\frac{\pi mz}{d}dz = \frac{\varepsilon}{2}|E_\omega|^2\left(\frac{Z}{Z_0}\right)^2\frac{d}{2}.$$

The resonator's energy loss (per unit time), due to the radiation through its interfaces, is the sum of the time-averaged Poynting vectors of the outcoming waves, whose amplitudes T_0 and R_0 are (in our approximation) equal to each other:

$$\frac{\mathscr{P}_{\text{loss}}}{A} = 2\overline{S}_T = \frac{1}{Z_0}|T_0E_\omega|^2,$$

so at the resonance ($T_0 = 1$), Eq. (7.227) yields

$$Q \equiv \omega_m\frac{U}{\mathscr{P}_{\text{loss}}} = \omega_m\frac{\varepsilon d}{4}Z_0\left(\frac{Z}{Z_0}\right)^2 = \frac{k_m}{4Z}Z_0\left(\frac{Z}{Z_0}\right)^2 = \frac{\pi m}{4}\frac{Z}{Z_0}. \qquad (*)$$

This formula is only valid if $Q \gg 1$, as it is in our case $Z \gg Z_0$.

(ii) The final result of problem 7.9 was the following expression for the (amplitude) transmission coefficient:

$$T_0 = \frac{4ZZ_0}{(Z+Z_0)^2\exp\{i(k_0 - k)d\} - (Z - Z_0)^2\exp\{i(k_0 + k)d\}} \equiv \frac{N}{D},$$

whose denominator D (a complex number) may be conveniently represented by the 2D vector diagrams reproduced in the figure below. (The right-hand panel of this figure confirms, in particular, that the resonance transmission ($T_0 = 1$) is reached at $k_md = \pi m$.)

[49] See, e.g. *Part CM* section 5.1, in particular the right-hand panel of figure 5.1.

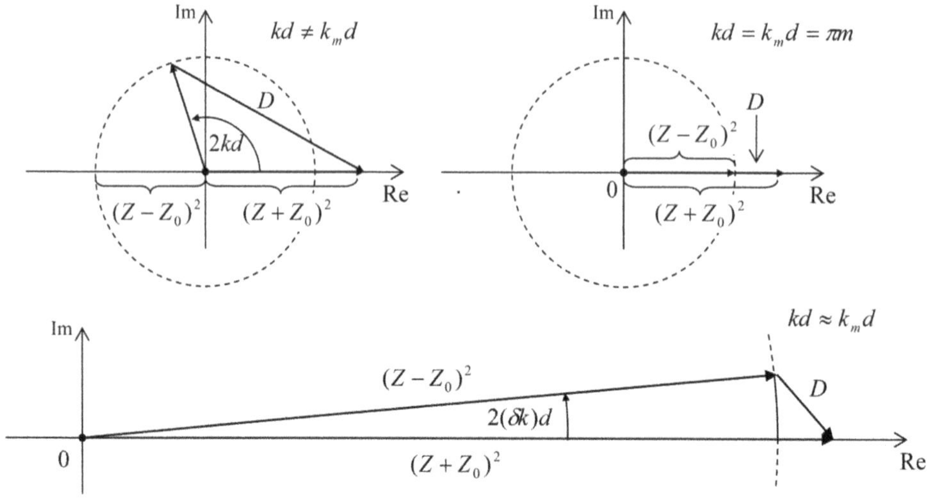

The last diagram shows that at a strong impedance mismatch ($Z \gg Z_0$), even a small deviation $\delta k \equiv k - k_m$ from an exact resonance leads to a fast growth of the denominator and hence to a fast suppression of $|T|$. As a result, in order to find the power transmission coefficient $|T|^2$ near the resonance, we may use the following approximation (see the bottom panel of the above figure):

$$|D|^2 \approx [(Z+Z_0)^2 - (Z-Z_0)^2]^2 + (Z-Z_0)^4[2(\delta k)d]^2 \approx [4ZZ_0]^2 + Z^4[2(\delta k)d]^2.$$

The denominator increases to twice its resonance value (and hence the transmitted power drops to one half of its resonance value) when the second term in this expression becomes equal to the first one, i.e. when

$$(\delta k)|_{1/2}d = \pm\frac{2Z_0}{Z}.$$

According to Eq. (7.229)[50], the Q factor of the resonator may be determined from the 'full-width half-maximum' (FWHM) bandwidth:

$$Q = \frac{\omega_m}{\Delta\omega} = \frac{k_m d}{(\Delta k)d} = \frac{\pi m}{2\,|\delta k|_{1/2}d} = \frac{\pi m Z}{4Z_0},$$

i.e. the exactly same result as given by Eq. (*).

References

[1] Jones P *et al* 2016 *Optical Tweezers: Principles and Applications* (Cambridge: Cambridge University Press)

[2] Tinkham M 1996 *Introduction to Superconductivity* 2nd edn (New York: McGraw-Hill)

[3] Dwight H 1949 *J. Math. Phys.* **27** 84

[50] Again, its (quite simple) derivation may be found, for example, in *Part CM* section 5.1.

IOP Publishing

Chapter 8

Radiation, scattering, interference, and diffraction

***Problem* 8.1.** In the electric dipole approximation, calculate the angular distribution and the total power of electromagnetic radiation using the following classical model of the hydrogen atom: an electron rotating, at a constant distance R, about a much heavier proton. Use the latter result to evaluate the classical lifetime of the atom, borrowing the initial value of R from quantum mechanics: $R(0) = r_B \approx 0.53 \times 10^{-10}$ m.

Solution: For the radial component of the instant Poynting vector, we can use Eq. (8.26) of the lecture notes:

$$S_r = \frac{Z}{(4\pi\ vr)^2}\ddot{p}^2 \sin^2 \Theta.$$

In our current problem, with the dipole moment $\mathbf{p} = q\mathbf{R} = -e\mathbf{R}$ (where $\mathbf{R}$ is the electron's radius-vector, with $R \ll r$) rotates, the angle Θ is a function of time—see the figure below. The time average of $\sin^2\Theta$ may be calculated in several equivalent ways, for example by returning to Eq. (8.24) to write

$$\ddot{p}^2 \sin^2 \Theta = |\mathbf{n} \times \ddot{\mathbf{p}}|^2,$$

and calculating this vector product in some immobile reference frame—for example, that shown in the figure below, with the axes x and y within the plane of particle's rotation, being turned about axis z so that the observation point $\mathbf{r}$ is within the plane $[x, z]$, i.e. for that point, $y = 0$. As the figure shows, in this frame,

$$\mathbf{n} = \mathbf{n}_x \sin\theta + \mathbf{n}_z \cos\theta, \quad \ddot{\mathbf{p}} = eR\omega^2(\mathbf{n}_x \cos\varphi + \mathbf{n}_y \sin\varphi), \quad \text{with } \varphi \equiv \omega t + \text{const},$$

where ω is the angular velocity of the electron rotation. Now the rule of the vector multiplication yields

$$\mathbf{n} \times \ddot{\mathbf{p}}(t) = eR\omega^2 \begin{vmatrix} \mathbf{n}_x & \mathbf{n}_y & \mathbf{n}_z \\ \sin\theta & 0 & \cos\theta \\ \cos\varphi & \sin\varphi & 0 \end{vmatrix}$$

$$= eR\omega^2(-\mathbf{n}_x \cos\theta \sin\varphi + \mathbf{n}_y \cos\theta \cos\varphi + \mathbf{n}_z \sin\theta \sin\varphi).$$

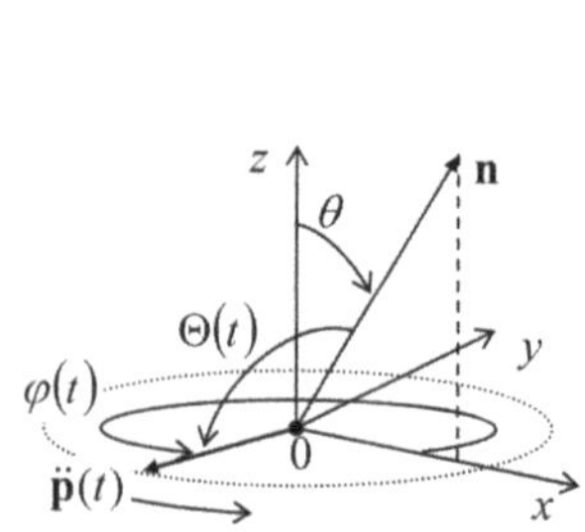

Averaging the square of this vector over the period of rotation, i.e. over the interval $\Delta\varphi = 2\pi$, we obtain

$$\overline{(\ddot{p}\sin\Theta)^2} = q^2R^2\omega^4\overline{(\cos^2\theta\sin^2\varphi + \cos^2\theta\cos^2\varphi + \sin^2\theta\sin^2\varphi)}$$
$$= e^2R^2\omega^4\left(\cos^2\theta + \frac{1}{2}\sin^2\theta\right).$$

This formula shows that the angular distribution of the average radiated power is indeed different from that produced by an oscillating dipole of fixed orientation: the radiation is strongest at $\theta = 0$ and $\theta = \pi$, i.e. along the direction normal to the charge rotation plane (in our notation, axis z), but is non-vanishing along any other direction. The total average power of the radiation (into the free space) is

$$\mathscr{P} = \int_{4\pi}\overline{S}_r r^2 d\Omega = Z_0\left(\frac{\omega^2 eR}{4\pi c}\right)^2 2\pi\int_0^{\pi}\left(\cos^2\theta + \frac{1}{2}\sin^2\theta\right)\sin\theta d\theta$$
$$= Z_0\left(\frac{\omega^2 eR}{4\pi c}\right)^2 2\pi\frac{4}{3} = \frac{Z_0\omega^4 e^2R^2}{6\pi c^2}.$$

Comparing this result with Eq. (8.29) for a fixed-direction dipole, we see that the radiation power by a rotating charge equals a sum of those from its two oscillating components $p_x(t)$ and $p_y(t)$, calculated independently. The reason of this independence (which is not immediately apparent and requires a proof, for example the one given above) is that the polarizations of the partial waves radiated by the component dipoles are mutually perpendicular.

Proceeding to the calculation of the classical lifetime of the classical atom, from the elementary classical mechanics we know that at the circular motion of a non-relativistic particle in the Coulomb attractive field (with the potential energy $U(R) = -e^2/4\pi\varepsilon_0 R < 0$), its kinetic energy $T \equiv mv^2/2$ equals $-U(R)/2 > 0$, so that the full energy $\mathscr{E} \equiv T + U$ equals $U(R)/2$, i.e. $T = -\mathscr{E}$. In our case, this means that

$$U = -\frac{e^2}{4\pi\varepsilon_0 R}, \qquad \mathscr{E} = -\frac{e^2}{8\pi\varepsilon_0 R}, \qquad T = \frac{e^2}{8\pi\varepsilon_0 R},$$

so that

$$\omega^2 = \frac{v^2}{R^2} = \frac{2T/m_e}{R^2} = \frac{e^2}{4\pi\varepsilon_0 m_e R^3}.$$

A weak radiation of power $\mathscr{P} \ll \omega\mathscr{E}$ causes a relatively slow energy reduction: $d\mathscr{E}/dt = -\mathscr{P}$, where $\mathscr{P}$ may be calculated (as this was done above) for a circular orbit. In our particular case this equation is

$$\frac{d\mathscr{E}}{dt} \equiv -\frac{e^2}{8\pi\varepsilon_0}\frac{d}{dt}\left(\frac{1}{R}\right) = -\frac{Z_0\omega^4 e^2 R^2}{6\pi c^2} \equiv -\frac{Z_0 e^6}{6\pi c^2(4\pi\varepsilon_0)^2 m_e^2}\frac{1}{R^4} < 0.$$

This equation, describing a monotonic *increase* of $(1/R)$, i.e. the electron's fall on the proton nucleus, may be rewritten in the dimensionless form

$$\tau\dot{\xi} = -\frac{1}{3\xi^2}, \qquad \text{i.e. } 3\xi^2 d\xi = -\frac{dt}{\tau},$$

where $\xi \equiv R(t)/R(0)$, and τ is the time scale of the decay process:

$$\tau \equiv \frac{4\pi^2 m_e^2 R^3(0)}{e^4 Z_0 \mu_0}.$$

The dimensionless differential equation may be readily integrated to give

$$\xi(t) = \left(1 - \frac{t}{\tau}\right)^{1/3},$$

showing that $R(t)$ vanishes (i.e. the electron drops on the proton) exactly at $t = \tau$. For an electron ($e \approx -1.6 \times 10^{-19}$ C, $m_e \approx 0.91 \times 10^{-30}$ kg), initially rotating at the Bohr radius, $R(0) = r_B \approx 0.53 \times 10^{-10}$ m, the above formula for τ yields approximately 1.57×10^{-11} s.

Such an atomic collapse is very fast on the human scale of events (let us thank quantum mechanics for preventing this disaster!), but since for our parameters $\omega \sim 4 \times 10^{16}$ s^{-1}, the same process is slow on the rotation period scale: $\omega\tau \sim 10^4 \gg 1$. This strong relation justifies our approach, which neglects the loss of energy (and hence of R) during one rotation period, at the average radiated power calculation stage. Another necessary sanity check is that the radiation wavelength scale, $\lambda = 2\pi c/\omega \sim 5 \times 10^{-8}$ m, is much larger than $R \sim 10^{-10}$ m, so that the radiating charge could indeed be treated as an electric dipole—as it was.

***Problem* 8.2.** A non-relativistic particle of mass m, with electric charge q, is placed into a uniform magnetic field $\mathbf{B}$. Derive the law of decrease of the particle's kinetic energy due to its electromagnetic radiation at the *cyclotron frequency* $\omega_c = qB/m$. Evaluate the rate of such radiation cooling for electrons in a magnetic field of 1 T, and estimate the electron energy interval in which this result is qualitatively correct.

Hint: The cyclotron motion is discussed in detail (for arbitrary particle's velocities $v \sim c$) in section 9.6 of the lecture notes, but I hope that the reader knows that in the non-relativistic case ($v \ll c$) the above formula for ω_c may be readily obtained by combining Newton's second law $mv_\perp^2/R = qv_\perp B$ for the circular motion of the particle under the effect of the magnetic component of the Lorentz force (5.10), and the geometric relation $v_\perp = R\omega_c$. (Here $\mathbf{v}_\perp$ is the particle's velocity within the plane normal to the vector $\mathbf{B}$.)

Solution: Let us assume that the particle's energy loss because of the radiation during one revolution is much smaller than the energy itself. (This condition has to be verified *a posteriori* for the application of the calculation result to any particular system.) Then its orbit in the plane normal to the field is approximately circular, and we may reuse the following intermediate result of the solution of the previous problem: the power of the electric-dipole radiation into the free space by a particle of charge q, moving on a circular orbit of radius R with angular velocity ω, averaged over the rotation period, is

$$\mathscr{P} = \frac{Z_0 q^2 \omega^4 R^2}{6\pi c^2} \equiv \frac{Z_0 q^2}{6\pi c^2}\omega^2(\omega R)^2.$$

For our current case, $\omega = \omega_c = qB/m$, and $\omega R = v_\perp$, so that[1]

$$\mathscr{P} = \frac{Z_0 q^2}{6\pi c^2}\left(\frac{qB}{m}\right)^2 v_\perp^2 = \frac{Z_0 q^4 B^2}{3\pi c^2 m^3}T, \qquad (*)$$

where $T \equiv m{v_\perp}^2/2$ is the kinetic energy of the particle's motion in the plane normal to the magnetic field's direction. The magnetic field force $q\mathbf{v} \times \mathbf{B}$ cannot do any work on the particle, i.e. does not create any potential energy, and the particle's motion along the magnetic field is not affected by it, so that we may represent the left-hand side of Eq. (*) as the kinetic energy decrease rate, $-dT/dt$. As a result, we get a very simple equation for energy evolution:

$$\frac{dT}{dt} = -\frac{T}{\tau}, \qquad \text{where } \tau \equiv \frac{3\pi c^2 m^3}{Z_0 q^4 B^2},$$

with the evident solution $T(t) = T(0)\exp\{-t/\tau\}$, so that the constant τ defined above gives a fair scale of the energy decay[2]. Note that in contrast with the previous problem, the decay time does not depend on the initial energy of the particle.

For an electron ($q = -e \approx 1.6 \times 10^{-19}$ C, $m = m_e \approx 0.91 \times 10^{-30}$ kg) in the field $B = 1$ T, the above expression yields $\tau \approx 2.5$ s, while $\omega_c \approx 1.8 \times 10^{11}$ s^{-1}. This means that the relative energy loss $\Delta T/T \approx 2\pi/\omega_c\tau$ during one rotation is indeed very small ($\sim 10^{-11}$), so that the applied approximation of constant energy at each revolution is indeed valid.

Another condition of validity of our result is imposed by the applicability of the electric dipole approximation: $k_c R \ll 1$, where $k_c \equiv \omega_c/c$ is the radiation's wave vector. Since $R = \omega_c/v_\perp$, this condition is reduced to $v_\perp \ll c$, i.e. to that the particle is non-relativistic, meaning that energy T has to be much *lower* than mc^2. (For an electron, $mc^2 \sim 0.5$ MeV.)

Finally, we have treated the electron as a classical particle. In the magnetic field, this treatment is only valid if T is much *higher* than the distance $\hbar\omega_c$ between the so-

[1] The generalization of this relation to the relativistic case (when this effect is referred to as the *synchrotron radiation*) will be discussed in section 10.3 of the lecture notes.

[2] Although particle's velocity $v_{||}$ along the magnetic field is not directly affected by the radiation, in many low-density systems, rare electron–electron interactions lead to their continuous 'thermalization', i.e. an approximate equilibration of all velocity components. In this case, the cyclotron radiation leads to the reduction of $v_\perp$ as well.

called *Landau levels*[3], where $\hbar \approx 1.055 \times 10^{-34}$ J·s is the Planck constant. For the above example, $\hbar\omega_c \approx 0.12$ meV. Thus, the above result is valid within at least nine orders of energy magnitude.

***Problem* 8.3.** Solve the dipole antenna radiation problem discussed in section 8.2 of the lecture notes (see figure 8.3, partly reproduced below) for its optimal length $l = \lambda/2$, assuming[4] that the current distribution in each of its arms is sinusoidal:

$$I(z, t) = I_0 \cos\frac{\pi z}{l}\cos\omega t.$$

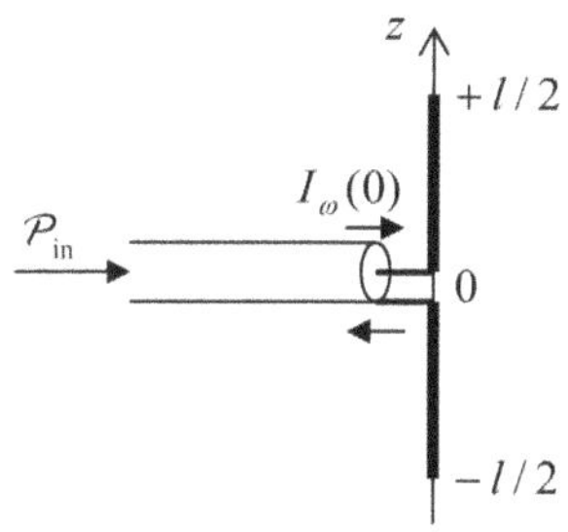

Solution: Since the main condition of the dipole approximation condition, $kl \ll 1$, is not satisfied for an antenna that long ($kl = (2\pi/\lambda)l = \pi$), we cannot use the formulas derived in section 8.2, and need to return to the general expressions (8.17) for the retarded potentials. However, we may use the experience of section 8.2, indicating that the fields in the far field zone ($kr \gg 1$) may be more readily calculated from the vector-potential than from the scalar potential. Integrating Eq. (8.17b) over the antenna wire's cross-section, we obtain the following expression for the vector-potential in the free space surrounding the antenna[5]:

$$\begin{aligned}\mathbf{A}(\mathbf{r}, t) &= \mathbf{n}_z\frac{\mu_0 I_0}{4\pi}\int_{-l/2}^{+l/2}\cos\frac{\pi z'}{l}\cos\left[\omega\left(t - \frac{R}{c}\right)\right]\frac{dz'}{R} \\ &\equiv \mathbf{n}_z\frac{\mu_0 I_0}{4\pi}\mathrm{Re}\left[\int_{-l/2}^{+l/2}\cos\frac{\pi z'}{l}\exp\left\{-i\omega\left(t - \frac{R}{c}\right)\right\}\frac{dz'}{R}\right],\end{aligned} \qquad (*)$$

where $R = [\rho^2 + (z - z')^2]^{1/2}$, with ρ and z being the cylindrical coordinates of the observation point $\mathbf{r}$ in the reference frame shown in the figure above. At large distances, $r \gg l$, the argument of the complex exponent may be approximated as

[3] See, e.g. *Part QM* section 3.2.

[4] As was emphasized in the lecture notes, this is a reasonable guess rather than a controllable approximation. The exact (rather involved!) theory shows that this assumption gives errors ~5%.

[5] Just as in section 8.2, we ignore the possible distortions of the field by the generator and/or transmission line used to create current $I(0, t)$ into the antenna. Since the TEM line's cross-section may be much less than λ, in many practical cases this is a very good approximation.

$$\omega\left(t - \frac{R}{c}\right) \equiv \omega t - k[\rho^2 + (z - z')^2]^{1/2}$$
$$\approx \omega t - k[\rho^2 + z^2 - 2zz']^{1/2}$$
$$\approx \omega t - kr\left(1 - \frac{zz'}{r^2}\right) \equiv (\omega t - kr) + kz' \cos\Theta,$$

where $k = \omega/c$ is the wave number, $r \equiv (\rho^2 + z^2)^{1/2}$, and $\Theta \equiv \cos^{-1}(z/r)$ is the angle between the direction toward the observation point and the antenna's axis z. Also, in the same limit, R in the denominator of Eq. (*) may be approximately replaced with r, and moved out of the integral over z'.[6] These simplifications allow us to reduce that expression, in the far field zone, to

$$\mathbf{A}(\mathbf{r}, t) \approx \mathbf{n}_z \frac{\mu_0 I_0}{4\pi r} \mathrm{Re}\left[\exp\{i(kr - \omega t)\} \int_{-l/2}^{+l/2} \cos\frac{\pi z'}{l} \exp\{-ikz' \cos\Theta\} dz'\right],$$
$$\text{at } kr \gg 1,$$

and to work out of the remaining integral analytically:

$$\int_{-l/2}^{+l/2} \cos\frac{\pi z'}{l} \exp\{-ikz'\cos\Theta\} dz' \equiv \int_{-\pi/2k}^{+\pi/2k} \cos kz' \; \exp\{-ikz'\cos\Theta\} dz'$$
$$\equiv \frac{1}{2}\sum_{\pm} \int_{-\pi/2k}^{+\pi/2k} \exp\{ik(\pm 1 - \cos\Theta)z'\} dz'$$
$$= \frac{1}{2}\sum_{\pm} \frac{\exp\{ik(\pm 1 - \cos\Theta)z'\}}{ik(\pm 1 - \cos\Theta)}\Bigg|_{kz' = -\pi/2}^{kz' = +\pi/2}$$
$$\equiv \sum_{\pm} \frac{\sin\left(\pm\frac{\pi}{2} - \frac{\pi}{2}\cos\Theta\right)}{k(\pm 1 - \cos\Theta)}$$
$$\equiv \sum_{\pm} \frac{\pm\cos\left(\frac{\pi}{2}\cos\Theta\right)}{k(\pm 1 - \cos\Theta)} \equiv 2\frac{\cos\left(\frac{\pi}{2}\cos\Theta\right)}{k\sin^2\Theta}.$$

Since the result is purely real, the vector-potential equals

$$\mathbf{A}(\mathbf{r}, t) = \mathbf{n}_z \frac{\mu_0 I_0}{2\pi} \frac{\cos\left(\frac{\pi}{2}\cos\Theta\right)}{\sin^2\Theta} \frac{\cos(\omega t - kr)}{kr}$$
$$\equiv \mathbf{n}_z \frac{\mu_0 I(0, t - r/c)}{2\pi kr} \frac{\cos\left(\frac{\pi}{2}\cos\Theta\right)}{\sin^2\Theta}.$$

[6] Note that these are essentially the same approximations as used at the derivation of Eq. (8.78), which expresses the Huygens principle, so the reader who needs more detailed explanations may be referred there.

The structure of this expression is exactly the same as that of (8.23), with the replacement

$$\dot{\mathbf{p}}(t) \rightarrow \mathbf{n}_z \frac{2I(0, t)}{k} \frac{\cos\left(\frac{\pi}{2}\cos\Theta\right)}{\sin^2\Theta},$$

and we may repeat all the transitions from that formula to Eq. (8.26) to obtain the following radiation power density, i.e. the radial component of the Poynting vector:

$$S_r = \frac{Z_0}{(4\pi\ cr)^2}\left(\frac{2\dot{I}(0, t)}{k}\right)^2 f(\Theta) = \frac{Z_0 I_0^2 \cos^2(\omega t - kr)}{4\pi^2 r^2} f(\Theta),$$

where

$$f(\Theta) \equiv \frac{\cos^2[(\pi/2)\cos\Theta]}{\sin^2\Theta}.$$

This expression shows that the angular distribution of the radiation, shown in the figure below, is somewhat (although not too much) different from that ($\sin^2\Theta$) of a radiating dipole—in particular, of a short dipole antenna analyzed in section 8.2. Because of that, its solid-angle integral,

$$J \equiv \oint_{4\pi} f(\Theta) d\Omega = 2\pi \int_0^{\pi} f(\Theta) \sin\Theta d\Theta,$$

is also slightly different—it equals approximately 7.658 instead of $8\pi/3 \approx 8.378$ for $\sin^2\Theta$. As a result, the total average radiation power is

$$\mathscr{P} \equiv r^2 \oint_{4\pi} \overline{S}_r d\Omega \approx Z_0 \frac{J}{4\pi^2} \frac{I_0^2}{2}.$$

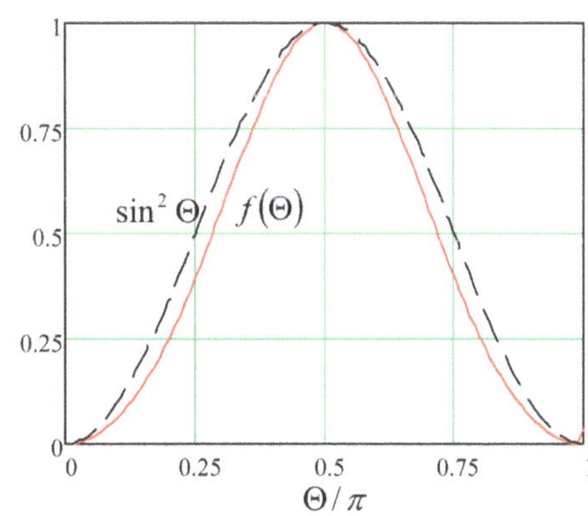

Just as was done in section 8.2 for the short antenna, we may recalculate this result into the antenna's impedance as 'seen' by the generator (or transmission line) feeding it:

$$\mathrm{Re}\, Z_A = \frac{J}{4\pi^2} Z_0 \approx \frac{7.658}{4\pi^2} Z_0 \approx 73.1 \text{ Ohms}.$$

This impedance is very close to one of the coaxial cable standard values (75 Ohms), making its matching with antenna straightforward. (If this statement is not clear, please revisit the discussion in the end of section 7.5 of the lecture notes.) Again, all these results are only valid with an accuracy of a few percent, because they are based on the sinusoidal (rather than self-consistently calculated) distribution of the current along the antenna's length.

***Problem* 8.4.** Use the Lorentz oscillator model of a bound charge, described by Eq. (7.30) of the lecture notes, to explore the transition between the two scattering limits discussed in section 8.3 of the lecture notes, in particular the *resonant scattering* taking place at $\omega \approx \omega_0$. In this context, discuss the contribution of the scattering into the oscillator's damping.

Solution: The differential equation (7.30), describing the charge dynamics, is linear; hence the forced oscillations $x(t)$ of charge's displacement, and hence of its dipole moment $p(t) = qx(t)$, induced by a monochromatic incident wave with frequency ω, are sinusoidal, with the same frequency. The average power of the wave radiated (in the case of scattering, re-radiated, i.e. scattered) by such a sinusoidally oscillating dipole into the free space is given by Eq. (8.28) of the lecture notes, with $Z = Z_0$ and $v = c$:

$$\overline{\mathscr{P}} = \frac{Z_0\omega^4}{12\pi c^2}|p_\omega|^2 = \frac{Z_0\omega^4}{12\pi c^2}q^2|x_\omega|^2, \qquad (*)$$

where x_ω is the complex amplitude of the charge's oscillations. Using Eq. (7.31) to express this amplitude via that of the electric field of the incident wave, we obtain

$$\overline{\mathscr{P}} = \frac{Z_0\omega^4 q^4}{12\pi mc^2}\frac{|E_\omega|^2}{(\omega_0^2 - \omega^2)^2 + (2\omega\delta_0)^2},$$

so that Eq. (8.39) yields the following full cross-section of scattering:

$$\sigma = \frac{Z_0^2 q^4}{6\pi mc^2}\frac{\omega^4}{(\omega_0^2 - \omega^2)^2 + (2\omega\delta_0)^2}. \qquad (**)$$

In the high-frequency limit $\omega/\omega_0 \to \infty$, this expression is reduced to the Thomson scattering formula (8.40)–(8.41), while in the opposite limit, $\omega/\omega_0 \to 0$, it leads to Eq. (8.45) with

$$\alpha \equiv \left.\frac{qx_\omega}{E_\omega}\right|_{\omega \ll \omega_0} \equiv \frac{q^2}{m\omega_0^2},$$

and hence to the Rayleigh scattering. However, Eq. (**) shows that on the way between these two limits, the scattering cross-section may have a very high peak at $\omega \approx \omega_0$, with a height limited only by the oscillator's damping:

$$\sigma|_{\omega=\omega_0} = \frac{Z_0^2 q^4}{6\pi c^2 m}\frac{\omega_0^2}{(2\delta_0)^2} \equiv Q^2\sigma|_{\omega\to\infty},$$

where Q is the Q-factor of the resonance, which may be much larger than 1.

The damping may be contributed not only by some internal losses inside the bound charge system, but by the scattering itself. Indeed, the average rate of the energy loss (i.e. the dissipation power) at forced oscillations of a harmonic oscillator is[7]

$$\overline{\mathscr{P}_{\text{loss}}} = 2m\,\overline{\dot{x}^2}\delta_0 = m\omega^2\,|x_\omega|^2\delta_0\,.$$

If the electromagnetic wave's (re-)radiation is the only energy loss mechanism of the oscillator, this expression has to be equal to the one given by Eq. (*), giving, at $\omega \approx \omega_0$,

$$Q^{-1} \equiv \frac{2\delta_0}{\omega} = \frac{Z_0\omega q^2}{6\pi c^2} \equiv \frac{2}{3}(Z_0 c\varepsilon_0)\left(\frac{\omega}{c}\right)\left(\frac{q^2}{4\pi\varepsilon_0 mc^2}\right) \equiv \frac{2}{3}kr_c,$$

where $k = \omega/c$ is the wave number, and r_c is the classical radius of the charge, defined by Eq. (8.41) of the lecture notes. For an electron, $r_c \sim 3 \times 10^{-15}$ m, so that the product kr_c is very small (i.e. Q is very large) at all frequencies up to hard γ rays. Hence, the resonant scattering may be much larger than the Thomson and Rayleigh scattering.

The quantum mechanics[8] not only confirms the above formulas for the harmonic oscillator, but also shows that the resonant scattering takes place for any scatterer, with energy levels E_n, as soon as the wave frequency ω approaches *any* of the quantum transition frequencies $\omega_{nn'} = (E_n - E_{n'})/\hbar$, with $n \neq n'$, provided that the corresponding matrix element of the electric dipole moment's operator is different from 0, and that at least one of the involved energy levels is occupied.

Problem* 8.5. A sphere of radius R, made of a material with a uniform permanent electric polarization $\mathbf{P}_0$ and a constant mass density ρ, is free to rotate about its center. Calculate the average total cross-section of scattering, by the sphere, of a linearly polarized electromagnetic wave of frequency $\omega \ll c/R$, propagating in free space, in the limit of a small wave amplitude, assuming that the initial orientation of the polarization vector is random.

Solution: As we know from the solution of problem 3.11, the electric field induced by a polarized sphere is identical to that of a point electric dipole, placed in sphere's center, with the moment given by a very simple and natural formula

$$\mathbf{p} = V\mathbf{P}_0 = \frac{4\pi}{3}R^3\mathbf{P}_0.$$

In the low-frequency limit $\omega \ll c/R$, the wave's electric field $\mathbf{E}$, on the distances of the order of R, is virtually uniform. According to Eq. (3.17) of the lecture notes, such a

[7] This expression may be readily obtained by interpreting the second term on the left-hand side of Eq. (7.30) as a viscous friction force, proportional to charge's velocity: $F_v = -2mv\delta$, giving the instant dissipation power $\mathscr{P}_{\text{loss}} = -F_v v = 2mv^2\delta$.

[8] See, e.g. *Part QM* chapters 6 and 9.

uniform electric field **E** exerts, on the dipole (i.e. on the sphere), the following mechanical torque:

$$\boldsymbol{\tau} = \mathbf{p} \times \mathbf{E}.$$

As we know from the classical mechanics basics[9], under the effect of such a torque the mechanical angular momentum **L** of the sphere evolves as

$$\dot{\mathbf{L}} = \boldsymbol{\tau}.$$

Classical mechanics also says[10] that the angular momentum **L** of a spherically symmetric body (frequently called the *spherical top*) equals $I\boldsymbol{\Omega}$, where $\boldsymbol{\Omega}$ is the instantaneous angular velocity vector and I is the (only) principal moment of inertia of such a body. (For a sphere with a constant mass density ρ, this moment is $I = (8\pi/15)\rho R^5$.)[11] Combining the above relations, we obtain the following equation of time evolution of the vector $\boldsymbol{\Omega}$:

$$I\dot{\boldsymbol{\Omega}} = \mathbf{p} \times \mathbf{E}. \qquad (*)$$

One more relation between the vectors $\boldsymbol{\Omega}$ and **p** follows from the kinematics of rotation of any vector of constant length, in particular **p**:[12]

$$\dot{\mathbf{p}} = \boldsymbol{\Omega} \times \mathbf{p}. \qquad (**)$$

A joint solution of two vector equations (*) and (**), with a time-dependent (oscillating) vector **E**, is not a trivial task, but may be simplified if the incident wave's amplitude is relatively weak: $E \to 0$. To see this, let us differentiate Eq. (**) over time, and then plug Eq. (*), and Eq. (**) again, into the resulting right-hand side:

$$\ddot{\mathbf{p}} = \dot{\boldsymbol{\Omega}} \times \mathbf{p} + \boldsymbol{\Omega} \times \dot{\mathbf{p}} = \frac{1}{I}(\mathbf{p} \times \mathbf{E}) \times \mathbf{p} + \boldsymbol{\Omega} \times (\boldsymbol{\Omega} \times \mathbf{p}). \qquad (***)$$

Since $p = \text{const}$, the first term in the last expression contains a component oscillating with wave's frequency ω, and proportional to its amplitude E_ω. On the other hand, Eq. (*) shows that the amplitude of such oscillations of the vector $\boldsymbol{\Omega}$ is also proportional to E_ω, so that, the fast oscillations of the term $\boldsymbol{\Omega} \times (\boldsymbol{\Omega} \times \mathbf{p})$ are proportional to $|E_\omega|^2$, i.e. are much smaller than those of the first term at $E_\omega \to 0$. (This reasoning is strict in the absence of the initial rotation of the sphere, but even if such a rotation is present, its angular velocity Ω_0 has to be as high as ω to affect the approximation used below.)

Dropping, on these grounds, the second term on the right-hand side of Eq. (***) (but only for a while—see below), we obtain

$$\ddot{\mathbf{p}} = -\frac{1}{I}\mathbf{p} \times (\mathbf{p} \times \mathbf{E}).$$

---

[9] See, e.g. *Part CM*, Eq. (1.33).

[10] See, e.g. *Part CM* sections 4.1 and 4.2, in particular Eqs. (4.31) and (4.32).

[11] If the reader has any difficulty with this well-known result, he or she may have a look, for example, at the model solution of problem 4.1 in *Part CM*.

[12] See, e.g. *Part CM* section 4.1, in particular Eq. (4.9).

According to this expression, if the incident wave is linearly polarized, i.e. the vector **E** is rapidly oscillating along one direction, the resulting vector $\ddot{\mathbf{p}}$ also oscillates with the same frequency along a line normal to vector **p** (see the figure below), with the magnitude

$$\ddot{p} = -\frac{p^2 E}{I} \sin\theta,$$

where θ is the angle between the vectors **E** and **p**. Plugging this expression into Eq. (8.27), for the average power of radiation into the free space (with $Z = Z_0$ and $v = c$), we obtain a frequency-independent result:

$$\overline{\mathscr{P}} = \frac{Z_0}{6\pi c^2}\left(\frac{p^2}{I}\right)^2 \overline{E^2(t)} \sin^2\theta = \frac{Z_0}{12\pi c^2}\left(\frac{p^2}{I}\right)^2 |E_\omega|^2 \sin^2\theta.$$

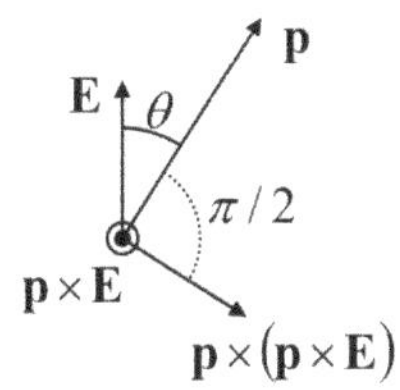

Here the power has been averaged over the period of the wave frequency ω. Now returning to Eq. (***), we have to notice that the second term on its right-hand side, neglected in the above calculation, also has a certain small (proportional to $|E_\omega|^2$) but non-vanishing average over the period. This average component of the second derivative of vector **p** results in a rather complex slow evolution of the direction of the vector, and hence to a change of the angle θ. Although slow on the scale of the wave frequency ω, this evolution may be noticeable on the scale of the scattering measurement experiment, and generally should be accounted for in theory. However, if the initial direction of sphere's polarization vector $\mathbf{P}_0$ is random, the evolution details are not important for the statistical average[13] of the radiation power, which may be calculated as

$$\langle \overline{\mathscr{P}} \rangle = \frac{1}{4\pi}\oint_{4\pi} \overline{\mathscr{P}}\, d\Omega = \frac{1}{4\pi} 2\pi \int_0^\pi \overline{\mathscr{P}} \sin\theta d\theta$$
$$= \frac{1}{2}\frac{Z_0}{12\pi c^2}\left(\frac{p^2}{I}\right)^2 |E_\omega|^2 \int_0^\pi \sin^2\theta \sin\theta d\theta.$$

[13] Such average is all we need if we measure the simultaneous, independent scattering of a wave by many similar spheres with random polarization directions—just as was discussed in section 8.3 for the Rayleigh scattering. (Please note the statistical averaging sign ⟨ ⟩, which differs from the time average sign; it will be used very frequently in *Part SM* of this series.)

With the variable substitution $\xi \equiv \cos\theta$, it is easy to see that the last integral is just

$$\int_0^{\pi}(1-\cos^2\theta)\sin\theta d\theta = \int_{-1}^{+1}(1-\xi^2)\,d\xi = \frac{4}{3},$$

so that, finally,

$$\langle \mathscr{P} \rangle = \frac{1}{2}\frac{Z_0}{12\pi c^2}\left(\frac{p^2}{I}\right)^2 |E_\omega|^2 \frac{4}{3} \equiv \frac{Z_0 p^4}{18\pi c^2 I^2}|E_\omega|^2.$$

Now, for the statistically averaged cross-section, Eq. (8.39) of the lecture notes yields

$$\langle \sigma \rangle = \frac{\langle \mathscr{P} \rangle}{|E_\omega|^2/2Z_0} = \frac{Z_0^2 p^4}{9\pi c^2 I^2},$$

and plugging in the expressions of p and I for a uniform sphere, discussed above, we obtain

$$\langle \sigma \rangle = \frac{Z_0^2}{9\pi c^2}\left(\frac{4\pi}{3}R^3 P_0\right)^4\left(\frac{8\pi}{15}\rho R^5\right)^{-2} \equiv \frac{100}{81}\left(\frac{P_0^2/\varepsilon_0}{\rho c^2}\right)^2 \pi R^2. \qquad (****)$$

It is curious that the scattering cross-section is proportional to the physical cross-section $\sigma_0 \equiv \pi R^2$ of the sphere; however, for any non-exotic matter it is much smaller. Indeed, as was mentioned in section 3.3 of the lecture notes, the highest values of the remnant polarization P_0 in ferroelectrics do not exceed $\sim 1\ \mathrm{C\,m^{-2}}$, while their density ρ is of the order of $10^4\ \mathrm{kg\,m^{-3}}$. Plugging these values (both in correct SI units) into our result, we see that the ratio $\langle\sigma\rangle/\sigma_0$ can hardly be higher than $\sim 10^{-20}$.[14]

This smallness is not occasional; indeed, the largest polarization of usual condensed matter corresponds to the displacement of a few outer-shell electrons by the distance of the order of interatomic distances a, which in turn are of the order of the Bohr radius r_B, so that the largest dipole moment per atom is $p_0 \sim er_B$, and the largest polarization is $P_0 \sim p_0/r_B^3 \sim e/r_B^2$. Plugging in the quantum-mechanical expression for the Bohr radius[15], $r_B = \hbar^2/m_e(e^2/4\pi\varepsilon_0)$, and the natural (although crude) estimate of the material's density, $\rho \sim Am_p/r_B^3$, where A is the atomic number, we see that the fraction in the last form of Eq. (****) is of the order of $(m_e/Am_p)\alpha^2 \ll 1$, where $\alpha \equiv e^2/4\pi\varepsilon_0\hbar c \approx 1/137$ is the fine-structure constant, which describes the weakness of electromagnetic interactions on the quantum-relativistic scale—see also the model solutions of problems 5.14 and 9.12.

***Problem* 8.6.** Use Eq. (8.56) of the lecture notes to analyze the interference/diffraction pattern produced by a plane wave's scattering on a set of N similar,

[14] This extreme smallness of the calculated effect makes it hard to detect on the masking background of other effects not accounted for in our simple model, such as small but non-vanishing differential ('dynamic') polarizability $\alpha_{dyn} \equiv dp_0/dE$ of the ferroelectric's atoms.

[15] It was already mentioned in section 2.1 of the lecture notes, and in the model solutions of problems 5.14 and 5.15, and will be discussed in detail in *Part QM* of this series)

equidistant point objects on a straight line, normal to the direction of the incident wave's propagation—see the figure below. Discuss the trend(s) of the pattern in the limit $N \to \infty$.

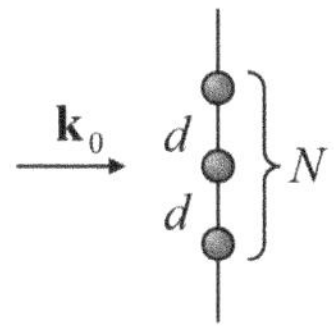

Solution: According to the discussion in section 8.4 of the lecture notes, in the case of similar small scatterers located at points $\mathbf{r}_j$, the scattered wave intensity is proportional (apart from the smooth polarization-related function $\sin^2\Theta$) to the scattering function $F = |I(\mathbf{q})|^2$, where $I(\mathbf{q})$ is the phase sum (8.57),

$$I(\mathbf{q}) \equiv \sum_j \exp\{-i\mathbf{q}\cdot\mathbf{r}_j\},$$

and $\mathbf{q} \equiv \mathbf{k} - \mathbf{k}_0$ is the wave vector change at scattering. Introducing the Cartesian coordinates just as in figure 8.6 (see the figure below), we obtain $\mathbf{r}_j = x_j\mathbf{n}_x$, with $x_j = jd$, so that for the observation point within the plane of drawing (the common for line of elementary scatterers and the wave vector $\mathbf{k}_0$), where $q_x = k\sin\theta$, we obtain

$$I(\mathbf{q}) = \sum_{j=1}^{N} \exp\{-ikjd\sin\theta\} \equiv \sum_{j=1}^{N} \exp\{-ij\zeta\}, \qquad (*)$$

where $\zeta \equiv kd\sin\theta$, so that

$$F(\zeta) = |I(\mathbf{q})|^2 \equiv I(\mathbf{q})I^*(\mathbf{q}) = \left(\sum_{j=1}^{N} \exp\{-ij\zeta\}\right)\left(\sum_{j'=1}^{N} \exp\{-ij'\zeta\}\right)^*$$

$$\equiv \sum_{j,j'=1}^{N} \exp\{i(j'-j)\zeta\}.$$

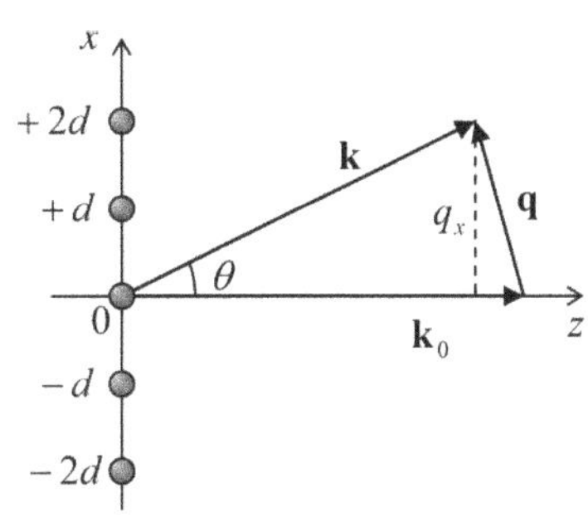

Three most significant features of the function $F(\zeta)$, describing an interplay between the interference of waves scattering from adjacent points, and the diffraction on the system as the whole, are immediately apparent from this formula:

(i) the function is even, because every term of the last sum is invariant with respect to the simultaneous replacements $\{\zeta \to -\zeta, j \leftrightarrow j'\}$, and the sums over j and j' are similar;
(ii) it is 2π-periodic, because each term of the sum is; and
(iii) the function reaches its major maximum[16], $F_{\max} = N^2$, at $\zeta = 0$ (and hence at all $\zeta = 2\pi n$, with any integer n), because at this point each term of the sum reaches its maximum, equal to 1.

These features are clearly visible on the numerical plots of the function for several values of N, shown in the figure below. (Its right panel zooms on the main peak near the origin.)

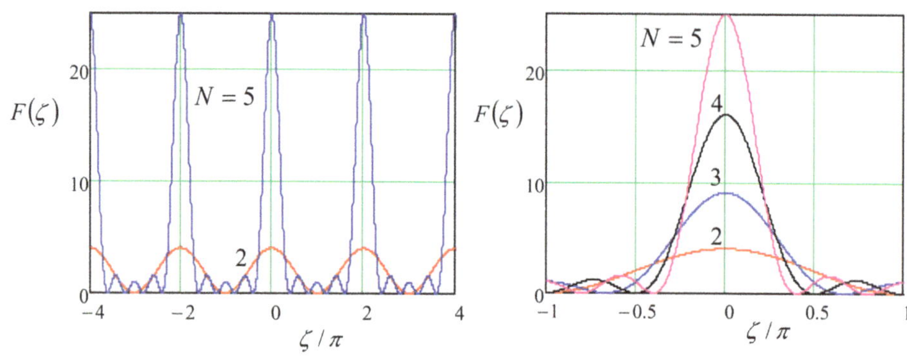

The plots show that as the number N of scatterers is increased, the width $\delta\zeta$ of the function's major maxima decreases as $\sim \pi/N$,[17] and that at $N \to \infty$ the function rapidly approaches a periodic set of the Fraunhofer diffraction patterns—see figure 8.8 of the lecture notes:

$$F(\zeta) \approx \left(N\frac{\sin \xi}{\xi}\right)^2 \equiv N^2 \text{sinc}^2 \xi,$$

$$\text{with} \quad \xi \equiv \frac{N(\zeta - 2\pi n)}{2} = \frac{k(Nd)\sin\theta}{2} - \pi N n, \qquad \text{for } N \gg 1.$$

This is natural, because at $N \to \infty$, the sum in Eq. (*) tends to the phase integral defined by Eq. (8.63) and given, for our case, by Eq. (8.64), with V replaced with N.

Note, however, that in terms of the observation angle θ, the transition $N \to \infty$ looks rather different in two ultimate cases.

[16] As the figure below shows, the function has $(N - 1)$ additional, lower maxima at each period.

[17] The exact coefficient in this relation depends on the accepted qualitative definition of peak's width.

(i) If the total length $a \equiv Nd$ of the scattering system is kept fixed (i.e. the distance d between the adjacent scatterers is decreased as $1/N$), then the width

$$\delta(\sin\theta) \equiv \frac{2}{ka}\delta\xi \equiv \frac{N}{ka}\delta\zeta \approx \frac{\pi}{ka}$$

of each diffraction peak does not depend on N, while the intervals between the peaks,

$$\Delta\xi \equiv \frac{ka}{2}\Delta(\sin\theta) = \frac{N}{2}\Delta\zeta = N\pi,$$

grow as N. As a result, since the largest magnitude of $\sin\theta$ is 1, in the typical case $ka \sim 1$, none of the diffraction peaks but one (with $n = 0$, i.e. $\theta \approx 0$) exist. This is exactly the case of diffraction, discussed in section 8.4 of the lecture notes.

(ii) In the opposite case, when the period d of the scattering structure is kept fixed, so that the increase of N leads to the proportional increase of its total length $a = Nd$, the intervals

$$\Delta(\sin\theta) \equiv \frac{2}{ka}\Delta\xi = \frac{N}{ka}\Delta\zeta = \frac{2\pi}{kd}$$

between the major maxima of the scattering intensity remain constant, so that they show up at the same values of the observation angle θ. In contrast, the width $\delta\theta$ of each maximum shrinks:

$$\delta(\sin\theta) \approx \frac{\pi}{kd}\frac{1}{N} \to 0,$$

i.e. the diffraction peaks become very sharp, so that the difference of their positions for different wavelengths $\lambda = 2\pi/k$ may be used for high-resolution spectroscopy. This is exactly the limit pursued in diffraction gratings—see section 8.8 of the lecture notes for a brief discussion of these important devices.

Finally notice that if the observation point is off the common plane shown in the figure above, the interference/diffraction pattern is quantitatively different—because the vector $\mathbf{q}$ also has an off-plane component, so that the scattering angle θ is now different from $(\pi/2 - \Theta)$—see, e.g. figure 8.5 in the lecture notes. However, in the most realistic case $kd > 1$, when the most important features of the pattern are located at small scattering angles, its main features described above remain qualitatively the same.

***Problem* 8.7.** Use the Born approximation to calculate the differential cross-section of the plane wave scattering by a non-magnetic, uniform dielectric sphere of an arbitrary radius R. In the limits $kR \ll 1$ and $1 \ll kR$ (where k is the wave number), analyze the angular dependence of the differential cross-section, and calculate the full cross-section of scattering.

Solution: According to Eqs. (8.62) and (8.63) of the lecture notes, the differential cross-section may be calculated as

$$\frac{d\sigma}{d\Omega} = \left(\frac{k^2}{4\pi}\right)^2 (\kappa - 1)^2 |I(\mathbf{q})|^2 \sin^2 \Theta, \qquad \text{with } I(\mathbf{q}) = \int_V \exp\{-i\mathbf{q}\cdot\mathbf{r}\} d^3r,$$

where the phase integral $I(\mathbf{q})$ is taken over the scatterer's volume, and $\mathbf{q} = \mathbf{k} - \mathbf{k}_0$ is the wave vector's change due to scattering.

According to figure 8.5 (reproduced, with additions, below), the magnitude of the vector $\mathbf{q}$ is related to the scattering angle θ (i.e. the angle between vectors $\mathbf{k}$ and $\mathbf{k}_0$, of the same length k) as $q = 2k\sin(\theta/2)$. As the figure also shows, the angle θ is related to the polarization angle Θ between the vectors $\mathbf{E}_\omega$ and $\mathbf{k}$, and the azimuthal angle ϕ, as follows

$$\sin^2 \Theta = \frac{k_y^2 + k_z^2}{k^2} = \sin^2 \theta \sin^2 \phi + \cos^2 \theta.$$

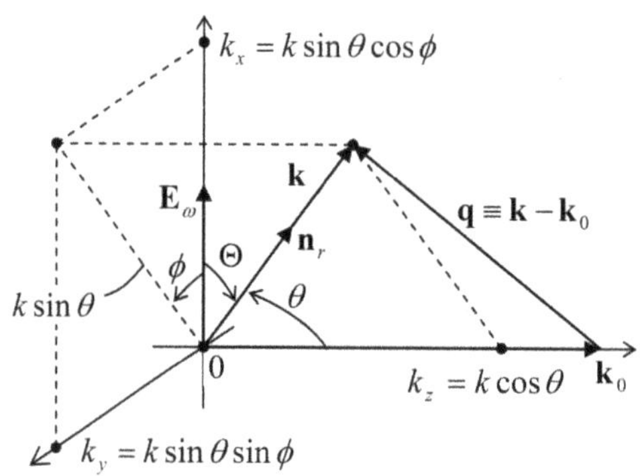

For a sphere, it is natural to work out the phase integral by taking the direction of the vector $\mathbf{q}$ for the polar axis, so that $\mathbf{q}\cdot\mathbf{r} = qr\cos\alpha$, and

$$\begin{aligned}
I(\mathbf{q}) &= 2\pi \int_0^R r^2 dr \int_0^\pi \exp\{-iqr\cos\alpha\} \sin\alpha d\alpha \\
&\equiv -2\pi \int_0^R r^2 dr \int_0^\pi \exp\{-iqr\cos\alpha\} d(\cos\alpha) \\
&= -2\pi \int_0^R r^2 dr \frac{1}{-iqr}(e^{iqr} - e^{-iqr}) \\
&\equiv \frac{4\pi}{q}\int_0^R r dr \sin qr = \frac{4\pi}{q^3}(\sin qR - qR\cos qR).
\end{aligned}$$

As a result, the differential cross-section is

$$\frac{d\sigma}{d\Omega} = (k^2R^3)^2 (\kappa - 1)^2 \sin^2 \Theta f(qR), \qquad (*)$$

where the function f, defined as

$$f(\xi) \equiv \frac{(\sin \xi - \xi \cos \xi)^2}{\xi^6},$$

behaves as shown by the red line in the figure below.

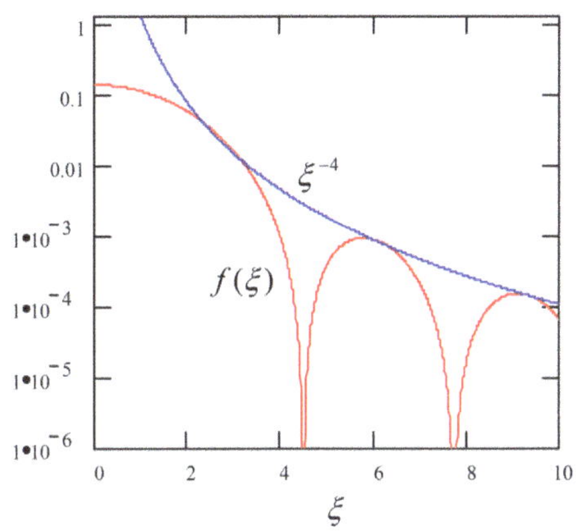

For a relatively large sphere, $kR \gg 1$, the function $f(qR)$ oscillates already at relatively small values of $q \sim 1/R$ (and hence of the scattering angle θ), so that $\Theta \approx \pi/2$, and $d\sigma/d\Omega$ is virtually independent of the azimuthal angle ϕ, and describes a round-ring diffraction pattern. As the number n of the intensity maxima is increased, their positions rapidly approach the following asymptotic values

$$q_n = \left(n + \frac{1}{2}\right)\frac{\pi}{R}, \quad n = 1, 2, \dots,$$

i.e. the scattering angles

$$\theta_n \approx \frac{q_n}{k} = \left(n + \frac{1}{2}\right)\frac{\pi}{kR} \ll 1.$$

As qR is increased, the oscillations' envelope is dropping very fast: as $(qR)^{-4}$—see the blue dashed line in the figure above. As a result, the integral (8.50), which gives the total cross-section, converges already at small scattering angles $\theta \ll 1$, where we may take $\sin^2\Theta \approx 1$, $q \approx k\theta$, and $\sin \theta \approx \theta$, so that

$$\sigma \approx \int_{4\pi} \frac{d\sigma}{d\Omega}\bigg|_{1/k \ll \alpha \ll 1} d\Omega = 2\pi k^4 R^6 (\kappa - 1)^2 \int_0^{1/k \ll \alpha \ll 1} f(kR\theta)\theta d\theta$$
$$= 2\pi k^2 R^4 (\kappa - 1)^2 \int_0^{\infty} f(\xi)\xi d\xi.$$

This is a table integral equal to ¼,[18] so that, finally,

$$\sigma = \frac{\pi}{2}(kR^2)^2(\kappa - 1)^2 \equiv \frac{1}{2}\sigma_0(kR)^2(\kappa - 1)^2, \qquad \text{for } kR \gg 1,$$

where $\sigma_0 \equiv \pi R^2$ is the geometric cross-section of the sphere. Since $kR \gg 1$, it is tempting to think that σ described by this formula may be larger than σ_0. However, a straightforward analysis of validity of the Born approximation for extended uniform objects[19] yields the condition $\sigma \ll \sigma_0$, so that the above result is only valid if $(\kappa - 1) \ll 1/kR$.

In the opposite limit, $kR \ll 1$, we can use Eq. (*) with $qR \ll 1$ and $f(qR) \approx f(0) = 1/9$ for all possible directions of vector $\mathbf{k}$ ($0 \leqslant q \leqslant 2k$), so that

$$\frac{d\sigma}{d\Omega} = \left(\frac{k^2R^3}{3}\right)^2(\kappa - 1)^2 \sin^2\Theta.$$

This result coincides with the general Eq. (8.53) of the lecture notes (with $V = (4\pi/3)R^3$) and shows that in this limit, there is no diffraction pattern to speak about. As we have already seen at the derivation of Eq. (8.27), the solid-angle integral of $\sin^2\Theta$ equals $8\pi/3$, so that the full cross-section in this limit is

$$\sigma = \frac{8\pi}{27}k^4R^6(\kappa - 1)^2 \equiv \frac{8}{27}\sigma_0(kR)^4(\kappa - 1)^2, \qquad \text{for } kR \ll 1.$$

This expression clearly shows that σ is much smaller than the geometric cross-section σ_0 of the sphere.

***Problem* 8.8.** A sphere of radius R is made of a uniform, non-magnetic, linear dielectric material, with an arbitrary dielectric constant. Derive an explicit expression for its full cross-section of scattering of a low-frequency monochromatic wave, with $k \ll 1/R$, and compare the result with the solution of the previous problem.

Solution: In the low-frequency limit $kR \ll 1$, i.e. when the wavelength $\lambda = 2\pi/k$ is much larger than sphere's radius R, the incident wave's field is essentially uniform at distances of the order of R. As the analysis in section 3.4 has shown, in this case the electric field created by the sphere's polarization outside it is exactly the same as that of a point electric dipole with the dipole moment (3.64)[20]:

$$\mathbf{p}_\omega = 4\pi\varepsilon_0 R^3 \frac{\kappa - 1}{\kappa + 2}\mathbf{E}_\omega. \qquad (*)$$

(Note that in the limit $\kappa \to 1$, this relation is reduced to Eq. (8.52) of the lecture notes.) This means, in particular, that the power re-radiated by the sphere, and hence the differential cross-section, have the same angular dependence ($\propto \sin^2\Theta$) as given

[18] See, e.g. Eq. (A.30b). Note that this formula allows the calculation of the full cross section, for an arbitrary ratio qR, in an analytical but bulky form—see the model solution of problem 3.6 in *Part QM*.

[19] See the model solution of problem 8.10.

[20] Here and below, for a material with frequency-dependent complex electric permittivity $\varepsilon(\omega)$, the dielectric constant κ has to be understood as the ratio $\varepsilon(\omega)/\varepsilon_0$.

by Eq. (8.26) of the lecture notes. Plugging Eq. (*) into Eqs. (8.28) and (8.39) of the lecture notes, we obtain

$$\sigma = \frac{8\pi}{3}(kR)^4\left(\frac{\kappa - 1}{\kappa + 2}\right)^2 R^2.$$

Comparing this expression with the solution of the previous problem (obtained in the Born approximation) in the corresponding frequency limit, we see that they indeed coincide at $\kappa \to 1$, when $(\kappa + 2)^2 \to 9$. On the other hand, if the dielectric constant is increased, the cross-section gradually approaches its largest possible value

$$\sigma \approx \sigma_{\max} = \frac{8\pi}{3}(kR)^4 R^2, \qquad \text{so that } \frac{\sigma_{\max}}{\pi R^2} = \frac{8}{3}(kR)^4, \qquad \text{at } \kappa \gg 1.$$

The last expression clearly shows that in the low-frequency limit ($kR \ll 1$) even this maximum value is much smaller than the geometric cross-section $\sigma_0 \equiv \pi R^2$ of the sphere.

***Problem* 8.9.** Use the Born approximation to calculate the differential cross-section of the plane wave scattering on a right, circular cylinder of length l and radius R, for an arbitrary angle of incidence.

Solution: In order to use Eq. (8.62) of the lecture notes, we have to calculate the phase integral (8.63):

$$\begin{aligned} I(\mathbf{q}) &\equiv \int_V \exp\{-i\mathbf{q}\cdot\mathbf{r}'\}d^3r' \\ &= \int_{-l/2}^{+l/2} dz \exp\left\{-iq_z z'\right\}\int_0^R \rho' d\rho' \int_0^{2\pi} d\varphi \exp\left\{-iq_\perp \rho' \cos\varphi\right\}, \end{aligned}$$

where q_z is the component of the scattering vector $\mathbf{q} = \mathbf{k} - \mathbf{k}_0$ along the cylinder's axis (taken for the coordinate axis z), $q_\perp$ is the magnitude of its component in the plane of cylinder's cross-section, and φ is the angle between that component and the vector $\boldsymbol{\rho}' \equiv \mathbf{r}' - \mathbf{n}_z z'$. The first integral is simple—see, e.g. Eqs. (8.64)–(8.66) and figure 8.7:

$$\int_{-l/2}^{+l/2} e^{-iq_z z'}dz' = l \ \text{sinc}\frac{q_z l}{2},$$

while the second one may be expressed via a Bessel function of the first kind, using their integral representation[21] and then the recurrence relation (2.143) for $n = 1$:

$$\begin{aligned} \int_0^R \rho' d\rho' \int_0^{2\pi} d\varphi\ e^{-iq_\perp \rho' \cos\varphi} &= 2\pi\int_0^R J_0(q_\perp \rho')\rho' d\rho' \equiv \frac{2\pi}{q_\perp^2}\int_0^{q_\perp R} J_0(\xi)\xi d\xi \\ &= \frac{2\pi}{q_\perp^2}[\xi J_1(\xi)]_0^{q_\perp R} \equiv \frac{2\pi R}{q_\perp}J_1(q_\perp R). \end{aligned}$$

[21] See, e.g. Eq. (A.42a) with $n = 0$.

With these substitutions, Eqs. (8.62) and (8.63) yield

$$\frac{d\sigma}{d\Omega} = \left(\frac{k^2 l R^2}{4}\right)^2 (\kappa - 1)^2 \sin^2\Theta \ \text{sinc}^2\frac{q_z l}{2}\left[\frac{J_1(q_\perp R)}{q_\perp R/2}\right]^2.$$

Here the factors inside the last square brackets have been grouped to emphasize that this fraction is very much similar to the sinc function, and in particular tends to 1 when its argument (in our case, $q_\perp R$) tends to zero—see the figure below.

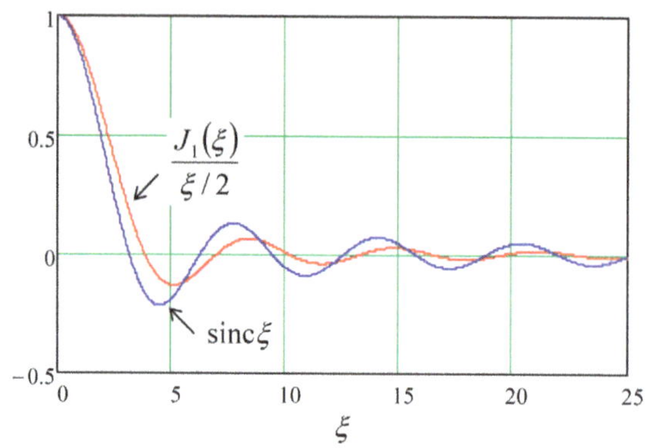

In this form, it is evident that for a very small cylinder with kl, $kR \ll 1$, our result reduces to Eq. (8.53) with $V \equiv \pi R^2 l$. However, for relatively large cylinders, our apparently simple analytical result describes a relatively complex angular pattern, because the relation between q_z, $q_\perp$, and Θ depends on the direction of the vectors **E**, **k**, and $\mathbf{k}_0$ relative to the cylinder's axis—see figure 8.5.

***Problem* 8.10.** Formulate the quantitative condition of the Born approximation's validity for a uniform linear-dielectric scatterer, with all linear dimensions of the order of the same scale a.

Solution: As was mentioned in the beginning of section 8.3 of the lecture notes, the general condition of Born approximation's validity is the weakness of the scattered wave in comparison with the incident wave $\mathbf{E}_{\text{in}}$ inside the scatterer. For a linear-dielectric, non-magnetic object, which scatters a plane, monochromatic incident wave with wave vector $\mathbf{k}_0$, this means that its electric polarization **P** at point $\mathbf{r}'$ is given by Eq. (3.45) with $\mathbf{E}(\mathbf{r}') = \mathbf{E}_{\text{in}}(\mathbf{r}') = \text{Re}[(\mathbf{E}_\omega)_{\text{in}}\exp\{i\mathbf{k}_0 \cdot \mathbf{r}'\}]$:

$$\mathbf{P}(\mathbf{r}') = (\kappa - 1)\varepsilon_0 \mathbf{E}_{\text{in}}(\mathbf{r}') = (\kappa - 1)\varepsilon_0 \,\text{Re}\,[(\mathbf{E}_\omega)_{\text{in}} e^{i\mathbf{k}_0\cdot\mathbf{r}'}].$$

Using this expression to calculate the dipole moment $d\mathbf{p}(\mathbf{r}') = \mathbf{P}(\mathbf{r}')\, d^3r'$ of the elementary volume of the scatterer, we can use Eq. (8.17b) to find the contribution of this elementary volume to the complex amplitude of the vector-potential of the scattered wave at a certain observation point **r**:

$$dA_\omega(\mathbf{r}) = \frac{\mu_0}{2\pi}\frac{j_\omega(\mathbf{r}')}{R}e^{i\mathbf{k}\cdot\mathbf{R}}d^3r' = \frac{\mu_0}{2\pi}\frac{(-i\omega d\mathbf{p}_\omega)}{R}e^{i\mathbf{k}\cdot\mathbf{R}}$$
$$= -i\frac{\mu_0\omega}{2\pi}(\kappa - 1)\varepsilon_0\mathbf{E}_\omega\frac{1}{R}e^{i\mathbf{k}\cdot\mathbf{R}}e^{i\mathbf{k}_0\cdot\mathbf{r}'}d^3r',$$

where $\mathbf{R} \equiv \mathbf{r} - \mathbf{r}'$ is the vector connecting the source and observation points. Summing up these contributions over all the scatterer's volume V, and taking into account that according to Eq. (6.7), the vector-potential's amplitude in the (plane, monochromatic) incident wave is $(\mathbf{A}_\omega)_{\text{in}} = (\mathbf{E}_\omega)_{\text{in}}\exp\{i\mathbf{k}_0 \cdot \mathbf{r}\}/(-i\omega)$, we obtain[22]

$$\left|\frac{(\mathbf{A}_\omega)_{\text{scat}}}{(\mathbf{A}_\omega)_{\text{in}}}\right| = \frac{\varepsilon_0\mu_0\omega^2}{2\pi}(\kappa - 1)\left|e^{-i\mathbf{k}_0\cdot\mathbf{r}}\int_V e^{i\mathbf{k}\cdot\mathbf{R}}e^{i\mathbf{k}_0\cdot\mathbf{r}'}\frac{d^3r'}{R}\right|$$
$$\equiv \frac{k^2}{2\pi}(\kappa - 1)\left|e^{-i\mathbf{k}_0\cdot\mathbf{r}}\int_V e^{i\mathbf{k}\cdot(\mathbf{r}-\mathbf{r}')}e^{i\mathbf{k}_0\cdot\mathbf{r}'}\frac{d^3r'}{R}\right|$$
$$\equiv \frac{k^2}{2\pi}(\kappa - 1)|e^{i(\mathbf{k}-\mathbf{k}_0)\cdot\mathbf{r}}|\left|\int_V e^{i\mathbf{k}\cdot\mathbf{r}'}e^{i\mathbf{k}_0\cdot\mathbf{r}'}\frac{d^3r'}{R}\right| \qquad (*)$$
$$= \frac{k^2}{2\pi}(\kappa - 1)\left|\int_V e^{-i\mathbf{q}\cdot\mathbf{r}'}\frac{d^3r'}{|\mathbf{r} - \mathbf{r}'|}\right|, \quad \text{with } \mathbf{q} \equiv \mathbf{k} - \mathbf{k}_0.$$

The Born approximation is valid if the right-hand side of this equality is much less than 1 at any point of the scatterer. For a sample much smaller than the wavelength, we can take $qr' \ll 1$ at any point, so that the integral in Eq. (*) is of the order of a^2,[23] and the approximation's validity condition is

$$\frac{(ka)^2}{2\pi}(\kappa - 1) \ll 1, \qquad \text{for } ka \ll 1. \qquad (**)$$

Comparing this expression with Eq. (8.53) of the lecture notes, which gives the full cross-section for this case,

$$\sigma \equiv \oint_{4\pi}\frac{d\sigma}{d\Omega}d\Omega = \left[\frac{k^2V}{4\pi}(\kappa - 1)\right]^2\oint_{4\pi}\sin^2\Theta d\Omega$$
$$= \left[\frac{k^2V}{4\pi}(\kappa - 1)\right]^2 2\pi\int_0^\pi \sin^3\Theta d\Theta$$
$$= \frac{4\pi}{3}\left[\frac{k^2V}{4\pi}(\kappa - 1)\right]^2,$$

[22] For distant observation points (R, $r \gg a$), this integral is just $I(\mathbf{q})/r$, where $I(\mathbf{q})$ is the phase integral (8.63), but for our current task, we need to consider points inside the scatterer: $r \sim a$.

[23] Let me remind the reader that the 3D singularity $1/r$ is integrable: $\int_{0\leqslant r<\varepsilon} d^3r/r = 4\pi\int_0^\varepsilon rdr = 2\pi\varepsilon^2 \to 0$ at $\varepsilon \to 0$.

with $V \sim a^3$, and with the evident estimate $\sigma_0 \sim a^2$ for the physical cross-section area of the scatterer, we see that Eq. (8.13) may be rewritten just as

$$\sigma \ll \sigma_0.$$

On the other hand, for an extended object, with $a \gg \lambda$, the function under the integral in Eq. (*) rapidly oscillates in the direction of the vector **q**, so that the contribution of this direction to the integral is accumulated only on distances $\sim\lambda \sim 1/k$ rather than a. As a result, a fairer estimate of the integral is ka, so that the Born approximation is valid if[24]

$$\frac{ka}{2\pi}(\kappa - 1) \ll 1, \qquad \text{for } 1 \ll ka. \tag{***}$$

Note, however, that estimates (**) and (***) are only valid if all linear dimensions of the scatterer are comparable, and should be modified, for example, for strongly elongated objects—say, long/thin rods. In such objects, for certain parameters, the Born approximation may simultaneously work well in some scattering directions and fail in others.

***Problem* 8.11.** If a scatterer absorbs some part of the incident wave's power, it may be characterized by an *absorption cross-section* σ_a defined similarly to Eq. (8.39) for the scattering cross-section:

$$\sigma_a \equiv \frac{\overline{\mathscr{P}_a}}{|E_\omega|^2/2Z_0},$$

where the nominator is the time-averaged power absorbed in the scatterer. Calculate σ_a for a very small sphere of radius $R \ll k^{-1}$, δ_s, made of a non-magnetic material with Ohmic conductivity σ, and with high-frequency permittivity $\varepsilon_{opt} = \varepsilon_0$. Can σ_a of such a sphere be larger than its geometric cross-section πR^2?

Solution: Due to the condition $R \ll k^{-1}$, the displacement currents are negligible, and the external field is virtually uniform at the spatial scale of the order of R, while due to the condition $R \ll \delta_s$, the magnetic field of the eddy currents is also negligible. At these conditions, we may calculate the electric field inside the sphere from the first of Eqs. (3.65) of the lecture notes, which may be rewritten as

$$\mathbf{E}_{int} = \frac{3\varepsilon_0}{\varepsilon + 2\varepsilon_0}\mathbf{E}_0,$$

with the replacements discussed in section 7.2: $\mathbf{E}_{int} \to (\mathbf{E}_{int})_\omega$, $\mathbf{E}_0 \to \mathbf{E}_\omega$, $\varepsilon \to \varepsilon(\omega)$. For an Ohmic conductor, at not very high frequencies, the complex electric permittivity may be calculated using Eq. (7.46), with $\varepsilon_{opt} = \varepsilon_0$. Combining these replacements, we obtain

$$(\mathbf{E}_{int})_\omega = \frac{3\varepsilon_0}{\varepsilon(\omega) + 2\varepsilon_0}\mathbf{E}_\omega = \frac{3\varepsilon_0}{(\varepsilon_0 + i\sigma/\omega) + 2\varepsilon_0}\mathbf{E}_\omega \equiv \frac{1}{1 + i/3\omega\tau_r}\mathbf{E}_\omega,$$

[24] For the analog of conditions (**) and (***) for the Born approximation in quantum mechanics, see *Part QM*, Eq. (3.77).

where $\tau_r \equiv \varepsilon_0/\sigma$ is the charge relaxation time of the sphere's material, which was discussed in section 4.2—see the second of Eqs. (4.10), with $\kappa = 1$.

Now calculating the time average of the Joule power loss (4.40) exactly as it was done for the Poynting vector in Eqs. (7.41) and (7.42) of the lecture notes, we obtain

$$\begin{aligned}\overline{\mathscr{P}_a} &= \sigma \int_V \overline{E_{\text{int}}^2} d^3r = \sigma \, \overline{\{\text{Re}\,[(E_{\text{int}})_\omega e^{-i\omega t}]\}^2} \; V \\ &= \sigma \frac{1}{2} \mid (E_{\text{int}})_\omega \mid^2 \frac{4\pi}{3} R^3 \\ &= \sigma \frac{1}{2} \left| \frac{1}{1 + i/3\omega\tau_r} E_\omega \right|^2 \frac{4\pi}{3} R^3 \\ &= \sigma \frac{1}{2} \frac{1}{1 + (1/3\omega\tau_r)^2} |E_\omega|^2 \frac{4\pi}{3} R^3,\end{aligned}$$

so that the absorption cross-section is

$$\sigma_a \equiv \frac{\overline{\mathscr{P}_a}}{|E_\omega|^2/2Z_0} = \frac{4\pi}{3} \frac{Z_0 \sigma R^3}{1 + (1/3\omega\tau_r)^2}. \qquad (*)$$

This result shows that in the limit of low frequencies, $\omega \ll 1/\tau_r$, the cross-section is proportional to the material's *resistivity* $1/\sigma$:

$$\sigma_a \approx \frac{4\pi}{3} \frac{Z_0 \sigma R^3}{(1/3\omega\tau_r)^2} \equiv 12\pi Z_0 \sigma R^3 \omega^2 \tau_r^2 \equiv 12\pi \frac{Z_0 \varepsilon_0^2 R^3 \omega^2}{\sigma}, \qquad \text{for } \omega\tau_r \ll 1,$$

and vanishes at $\omega \to 0$. This is natural, because the low-frequency field is screened out of any conductor, unless its size is smaller than the (extremely small) screening length—see section 2.1 of the lecture notes.

As the wave's frequency is increased, the σ_a given by Eq. (*) grows, but then saturates at a frequency-independent value proportional to the sphere material's *conductivity* σ:

$$\sigma_a = \frac{4\pi}{3} Z_0 \sigma R^3, \qquad \text{so that } \frac{\sigma_a}{\pi R^2} = \frac{4}{3} Z_0 (\sigma R), \qquad \text{for } \omega\tau_r \gg 1.$$

(Note that, as the elementary Eq. (4.21) shows, the right-hand side of the last relation, besides a numerical factor of the order of unity, is equal to the ratio of the free-space's wave impedance $Z_0 \approx 377\ \Omega$ to its dc resistance[25].) Plugging into this

[25] Assuming that the resistance is measured between two electrodes reasonably distributed on the opposite sides of a sphere's diameter.

result the high-frequency condition in the form $\sigma \ll \varepsilon_0\omega$, and the first condition on R imposed in the assignment, $R \ll 1/k \equiv c/\omega$, we obtain

$$\frac{\sigma_a}{\pi R^2} \sim Z_0\sigma R \ll Z_0(\varepsilon_0\omega)\frac{c}{\omega} \equiv Z_0\varepsilon_0 c \equiv 1,$$

so that our result is valid only if the ratio is much less than one. The second condition, $R \ll \delta_s \equiv (2/\mu_0\sigma\omega)^{1/2}$, also together with the high-frequency condition $\sigma \ll \varepsilon_0\omega$, leads to a similar restriction:

$$\frac{\sigma_a}{\pi R^2} \sim Z_0\sigma R \ll Z_0\left(\frac{\sigma}{\mu_0\omega}\right)^{1/2} \ll Z_0\left(\frac{\varepsilon_0\omega}{\mu_0\omega}\right)^{1/2} \equiv 1.$$

This means that σ_a, within the specified restrictions $R \ll k^{-1}$, δ_s, cannot be larger than the geometric cross-section πR^2 of the sphere.

***Problem* 8.12.** Use the Huygens principle to calculate the wave's intensity at the symmetry plane of the slit diffraction experiment (i.e. at $x = 0$ in figure 8.12 of the lecture notes), for an arbitrary ratio z/ka^2.

Solution: The general solution of the problem is given by Eqs. (8.105)–(8.107) of the lecture notes. For our particular case $x = 0$, Eq. (8.107) becomes

$$\begin{aligned} I_x &= \int_{-a/2}^{+a/2} \exp\frac{ikx'^2}{2z}dx' \\ &= 2\int_0^{+a/2}\left[\cos\frac{kx'^2}{2z} + i\sin\frac{kx'^2}{2z}\right]dx' \\ &= 2\left(\frac{2z}{k}\right)^{1/2}\int_0^{\xi}[\cos(\zeta^2) + i\sin(\zeta^2)]\,d\zeta \\ &= 2\left(\frac{\pi z}{k}\right)^{1/2}[\mathcal{C}(\xi) + i\mathcal{S}(\xi)], \qquad \text{where } \xi \equiv \left(\frac{k}{8z}\right)^{1/2} a \equiv \left(\frac{\pi}{4}\frac{a^2}{\lambda z}\right)^{1/2}, \end{aligned}$$

and $\mathcal{C}(\xi)$ and $\mathcal{S}(\xi)$ are the Fresnel integrals defined by Eqs. (8.112). Combining this relation with Eqs. (8.104) and (8.105), for the relative wave intensity we obtain

$$\frac{\overline{S}(0, z)}{S_0} = \left|\frac{f_\omega(0, z)}{f_0}\right|^2 = 2[\mathcal{C}^2(\xi) + \mathcal{S}^2(\xi)].$$

This result is shown with the solid red curve in the figure below. As the distance z of the observation point from the slit is increased (i.e. the argument ξ of the Fresnel integral is decreased), the intensity first oscillates, with an increasing amplitude, near the average value of S_0, but just beyond the Fresnel-to-Fraunhofer diffraction crossover, the intensity starts a monotonic drop (see the dashed line):

$$\frac{\overline{S}(0, z)}{S_0} \approx \frac{4}{\pi}\xi^2 \equiv \frac{a^2}{\lambda z} \ll 1, \qquad \text{at } z \gg \frac{a^2}{\lambda}.$$

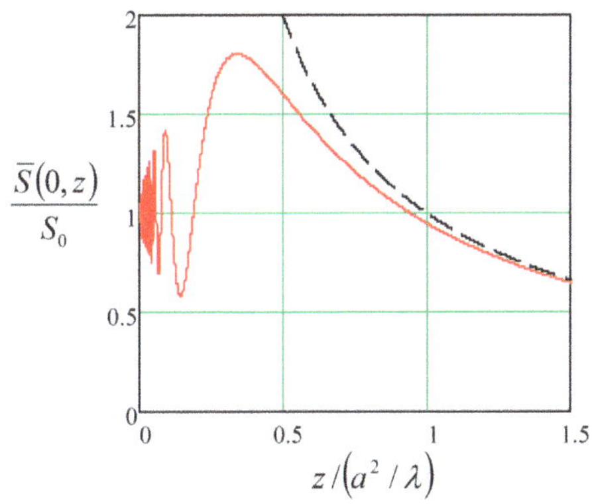

Qualitatively, this behavior is similar to that for the diffraction at a circular orifice (see the next problem), but the intensity oscillations with z are less pronounced (with the maximum intensity never reaching $2S_0$, and the minimal intensity not hitting zero), and its drop at large distances is not so fast—all due to the fact that in our current case the diffraction takes place is only one dimension.

***Problem* 8.13.** A plane wave with wavelength λ is normally incident on an opaque, plane screen, with a round orifice of radius $R \gg \lambda$. Use the Huygens principle to calculate the passing wave's intensity distribution along the system's symmetry axis, at distances $z \gg R$ from the screen (see the figure below) and analyze the result.

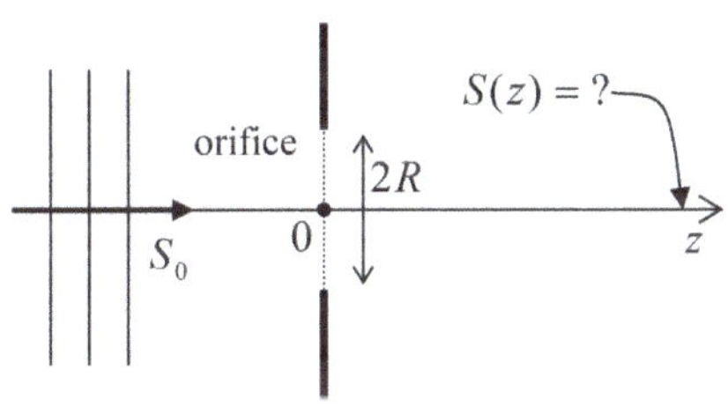

Solution: According to Eq. (8.91) of the lecture notes, the complex amplitude of the wave in the observation point is

$$f_\omega(z) = f_0 \frac{k}{2\pi i} \int_0^R \rho' d\rho' \int_0^{2\pi} d\varphi' \frac{\exp\{ik(\rho'^2 + z^2)^{1/2}\}}{(\rho'^2 + z^2)^{1/2}}.$$

Using the axial symmetry of the problem, and the conditions $\lambda \ll R \ll z$, this expression may be simplified, just as has been done in section 8.5—see, e.g. Eq. (8.86):

$$f_\omega(z) \approx f_0 \frac{k}{iz} e^{ikz} \int_0^R \exp\left\{i\frac{k\rho'^2}{2z}\right\} \rho' d\rho' = f_0 \frac{k}{iz} e^{ikz} \frac{z}{k} \int_0^{kR^2/2z} \exp\{i\xi\} d\xi$$

$$= -f_0 e^{ikz}\left(\exp\left\{i\frac{kR^2}{2z}\right\} - 1\right),$$

so that the wave's intensity is

$$S(z) = S_0 \left| \exp\left\{ i\frac{kR^2}{2z} \right\} - 1 \right|^2 = 2S_0\left(1 - \cos\frac{kR^2}{2z}\right) \equiv 2S_0\left(1 - \cos\frac{\pi R^2}{z\lambda}\right).$$

The result shows that the intensity oscillates, reaching zero at points $z = z_n$ where $\cos(\pi R^2/z_n\lambda) = 1$, i.e. at $\pi R^2/z_n\lambda = 2\pi n$ (with $n = 1, 2, \ldots$), so that

$$z_n = \frac{R^2}{2\lambda n}.$$

Between each pair of the minima, the intensity reaches a maximum (twice larger than the incident wave's intensity!) at points $z = z'_n$ where $\cos(\pi R^2/z'_n\lambda) = -1$, i.e. $\pi R^2/z'_n\lambda = 2\pi n - \pi$, so that[26]

$$z'_n = \frac{R^2}{2\lambda(n - 1/2)}.$$

Note that the first few maxima and minima (most distant from the screen) correspond to the point of transition from the Fresnel to Fraunhofer diffraction limits: $z/R \sim R/\lambda$, and that in the latter limit, the intensity does not oscillate, but rather gradually vanishes at $z \to \infty$:

$$S(z) \to S_0\left(\frac{\pi R^2}{z\lambda}\right)^2 \propto \frac{1}{z^2}, \qquad \text{at } z \gg \frac{R^2}{\lambda}. \qquad (*)$$

This decrease may be readily explained as the wave's taking a spherical form—the behavior natural for large distances, from which the orifice looks like a point source.

***Problem* 8.14.** A plane monochromatic wave is now normally incident on an opaque circular disk of radius $R \gg \lambda$. Use the Huygens principle to calculate the wave's intensity at distance $z \gg R$ behind the disk's center (see the figure below). Discuss the result.

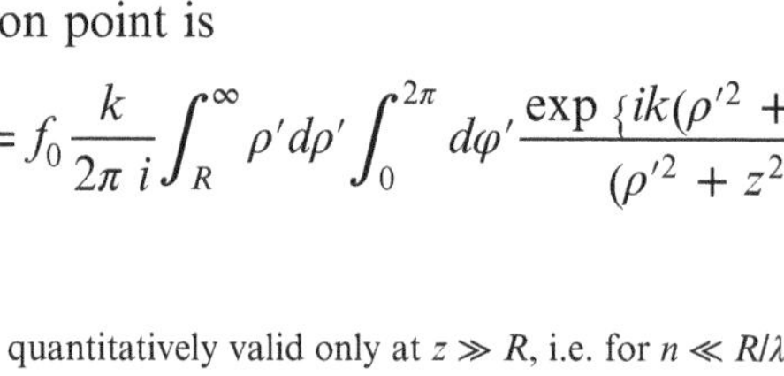

Solution: According to Eq. (8.91) of the lecture notes, the complex amplitude of the wave in the observation point is

$$f_\omega(z) = f_0 \frac{k}{2\pi i}\int_R^\infty \rho' d\rho' \int_0^{2\pi} d\varphi' \frac{\exp\{ik(\rho'^2 + z^2)^{1/2}\}}{(\rho'^2 + z^2)^{1/2}}.$$

[26] Note that these results are quantitatively valid only at $z \gg R$, i.e. for $n \ll R/\lambda$; however, since that last ratio has to be large for the Huygens principle to be valid, the prediction of many maxima and minima is a sound one, and may be readily verified experimentally.

Using the axial symmetry of the problem, and the Huygens principle conditions (8.83), $z \gg R \gg \lambda$, this expression may be simplified, just as has been done in section 8.5—see, e.g. Eq. (8.86):

$$f_\omega(z) \approx f_0 \frac{k}{iz} e^{ikz} \int_R^\infty \exp\left\{i\frac{k\rho'^2}{2z}\right\}\rho' d\rho' = f_0 \frac{k}{iz} e^{ikz} \frac{z}{k} \int_{kR^2/2z}^\infty \exp\{i\xi\} d\xi$$
$$= -f_0 e^{ikz} \exp\{i\xi\} \Big|_{kR^2/2z}^{\infty}.$$

Since $\exp\{i\xi\} = \cos\xi + i\sin\xi$ is an oscillating rather than a decaying function, the result of the upper-limit substitution may not be apparent. However, since this is an analytical function, we can always replace $i\xi$ with $(i - \delta)\xi$, where δ is a small real, positive number. As a result, the upper-limit substitution disappears and now we can make, in the lower-limit substitution, transition $\delta \to 0$. As a result, we obtain

$$f_\omega(z) = f_0 e^{ikz} \exp\left\{i\frac{kR^2}{2z}\right\},$$

leading to a paradoxical answer

$$S = S_0.$$

It is curious that this result was first obtained in 1818 by Poisson who used it as an argument *against* the theory of diffraction that had been developed by Fresnel—and against the wave theory of light as the whole. The following experiments have shown, however, that the pattern of diffraction on a disk with $R \ll z$ indeed features a light spot in the center.

Note also that in the Fraunhofer diffraction limit (i.e. at $z \gg kR^2$), this result may be obtained directly from Eq. (*) in the model solution of the previous problem, and the Babinet principle (8.129).

***Problem* 8.15.** Use the Huygens principle to analyze the Fraunhofer diffraction of a plane wave normally incident on a square-shaped hole, of size $a \times a$, in an opaque screen. Sketch the diffraction pattern you would observe at a sufficiently large distance, and quantify expression 'sufficiently large' for this case.

Solution: As was discussed in section 8.8 of the lecture notes, in the Fraunhofer limit the diffraction pattern is just a 2D Fourier transform of the initial wave amplitude distribution. In our case this means that the complex amplitude of the scattered wave may be calculated as

$$f(x, y) = \text{const} \times \int_{-a/2}^{+a/2} dx' \int_{-a/2}^{+a/2} dy' \exp\{i(k_x x' + k_y y')\},$$

with $k_x = k(x/z)$, $k_y = k(y/z)$, $k = 2\pi/\lambda$. This integral is just a product of two 1D elementary integrals of complex-exponential (i.e. sinusoidal) functions, giving finally

$$f(x, y) = \text{const} \times \frac{\sin(k_x a/2)}{(k_x a/2)} \frac{\sin(k_y a/2)}{(k_y a/2)} = \text{const} \times \text{sinc}\left(\frac{kax}{2z}\right)\text{sinc}\left(\frac{kay}{2z}\right).$$

The figure below shows lines of equal intensity (proportional to f^2), drawn at each 1% of the maximum value. Note that though the secondary maxima, centered at points $\{k_x = n\pi/a,\ k_y = m\pi/a\}$ do exist for all integer n and m, they are extremely weak unless one of these numbers equals 1, while the other one is not much larger than 1.

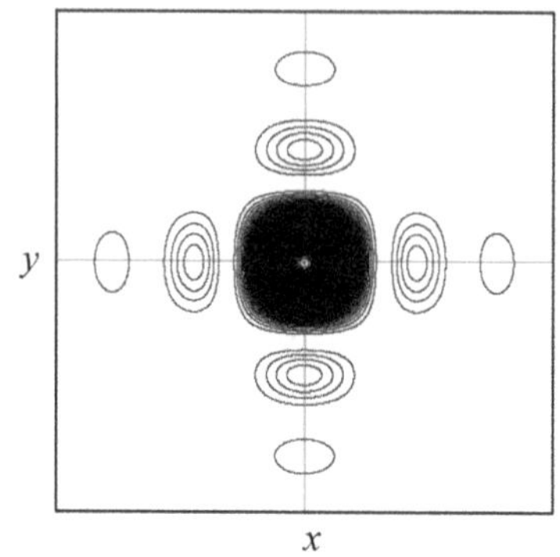

This result is valid for 'sufficiently large' distances z in the sense that they are well beyond the crossover from the Fresnel to Fraunhofer diffraction (see section 8.6): $z \gg a^2/\lambda \sim ka^2$.

***Problem* 8.16.** Use the Huygens principle to analyze the propagation of a monochromatic Gaussian beam, described by Eq. (7.181) of the lecture notes, with the initial characteristic width $a_0 \gg \lambda$, in a uniform, isotropic medium. Use the result for a semi-quantitative derivation of the so-called *Abbe limit*[27] for the spatial resolution of an optical system,

$$w_{\min} = \frac{\lambda}{2\sin\theta},$$

where θ is the half-angle of the wave cone propagating from the object, and captured by the system.

Solution: Plugging Eq. (7.181) into Eq. (8.91), and using the Cartesian coordinates similar to those shown in figure 8.11 of the lecture notes (see the figure below), we obtain

$$\begin{aligned} f_\omega(\mathbf{r}) &= \frac{k}{2\pi i} f_0 \int_{z'=0} \exp\left\{-\frac{\zeta\rho'^2}{2}\right\}\frac{e^{ikR}}{R} d^2r' \\ &\equiv \frac{k}{2\pi i} f_0 \int_{-\infty}^{+\infty} dx' \int_{-\infty}^{+\infty} dy' \exp\left\{-\frac{\zeta(x'^2+y'^2)}{2}\right\}\frac{e^{ikR}}{R}, \end{aligned} \qquad (*)$$

[27] Reportedly, due to not only E Abbe (1873), but also to H von Helmholtz (1874).

where $R = [(x - x')^2 + (y - y')^2 + z^2]^{1/2}$, and $\zeta \equiv 1/a_0^2$.

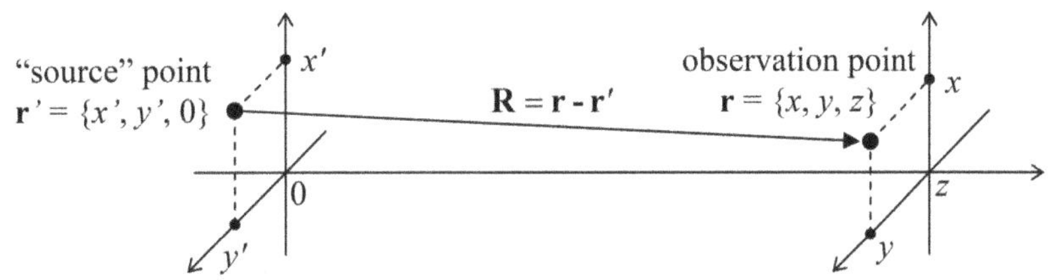

Since the Huygens principle, in its simple form (8.91), is only valid if $z^2 \gg x^2, y^2, x'^2, y'^2 \sim a^2$, we may use the approximations similar to those used at the derivation of Eq. (8.87) of the lecture notes, and reduce Eq. (*) to

$$f_\omega(\mathbf{r}) = \frac{k}{2\pi i} f_0 \frac{e^{ikz}}{z} I_x I_y,$$

where I_x and I_y are two similar integrals, e.g.

$$I_x \equiv \int_{-\infty}^{+\infty} \exp\left\{-\frac{\zeta x'^2}{2} + \frac{ik}{2z}(x - x')^2\right\} dx'.$$

This is a typical *Gaussian integral*[28]; it may be readily worked out by representing the exponent's argument as a full square of the difference (x' – const), plus another term independent of x':

$$\begin{aligned}-\frac{\zeta x'^2}{2} + \frac{ik}{2z}(x - x')^2 &\equiv -\frac{\zeta x'^2}{2} + \frac{ik}{2z}x^2 + \frac{ik}{z}xx' + \frac{ik}{2z}x'^2 \\ &\equiv -\left[\left(\frac{\zeta}{2} - \frac{ik}{2z}\right)^{1/2} x' - \frac{ik}{2z}\left(\frac{\zeta}{2} - \frac{ik}{2z}\right)^{-1/2} x\right]^2 \\ &\quad + \left[\frac{ik}{2z} - \left(\frac{k}{2z}\right)^2\left(\frac{\zeta}{2} - \frac{ik}{2z}\right)^{-1}\right] x^2.\end{aligned}$$

Now we may argue that although the coefficient before x' is a complex number, its real part is positive and the function under the integral is analytical, so that it may be worked out exactly as that of a real argument[29]:

[28] Such integrals are frequently met in various fields of physics (and other qualitative sciences). For example, they naturally arise at the analysis of the dispersion of 1D wave packets—see, e.g. *Part QM* section 2.2.
[29] Since the integration limits are infinite, the constant offset of x', expressed by the second term in the square brackets, does not affect the integral.

$$I_x = \exp\left\{\left[\frac{ik}{2z} - \left(\frac{k}{2z}\right)^2\left(\frac{\zeta}{2} - \frac{ik}{2z}\right)^{-1}\right]x^2\right\}$$

$$\int_{-\infty}^{+\infty} \exp\left\{-\left[\left(\frac{\zeta}{2} - \frac{ik}{2z}\right)^{1/2} x' - \frac{ik}{2z}\left(\frac{\zeta}{2} - \frac{ik}{2z}\right)^{-1/2} x\right]^2\right\}dx'$$

$$\equiv \exp\left\{\left[\frac{ik}{2z} - \left(\frac{k}{2z}\right)^2\left(\frac{\zeta}{2} - \frac{ik}{2z}\right)^{-1}\right]x^2\right\}\left(\frac{\zeta}{2} - \frac{ik}{2z}\right)^{-1/2}\int_{-\infty}^{+\infty} \exp\{-\xi^2\}d\xi.$$

The last, table integral[30] equals $\pi^{1/2}$, so that using the absolutely similar integral I_y, we finally obtain[31]

$$f_\omega(\mathbf{r}) = \frac{k}{2i}f_0\frac{e^{ikz}}{z}\left(\frac{\zeta}{2} - \frac{ik}{2z}\right)^{-1}\exp\left\{\left[\frac{ik}{2z} - \left(\frac{k}{2z}\right)^2\left(\frac{\zeta}{2} - \frac{ik}{2z}\right)^{-1}\right](x^2 + y^2)\right\}$$

$$\equiv f_0 e^{ikz}\left(1 + \frac{i\zeta z}{k}\right)^{-1}\exp\left\{\left(\frac{i\zeta^2 z}{2k} - \frac{\zeta}{2}\right)\rho^2 \Big/ \left(1 + \frac{\zeta^2 z^2}{k^2}\right)\right\}.$$

The wave's intensity is proportional to

$$|f_\omega(\mathbf{r})|^2 = |f_0|^2\left(1 + \frac{\zeta^2 z^2}{k^2}\right)^{-1}\exp\left\{-\zeta\rho^2 \Big/ \left(1 + \frac{\zeta^2 z^2}{k^2}\right)\right\}$$

$$\equiv |f_0|^2\left(1 + \frac{z^2}{k^2 a_0^4}\right)^{-1}\exp\left\{-\frac{\rho^2}{a_0^2} \Big/ \left(1 + \frac{z^2}{k^2 a_0^4}\right)\right\} \qquad (**)$$

$$\equiv |f_0|^2\left(\frac{a}{a_0}\right)^{-2}\exp\left\{-\frac{\rho^2}{a^2}\right\},$$

where a is the beam's effective radius at the distance z from the initial plane[32]:

$$a^2 = a_0^2 + \frac{z^2}{k^2 a_0^2} \equiv a_0^2 + \frac{z^2\lambda^2}{4\pi^2 a_0^2}. \qquad (***)$$

So, the beam retains its Gaussian shape at all distances, but gradually broadens. The most interesting feature of this effect is that the narrower the initial beam, the

[30] See, e.g. Eq. (A.36).

[31] Alternatively, this result may be obtained by solving the approximate 2D *paraxial* (or 'parabolic') *wave equation*,

$$\partial a/\partial z = (1/2ik)\nabla_{x,y}^2 a,$$

for the complex amplitude $a(\mathbf{r})$ of the field represented in the form $f(\mathbf{r}) = a(\mathbf{r})e^{ikz}$, with appropriate boundary conditions. The paraxial approximation follows from the Helmholtz equation (7.204) in the same limit ($|\nabla a| \ll k$; $|x|$, $|y| \ll z$) as Eq. (8.91), which is essentially one of its solutions.

[32] Note also the pre-exponential factor in Eq. (**), which ensures that the total power of the beam, proportional to $|f_\omega|^2 a^2$, is independent of z—as it should be.

faster it broadens[33]. Actually, this is not quite surprising, and is just the standard feature of the Fraunhofer diffraction, as discussed in section 8.6 of the lecture notes. Indeed, the finite width a_0 of the initial beam plays the same role as the finite width of the orifice (or slit) at the usual diffraction—see figure 8.12. At relatively small distances, $z/a_0 \ll a_0/\lambda$, the beam is the subject of the Fresnel diffraction. As we know, this diffraction is limited to small vicinities of sharp edges of the wave distribution. The Gaussian beam does not have such edges, and hence remains virtually intact. On the other hand, at large distances, where $z/a_0 \gg a_0/\lambda$, the beam is affected by the Fraunhofer diffraction, at which the effective half-angle $\theta \equiv a/z$ of the pattern's divergence is independent of z—see, e.g. Eq. (8.110). Similarly, in this limit Eq. (***) yields $a = z\lambda/2\pi a_0$, i.e.

$$\theta \equiv \frac{a}{z} = \frac{\lambda}{2\pi a_0}.$$

Note that the smaller a_0, the larger this angle (though the Huygens approximation is valid only if $a_0 \gg \lambda$, i.e. if the diffraction angle is small, $\theta \ll 1$).

Now rewriting the same result as

$$a_{\min} \equiv a_0 = \frac{\lambda}{2\pi\theta},$$

we see that at $\theta \ll 1$ (when our calculation is valid), it has the same functional form as the Abbe limit's formula[34]. (Some difference of the numerical coefficients is due to the convention in definition of w, which is closer to the full width of the beam than to its effective width a_0.)

***Problem* 8.17.** Within the Fraunhofer approximation, analyze the pattern produced by a 1D diffraction grating with the periodic transparency profile shown in the figure below, for the normal incidence of a plane, monochromatic wave.

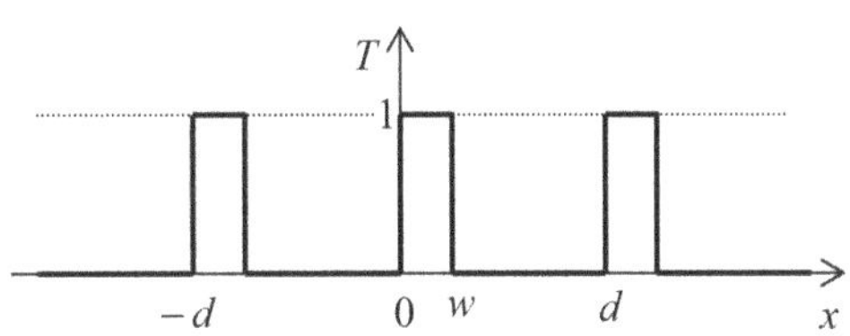

Solution: As was discussed in section 8.8, within the Fraunhofer approximation (valid when the distance z between the lattice and the observer is much larger than

[33] The similar behavior, described by a very similar math, is pertinent to the 1D wave packets propagating in a dispersive medium—see, e.g. *Part QM* section 2.2.

[34] Note that usually the Abbe limit is represented as $w_{\min} = \lambda_0/2n\sin\theta$, where n is the refraction index defined by Eq. (7.84), and $\lambda_0 = n\lambda$ is the free-space wavelength.

d^2/λ), the diffracted wave's amplitude $f_\omega(\rho)$ is just the spatial Fourier transform of the incident wave's amplitude $f_\omega(\rho')$ just behind the lattice, i. e. of its transparency $T(\rho') \equiv f_\omega(\rho')/f_0$. In our 1D case we have

$$f_\omega(\rho) \to f_\omega(x) = \text{const} \times \int_{-\omega}^{+\infty} T(x')\exp\{ik_t x'\}dx'$$
$$= \text{const} \times \int_{-\omega}^{+\infty} T(x')\exp\left\{i\frac{kx}{z}x'\right\}dx'.$$

Since in our case the transparency is periodic, $T(x' + d) = T(x')$, the intensity is non-vanishing only for a discrete set of equidistant values

$$k_n = \frac{2\pi}{d}n, \qquad x_n = \frac{z}{k}k_n = \frac{2\pi z}{kd}n, \qquad \text{with } n = 0, \pm 1, \pm 2, \ldots$$

which produce on a screen as a set of thin 'interference stripes' (or 'fringes') parallel to those on the diffraction lattice. For these values,

$$f_n = \text{const} \times \int_0^d T(x')\exp\left\{2\pi i\frac{x'}{d}n\right\}dx' \propto \int_0^w \exp\left\{2\pi i\frac{x'}{d}n\right\}dx'$$
$$= \frac{d}{2\pi i n}\left(\exp\left\{2\pi i\frac{x'}{d}n\right\}\right)_{x'=0}^{x'=w} = \frac{d}{2\pi i n}\left(\exp\left\{2\pi i\frac{w}{d}n\right\} - 1\right).$$

For analysis, it is convenient to represent this complex amplitude as a product of a real function by a phase factor that does not affect the wave's intensity:

$$f_n \propto \frac{d}{\pi n}\sin\frac{\pi w n}{d}\exp\left\{i\pi\frac{w}{d}n\right\} = a\,\text{sinc}\frac{\pi w n}{d}\exp\left\{i\pi\frac{w}{d}n\right\}.$$

Thus the stripe intensity follows the square of the sinc function,

$$|f_n|^2 \propto \text{sinc}^2\frac{\pi w n}{d},$$

see the solid red line in figure 8.8 of the lecture notes. Note that if d is a multiple of w, some interference stripes disappear; a convenient choice is $d = 2w$, leading to the suppression of each other interference fringe, and thus doubling the wavelength range available for unperturbed spectroscopic analysis of the diffracting wave.

***Problem* 8.18.** N equal point charges are attached, at equal intervals, to a circle rotating with a constant angular velocity about its center—see the figure below. For what values of N does the system emit:

(i) the electric dipole radiation?
(ii) the magnetic dipole radiation?
(iii) the electric quadrupole radiation?

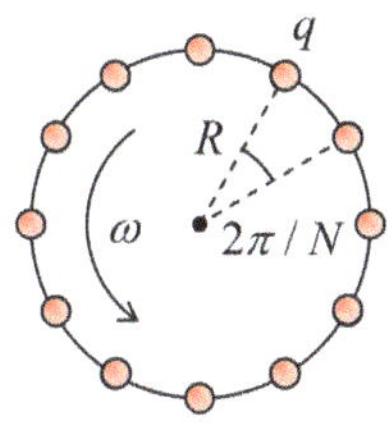

Solutions: Let us number the charges sequentially, as $k = 1, 2, \dots N$; then the angular position of the kth charge is

$$\varphi_k = \frac{2\pi}{N}k + \varphi_0, \qquad \text{where } \varphi_0 = \omega t + \text{const},$$

so that in the reference frame with the origin in circle's center and axis z normal to its plane the Cartesian coordinates of the charges are

$$\begin{aligned} x_k &= R\cos\varphi_k = R\cos\left(\frac{2\pi}{N}k + \varphi_0\right), \\ y_k &= R\cos\varphi_k = R\sin\left(\frac{2\pi}{N}k + \varphi_0\right), \qquad z_k = 0. \end{aligned} \tag{*}$$

(i) According to the definition (3.6) of the electric dipole moment $\mathbf{p}$, for this system its Cartesian components are

$$p_x = qR\sum_{k=1}^{N}\cos\left(\frac{2\pi}{N}k + \varphi_0\right) \equiv qR[\Sigma_c(N)\cos\varphi_0 - \Sigma_s(N)\sin\varphi_0],$$

$$p_y = qR\sum_{k=1}^{N}\sin\left(\frac{2\pi}{N}k + \varphi_0\right) \equiv qR[\Sigma_s(N)\cos\varphi_0 + \Sigma_c(N)\sin\varphi_0], \quad p_z = 0,$$

where

$$\Sigma_c(N) \equiv \sum_{k=1}^{N}\cos\frac{2\pi}{N}k, \qquad \Sigma_s(N) \equiv \sum_{k=1}^{N}\sin\frac{2\pi}{N}k\,. \tag{**}$$

Mathematics tells us[35] that both these sums equal 0 for any integer $N > 0$, with the sole exception of $\Sigma_c(1) = \cos 2\pi = 1$. Hence only the system with one rotating charge can emit electric dipole radiation. (Its calculation was the task of problem 8.1.) This result may be interpreted as the mutual compensation of the elementary dipole moments $q\mathbf{r}_k$ of $N > 1$ charges, due to the different orientation of their radius-vectors $\mathbf{r}_k$.

[35] See, e.g. Eq. (A.21) with $n = N$ and $\xi = 2\pi/N$. A simple alternative way to prove this property is to represent the vectors $\mathbf{R}_k$, connecting the adjacent vertices of a regular N-side polygon (so that $R_k = \text{const}$), by complex numbers $R_k\exp\{i(2\pi k/N + \text{const})\}$, and plug this representation into the condition that the polygon is a closed line: $\Sigma_k\mathbf{R}_k = 0$. (Evidently, this trick does not work for the special case $N = 1$.)

(ii) For a system of N point charges q, moving with velocities $\mathbf{v}_k$, the definition (5.91) of the magnetic dipole moment takes the form

$$\mathbf{m} = \frac{q}{2}\sum_{k=1}^{N}\mathbf{r}_k \times \mathbf{v}_k.$$

For our system, the velocity vector $\mathbf{v}_k$ of each charge is normal to both its radius-vector $\mathbf{r}_k$ and the axis z, so that all vector products $\mathbf{r}_k \times \mathbf{v}_k$, and hence the net moment $\mathbf{m}$, are directed along that axis:

$$\mathbf{m} = m\mathbf{n}_z, \qquad \text{with } m = \frac{q}{2}\sum_{k=1}^{N} r_k v_k = \frac{qN}{2}R^2\omega.$$

Although this moment is not equal to zero for any $N > 0$, its time derivative is, so that according to Eqs. (8.137)–(8.139), the system does not emit the magnetic-dipole radiation at any N. (Note that the answer might be different if the angular velocity ω depended on time.)

(iii) Plugging Eqs (*) into Eq. (8.142) for the electric quadrupole tensor, with all $q_k = q$, we obtain

$$\mathscr{Q} = qR^2\sum_{k=1}^{N}\begin{pmatrix} 2\cos^2\varphi_k - \sin^2\varphi_k & 2\sin\varphi_k\cos\varphi_k & 0 \\ 2\sin\varphi_k\cos\varphi_k & 2\sin^2\varphi_k - \cos^2\varphi_k & 0 \\ 0 & 0 & -\cos^2\varphi_k - \sin^2\varphi_k \end{pmatrix}$$

$$\equiv qR^2\sum_{k=1}^{N}\begin{pmatrix} (1 + 3\cos 2\varphi_k)/2 & \sin 2\varphi_k & 0 \\ \sin 2\varphi_k & (1 - 3\cos 2\varphi_k)/2 & 0 \\ 0 & 0 & -1 \end{pmatrix}.$$

Besides the time-independent terms (which, according to Eqs. (8.143) and (8.147) of the lecture notes, do not provide any radiation), the tensor elements include the following sums:

$$\sum_{k=1}^{N}\cos 2\varphi_k \equiv \Sigma_c'(N)\cos 2\varphi_0 - \Sigma_s'(N)\sin 2\varphi_0,$$

$$\sum_{k=1}^{N}\sin 2\varphi_k \equiv \Sigma_s'(N)\cos 2\varphi_0 + \Sigma_c'(N)\sin 2\varphi_0,$$

where the trigonometric function sums differ from those defined by Eq. (8.20) by an additional factor 2 before the argument:

$$\Sigma_c'(N) \equiv \sum_{k=1}^{N}\cos\frac{4\pi}{N}k, \qquad \Sigma_s'(N) \equiv \sum_{k=1}^{N}\sin\frac{4\pi}{N}k\ .$$

Due to this difference, the first sum does not vanish not only for $N = 1$, but also for $N = 2$:

$$\Sigma_c'(1) = \cos 4\pi = 1, \qquad \Sigma_c'(2) = \cos 2\pi + \cos 4\pi = 2,$$

so that the electric-quadrupole radiation is emitted not only by a single rotating charge (for which it is much weaker than the dominant electric-dipole radiation), but also for a pair of equal charges on the opposite ends of the same circle's diameter (which does not emit the electric-dipole radiation).

Completing the calculation of the electric-quadrupole radiation for the latter case, using the above expression for the tensor $\mathscr{Q}$, together with Eqs. (8.143) and (8.147), is a simple but useful additional exercise, highly recommended to the reader.

***Problem* 8.19.** What general statements can you made about:

(i) the electric dipole radiation, and
(ii) the magnetic dipole radiation,

due to a collision of an arbitrary number of similar classical, non-relativistic particles?

Solutions:

(i) According to Eq. (8.26) of the lecture notes, the electric dipole radiation power is proportional to the square of the second time derivative of the net electric dipole moment of the system,

$$\mathbf{p} \equiv \sum_k q_k \mathbf{r}_k.$$

But for similar particles, with equal masses and charges, the moment is proportional to the total linear momentum of the system:

$$\mathbf{p} = q\sum_k \mathbf{r}_k = \frac{q}{m}\sum_k m\mathbf{r}_k.$$

Classical mechanics says[36] that the momentum cannot change in time as a result of internal interactions between the components of the system, i.e. $d\mathbf{p}/dt = 0$, and hence and $d^2\mathbf{p}/dt^2 = 0$ as well, so that the collision cannot induce any electric dipole radiation.

(ii) Very similarly, according to Eq. (8.139), the magnetic dipole radiation power is proportional to the square of the second derivative of the net magnetic dipole moment $\mathbf{m}$ of the system. But according to Eq. (5.95), for a system of particles with the same masses and electric charges, $\mathbf{m}$ is proportional to the angular momentum $\mathbf{L}$, and according to classical mechanics[37], cannot change in time as a result of internal

[36] See, e.g. *Part CM*, Eq. (1.10).
[37] See, e.g. *Part CM*, Eq. (1.38).

interactions, i.e. $d\mathbf{m}/dt = 0$, and hence $d^2\mathbf{m}/dt^2 = 0$, so that the collision cannot induce any magnetic dipole radiation either.

Note that at such conditions, the electric quadrupole radiation can still take place—see, e.g. the two-particle example given in the end of section 8.9.

***Problem* 8.20.** Calculate the angular distribution and the total power radiated by a small round loop antenna with radius R, fed by ac current $I(t)$ with frequency ω and amplitude I_0, into the free space.

Solution: If the antenna's radius is sufficiently small, $R \ll \lambda \equiv 2\pi/k = 2\pi c/\omega$, the multipole expansion of loop's magnetic field is dominated by the effect of its dipole moment $\mathbf{m}(t)$, directed normally to the loop area, with the magnitude given by Eq. (5.97):

$$m(t) = \pi R^2 I(t) = \pi R^2 I_0 \cos(\omega t + \varphi_0), \qquad \text{so that}$$
$$\ddot{m}(t) = \pi R^2 \ddot{I}(t) = -\pi\omega^2 R^2 I_0 \cos(\omega t + \varphi_0),$$

where φ_0 is a constant. Hence the angular distribution of the instant power of radiation into free space may be calculated using Eq. (8.139) of the lecture notes, with $Z = Z_0$ and $v = c$:

$$\frac{d\mathscr{P}}{d\Omega} = r^2 S_r = \frac{Z_0}{(4\pi c^2)^2}\ddot{m}^2(t)\sin^2\Theta = \frac{Z_0}{(4\pi c^2)^2}(\pi\omega^2 R^2 I_0)^2\cos^2(\omega t + \varphi_0)\sin^2\Theta,$$

where Θ is the angle between the direction of vector $\mathbf{m}(t)$ (i.e. the symmetry axis of the antenna loop) and the direction toward the observation point.

Averaging this expression over time, we obtain the average power per unit solid angle:

$$\frac{d\overline{\mathscr{P}}}{d\Omega} = r^2 \overline{S_r} = \frac{1}{2}\frac{Z_0}{(4\pi c^2)^2}(\pi\omega^2 R^2 I_0)^2\sin^2\Theta \equiv \frac{1}{32}Z_0 I_0^2 (kR)^4 \sin^2\Theta,$$

and an easy integration (similar to that carried out for the electric dipole radiation in section 8.2) yields the total average power

$$\overline{\mathscr{P}} = \frac{1}{32}Z_0 I_0^2 (kR)^4 \oint_{4\pi} \sin^2\Theta d\Omega = \frac{1}{32}Z_0 I_0^2 (kR)^4 \frac{8\pi}{3} \equiv \frac{\pi}{12}Z_0 I_0^2 (kR)^4.$$

The power grows very fast with the loop's radius R, until it becomes comparable with the wavelength $\lambda = 2\pi/k = 2\pi c/\omega$ of the emitted radiation. (The analysis of the radiation at $R \sim \lambda$ is much more difficult, because even the assumption of the ac current constancy along the loop's length becomes unrealistic.)

***Problem* 8.21.** The orientation of a magnetic dipole $\mathbf{m}$, of a fixed magnitude, is rotating about a certain axis with angular velocity ω, with the angle α between them staying constant. Calculate the angular distribution and the average power of its radiation (into free space).

Solution: The rotating dipole moment $\mathbf{m}(t)$ may be represented by a sum of its two components: a time-independent component $m_z = m\cos\alpha$ along the rotation axis, and the perpendicular component

$$\mathbf{m}_\perp(t) = m \sin\alpha\left(\mathbf{n}_x \cos\varphi + \mathbf{n}_y \sin\varphi\right), \qquad \text{with } \varphi \equiv \omega t + \text{const},$$

rotating about this axis—see the figure below, so that

$$\ddot{\mathbf{m}}(t) = \ddot{\mathbf{m}}_\perp(t) = -m\omega^2 \sin\alpha\left(\mathbf{n}_x \cos\varphi + \mathbf{n}_y \sin\varphi\right).$$

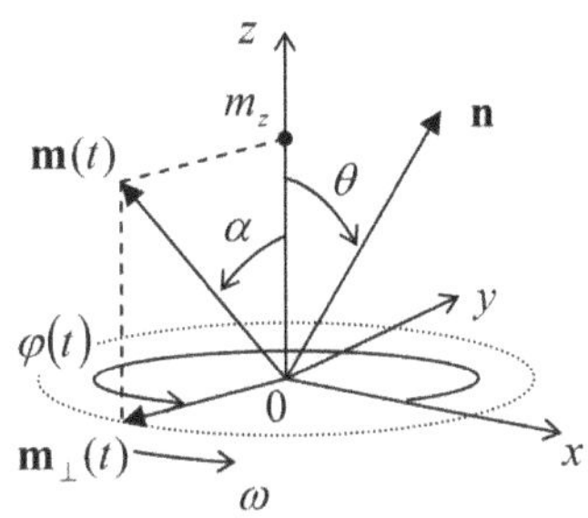

This expression is similar to the formula for $\ddot{\mathbf{p}}(t)$ used in the model solution of problem 8.1, with the replacement

$$eR \rightarrow -m \sin\varphi.$$

Now we may use the fact that Eq. (8.139) for the magnetic dipole radiation is similar to Eq. (8.26) for the electric dipole radiation, with the following replacement:

$$\ddot{\mathbf{p}} \rightarrow \frac{1}{c}\ddot{\mathbf{m}}.$$

Hence, the angular distribution of the average radiation power in these two problems is also similar:

$$S_r \propto \cos^2\theta + \frac{1}{2}\sin^2\theta,$$

where θ is the angle between the axis of moment rotation and the direction to the observation point—see the figure above. Performing the above replacements in the expression for the total average power obtained in the solution of problem 8.1,

$$\overline{\mathscr{P}}_{\text{electric}} = \frac{Z_0\omega^4 e^2 R^2}{6\pi c^2},$$

we obtain

$$\overline{\mathscr{P}}_{\text{magnetic}} = \frac{Z_0\omega^4 m^2 \sin^2\alpha}{6\pi c^4}. \qquad (*)$$

Note that such rotation of the magnetic moment may be caused, in particular, by a constant external magnetic field $\mathbf{B}_{ext}$ parallel to axis z, due to the torque $\boldsymbol{\tau} = \mathbf{m} \times \mathbf{B}_{ext}$ it exerts—see Eq. (5.101), and a discussion of the torque-induced precession[38]. In this case, the radiation power calculated above leads to a gradual reduction of the potential energy (5.100), $U = -\mathbf{m} \cdot \mathbf{B}_{ext} = -mB_{ext}\cos\alpha$ of the dipole–field interaction, via a gradual increase of the angle α, eventually leading to dipole's alignment with the field ($\alpha \to \pi$, $\sin\alpha \to 0$) and, according to Eq. (*), the corresponding expiration of radiation's emission.

***Problem* 8.22.** Solve problem 8.8 (also in the low-frequency limit $kR \ll 1$), for the case when the sphere's material has a frequency-independent Ohmic conductivity σ, and $\varepsilon_{opt} = \varepsilon_0$, in two limits:

(i) of a very large skin depth ($\delta_s \gg R$), and
(ii) of a very small skin depth ($\delta_s \ll R$).

Solutions:

(i) In the limit $\delta_s \gg R$, the eddy currents, which are responsible for the skin effect, may be neglected—and hence their magnetic moment as well. For the electric dipole moment, we may use Eq. (3.64) of the lecture notes with the replacements $\mathbf{p} \to \mathbf{p}_\omega$, $\mathbf{E}_0 \to \mathbf{E}_\omega$, and $\kappa \to \varepsilon(\omega)/\varepsilon_0$:

$$\mathbf{p}_\omega = 4\pi\varepsilon_0 R^3 \frac{\varepsilon(\omega)/\varepsilon_0 - 1}{\varepsilon(\omega)/\varepsilon_0 + 2}\mathbf{E}_\omega.$$

For an Ohmic conductor, the complex electric permittivity $\varepsilon(\omega)$ may be calculated using Eq. (7.46), so that with $\varepsilon_{opt} = \varepsilon_0$, we obtain[39]

$$\mathbf{p}_\omega = 4\pi\varepsilon_0 R^3 \frac{(\varepsilon_0 + i\sigma/\omega)/\varepsilon_0 - 1}{(\varepsilon_0 + i\sigma/\omega)/\varepsilon_0 + 2}\mathbf{E}_\omega \equiv 4\pi\varepsilon_0 R^3 \frac{1}{1 - 3i\omega\tau_r}\mathbf{E}_\omega,$$

where $\tau_r \equiv \varepsilon_0/\sigma$ is the charge relaxation time constant of the material—see Eq. (4.10) of the lecture notes[40]. Now Eqs. (8.28) and (8.39) of the lecture notes yield

$$\sigma = \frac{8\pi}{3}(kR)^4 \left| \frac{1}{1 - 3i\omega\tau_r} \right|^2 R^2$$
$$\equiv \frac{8}{3}\frac{(kR)^4}{1 + 9\omega^2\tau_r^2}\sigma_0, \qquad \text{where } \sigma_0 \equiv \pi R^2.$$

This result shows that if $\omega\tau_r \ll 1$, i.e. if the conductivity of sphere's material is relatively high, the full cross-section of scattering is the same as σ_{max} in the solution

[38] See, e.g. *Part CM* section 4.5.
[39] See also the solution of problem 8.11.
[40] The Ohmic conductivity σ participating in the above expression should not be confused with the full cross-section of scattering. Let me hope that these notions may be readily distinguished from the context.

of problem 8.8. The cross-section first grows fast (as ω^4) with the wave's frequency, but as ω reaches the reciprocal relaxation time, this increase is tempered by the factor ω^2 in the denominator. (Note that the imposed conditions $R \ll 1/k$, δ_s make this crossover reachable only if $R \ll 1/Z_0\sigma$; for a good conductor like copper, this condition requires atomic dimensions.)

(ii) In this limit, the skin effect does not allow the electromagnetic wave's fields to penetrate into the sphere's bulk. As we know from the discussion in sections 3.4 and 5.6 of the lecture notes (which are applicable to our current problem because of the low-frequency condition $kR \ll 1$), in this case the electric and magnetic fields induced by the sphere outside it coincide exactly with those of the point electric and magnetic dipoles. Their moments are given, respectively, by Eq. (3.64) with $\kappa \to \infty$,

$$\mathbf{p} = 4\pi\varepsilon_0 R^3\mathbf{E},$$

and Eq. (5.125) with $\mu \to 0$:

$$\mathbf{m} = -2\pi R^3\mathbf{H},$$

where **E** and **H** are the unperturbed fields of the incident wave. Since in a traveling wave, these fields change in time synchronously (see Eqs. (7.6)–(7.8) of the lecture notes),

$$\mathbf{E}(t) = -Z_0\mathbf{n}_0 \times \mathbf{H}(t),$$

so do the induced electric and magnetic moments, and their radiation fields—see Eqs. (8.24) and (8.137), both with $\varepsilon = \varepsilon_0$, $\mu = \mu_0$, and hence $v = c$:

$$\begin{aligned}\mathbf{B}_e = -\frac{\mu_0}{4\pi rc}\mathbf{n} \times \ddot{\mathbf{p}} &= -\frac{\mu_0}{4\pi rc}4\pi\varepsilon_0 R^3\mathbf{n} \times \ddot{\mathbf{E}} \\ &= \frac{\mu_0}{4\pi rc}4\pi\varepsilon_0 R^3 Z_0\mathbf{n} \times (\mathbf{n}_0 \times \ddot{\mathbf{H}}) \\ &\equiv \frac{\mu_0 R^3}{c^2 r}\mathbf{n} \times (\mathbf{n}_0 \times \ddot{\mathbf{H}}),\end{aligned}$$

$$\begin{aligned}\mathbf{B}_m = \frac{\mu_0}{4\pi rc^2}\mathbf{n} \times (\mathbf{n} \times \ddot{\mathbf{m}}) &= -\frac{\mu_0}{4\pi rc^2}2\pi R^3\mathbf{n} \times (\mathbf{n} \times \ddot{\mathbf{H}}) \\ &\equiv -\frac{1}{2}\frac{\mu_0 R^3}{c^2 r}\mathbf{n} \times (\mathbf{n} \times \ddot{\mathbf{H}}).\end{aligned}$$

(Here $\mathbf{n}_0 \equiv \mathbf{k}_0/k$ and $\mathbf{n} \equiv \mathbf{k}/k$ are the unit vectors directed along the propagation directions of, respectively, the incident and scattered waves, and the retarded time is implied.) These expressions show that the induced fields are not only coherent, but also comparable in magnitude, so that in order to calculate the radiation power correctly, we have to add up the fields, rather than their partial powers:

$$\mathbf{B} = \frac{\mu_0 R^3}{c^2 r}\left[\mathbf{n} \times (\mathbf{n}_0 \times \ddot{\mathbf{H}}) - \frac{1}{2}\mathbf{n} \times (\mathbf{n} \times \ddot{\mathbf{H}})\right].$$

To evaluate the expression in the square brackets, let us use the coordinate system shown in the figure below, so that the unit vector $\mathbf{n}_0$ is directed along the axis z, the vector $\ddot{\mathbf{H}}$(normal to $\mathbf{n}_0$) is within the [x, y] plane, and the vector $\mathbf{n}$ is within the [x, z] plane. In these coordinates, each of the vectors has one zero component,

$$\mathbf{n} = \{\sin\theta,\ 0,\ \cos\theta\}, \qquad \mathbf{n}_0 = \{0,\ 0,\ 1\}, \qquad \ddot{\mathbf{H}} = \ddot{H}\ \{\cos\varphi,\ \sin\varphi,\ 0\},$$

so that the result of the (straightforward) calculation of the double vector products not too cumbersome:

$$\mathbf{n}\times(\mathbf{n}_0\times\ddot{\mathbf{H}}) - \frac{1}{2}\mathbf{n}\times(\mathbf{n}\times\ddot{\mathbf{H}}) = \ddot{H}\left\{-\cos\theta\cos\varphi\left(1-\frac{1}{2}\cos\theta\right),\right.$$

$$\left.\sin\varphi\left(\frac{1}{2}-\cos\theta\right),\quad \sin\theta\cos\varphi\left(1-\frac{1}{2}\cos\theta\right)\right\},$$

giving

$$\left[\mathbf{n}\times(\mathbf{n}_0\times\ddot{\mathbf{H}}) - \frac{1}{2}\mathbf{n}\times(\mathbf{n}\times\ddot{\mathbf{H}})\right]^2 = \ddot{H}^2\left[\left(1-\frac{1}{2}\cos\theta\right)^2(\cos^2\theta\cos^2\varphi + \sin^2\theta\sin^2\varphi)\right.$$

$$\left.+\left(\frac{1}{2}-\cos\theta\right)^2\sin^2\varphi\right].$$

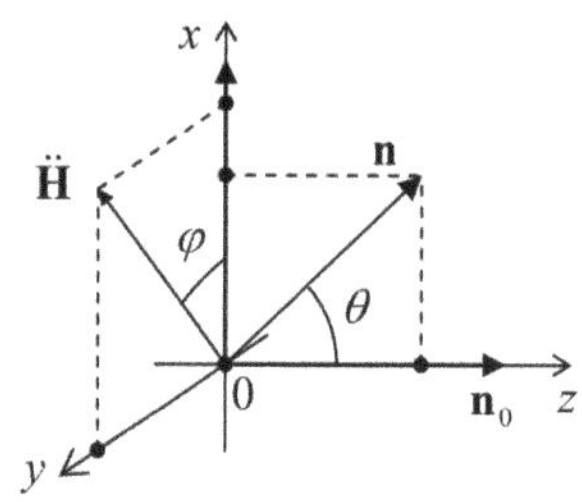

Now we may use Eq. (7.9b) to calculate the instant total power of the radiation:

$$\mathscr{P} = r^2\oint_{4\pi} S_r d\Omega = Z_0 r^2\oint_{4\pi}\left(\frac{B}{\mu_0}\right)^2 d\Omega$$

$$= \frac{Z_0 R^6}{c^4}\oint_{4\pi}\left[\mathbf{n}\times(\mathbf{n}_0\times\ddot{\mathbf{H}}) - \frac{1}{2}\mathbf{n}\times(\mathbf{n}\times\ddot{\mathbf{H}})\right]^2 d\Omega.$$

Here the integration is nominally over all directions of the vector $\mathbf{n}$, at a fixed direction of the vector $\ddot{\mathbf{H}}$, but due to the spherical symmetry of the scatterer, it is equivalent to the usual integration over the spherical angles θ and φ (see the figure above):

$$\mathscr{P} = \frac{Z_0 R^6}{c^4}\ddot{H}^2 \int_0^{\pi} \sin\theta d\theta \int_0^{2\pi} d\varphi \left[\left(1 - \frac{1}{2}\cos\theta\right)^2 (\cos^2\theta\cos^2\varphi + \sin^2\theta\sin^2\varphi) + \left(\frac{1}{2} - \cos\theta\right)^2 \sin^2\varphi \right]$$

$$= \frac{Z_0 R^6}{c^4}\ddot{H}^2 \pi \int_0^{\pi} \left[\left(1 - \frac{1}{2}\cos\theta\right)^2 + \left(\frac{1}{2} - \cos\theta\right)^2 \right] \sin\theta d\theta$$

$$= \frac{Z_0 R^6}{c^4}\ddot{H}^2 \pi \frac{4}{3}.$$

Finally, averaging this power over time (using the sinusoidal time dependence of the incident field's H), so that

$$\overline{\ddot{H}^2} = \frac{\omega^4}{2}|H_\omega|^2,$$

and using the definition Eq. (8.39) of the full cross-section (with $E_\omega = H_\omega Z_0$), we obtain

$$\sigma = \frac{4\pi}{3}\frac{R^6}{c^4}\omega^4 = \frac{4\pi}{3}k^4 R^6 \equiv \frac{4}{3}(kR)^4\sigma_0, \qquad \text{where } \sigma_0 \equiv \pi R^2.$$

This result should be compared with that of problem 8.8 for a non-conducting, dielectric sphere, in the corresponding limit $\kappa \to \infty$ (when the sphere's interior is also fully screened from the penetration of the incident field):

$$\sigma = \frac{8}{3}(kR)^4\sigma_0.$$

The comparison shows that in contrast to electrostatics, where a conductor may be adequately described as a dielectric with an infinite dielectric constant, in dynamics the ac eddy currents flowing on the surface interfere with scattering, giving (for a spherical geometry) a 50% decrease of the full cross-section, and also changing the angular distribution of the scattered wave.

***Problem* 8.23.** Complete the solution of the problem started in section 8.9 of the lecture notes, by calculating the full power of radiation of the system of two charges oscillating in antiphase along the same straight line—see figure 8.16, partly reproduced below. Also, calculate the average radiation power for the particular case of harmonic oscillations, $d(t) = a\cos\omega t$, compare it with the case of a single charge performing similar oscillations, and interpret the difference.

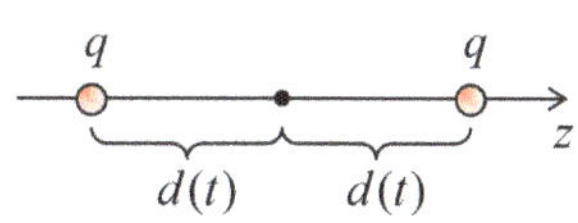

Solution: According to Eq. (8.147), the full power of the electric-quadrupole radiation into the free space is

$$\mathscr{P} = \frac{Z_0}{1440\pi c^4}\sum_{j,j'=1}^{3}\left(\dddot{\mathscr{Q}}_{jj'}\right)^2.$$

Using Eqs. (8.144) for the components of the quadrupole tensor, we obtain

$$\mathscr{P} = \frac{Z_0}{1440\pi c^4}\left\{2\left[\frac{d^3}{dt^3}(-2qd^2)\right]^2 + \left[\frac{d^3}{dt^3}(4qd^2)\right]^2\right\} = \frac{Z_0q^2}{60\pi c^4}\left[\frac{d^3}{dt^3}d^2(t)\right]^2. \qquad (*)$$

For the particular case of harmonic oscillations, $d(t) = a\cos\omega t$, the time derivative is

$$\frac{d^3}{dt^3}d^2(t) = \frac{a^2}{2}\frac{d^3}{dt^3}(1 + \cos 2\omega t) = 4a^2\omega^3 \sin 2\omega t, \qquad (**)$$

so that the average radiation power is

$$\overline{\mathscr{P}} = \frac{Z_0q^2}{60\pi c^4}(4a^2\omega^3)^2\,\overline{\sin^2 2\omega t} = \frac{2Z_0q^2a^4\omega^6}{15\pi c^4}.$$

For the similar motion of a single charge q, the electric dipole radiation dominates. For it, we may use Eq. (8.29) of the lecture notes:

$$\overline{\mathscr{P}_1} = \frac{Z_0q^2a^2\omega^4}{12\pi c^2}.$$

The ratio of these two powers is

$$\frac{\overline{\mathscr{P}}}{\overline{\mathscr{P}_1}} = \frac{8}{5}\left(\frac{a\omega}{c}\right)^2.$$

The fraction in the parentheses is just the ratio $v_{\max}/c$, where $v_{\max}$ is the amplitude of velocity oscillations. All expressions of chapter 8, starting from section 8.2, and hence the above results, are only valid for non-relativistic particles moving with velocity $v \ll c$. In this case, the calculated electric-quadrupole radiation power is much smaller than the electric-dipole radiation power.

This difference is natural, because the two charges moving coherently in antiphase kill the main dipole components of each other's radiation, and the quadrupole radiation is just what remains of the total radiation after this cancellation. Note that according to Eq. (**) the quadrupole system of two charges radiates at frequency 2ω, while the potential dipole radiation of each charge at the basic frequency ω is completely cancelled.

Chapter 9

Special relativity

***Problem* 9.1.** Use the non-relativistic picture of the Doppler effect, in which light propagates with velocity c in a Sun-bound aether, to derive Eq. (9.4) of the lecture notes.

Solution: As observed in the interferometer-bound reference frame, the light in its longitudinal arm (i.e. the arm directed along the aether's velocity v) of length l_l moves with speed $(c - v)$ in one direction and speed $(c + v)$ in the opposite direction, so that the roundtrip travel time is

$$t_l = \frac{l_l}{c - v} + \frac{l_l}{c + v} = \frac{2l_l}{c}\frac{1}{1 - v^2/c^2}.$$

The calculation of the roundtrip time t_t in the transverse arm, of length l_t, is easier in the aether-bound reference frame. In this frame, during a time interval Δt, the interferometer passes distance $-v\Delta t$ in the longitudinal direction, so that the total length of the light's roundtrip (see the figure below) is

$$l = 2\left[l_t^2 + \left(\frac{vt_t}{2}\right)\right]^{1/2}.$$

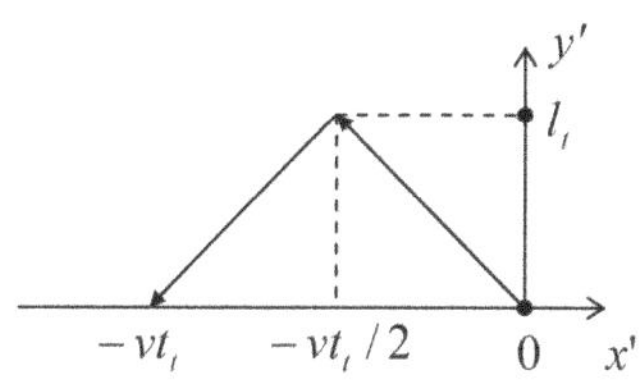

In this frame, by the aether's definition, the light's velocity equals c, so we have to require that $l = ct_t$. Solving the resulting equation for t_t, we obtain

$$t_t = \frac{2l_t}{c}\frac{1}{(1 - v^2/c^2)^{1/2}}.$$

doi:10.1088/978-0-7503-1407-7ch9

For the roundtrip time difference $\Delta t \equiv t_l - t_t$, these expressions yield the first form of Eq. (9.4):

$$\Delta t = \frac{2}{c}\left(\frac{l_l}{1 - v^2/c^2} - \frac{l_t}{(1 - v^2/c^2)^{1/2}}\right).$$

Now Taylor-expanding both fractions on the right-hand side of this relation in small parameter $(v/c)^2 \sim 10^{-8}$, in the linear approximation we obtain

$$\Delta t \approx \frac{2}{c}\left[l_l\left(1 + \frac{v^2}{c^2}\right) - l_t\left(1 + \frac{v^2}{2c^2}\right)\right] \equiv \frac{2(l_l - l_t)}{c} + \frac{(2l_l - l_t)}{c}\left(\frac{v}{c}\right)^2.$$

Although at $l_l \neq l_t$ the first term may be much larger than the second one, it does not change as a result of the interferometer's rotation relative to the aether's motion direction, and may be readily filtered out from the data. In contrast, the second term, which does not vanish even at $l_l = l_t$, describes the effect we are looking for. For a nearly symmetric instrument, with $l_l = l_t \equiv l$, it yields the second form of Eq. (9.4):

$$\Delta t \approx \frac{l}{c}\left(\frac{v}{c}\right)^2.$$

***Problem* 9.2.** Show that two successive Lorentz space/time transforms in the same direction, with velocities u' and v, are equivalent to a single transform, with the velocity u given by Eq. (9.25) of the lecture notes.

Solution: Let us denote the normalized velocity u'/c as β_1, and v/c as β_2, with the corresponding indices of the Lorentz factor γ —see Eq. (9.17). Then the first transform, given by Eq. (9.19b), yields

$$x_1 = \gamma_1(x - \beta_1 ct), \qquad ct_1 = \gamma_1(ct - \beta_1 x),$$

while for the second transform we obtain (from the same relation):

$$\begin{aligned} x_2 &= \gamma_2(x_1 - \beta_2 ct_1) = \gamma_1\gamma_2\left[(x - \beta_1 ct) - \beta_2(ct - \beta_1 x)\right] \\ &\equiv \gamma_1\gamma_2\left[(1 + \beta_1\beta_2)x - (\beta_1 + \beta_2)ct\right], \\ ct_2 &= \gamma_2(ct_1 - \beta_2 x_1) = \gamma_1\gamma_2\left[(ct - \beta_1 x) - \beta_2(x - \beta_1 ct)\right] \\ &\equiv \gamma_1\gamma_2\left[(1 + \beta_1\beta_2)ct - (\beta_1 + \beta_2)x\right]. \end{aligned} \qquad (*)$$

Now let us spell out the product $\gamma_1\gamma_2$ involved in this result:

$$\begin{aligned} \gamma_1\gamma_2 &= \frac{1}{\left(1 - \beta_1^2\right)^{1/2}}\frac{1}{\left(1 - \beta_2^2\right)^{1/2}} = \frac{1}{\left[\left(1 - \beta_1^2\right)\left(1 - \beta_2^2\right)\right]^{1/2}} \\ &= \frac{1}{\left(1 + \beta_1^2\beta_2^2 - \beta_1^2 - \beta_2^2\right)^{1/2}}. \end{aligned} \qquad (**)$$

On the other hand, in our notation, Eq. (9.25) takes the form

$$\beta = \frac{\beta_1 + \beta_2}{1 + \beta_1\beta_2}, .$$

so that the corresponding Lorentz factor is[1]

$$\gamma = \frac{1}{(1-\beta^2)^{1/2}} = \frac{1}{\left[1 - \dfrac{(\beta_1+\beta_2)^2}{(1+\beta_1\beta_2)^2}\right]^{1/2}} \equiv \frac{1+\beta_1\beta_2}{\left(1 + \beta_1^2\beta_2^2 - \beta_1^2 - \beta_2^2\right)^{1/2}},$$

and using Eq. (**), we may reduce the last relation to

$$\gamma = \gamma_1\gamma_2(1 + \beta_1\beta_2).$$

As a result, the direct transform from $\{x, t\}$ to $\{x_2, t_2\}$, using these values of β and γ, is

$$x_2 = \gamma(x - \beta ct) = \gamma_1\gamma_2(1+\beta_1\beta_2)\left(x - \frac{\beta_1+\beta_2}{1+\beta_1\beta_2}ct\right),$$

$$ct_1 = \gamma(ct - \beta x) = \gamma_1\gamma_2(1+\beta_1\beta_2)\left(ct - \frac{\beta_1+\beta_2}{1+\beta_1\beta_2}x\right),$$

giving the same results as Eqs. (*) following from two sequential transforms.

***Problem* 9.3.** $N + 1$ reference frames, numbered by index n (taking values 0, 1, …, N), move in the same direction as a particle. Express the particle's velocity in the frame number 0 via its velocity u_N in the frame number N, and the set of velocities v_n of the frame number n relative to the frame number $(n - 1)$.

Solution: According Eq. (9.25) of the lecture notes, applied to reference frames number $(n - 1)$ and number n,

$$u_{n-1} = \frac{u_n + v_n}{1 + u_n v_n/c^2}, \quad \text{i.e.} \quad \beta_{n-1} = \frac{\beta_n + \mathrm{\beta}_{\mathrm{n}}}{1 + \beta_n \mathrm{\beta}_{\mathrm{n}}}, \quad \text{where } \beta_n \equiv \frac{u_n}{c}, \quad \text{while } \mathrm{\beta}_{\mathrm{n}} = \frac{v_n}{c}, \qquad (*)$$

where u_n is the velocity of the particle, measured in the nth reference frame. A sequential application of N such formulas, starting from $n = N$ all the way down to $n = 1$, would give us the required answer for u_0, but at large N the formulas would be extremely bulky. However, using the definition of rapidity φ,

$$\tanh\varphi \equiv \beta, \qquad (**)$$

[1] Note that γ is very different from $\gamma_1\gamma_2$, so that the length contraction and time dilation formulas (9.20) and (9.21) cannot be applied sequentially! (Explain why.)

we may rewrite Eq. (*) as

$$\tanh \varphi_{n-1} = \frac{\tanh \varphi_n + \tanh \phi_n}{1 + \tanh \varphi_n \tanh \phi_n}, \quad \text{where } \tanh \varphi_n \equiv \beta_n \equiv \frac{u_n}{c}, \quad \tanh \phi_n \equiv \beta_n \equiv \frac{v_n}{c}.$$

But the first of these relations is just the algebraic identity for $\tanh(\varphi_n + \phi_n)$, so that we may write simply

$$\varphi_{n-1} = \varphi_n + \phi_n.$$

(This is why the notion of rapidity is so useful; these parameters add up as non-relativistic velocities even in the relativistic case.) Now applying this formula N times, starting from $n = N$, we obtain

$$\begin{aligned}
&\varphi_{N-1} = \varphi_N + \phi_N, \\
&\varphi_{N-2} = \varphi_{N-1} + \phi_{N-1} = (\varphi_N + \phi_N) + \phi_{N-1} \equiv \varphi_N + (\phi_N + \phi_{N-1}), \\
&\ldots \\
&\varphi_0 = \varphi_N + (\phi_N + \phi_{N-1} + \ldots + \phi_1) \equiv \varphi_N + \sum_{n-1}^{N} \phi_n .
\end{aligned}$$

What remains is to use Eq. (**) to rewrite the result for φ_0 in terms of the usual velocities:

$$u_0 = c \tanh\left(\tanh^{-1}\frac{u_N}{c} + \sum_{n=1}^{N} \tanh^{-1}\frac{v_n}{c}\right).$$

For relatively low (non-relativistic) velocities we may use the fact that both involved functions, tanh and $\tanh^{-1}$, are well approximated by their arguments, so that our general result is reduced to the evident equality

$$u_0 \approx u_N + \sum_{n=1}^{N} v_n.$$

In the opposite limit, when one or several of the involved velocities approach c, so that the corresponding $\tanh^{-1}$ functions become very large, the external tanh tends to 1, so that we obtain the natural answer

$$u_0 \to c.$$

***Problem* 9.4.** A spaceship, moving with constant velocity v directly from the Earth, sends back brief flashes of light with a period Δt_S—as measured by the spaceship's clock. Calculate the period with which Earth's observers may receive the signals—as measured by the Earth's clock.

Solution: Due to the time dilation effect, described by Eq. (9.21), the period between the light flash emitting events, as measured by the Earth's clocks, is $\Delta t_E = \gamma \Delta t_S$, where $\gamma \equiv (1 - v^2/c^2)^{1/2} \geqslant 1$. During such a time interval, the spaceship moves farther from the Earth by the space interval $\Delta x = v\Delta t_E = \gamma v \Delta t_S$—again, as measured in

Earth's reference frame. Now, by the Earth's clock again, it will take the signal, propagating with the speed of light, an additional time $\Delta t_A = \Delta x/c = \gamma v \Delta t_S/c = \beta\gamma\Delta t_S$ (where $\beta \equiv v/c$) to return to the point where the previous flash has been emitted, so that the total interval between the received signals is

$$\Delta t_R = \Delta t_E + \Delta t_A = \Delta t_E + \beta\Delta t_E \equiv (1+\beta)\Delta t_E = (1+\beta)\gamma\Delta t_S \equiv \left(\frac{1+\beta}{1-\beta}\right)^{1/2}\Delta t_S.$$

Note that the ratio $\Delta t_R/\Delta t_S$ is expressed by the fraction reciprocal to that describing the Doppler effect for frequencies—see Eq. (9.44) of the lecture notes—with the relevant (upper) sign, corresponding to the motion of the wave source. This is natural, because $1/\Delta t$ is just the (cyclic) frequency of the periodic flashes. (In such a dispersion-free 'medium' as free space, the group velocity of signal propagation and the phase velocity of phase propagation, coincide.) In the non-relativistic limit, $v \ll c$, the ratio $\Delta t_R/\Delta t_S$ approaches $(1+\beta) \to 1$, while at $v \to c$, the ratio grows as $\sim 2\gamma \to \infty$.

***Problem* 9.5.** From the point of view of observers in a 'moving' reference frame 0′, a straight thin rod, parallel to axis x', is moving, without rotation, with a constant velocity $\mathbf{u}'$ directed along axis y'. The reference frame 0′ is itself moving relative to another ('lab') reference frame 0 with a constant velocity $\mathbf{v}$ along axis x, also without rotation—see the figure below. Calculate:

(i) the direction of the rod's velocity, and
(ii) the orientation of the rod on the $[x, y]$ plane,

both as observed from the lab reference frame. Is the velocity, in this frame, perpendicular to the rod?

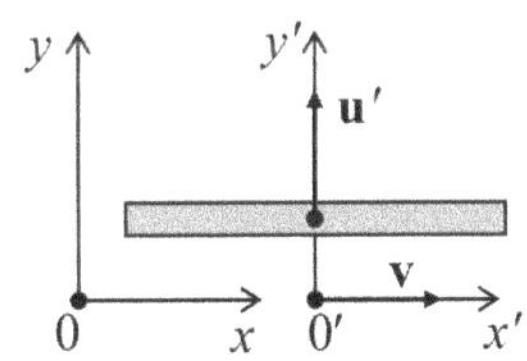

Solutions:

(i) To calculate the components of the rod's velocity $\mathbf{u}$ in the frame 0, we may use Eqs. (9.23) of the lecture notes, for the particular case $u'_x = 0$ and $u'_y = u'$. The result is

$$u_x = v, \qquad u_y = \frac{u'}{\gamma}, \qquad \text{where} \quad \gamma \equiv \frac{1}{(1 - v^2/c^2)^{1/2}},$$

so that the velocity's direction may be characterized by angle $\varphi_\mathbf{u}$ (see the figure below) with

$$\tan\varphi_\mathbf{u} \equiv \frac{u_y}{u_x} = \frac{u'}{\gamma v}. \qquad (*)$$

(ii) The simplest way to calculate the rod's orientation in the frame 0 is to select the space and time origins of both frames to coincide with each other, and the middle of the rod, at $t = t' = 0$. Then we may use Eqs. (9.19) of the lecture notes, with $\{x', y' = u't'\}$ being the Cartesian coordinates of another point of the rod, displaced by x' from its center, as measured in the 'moving' frame 0′. In order to find the rod's direction in the 'lab' frame 0, we need to calculate the coordinates $\{x, y\}$ of that point in this frame at a certain fixed 'lab' time t, for example $t = 0$. With this substitution, Eqs. (9.19a) take the form

$$x = \gamma(x' + \beta ct'), \qquad y = u't', \qquad 0 = ct' + \beta x', \quad \text{with} \quad \beta \equiv \frac{v}{c},$$

so that the elementary substitution of the last result, $t' = -\beta x'/c \equiv -vx'$, into the first two relations yields

$$x = \gamma(x' - \beta^2 x') \equiv \frac{x'}{\gamma}, \qquad y = \frac{u'}{c}ct' = -\frac{u'}{c}\beta x'.$$

so that from here, for the angle describing the direction of the vector $\mathbf{n}$ normal to the rod (see the figure above) we obtain

$$\tan\varphi_\mathbf{n} \equiv -\frac{x}{y} = \frac{c}{\gamma\beta u'} \equiv \frac{c^2}{\gamma u'v}. \qquad (**)$$

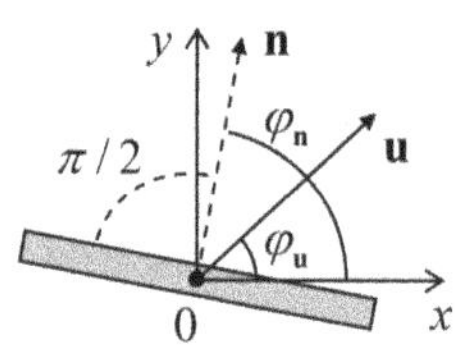

A comparison of Eqs. (*) and (**) shows that generally the angles $\varphi_\mathbf{u}$ and $\varphi_\mathbf{n}$ are rather different, even for non-relativistic velocities, and coincide (so that the directions of the vectors **u** and **n** are the same) only asymptotically, in the limits when either $v \to 0$ (so that $\tan\varphi_\mathbf{u} \to \tan\varphi_\mathbf{n} \to \infty$, and hence $\varphi_\mathbf{u} \to \varphi_\mathbf{n} \to \pi/2$), or $u' \to c$ (so that $\tan\varphi_\mathbf{u} \to \tan\varphi_\mathbf{n} \to c/\gamma v \equiv 1/\beta\gamma$).

***Problem* 9.6.** Starting from the rest at $t = 0$, a spaceship moves directly from the Earth, with a constant acceleration as measured in its instantaneous rest frame. Find its displacement $x(t)$ from the Earth as measured from the Earth's reference frame, and interpret the result.

Hint: The instantaneous rest frame of a moving particle is the inertial reference frame that, at the considered moment of time, has the same velocity as the particle.

Solution: Let us apply Eq. (9.25) of the lecture notes,

$$u = \frac{u' + v}{1 + u'v/c^2}, \qquad (*)$$

to the velocity of the spaceship at moment $(t + dt)$, as measured, respectively, in the lab frame (u) and in the instantaneous rest frame of the spaceship (u'), whose velocity, as measured in the lab frame at moment t, is v. By this definition, at $dt = 0$, $v = u$, i.e. $u' = 0$, so that for a non-zero but small dt, $u' = du'$ is also small, and we may Taylor-expand both parts of Eq. (*) in small deviations proportional to dt, at constant t, i.e. at constant $v = u(t)$. Keeping only the two leading terms of the right-hand side's expansion,

$$u(t) + du = \frac{du' + u(t)}{1 + du'u(t)/c^2} \equiv u(t)\frac{1 + du'/u(t)}{1 + du'u(t)/c^2} \approx u(t)\left[1 + du'\left(\frac{1}{u(t)} - \frac{u(t)}{c^2}\right)\right]$$
$$\equiv u(t) + du'\left[1 - \frac{u^2(t)}{c^2}\right],$$

we obtain

$$du = du'\left[1 - \frac{u^2(t)}{c^2}\right] \equiv \frac{du'}{\gamma^2}, \qquad (**)$$

where γ is the Lorentz factor (9.17) of the spaceship at moment t. By definition, the acceleration a' measured in the instantaneous frame of the spaceship (frequently called the *proper acceleration*) is equal to $du'/d\tau$, where $d\tau = dt/\gamma$ is the proper time interval corresponding to dt—see Eq. (9.60) of the lecture notes. On the other hand, the acceleration a measured in the Earth's frame is evidently du/dt, so that Eq. (**) yields

$$a \equiv \frac{du}{dt} = \frac{1}{\gamma^2}\frac{du'}{dt} = \frac{1}{\gamma^2}\frac{du'}{\gamma d\tau} = \frac{1}{\gamma^3}a' \equiv \left(1 - \frac{u^2}{c^2}\right)^{3/2} a'.$$

Now we may integrate the resulting equation for the function $u(t)$,

$$\frac{du}{dt} = \left(1 - \frac{u^2}{c^2}\right)^{3/2} a',$$

with $a' = \text{const}$, over time, in a straightforward way:

$$\int \frac{d(u/c)}{(1 - u^2/c^2)^{3/2}} = \frac{a'}{c}\int dt + \text{const}.$$

Using the substitution $u/c \equiv \sin\xi$, so that $d(u/c) = d(\sin\xi) = \cos\xi\, d\xi$ and $(1 - u^2/c^2)^{3/2} = \cos^3\xi$, we obtain (with our initial condition $u(0) = 0$):

$$\int_0^{\sin^{-1}(u/c)} \frac{d\xi}{\cos^2\xi} \equiv \tan\left(\sin^{-1}\frac{u}{c}\right) = \frac{a't}{c},$$

giving us the following equation

$$u \equiv \frac{dx}{dt} = c\sin\left(\tan^{-1}\frac{a't}{c}\right)$$

for the distance $x(t)$ we are seeking. From here, with our initial condition, $x(0) = 0$, we obtain

$$x = c\int_0^t \sin\left(\tan^{-1}\frac{a't'}{c}\right)dt' \equiv \frac{c^2}{a'}\int_0^{a't/c} \sin(\tan^{-1}\xi)d\xi.$$

This integral may look intimidating, but in fact is rather simple. Indeed, introducing $\varphi \equiv \tan^{-1}\xi$, i.e. $\xi \equiv \tan\varphi$, and using the trigonometric identity $\sin\varphi = \tan\varphi/(1 + \tan^2\varphi)^{1/2}$, we obtain: $\sin(\tan^{-1}\xi) = \xi/(1 + \xi^2)^{1/2}$. Hence the integral is just

$$\int \sin(\tan^{-1}\xi)d\xi = \int\frac{\xi d\xi}{(1+\xi^2)^{1/2}} \equiv \frac{1}{2}\int\frac{d(1+\xi^2)}{(1+\xi^2)^{1/2}} = (1+\xi^2)^{1/2}.$$

Using this formula, we finally obtain

$$x(t) = \frac{c^2}{a'}\left\{\left[1 + \left(\frac{a't}{c}\right)^2\right]^{1/2} - 1\right\}.$$

This relation may be represented in a space–time symmetric form[2],

$$(x + x_0)^2 - (ct)^2 = \left(\frac{c^2}{a'}\right)^2, \quad \text{with} \quad x_0 \equiv \frac{c^2}{a'}, \qquad (***)$$

which describes, on the $[x,ct]$ plane, a hyperbola with ±45° asymptotes, offset by x_0 from the origin—see the sketch in the figure.

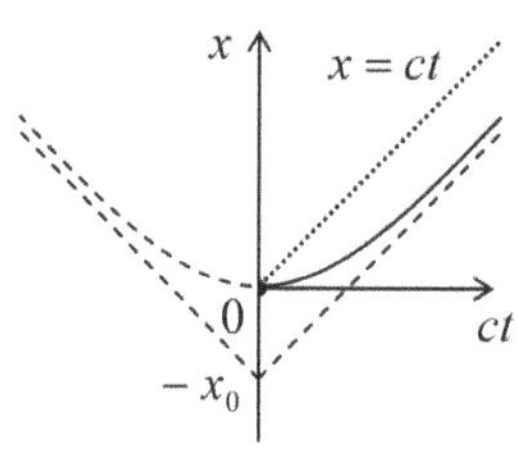

From this plot, x_0 may be clearly interpreted as the travel distance 'lost' (at a fixed, large travel time $t \gg x_0/c$) because of the final value of acceleration. For the Earth-surface acceleration $a' = g \approx 9.8$ m s^{-2} (creating perhaps the best balance between the comfort and health safety for a human astronaut :-), $x_0 = c^2/a' \approx 10^{16}$ m—almost

[2] This result is frequently derived in a different, less direct way, by first integrating Eq. (**), multiplied by $\gamma^2 \equiv 1/(1 - u^2/c^2)$, namely $du/(1 - u^2/c^2) = du' \equiv a'd\tau$, to obtain $u = (c/a')\sinh(a'\tau/c)$. Now this expression may be plugged into the definition of γ, giving $\gamma = \cosh(a'\tau/c)$. Finally, the expressions for u and γ may be plugged into the relation $dx/d\tau \equiv (dx/dt)(dt/d\tau) \equiv \gamma u$, giving an easy differential equation for x as a function of τ. Its solution yields $x = (c^2/a')\cosh(a'\tau/c) - x_0$, so that the identity $\cosh^2\xi - \sinh^2\xi = 1$, applied to $\xi = a'\tau/c$, coincides with Eq. (***). A useful by-product of this approach is the explicit relation between x and τ, which may be useful, in particular, for quantifying the twin paradox within a simple model of the roundtrip, with three constant-acceleration stages.

exactly 1 light-year. Hence, for hypothetical multi-light-year space missions, much higher accelerations would not create any substantial distance benefits, and just exacerbate the hardest problem of the interstellar travel—that of the initial mass of the rocket fuel, which is extremely challenging even for non-relativistic velocities.

***Problem* 9.7.** Calculate the first relativistic correction to the frequency of a harmonic oscillator as a function of its amplitude.

Solution: Combining the quadratic potential energy of the oscillator,

$$U = \frac{\kappa}{2}x^2 \equiv \frac{m\omega_0^2}{2}x^2,$$

with the Taylor expansion of its kinetic energy in small u^2/c^2, and keeping just two first terms (see Eq. (9.74) of the lecture notes), we would describe the usual harmonic oscillator, with frequency ω_0 independent of the oscillation amplitude A. In order to obtain the first relativistic correction to this result, we need to extend that expansion by one more term:

$$\frac{mc^2}{(1 - u^2/c^2)^{1/2}} \approx mc^2 + \frac{mu^2}{2} + \frac{3mu^4}{8c^2},$$

so that (ignoring the motion-independent term mc^2) we may approximate the full oscillator's energy $\mathscr{E}$ as

$$\mathscr{E} \approx \frac{mu^2}{2} + \frac{3mu^4}{8c^2} + \frac{m\omega_0^2x^2}{2}.$$

This energy is an integral of motion[3], and hence is always equal to its value at maximum deviations $x = \pm A$ from the equilibrium, when $u \equiv dx/dt = 0$:

$$\mathscr{E} = \frac{m\omega_0^2A^2}{2}.$$

Equating these two expressions for $\mathscr{E}$ and dividing all terms by $m/2$, we obtain the following relation between u and x:

$$u^2\left(1 + \frac{3u^2}{4c^2}\right) = \omega_0^2(A^2 - x^2).$$

With our assumption of the almost-non-relativistic motion, the second term in the first parentheses is small, so that we may plug into it the velocity calculated in the

[3] This (almost evident) fact may be strictly proved using the full Lagrangian function $\mathscr{L}$ of the oscillator, which may be formed by subtracting U from the free-particle's Lagrangian given by Eq. (9.68). According to analytical mechanics (see, e.g. *Part CM* section 2.3), since the total $\mathscr{L}$ does not depend on time explicitly ($\partial\mathscr{L}/\partial t = 0$), the Hamiltonian function $\mathscr{H}$ of the system is an integral of motion: $d\mathscr{H}/dt = 0$. Now calculating $\mathscr{H}$ exactly as was done at the derivation of Eq. (9.72) of the lecture notes, we may see that in such a system (even for an arbitrary ratio u/c, and any time-independent U), $\mathscr{H}$ and $\mathscr{E}$ coincide, so that $d\mathscr{E}/dt = 0$ as well.

zeroth (i.e. non-relativistic) approximation, $(u^2)_0 = \omega_0^2(A^2 - x^2)$, and obtain the first relativistic approximation

$$u^2 \approx \frac{\omega_0^2(A^2 - x^2)}{1 + (3\omega_0^2/4c^2)(A^2 - x^2)}.$$

Plugging this approximation into the well-known expression for the period of oscillations in a symmetric confining potential[4],

$$\mathcal{T} = \int_{\text{all period}} dt \equiv \int_{\text{all period}} \frac{dx}{dx/dt} \equiv \int_{\text{all period}} \frac{dx}{u} = 4\int_0^A \frac{dx}{|u|},$$

and then introducing a new, dimensionless integration variable $\xi \equiv x/A$, we obtain

$$\begin{aligned}
\mathcal{T} &\approx \frac{4}{\omega_0}\int_0^A \frac{[1 + (3\omega_0^2/4c^2)(A^2 - x^2)]^{1/2}}{(A^2 - x^2)^{1/2}}dx \equiv \frac{4}{\omega_0}\int_0^1 \frac{[1 + (3\omega_0^2A^2/4c^2)(1 - \xi^2)]^{1/2}}{(1 - \xi^2)^{1/2}}d\xi \\
&\approx \frac{4}{\omega_0}\int_0^1 \frac{1}{(1 - \xi^2)^{1/2}}\left[1 + \frac{3\omega_0^2A^2}{8c^2}(1 - \xi^2)\right]d\xi \\
&\equiv \frac{4}{\omega_0}\int_0^1 \frac{d\xi}{(1 - \xi^2)^{1/2}} + \frac{4}{\omega_0}\frac{3\omega_0^2A^2}{8c^2}\int_0^1 (1 - \xi^2)^{1/2}d\xi.
\end{aligned}$$

Both these integrals may be readily worked out by the same substitution $\xi \equiv \sin\alpha$ (so that $(1 - \xi^2)^{1/2} = \cos\alpha$), giving

$$\int_0^1 \frac{d\xi}{(1 - \xi^2)^{1/2}} = \int_0^{\pi/2} d\alpha = \frac{\pi}{2},$$

$$\int_0^1 (1 - \xi^2)^{1/2}d\xi = \int_0^{\pi/2} \cos^2\alpha \; d\alpha = \frac{1}{2}\int_0^{\pi/2} (1 + \cos 2\alpha)\; d\alpha = \frac{\pi}{4},$$

so that, finally, the oscillation period is

$$\mathcal{T} \approx \frac{2\pi}{\omega_0} + \frac{2\pi}{\omega_0}\frac{3\omega_0^2A^2}{16c^2} \equiv \frac{2\pi}{\omega_0}\left(1 + \frac{3\omega_0^2A^2}{16c^2}\right),$$

and the oscillation frequency

$$\omega \equiv \frac{2\pi}{\mathcal{T}} \approx \omega_0\left(1 + \frac{3\omega_0^2A^2}{16c^2}\right)^{-1} \approx \omega_0\left(1 - \frac{3\omega_0^2A^2}{16c^2}\right). \qquad (*)$$

This formula (which is quantitatively correct only if the second term in parentheses is much smaller than 1) shows that the oscillation frequency *reduces* with A—the natural result of the *growth* of particle's relativistic mass $M = m/(1 - u^2/c^2)^{1/2}$ with the increase of its velocity amplitude $u_{max} \approx \omega_0 A$.

---

[4] See, e.g. *Part CM* section 3.3.

Note that alternatively, Eq. (*) could be obtained by using the Lagrangian function (9.183) of the system,

$$\mathscr{L} = -\frac{mc^2}{\gamma} - \frac{m\omega_0^2 x^2}{2} \approx -mc^2 + \frac{mu^2}{2} - \frac{3mu^4}{8c^2} - \frac{m\omega_0^2 x^2}{2},$$

to derive its (approximate) Lagrange equation of motion,

$$\frac{d}{dt}\left(mu - \frac{3mu^3}{2c^2}\right) + m\omega_0^2 x = 0, \quad \text{i.e.} \quad \ddot{x} + \omega_0^2 x = \frac{d}{dt}\left(\frac{3\dot{x}^3}{2c^2}\right), \tag{**}$$

with the right-hand side considered small, and then solving it using the standard small-parameter method (also called the 'van der Pol method' or the 'rotating-wave approximation')[5]. However, for our particular case, when the right-hand side of Eq. (**) is a function of $u \equiv dx/dt$ alone, the approach spelled out above is a bit simpler.

***Problem* 9.8.** An atom, with an initial rest mass m, has been excited to an internal state with an additional energy $\Delta\mathscr{E}$, still being at rest. Next, it returns into its initial state, emitting a photon. Calculate the photon's frequency, taking into account the relativistic recoil of the atom.

Hint: In this problem, and in problems 9.11–9.13, treat photons as ultra-relativistic point particles with zero rest mass, energy $\mathscr{E} = \hbar\omega$, and momentum $\mathbf{p} = \hbar\mathbf{k}$.

Solution: Combining the energy conservation law,

$$mc^2 + \Delta\mathscr{E} = \hbar\omega + [(mc^2)^2 + p^2c^2]^{1/2},$$

and the momentum conservation law

$$0 = \frac{\hbar\omega}{c} - p$$

(both in the lab frame, which is also the center-of-mass frame in this problem), we readily obtain

$$\omega = \frac{\Delta\mathscr{E}}{\hbar}\,\frac{1 + \Delta\mathscr{E}/2mc^2}{1 + \Delta\mathscr{E}/mc^2} < \frac{\Delta\mathscr{E}}{\hbar}.$$

For real atoms ($m \sim 10^{-26}$ kg, $\Delta\mathscr{E} \sim 1$ eV), $\Delta\mathscr{E} \ll mc^2$, when the above expression is reduced to

$$\omega \approx \frac{\Delta\mathscr{E}}{\hbar}\left(1 - \frac{\Delta\mathscr{E}}{2mc^2}\right),$$

the recoil correction of frequency is relatively small ($\sim 10^{-10}$), but still may be measurable for quantum transitions with small natural linewidth. This correction is

[5] See, e.g. *Part CM* section 5.2.

dramatically reduced in the *Mössbauer effect*[6], in which the role of m is a played by the mass of the whole crystal lattice. This effect enables the important technique of high-resolution *Mössbauer spectroscopy.*

***Problem* 9.9.** A particle of mass m, initially at rest, decays into two particles, with rest masses m_1 and m_2. Calculate the total energy of the first decay product, in the reference frame moving with that particle.

Solution: In the rest frame of the decayed particle, its momentum p was zero, and its energy was mc^2. Hence, due to the energy and momentum conservation at the decay, the four-momenta of the decay products may be represented as

$$p_1^\alpha = \left\{\frac{\mathscr{E}_1}{c}, \mathbf{p}_1\right\}, \qquad p_2^\alpha = \left\{\frac{mc^2 - \mathscr{E}_1}{c}, -\mathbf{p}_1\right\}.$$

Due to the Lorentz invariance of the scalar products of four-vectors, the norm of the first four-vector has to be equal to its value in the rest frame of particle 1, i.e. $(m_1c)^2$, and that of the second vector should be equal to $(m_2c)^2$. This gives us a system of two equations for $\mathscr{E}_1$ and p_1:

$$\frac{\mathscr{E}_1^2}{c^2} - p_1^2 = (m_1c)^2, \qquad \frac{(mc^2 - \mathscr{E}_1)^2}{c^2} - p_1^2 = (m_2c)^2.$$

Now the elementary elimination of p_1 yields

$$\mathscr{E}_1 = \frac{m^2 + m_1^2 - m_2^2}{2m}c^2.$$

***Problem* 9.10.** A relativistic particle with a rest mass m, moving with velocity u, decays into two particles with zero rest mass.

(i) Calculate the smallest possible angle between the decay product velocities (in the lab frame, in which the velocity u is measured).
(ii) What is the largest possible energy of a product particle?

Solutions:

(i) In the reference frame moving with the velocity $\mathbf{v} = \mathbf{u} = u\mathbf{n}_x$ of the initial particle (i.e. that of the center of mass of the system), that particle does not move, so that the total momentum conservation requires the similar decay products to move in opposite directions along the same straight line. Since according to the basic equation (9.78) of the lecture notes, the speed of a massless particle with a non-zero energy has to be equal to c, the Cartesian components of the product particle velocities, as measured in this center-of-mass frame, depend on just one free parameter, say angle θ—see the figure below. (The plane of the drawing is the common plane of the initial and final particle velocities.)

[6] See, e.g. [1].

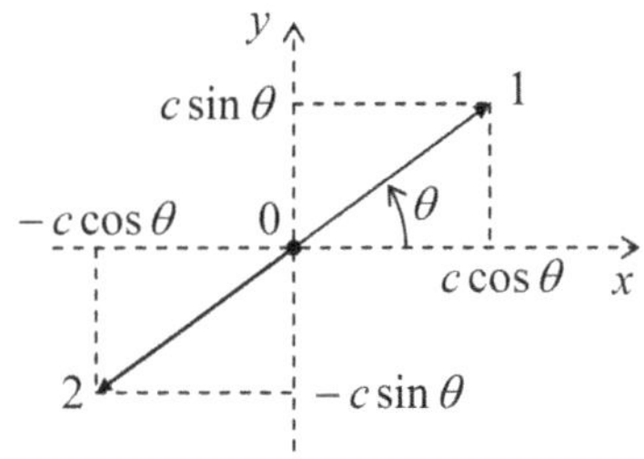

Now we may use Eqs. (9.23) of the lecture notes to Lorentz-transform the velocity components into the lab frame:

$$v_{x1} = \frac{c\cos\theta + u}{1 + u\cos\theta/c}, \quad v_{y1} = \frac{1}{\gamma}\frac{c\sin\theta}{1 + u\cos\theta/c},$$
$$v_{x2} = \frac{-c\cos\theta + u}{1 - u\cos\theta/c}, \quad v_{y2} = -\frac{1}{\gamma}\frac{c\sin\theta}{1 - u\cos\theta/c}.$$

Since the speed of both particles in the lab system is still c, the trigonometric functions of angles $\theta_{1,2}$ of their propagation (relative to the axis x) may be calculated as $\cos\theta_1 = v_{x1}/c$, etc, giving

$$\cos\theta_1 = \frac{\cos\theta + \beta}{1 + \beta\cos\theta}, \quad \sin\theta_1 = \frac{1}{\gamma}\frac{\sin\theta}{1 + \beta\cos\theta};$$
$$\cos\theta_2 = \frac{-\cos\theta + \beta}{1 - \beta\cos\theta}, \quad \sin\theta_2 = -\frac{1}{\gamma}\frac{\sin\theta}{1 - \beta\cos\theta}, \qquad (*)$$

where $\beta \equiv u/c$. Now we may calculate the angle $\theta_d \equiv \theta_1 - \theta_2$ between the two particles' velocities in the lab frame:

$$\cos\theta_d = \cos(\theta_1 - \theta_2) \equiv \cos\theta_1\cos\theta_2 + \sin\theta_1\sin\theta_2$$
$$= \frac{\beta^2 - \cos^2\theta}{1 - \beta^2\cos^2\theta} - \frac{\sin^2\theta}{\gamma^2(1 - \beta^2\cos^2\theta)}$$
$$\equiv \frac{\beta^2\sin^2\theta - (1 - \beta^2)}{\beta^2\sin^2\theta + (1 - \beta^2)} \equiv \frac{\xi - 1}{\xi + 1}, \quad \text{with } \xi \equiv \frac{\beta^2\sin^2\theta}{1 - \beta^2}.$$

Since the function $f(\xi) \equiv (\xi - 1)/(\xi + 1)$ grows monotonically at the segment $\xi = [0,1]$, $\cos\theta_d$ is largest, and hence θ_d (if defined on the segment $[0, \pi]$) is smallest, at $\sin^2\theta = 1$—i.e. at the product particles' propagation along the axis y, as observed in the center-of-mass frame. At that point, we obtain[7]

[7] The second step in the following calculation employs the trigonometric identity $\cos 2\xi = 2\cos^2\xi - 1$—see, e.g. the first of Eqs. (A.18d).

$$(\cos\theta_d)_{\max} = \frac{\beta^2 - (1-\beta^2)}{\beta^2 + (1-\beta^2)} \equiv 2\beta^2 - 1,$$

$$(\theta_d)_{\min} = \cos^{-1}(2\beta^2 - 1) \equiv 2\cos^{-1}\beta \equiv 2\cos^{-1}\frac{u}{c}.$$

So, at low velocities of the initial particle, the smallest angle θ_d tends to π, i.e. the value this angle always has in the center-of-mass frame, while at $u \to c$, it gradually approaches zero.

(ii) Evidently, the largest difference between the product energies is achieved at $\theta = 0$, when $\theta_1 = 0$ and $\theta_2 = \pi$, i.e. the particles move along the same line (in the figure above, axis x), but in opposite directions[8]. In this case we may use the conservation of energy and the x-component of momentum of the system in the lab frame, using the basic equations (9.70) and (9.73), with $p = \mathscr{E}/c$ for each of the product particles:

$$m\gamma u = \frac{\mathscr{E}_1}{c} - \frac{\mathscr{E}_2}{c}, \qquad m\gamma c^2 = \mathscr{E}_1 + \mathscr{E}_2.$$

Solving this simple system of equations, we obtain[9]

$$\mathscr{E}_{1,2} = \frac{m\gamma c^2}{2}\left(1 \pm \frac{u}{c}\right) \equiv \frac{mc^2}{2}\frac{1 \pm \beta}{(1-\beta^2)^{1/2}} \equiv \frac{mc^2}{2}\left(\frac{1 \pm \beta}{1 \mp \beta}\right)^{1/2}.$$

The result shows that if the initial particle is non-relativistic ($\beta \to 0$), its energy (in this limit close to the rest energy mc^2) is always equally divided between those of the decay products, but if it is ultra-relativistic ($\beta \to 1$), almost all its energy may go to just one product particle.

***Problem* 9.11.** A relativistic particle, propagating with velocity $\mathbf{u}$, in the absence of external fields, decays into two photons[10]. Calculate the angular dependence of the probability of photon detection, as measured in the lab frame.

Solution: Let θ be the angle between the direction of propagation of a detected photon and the vector of the initial velocity of the particle, as measured in the lab frame, while θ' being the similar angle as measured in the reference frame moving with the initial particle, i.e. with the velocity $\mathbf{v} = \mathbf{u}$. Then they are related by Eq. (9.27) of the lecture notes, with u' (the photon's speed) equal to c:

$$\tan\theta = \frac{\sin\theta'}{\gamma(\cos\theta' + \beta)}, \quad \text{with } \beta \equiv \frac{v}{c}, \quad \gamma \equiv \frac{1}{(1-\beta^2)^{1/2}}.$$

[8] In contrast, as follows from Eq. (*), at $\theta = \pi/2$, i.e. at $\cos\theta = 0$, and $\sin\theta = 1$, the product particles propagate at equal angles to the initial direction ($\theta_1 = -\theta_2$), and hence have equal energies.

[9] Note a substantial analogy between this expression and Eq. (9.44) describing the longitudinal Doppler effect.

[10] Such a decay may happen, for example, with a neutral pion.

For our purposes, it is beneficial to represent this relation as

$$\cos\theta \equiv \frac{1}{(1+\tan^2\theta)^{1/2}} = \frac{1}{[1+\sin^2\theta'/\gamma^2(\cos\theta'+\beta)^2]^{1/2}}$$
$$\equiv \frac{\cos\theta'+\beta}{[(\cos\theta'+\beta)^2+\sin^2\theta'(1-\beta^2)]^{1/2}}$$
$$\equiv \frac{\cos\theta'+\beta}{[(\cos\theta'+\beta)^2+(1-\cos^2\theta')(1-\beta^2)]^{1/2}}$$
$$\equiv \frac{\cos\theta'+\beta}{(1+2\beta\cos\theta'+\beta^2\cos^2\theta')^{1/2}} \equiv \frac{\cos\theta'+\beta}{1+\beta\cos\theta'},$$

with the reciprocal relation

$$\cos\theta' = \frac{\cos\theta-\beta}{1-\beta\cos\theta}. \qquad (*)$$

In the reference frame moving with the initial particle, the particle was at rest, so that in the absence of any external polarizing field, the angular distribution of photons has to be isotropic. This means that of the total $N \gg 1$ photons resulting from a large set ('statistical ensemble') of similar but independent experiments, the number of photons going into any small solid angle interval $d\Omega' = \sin\theta' d\theta' d\varphi'$ is the same: $dN = N(d\Omega'/4\pi) \equiv N(\sin\theta' d\theta' d\varphi'/4\pi)$. From the point of view of the lab frame, these dN photons go into the interval $d\Omega = \sin\theta d\theta d\varphi$, with $\varphi = \varphi'$ due to the axial symmetry of the problem. Hence the probability density of the photon propagation at angle θ is

$$w(\theta) \equiv \frac{1}{N}\frac{dN}{d\Omega} = \frac{1}{4\pi}\frac{\sin\theta' d\theta'}{\sin\theta d\theta} = \frac{1}{4\pi}\frac{d(\cos\theta')}{d(\cos\theta)}.$$

This relation is general (for an axially symmetric system); now using the particular Eq. (*) to calculate the derivative, we finally obtain

$$w(\theta) = \frac{1}{4\pi}\frac{1-\beta^2}{(1-\beta\cos\theta)^2}.$$

(As a sanity check, the integral of this function over the full solid angle, i.e. the total probability of a photon propagation in *some* direction, equals 1.)

This result shows that the photon detection probability is largest in the direction of the initial particle's propagation, $\theta = 0$:

$$w(0) = \frac{1}{4\pi}\frac{1+\beta}{1-\beta},$$

and that for ultra-relativistic particles ($\gamma \gg 1$), this enhancement may be very significant, scaling as γ^2. Note that this fact is in a qualitative agreement with the solution of the previous problem, devoted to a similar situation.

***Problem* 9.12.** A photon with wavelength λ is scattered by an electron, initially at rest. Calculate the wavelength λ' of the scattered photon as a function of the scattering angle α—see the figure below[11].

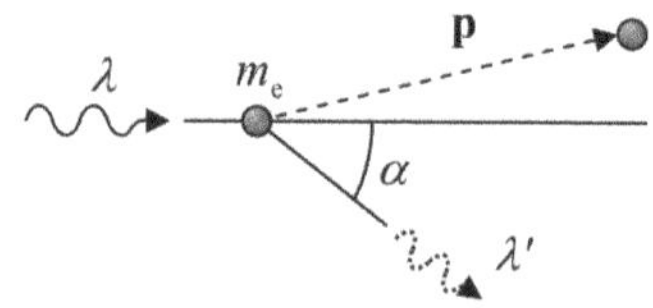

Solution: The problem may be most simply solved by writing the laws of conservation of the total relativistic momentum (9.70) and relativistic energy (9.73), both in the lab reference frame:

$$\hbar\mathbf{k} = \hbar\mathbf{k}' + \mathbf{p}, \qquad (*)$$

$$\hbar\omega + m_e c^2 = \hbar\omega' + [(m_e c^2)^2 + p^2c^2]^{1/2}, \qquad (**)$$

where $\mathbf{p}$ is the momentum of the electron after the collision (see the figure above), and $k = \omega/c = 2\pi/\lambda$. From Eq. (*), $\mathbf{p} = \hbar(\mathbf{k} - \mathbf{k}')$, so that[12]

$$p^2 = \hbar^2(\mathbf{k} - \mathbf{k}')^2 = \hbar^2(k^2 + k'^2 - 2kk'\cos\alpha).$$

Plugging this expression into Eq. (**), and solving it for λ', we obtain

$$\lambda' = \lambda + \frac{2\pi\hbar}{m_e c}(1 - \cos\alpha) \geqslant \lambda.$$

The only constant, with the dimension of length, participating in this result,

$$\frac{2\pi\hbar}{m_e c} \approx 2.43 \times 10^{-12}\text{m},$$

is called the *Compton wavelength*[13]. Note the relation between this length and the electron's classical radius r_e, given by Eq. (8.41):

$$\frac{r_e}{2\pi\hbar/m_e c} \equiv \frac{e^2}{4\pi\varepsilon_0}\frac{1}{m_e c^2}\frac{m_e c}{2\pi\hbar} \equiv \frac{1}{2\pi}\left(\frac{1}{4\pi\varepsilon_0}\frac{e^2}{c\hbar}\right) \equiv \frac{\alpha}{2\pi} \sim 10^{-3},$$

where

$$\alpha \equiv \frac{1}{4\pi\varepsilon_0}\frac{e^2}{c\hbar} \approx \frac{1}{137}$$

is the fine structure constant[14], which characterizes the strength (or rather the weakness) of electromagnetic interactions relative to the quantum effects. (The fact

[11] This the famous *Compton scattering* problem—see, e.g. *Part QM* section 1.1.
[12] Note this useful (and frequently used) approach of avoiding the introduction of the direction of the vector **p**.
[13] This historic name is rather misleading: no particle in the problem has this wavelength; neither did Dr A Compton, who discovered this effect in 1923, had such a wavelength (to the best of my knowledge :-).
[14] See also problems 5.14 and 8.5.

that $\alpha/2\pi \sim 10^{-3}$ is so much smaller than 1 enables the vast field of perturbative quantum electrodynamics.)

***Problem* 9.13.** Calculate the threshold energy of a γ-photon for the reaction

$$\gamma + \mathrm{p} \rightarrow \mathrm{p} + \pi^0,$$

if the proton was initially at rest.

Hint: For protons $m_\mathrm{p}c^2 \approx 938$ MeV, while for neutral pions $m_\pi c^2 \approx 135$ MeV.

Solution: Similarly to the analysis of the $\mathrm{p} + \mathrm{p} \rightarrow \mathrm{p} + \mathrm{p} + \mathrm{p} + \bar{\mathrm{p}}$ reaction in section 9.4 of the lecture notes, we may represent the conservation of the net four-momentum of the system as[15]

$$(\gamma + p)_\alpha(\gamma + p)^\alpha = (p + \pi^0)_\alpha(p + \pi^0)^\alpha, \qquad (*)$$

where the particle names are used to denote the corresponding partial four-momenta. After opening the parentheses, we can evaluate the Lorentz-invariant scalar product $p_\alpha p^\alpha$ on the left-hand side of Eq. (*) as $m_\mathrm{p}{}^2c^2$, and the similar product the γ-photon as zero, because the photon does not have rest mass. Similarly, the right-hand side of Eq. (*) equals $(m_\mathrm{p} + m_\pi)^2c^2$, because at the reaction threshold, its products have zero velocity in the center-of-mass frame. As a result, Eq. (*) reads

$$2\gamma_\alpha p^\alpha + m_\mathrm{p}^2c^2 = (m_\mathrm{p} + m_\pi)^2\, c^2, \quad \text{so that } 2\gamma_\alpha p^\alpha = \left[(m_\mathrm{p} + m_\pi)^2 - m_\mathrm{p}^2\right] c^2.$$

In the lab system, in which the initial proton rests, its four-vector p^α has only one non-vanishing component, $p_0 = m_\mathrm{p}c$, while the corresponding component γ_0 for the photon is just $\mathscr{E}_\gamma/c$. As a result, we obtain

$$\mathscr{E}_\gamma = \frac{1}{2m_\mathrm{p}}\left[(m_\mathrm{p} + m_\pi)^2 - m_\mathrm{p}^2\right] c^2 \equiv m_\pi c^2\left(1 + \frac{m_\pi}{2m_\mathrm{p}}\right).$$

The second term in the parentheses represents the energy price we have to pay for not having the center of mass of the initial particles at rest. With the given rest masses of the involved particles, the correction is close to 7% (quite noticeable!), and we finally obtain $\mathscr{E}_\gamma \approx 145$ MeV.

***Problem* 9.14.** A relativistic particle with energy $\mathscr{E}$ and rest mass m collides with a similar particle, initially at rest in the laboratory reference frame. Calculate:

(i) the final velocity of the center of mass of the system, in the lab frame,
(ii) the total energy of the system, in the center-of-mass frame, and
(iii) the final velocities of both particles (in the lab frame), if they move along the same direction.

[15] Here, just as in Eqs. (9.88) and (9.89) of the lecture notes, I am using italic font to denote the components of the four-momentum of each particle, leaving the roman font to denote the particle as such in the reaction's description.

Solutions:

(i) Before the collision, the total momentum p_s of the system in the lab frame equals that (p) of the only moving particle, which may be calculated from the basic Eq. (9.78) of the lecture notes:

$$p_s = p = [(\mathscr{E}/c)^2 - (mc)^2]^{1/2},$$

while the total energy of the system is

$$\mathscr{E}_s = \mathscr{E} + mc^2.$$

By definition, the center of mass of a system is an imaginary particle having the same total momentum p_s and total mass M_s as the whole system. (In relativity[16], the latter is the dynamic mass $M_s = \mathscr{E}_s/c^2$.) Applying to the center of mass the general formula $\mathbf{p} = M\mathbf{u}$, giving $\boldsymbol{\beta} = \mathbf{p}/Mc = \mathbf{p}c/\mathscr{E}$, we obtain

$$\beta_{\text{com}} = \frac{[(\mathscr{E}/c)^2 - (mc)^2]^{1/2}c}{\mathscr{E} + mc^2} \equiv \left(\frac{\mathscr{E} - mc^2}{\mathscr{E} + mc^2}\right)^{1/2}, \text{ so that}$$
$$\gamma_{\text{com}} \equiv \frac{1}{\left(1 - \beta^2_{\text{com}}\right)^{1/2}} = \left(\frac{\mathscr{E} + mc^2}{2mc^2}\right)^{1/2}. \qquad (*)$$

Note that despite the fact that the particles are similar, $\beta_{\text{com}} \to \beta/2$ only in the non-relativistic limit; in the opposite, ultra-relativistic limit ($v \to c$, $\beta \to 1$, $\mathscr{E} \gg mc^2$), our result yields $\beta_{\text{com}} \to 1$, i.e. the center-of-mass velocity is also close to c, while $\gamma_{\text{com}} \approx (\mathscr{E}/2mc^2)^{1/2} \gg 1$.

(ii) Applying the Lorentz transform formula, valid for any four-vector, to the zeroth component of the total four-momentum of the system, $p_s^\alpha = \{\mathscr{E}_s/c, p_s, 0, 0\}$, we obtain

$$\frac{(\mathscr{E}_s)_{\text{com}}}{c} = \gamma_{\text{com}}\left(\frac{\mathscr{E}_s}{c} - \beta_{\text{com}} p_s\right) = \gamma_{\text{com}}\left\{\frac{\mathscr{E} + mc^2}{c} - \beta_{\text{com}}\left[\left(\frac{\mathscr{E}}{c}\right)^2 - (mc)^2\right]^{1/2}\right\}.$$

Plugging in the expressions for β_{com} and γ_{com} from Eq. (*), we finally obtain[17]

$$(\mathscr{E}_s)_{\text{com}} = 2\gamma_s mc^2 = [2mc^2(\mathscr{E} + mc^2)]^{1/2}. \qquad (**)$$

[16] See the footnote at the end of section 9.3 of the lecture notes.

[17] Another (even simpler) way to derive Eq. (**) is to require the norm of the total four-momentum of the system, $p_s^\alpha p_{s\alpha}$, to be Lorentz invariant:

$$\frac{(\mathscr{E}_s)^2_{\text{com}}}{c^2} - (p_s)^2_{\text{com}} = \frac{(\mathscr{E}_s)^2}{c^2} - p_s^2,$$

and then notice that the net momentum in the center-of-mass frame, $(p_s)_{\text{com}}$, is zero.

Note that for ultra-relativistic particles, $\mathscr{E} \gg mc^2$, this energy (which is the value crucial for new particle generation) grows only as the square root of that in the lab frame; this is why particle colliders are more popular nowadays than resting-target accelerators. (The caveat is that colliders should have a very high density of particle beams to ensure a sufficiently high collision rate.)

(iii) By the definition of the center-of-mass reference frame, the velocities of two similar particles, observed from the frame should be equal and opposite, and since if the total kinetic energy is conserved at collision, their magnitudes β' should be the same both before and after it. If the particles, after the collision, move along the same direction as before it, this means that particles just exchange their velocities—see the figure below.

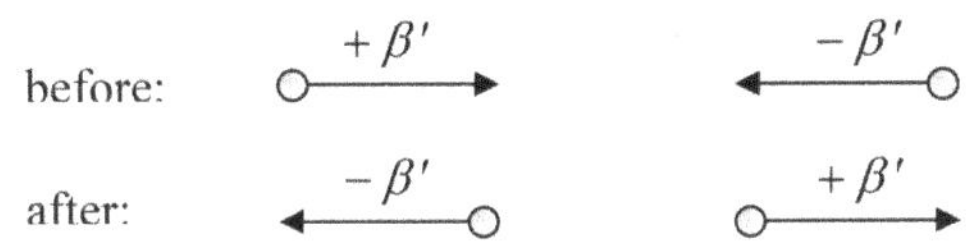

But this particular fact (of the particles exchanging velocities) should be true in *any* reference frame, because the Lorentz transform of a velocity does not depend on the particle it is pertinent to. Hence, in the lab system the final velocities of the particles are also the same as the initial ones:

$$\beta_+ = \beta = \frac{cp}{\mathscr{E}} = \frac{1}{\mathscr{E}}[\mathscr{E}^2 - (mc^2)^2]^{1/2}, \qquad \beta_- = 0.$$

***Problem* 9.15.** A 'primed' reference frame moves, with the reduced velocity $\boldsymbol{\beta} \equiv \mathbf{v}/c = \mathbf{n}_x\beta$, relative to the 'lab' frame. Use Eq. (9.109) of the lecture notes to express the components T'^{00} and T'^{0j} (with $j = 1, 2, 3$) of an arbitrary contravariant four-tensor $T^{\gamma\delta}$ via its components in the lab frame.

Solution: For the relative motion of the frames directed along axis x, the Lorentz transform equations have their traditional form (9.19), so that the mixed Lorentz tensor has the form (9.98):

$$\frac{\partial x'^\alpha}{\partial x^\beta} = L^\alpha_\beta = \begin{pmatrix} \gamma & -\beta\gamma & 0 & 0 \\ -\beta\gamma & \gamma & 0 & 0 \\ 0 & 0 & 1 & 0 \\ 0 & 0 & 0 & 1 \end{pmatrix}, \quad \text{where } \gamma \equiv \frac{1}{(1 - \beta^2)^{1/2}}.$$

Due to such an abundance of zero elements, of the 16 terms in each implicit sum over the indices γ and δ in the second of Eqs. (9.109), there are only four non-vanishing terms for each of T'^{00} and T'^{01}:

$$T'^{00} = \sum_{\gamma,\delta=0}^{3} \frac{\partial x'^0}{\partial x^\gamma}\frac{\partial x'^0}{\partial x^\delta} T^{\gamma\delta}$$
$$= \frac{\partial x'^0}{\partial x^0}\frac{\partial x'^0}{\partial x^0} T^{00} + \frac{\partial x'^0}{\partial x^0}\frac{\partial x'^0}{\partial x^1} T^{01} + \frac{\partial x'^0}{\partial x^1}\frac{\partial x'^0}{\partial x^0} T^{10} + \frac{\partial x'^0}{\partial x^1}\frac{\partial x'^0}{\partial x^1} T^{11}$$
$$= \gamma\gamma T^{00} + \gamma(-\beta\gamma)T^{01} + (-\beta\gamma)\gamma T^{01} + (-\beta\gamma)(-\beta\gamma)T^{11}$$
$$\equiv \gamma^2[T^{00} - \beta(T^{01} + T^{10}) + \beta^2 T^{11}],$$
$$T'^{01} = \sum_{\gamma,\delta=0}^{3} \frac{\partial x'^0}{\partial x^\gamma}\frac{\partial x'^1}{\partial x^\delta} T^{\gamma\delta}$$
$$= \frac{\partial x'^0}{\partial x^0}\frac{\partial x'^1}{\partial x^0} T^{00} + \frac{\partial x'^0}{\partial x^0}\frac{\partial x'^1}{\partial x^1} T^{01} + \frac{\partial x'^0}{\partial x^1}\frac{\partial x'^1}{\partial x^0} T^{10} + \frac{\partial x'^0}{\partial x^1}\frac{\partial x'^1}{\partial x^1} T^{11}$$
$$= \gamma(-\beta\gamma)T^{00} + \gamma\gamma T^{01} + (-\beta\gamma)(-\beta\gamma)T^{10} + (-\beta\gamma)\gamma T^{11}$$
$$\equiv \gamma^2[T^{01} - \beta(T^{00} + T^{11}) + \beta^2 T^{10}],$$

and only two non-vanishing term for each of T'^{02} and T'^{03}:

$$T'^{02} = \sum_{\gamma,\delta=0}^{3} \frac{\partial x'^0}{\partial x^\gamma}\frac{\partial x'^2}{\partial x^\delta} T^{\gamma\delta} = \frac{\partial x'^0}{\partial x^0}\frac{\partial x'^2}{\partial x^2} T^{02} + \frac{\partial x'^0}{\partial x^1}\frac{\partial x'^2}{\partial x^2} T^{12}$$
$$= \gamma T^{02} + (-\beta\gamma)T^{12} \equiv \gamma(T^{02} - \beta T^{12}),$$

and absolutely similarly, with the replacement $2 \to 3$:

$$T'^{03} = \gamma(T^{03} - \beta T^{13}).$$

As the simplest sanity check, at $\beta \to 0$ (when $\gamma \to 1$), all the tensor elements are the same in both reference frames.

***Problem* 9.16.** Static fields **E** and **B** are uniform but arbitrary (both in magnitude and in direction). What should be the velocity of an inertial reference frame in order to have the vectors **E**′ and **B**′, observed from that frame, parallel? Is this solution unique?

Solution: The only special direction **n**, defined by vectors **E** and **B**, is the one perpendicular to both these vectors:

$$\mathbf{n} = \frac{\mathbf{E}\times\mathbf{B}}{|\mathbf{E}\times\mathbf{B}|}. \qquad (*)$$

It is natural to assume that at least one of the reference frames 0′, in which **E** || **B**, should move along that direction: $\mathbf{v} = \mathbf{n}v$. According to Eq. (*), for that direction, in the lab frame, $\mathbf{E}_{||} = 0$, $\mathbf{E}_\perp = \mathbf{E}$, and $\mathbf{B}_{||} = 0$, $\mathbf{B}_\perp = \mathbf{B}$, so that according to Eq. (9.135) of the lecture notes, the fields measured in the moving frame are

$$\mathbf{E}'_{||} = \mathbf{E} = 0, \qquad \mathbf{B}'_{||} = \mathbf{B} = 0,$$
$$\mathbf{E}'_\perp = \gamma(\mathbf{E} + \mathbf{v}\times\mathbf{B}), \qquad \mathbf{B}'_\perp = \gamma(\mathbf{B} - \mathbf{v}\times\mathbf{E}/c^2),$$

The requirement to have vectors $\mathbf{E}'$ and $\mathbf{B}'$ parallel may be written as $\mathbf{E}' \times \mathbf{B}' = 0$, so that for the system's velocity we obtain the following equation:

$$(\mathbf{E} + \mathbf{v} \times \mathbf{B}) \times (\mathbf{B} - \mathbf{v} \times \mathbf{E}/c^2) = 0. \qquad (**)$$

Opening the parentheses, we can notice that all four operands of the result are vectors directed along the vector $\mathbf{n}$. Also, $\mathbf{E} \times (\mathbf{v} \times \mathbf{E}) = -vE^2\mathbf{n}$, because the vector $\mathbf{v} \times \mathbf{E}$ is perpendicular to $\mathbf{E}$. Similarly, $(\mathbf{v} \times \mathbf{B}) \times \mathbf{B} = vB^2\mathbf{n}$. Finally, $(\mathbf{v} \times \mathbf{B}) \times (\mathbf{v} \times \mathbf{E}) = v^2\mathbf{B} \times \mathbf{E}$, because the angle between vectors $\mathbf{v} \times \mathbf{B}$ and $\mathbf{v} \times \mathbf{E}$ equals that between vectors $\mathbf{B}$ and $\mathbf{E}$. As a result, Eq. (**) is reduced to a quadratic equation for v. This equation looks particularly simple for the reduced velocity $\beta \equiv v/c$:

$$\beta^2 - \frac{2}{\kappa}\beta + 1 = 0, \quad \text{where } \boldsymbol{\kappa} \equiv \frac{2(\mathbf{E}/c) \times \mathbf{B}}{(E/c)^2 + B^2}.$$

Formally, this equation has two solutions:

$$\beta_\pm = \kappa^{-1} \pm (\kappa^{-2} - 1)^{1/2}.$$

For any values and directions of vectors $\mathbf{E}$ and $\mathbf{B}$, the parameter κ, by its definition, is restricted to segment $[-1, +1]$, so that one of the solutions for β (plotted as functions of κ in the figure below) is always below 1, and hence the required reference frame exits for any set of fields.

Now we may notice that since $\mathbf{E}' \parallel \mathbf{B}'$, according to the Lorentz transform (9.135), in any other reference frame $0''$ moving relative to the frame $0'$ with a velocity parallel to these two vectors, the transverse fields would not appear, i.e. any such system also is a solution to our problem.

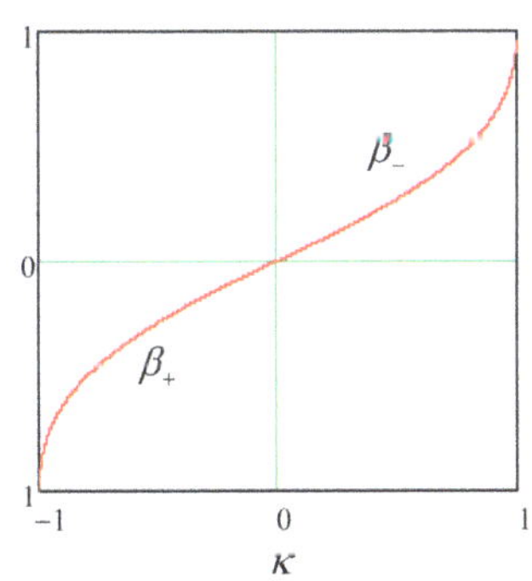

***Problem* 9.17.** Two charged particles, moving with the same constant velocity $\mathbf{u}$, are offset by distance vector $\mathbf{R} = \{a, b\}$ (see the figure below), as measured in the lab frame. Calculate the forces acting between the particles—also in the lab frame.

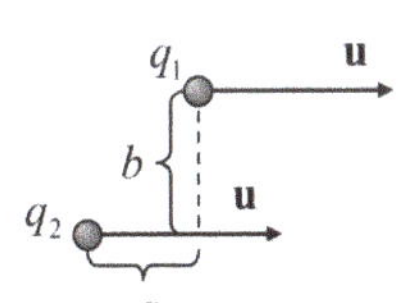

Solution: Let us first calculate the electric and magnetic fields created by charge q_2 at the location of particle 1, as measured in the lab frame. Since these fields do not depend on whether particle 1 moves or not, this task is equivalent to the problem solved in the end of section 9.5 of the lecture notes (see figure 9.11a), with the evident replacements $ut \to -a$ and $q \to q_1$. With these replacements, and with the same choice of Cartesian coordinates (see the figure below), the final forms of Eqs. (9.139) and (9.140) take the form

$$E_x = \frac{q_2}{4\pi\varepsilon_0}\frac{\gamma a}{(\gamma^2 a^2 + b^2)^{3/2}}, \quad E_y = \frac{q_2}{4\pi\varepsilon_0}\frac{\gamma b}{(\gamma^2 a^2 + b^2)^{3/2}}, \quad E_z = 0,$$

$$B_x = 0, \quad B_y = 0, \quad B_z = \frac{u}{c^2}\frac{q_2}{4\pi\varepsilon_0}\frac{\gamma b}{(\gamma^2 a^2 + b^2)^{3/2}} \equiv \frac{u}{c^2}E_y, \quad \text{where } \gamma \equiv \frac{1}{(1-u^2/c^2)^{1/2}}.$$

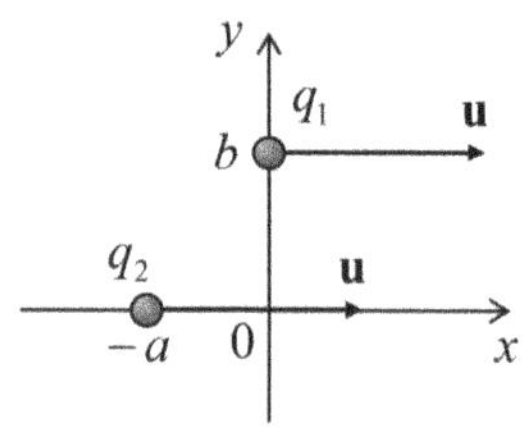

Now we may calculate both components of the Lorentz force (5.10), $\mathbf{F}_{12} = q_1(\mathbf{E} + \mathbf{u} \times \mathbf{B})$, exerted by these fields on particle 1. Since at our choice of coordinates, the calculated magnetic field $\mathbf{B}$ has only one Cartesian component, B_z, and the velocity $\mathbf{u}$ also only one component, $u_x = u$, the vector product $\mathbf{u} \times \mathbf{B}$ also has just one component, $(\mathbf{u} \times \mathbf{B})_y = -uB_z = -(u^2/c^2)E_y$, so that the full Lorentz force has just two non-vanishing components:

$$F_x = \frac{q_1 q_2}{4\pi\varepsilon_0}\frac{\gamma a}{(\gamma^2 a^2 + b^2)^{3/2}}, \quad F_y = \frac{q_1 q_2}{4\pi\varepsilon_0}\frac{\gamma b}{(\gamma^2 a^2 + b^2)^{3/2}}\left(1 - \frac{u^2}{c^2}\right) \equiv \frac{q_1 q_2}{4\pi\varepsilon_0}\frac{b}{\gamma(\gamma^2 a^2 + b^2)^{3/2}}.$$

The calculation of the reciprocal force $\mathbf{F}_{21}$ is absolutely similar, with the replacements $a \to -a$ and $b \to -b$, giving the same result as above, but with opposite signs for both Cartesian components—thus confirming Newton's third law.

Note, however, that the ratio F_x/F_y equals $\gamma^2 a/b$, i.e. coincides with the ratio a/b only in the non-relativistic limit, so that generally (in the lab frame) the forces between the charges are *not* directed along the line connecting them. This is a natural result of the finite speed of propagation electromagnetic interactions: one particle 'does not know' where the other particle is at this particular instant.

***Problem* 9.18.** Each of two thin, long, parallel particle beams of the same velocity $\mathbf{u}$, separated by distance d, carries electric charge with a constant density λ per unit length, as measured in the coordinate frame moving with the particles.

(i) Calculate the distribution of the electric and magnetic fields in the system (outside the beams), as measured in the lab frame.
(ii) Calculate the interaction force between the beams (per particle) and the resulting acceleration, both in the lab frame and in the system moving with the electrons. Compare the results and give a brief discussion of the comparison.

Solutions:

(i) In the reference frame moving with electrons, they are static, so there is no magnetic field: $\mathbf{B}' = 0$. The electric field observed in that frame may be represented as a sum of two fields (each created by one beam),

$$\mathbf{E}' = \mathbf{E}'_+ + \mathbf{E}'_-,$$

and each of these components may be readily found, say, from the Gauss theorem applied to a round cylinder of radius $\rho_\pm$ (see the figure below, in which the beams propagate along the axis x, i.e. perpendicular to the plane of drawing), with the corresponding beam serving as cylinder's axis:

$$\mathbf{E}'_\pm = \frac{\lambda \boldsymbol{\rho}_\pm}{2\pi\varepsilon_0 \rho_\pm^2},$$

or, with the coordinate choice shown in the figure below,

$$E'_{\pm x} = 0, \quad E'_{\pm y} = \frac{\lambda(y \mp d/2)}{2\pi\varepsilon_0[(y \mp d/2)^2 + z^2]}, \quad E'_{\pm z} = \frac{\lambda z}{2\pi\varepsilon_0[(y \mp d/2)^2 + z^2]}.$$

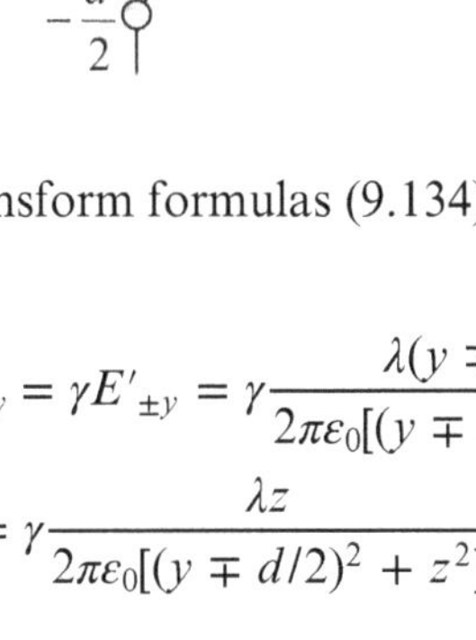

Now using the Lorentz transform formulas (9.134), in the lab system we also can write $\mathbf{E} = \mathbf{E}_+ + \mathbf{E}_-$, with[18]

$$E_{\pm x} = 0, \quad E_{\pm y} = \gamma E'_{\pm y} = \gamma \frac{\lambda(y \mp d/2)}{2\pi\varepsilon_0[(y \mp d/2)^2 + z^2]},$$

$$E_{\pm z} = \gamma E'_{\pm z} = \gamma \frac{\lambda z}{2\pi\varepsilon_0[(y \mp d/2)^2 + z^2]},$$

[18] The Lorentz transform does not change the length perpendicular to the relative velocity, so that $y' = y$, $z' = z$.

$$B_{\pm x} = 0, \quad B_{\pm y} = -\frac{\gamma u}{c^2} E'_{\pm z} = -\frac{\gamma u}{c^2} \frac{\lambda z}{2\pi\varepsilon_0[(y \mp d/2)^2 + z^2]},$$
$$B_{\pm z} = \frac{\gamma u}{c^2} E'_{\pm y} = \frac{\gamma u}{c^2} \frac{\lambda(y - d/2)}{2\pi\varepsilon_0[(y \mp d/2)^2 + z^2]}.$$

(ii) The Lorenz force acting on one beam (say, the one located at $y = +d/2$ in the figure above) comes only from the fields created by the other beam (located at $y' = -d/2$).[19] As a result, in the frame moving with the particles,

$$F'_y = qE'_{-y}|_{y=+d/2,\ z=0} = \frac{q\lambda}{2\pi\varepsilon_0 d}, \qquad F'_z = 0,$$

while in the lab frame the magnetic field contributes to the force as well:

$$\mathbf{F} = q(\mathbf{E}_- + \mathbf{u} \times \mathbf{B}_-)_{y=+d/2,\ z=0},$$

and we obtain

$$F_y = q\left(E_{-y} - uB_{-z}\right) = q\gamma\frac{\lambda}{2\pi\varepsilon_0 d}\left(1 - \frac{u^2}{c^2}\right) = \frac{1}{\gamma}\frac{q\lambda}{2\pi\varepsilon_0 d}, \qquad F_z = q\left(E_{-z} + uB_{-y}\right) = 0.$$

The resulting vertical acceleration in the moving frame (where $M' = m$) is

$$a'_y = \frac{F'_y}{m} = \frac{q\lambda}{2\pi\varepsilon_0 dm},$$

while in the lab frame, where $M = \gamma m$, it is

$$a_y = \frac{F_y}{M} = \frac{1}{\gamma^2}\frac{q\lambda}{2\pi\varepsilon_0 dm} \equiv \frac{a'_y}{\gamma^2}.$$

At $u \to c$, the results for F_y and a_y, obtained to the two reference frames, are dramatically different, but they are actually consistent, because time runs differently in the two frames. For example, if the acceleration produces a small beam shift $\Delta y \ll d$, with the vertical velocity still much below c, we can write

$$\Delta y' = \frac{a'_y}{2}(\Delta t')^2 = \frac{a'_y}{2}(\Delta\tau)^2, \qquad \Delta y = \frac{a_y}{2}(\Delta t)^2 = \frac{a'_y}{2\gamma^2}(\Delta t)^2.$$

Since, according to Eq. (9.21) of the lecture notes, the 'proper' time interval $\Delta t' = \Delta\tau$ and the lab frame interval Δt are related as $\Delta t = \gamma\Delta\tau$ (the time dilation), we obtain that $\Delta y' = \Delta y$—as it should due to the relation $y' = y$.

Note that alternatively, this problem might also be solved by the summation (actually, integration) of the result of the previous problem over all particles of a beam.

[19] Of course, different particles within each beam do interact, but the forces of those interactions are directed along the beam, and (in a uniform beam) mutually cancel.

***Problem* 9.19.** Spell out the Lorentz transform of the scalar potential and the vector-potential components, and use the result to calculate the potentials of a point charge q, moving with a constant velocity $\mathbf{u}$, as measured in the lab reference frame.

Solution: As was discussed in section 9.5 of the lecture notes, the scalar potential ϕ (divided by c) and the three Cartesian components of the vector potential $\mathbf{A}$ form a legitimate four-vector (9.116), whose contravariant form is

$$A^{\alpha} = \left\{\frac{\phi}{c}, \mathbf{A}\right\} \equiv \left\{\frac{\phi}{c}, A_x, A_y, A_z\right\}.$$

Hence the Lorentz transform of the components of this tensor follow the regular rule (9.91):

$$A^{\alpha} = L^{\alpha}_{\beta} A'^{\beta},$$

(with the implied summation over all four values of the repeated index β), where $L_{\beta}{}^{\alpha}$ is the (mixed) Lorentz tensor. If we, as usual, direct the axis x along the vector $\mathbf{v}$ of velocity of the 'primed' (i.e. 'moving') frame—as shown, for example in figure 9.1 of the lecture notes, the Lorentz tensor has the standard form (9.92):

$$L^{\alpha}_{\beta} = \begin{pmatrix} \gamma & \beta\gamma & 0 & 0 \\ \beta\gamma & \gamma & 0 & 0 \\ 0 & 0 & 1 & 0 \\ 0 & 0 & 0 & 1 \end{pmatrix}, \quad \text{with } \beta \equiv \frac{v}{c}, \quad \text{and } \gamma \equiv \frac{1}{(1-\beta^2)^{1/2}}.$$

Performing the usual matrix-by-vector multiplication, we obtain an explicit form of the required transform:

$$\frac{\phi}{c} = \gamma\frac{\phi'}{c} + \beta\gamma A'_x, \quad A_x = \beta\gamma\frac{\phi'}{c} + \gamma A'_x, \quad A_y = A'_y, \quad A_z = A'_z.$$

Frequently, it is more convenient to use these relations rewritten in the semi-vector form similar to Eqs. (9.135) for the fields:

$$\frac{\phi}{c} = \gamma\frac{\phi'}{c} + \beta\gamma A'_{||}, \quad A_{||} = \beta\gamma\frac{\phi'}{c} + \gamma A'_{||}, \quad \mathbf{A}_{\perp} = \mathbf{A}'_{\perp}, \qquad (*)$$

where the indices $||$ and $\perp$ mean the vector components, respectively, parallel and normal to the vector $\mathbf{v}$.[20]

Let us use Eq. (*) to calculate ϕ and $\mathbf{A}$ of a point charge $\mathbf{q}$ moving with velocity $\mathbf{u}$ relative to the ('lab') frame in which the potentials are measured. In the 'primed' frame, moving together with the charge in its origin, i.e. with velocity $\mathbf{v} = \mathbf{u}$ relative to the lab frame, the charge's field is purely electrostatic—see Eq. (1.35):

$$\phi' = \frac{q}{4\pi\varepsilon_0 r'}, \qquad \mathbf{A}' = 0,$$

[20] The reciprocal transform may be obtained either by solving the system of equations (*) for the primed potentials, or using Eq. (9.97b) with the reciprocal Lorentz matrix (9.98), or just by the reversal the velocity sign.

where $\mathbf{r}'$ is the observation point (as measured in the moving reference frame), so that Eqs. (*) yield

$$\phi = \gamma\phi' = \gamma\frac{1}{4\pi\varepsilon_0}\frac{q}{r'}, \quad A_{||} = \beta\gamma\frac{\phi'}{c} = \gamma\frac{v}{c^2}\frac{1}{4\pi\varepsilon_0}\frac{q}{r'}, \quad \mathbf{A}_{\perp} = 0. \qquad (**)$$

In a certain sense, this is already an answer, but for its applications we may need to spell out the potentials as explicit functions of observation point's position $\mathbf{r}$ and time t, as measured in the lab frame. For that, we should decompose the radius-vector $\mathbf{r}'$ of the field observation point into the parallel and normal components as well: $\mathbf{r}' = \mathbf{r}'_{||} + \mathbf{r}'_{\perp}$, and then use the first three of Eqs. (9.19b), which may be rewritten in a more compact, semi-vector form similar to that used in Eq. (*):

$$\mathbf{r}'_{||} = \gamma\left(\mathbf{r}_{||} - \boldsymbol{\beta}ct\right) \equiv \gamma(\mathbf{r}_{||} - \mathbf{u}t), \quad \mathbf{r}'_{\perp} = \mathbf{r}_{\perp},$$

so that, with $\mathbf{r}_{||} = \mathbf{n}_x x$ and $\mathbf{r}_{\perp} = \mathbf{n}_y y + \mathbf{n}_z z$, for the length r' of the vector $\mathbf{r}'$ we obtain

$$r' \equiv |\mathbf{r}'| = \left(r'^2_{||} + r'^2_{\perp}\right)^{1/2} = [\gamma^2(x - ut)^2 + y^2 + z^2]^{1/2}. \qquad (***)$$

If we select the time origin as was done in figure 9.11a of the lecture notes (partly reproduced below), just as it was done in section 9.5 for the derivation of expressions (9.139) and (9.140) for the electric and magnetic fields in the same situation, namely to provide $x = 0$ for the observation point, then Eq. (***) is simplified, and Eq. (**) yields

$$\phi = \gamma\frac{1}{4\pi\varepsilon_0}\frac{q}{(\gamma^2u^2t^2 + b^2)^{1/2}}, \quad A_{||} = \gamma\frac{u}{c^2}\frac{1}{4\pi\varepsilon_0}\frac{q}{(\gamma^2u^2t^2 + b^2)^{1/2}}, \quad \mathbf{A}_{\perp} = 0,$$

where $b \equiv (y^2 + z^2)^{1/2}$ is the 'impact parameter', i.e. the closest distance of the observation point from charge's trajectory.

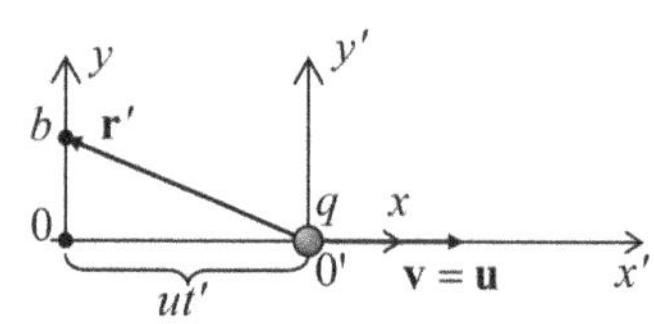

Comparing this result with Eqs. (9.139) and (9.140), we see that they involve the same denominator, but the expressions for the potentials are substantially simpler than those for the fields; in particular, they do not depend on the exact choice of the axes y and z within the plane normal to the vector $\mathbf{n}_x$ (and hence $\mathbf{u}$). However, we would need to make such a choice (for example, taking $y = b$, $z = 0$, as in the figure above) in order to derive the field formulas (9.139) and (9.140) from our result[21].

[21] For that, we would need to plug Eqs. (**), with r' given by Eq. (***), into Eqs. (9.121) of the lecture notes, spell them out in Cartesian components, perform the partial differentiation over x, y, z, and t, and only after that set $x = 0$, $y = b$, and $z = 0$. This calculation is more math than physics, but it still is recommended to the reader as an optional exercise.

***Problem* 9.20.** Calculate the scalar and vector potentials created by a time-independent electric dipole **p**, as measured in a reference frame that moves relatively to the dipole with a constant velocity **v**, with the shortest distance ('impact parameter') equal to b.

Solution: In the lab frame, in whose origin the dipole rests, its scalar potential is given by Eq. (3.7) of the lecture notes, while its magnetic field (and hence the vector potential) equals zero:

$$\phi = \frac{1}{4\pi\varepsilon_0}\frac{\mathbf{p}\cdot\mathbf{r}}{r^3}, \qquad \mathbf{A} = 0.$$

Applying to these potentials the Lorenz transform reciprocal to Eq. (*) of the previous problem's solution,

$$\frac{\phi'}{c} = \gamma\frac{\phi}{c} - \beta\gamma A_{||}, \quad A'_{||} = -\beta\gamma\frac{\phi}{c} + \gamma A_{||}, \quad \mathbf{A}'_{\perp} = \mathbf{A}_{\perp},$$

we obtain

$$\begin{aligned}\phi' &= \gamma\phi = \gamma\frac{1}{4\pi\varepsilon_0}\frac{\mathbf{p}\cdot\mathbf{r}}{r^3},\\ A'_{||} &= A'_x = -\beta\gamma\frac{\phi}{c} = -\gamma\frac{v}{c^2}\frac{1}{4\pi\varepsilon_0}\frac{\mathbf{p}\cdot\mathbf{r}}{r^3}, \qquad A'_y = A'_z = 0.\end{aligned} \qquad (*)$$

Just as in Eq. (**) in the previous problem's solution, this is already an answer, but it is natural to express the potentials explicitly via the time t' and the observation point's position $\mathbf{r}'$, as measured in the moving system—in particular since the vector $\mathbf{r}$ in the above formulas is an implicit function of time. For that, we may use the coordinate transform given by Eq. (9.19a) of the lecture notes:

$$x = \gamma(x' + vt'), \quad y = y', \quad z = z'.$$

From here,

$$\begin{aligned}r^2 &\equiv x^2 + y^2 + z^2 = \gamma^2(x' + vt')^2 + y'^2 + z'^2, \text{ and}\\ \mathbf{p}\cdot\mathbf{r} &\equiv p_x x + p_y y + p_z z = p_x\gamma(x' + vt') + p_y y' + p_z z',\end{aligned}$$

so that at the observation point with $x' = 0$,[22] $y' = b$, and $z' = 0$ (see figure 9.11 of the lecture notes), we obtain

$$\begin{aligned}\phi' &= \gamma\frac{1}{4\pi\varepsilon_0}\frac{p_x\gamma vt' + p_y b}{(\gamma^2 v^2 t'^2 + b^2)^{3/2}},\\ A'_x &= -\beta\gamma\frac{\varphi}{c} = -\gamma\frac{v}{c^2}\frac{1}{4\pi\varepsilon_0}\frac{p_x\gamma vt' + p_y b}{(\gamma^2 v^2 t'^2 + b^2)^{3/2}}, \qquad A'_y = A'_z = 0.\end{aligned}$$

Note that the Cartesian components p_x and p_y of the dipole moment, participating in this expression, are still as measured in the lab frame, rather than in the moving frame.

[22] As in section 9.5 of the lecture notes, and in the previous problem (see the figure in the model solution), this choice of x' implies such an origin of time t' that the distance between the observation point and the dipole is smallest (equal to b) at $t' = 0$.

One may run into claims that in the limit $v \ll c$, when $\gamma \approx 1$, this result is reduced to the sum of the statically calculated field of the initial electric dipole $\mathbf{p}$, and that of a co-located magnetic dipole $\mathbf{m} = \mathbf{v} \times \mathbf{p}$. However, this claim is incorrect. This may be proved, for example, by applying the *bac minus cab* rule of the vector algebra[23] to the double vector product $(\mathbf{v} \times \mathbf{p}) \times \mathbf{r}$, which would participate in this case in Eq. (5.90) of the lecture notes:

$$\mathbf{A}_{\mathrm{m}} = \frac{\mu_0}{4\pi}\frac{\mathbf{m} \times \mathbf{r}}{r^3} = \frac{\mu_0}{4\pi}\frac{(\mathbf{v} \times \mathbf{p}) \times \mathbf{r}}{r^3} \equiv -\frac{\mu_0}{4\pi}\frac{\mathbf{v}\,(\mathbf{p} \cdot \mathbf{r}) - \mathbf{p}\,(\mathbf{r} \cdot \mathbf{v})}{r^3},$$

and comparing the result with the expression $-\mathbf{v}(\mathbf{p} \cdot \mathbf{r})$ which participates in the three last Eqs. (*) in this limit, represented in the vector form:

$$\mathbf{A} \approx -\frac{\mathbf{v}}{c^2}\frac{1}{4\pi\varepsilon_0}\frac{\mathbf{p} \cdot \mathbf{r}}{r^3} \equiv -\frac{\mu_0}{4\pi}\frac{\mathbf{v}\,(\mathbf{p} \cdot \mathbf{r})}{r^3},$$

so that $\mathbf{A} \neq \mathbf{A}_{\mathrm{m}}$, unless $\mathbf{r} \perp \mathbf{v}$ (But: see the next problem.)

***Problem* 9.21.** Solve the previous problem, in the limit $v \ll c$, for a time-independent magnetic dipole $\mathbf{m}$.

Solution: In the lab frame, in whose origin the magnetic dipole rests, its electric field, and hence the scalar potential, equal zero, while its vector potential is given by Eq. (5.90) of the lecture notes:

$$\phi = 0, \qquad \mathbf{A} = \frac{\mu_0}{4\pi}\frac{\mathbf{m} \times \mathbf{r}}{r^3}. \tag{*}$$

In the leading, linear approximation in $\beta \ll 1$, the Lorenz factor $\gamma = 1/(1 - v^2/c^2)^{1/2} \approx 1 + v^2/2c^2$ equals just 1, so that the Lorentz transform relations used in the model solution of the previous problem,

$$\frac{\phi'}{c} = \gamma\frac{\phi}{c} - \beta\gamma A_{||}, \quad A'_{||} = -\beta\gamma\frac{\phi}{c} + \gamma A_{||}, \quad \mathbf{A}'_{\perp} = \mathbf{A}_{\perp},$$

are reduced to just

$$\frac{\phi'}{c} = \frac{\phi}{c} - \beta A_{||}, \quad A'_{||} = -\beta\frac{\phi}{c} + A_{||}, \quad \mathbf{A}'_{\perp} = \mathbf{A}_{\perp}. \tag{**}$$

Since in our current problem $\phi = 0$, the two last relations (**) may be summarized as $\mathbf{A}' = \mathbf{A}$, i.e. the vector potential is described by the second of Eqs. (*) even in the moving system, although we should remember that, just as in the previous problem, if $\mathbf{r}'$ is fixed, the longitudinal component of vector $\mathbf{r}$, and hence this vector as a whole, are now functions of time:

$$r_{||} = \gamma(r'_{||} + vt') \approx r'_{||} + vt', \qquad \mathbf{r}'_{\perp} = \mathbf{r}_{\perp}.$$

[23] See, e.g. Eq. (A.47).

In contrast, as the first of Eqs. (**) shows, the scalar potential ϕ' does not vanish in the linear approximation:

$$\phi' = -c\beta A_{||} = -v\frac{\mu_0}{4\pi}\frac{(\mathbf{m}\times\mathbf{r})_{||}}{r^3}.$$

Let us use the fact that the vector **v** has only one, longitudinal component to rewrite the last expression as

$$\phi' = -\frac{\mu_0}{4\pi}\frac{\mathbf{v}\cdot(\mathbf{m}\times\mathbf{r})}{r^3},$$

and then use the operand rotation rule of the vector algebra[24] to transform it to

$$\phi' = -\frac{\mu_0}{4\pi}\frac{\mathbf{r}\cdot(\mathbf{v}\times\mathbf{m})}{r^3} \equiv -\frac{1}{4\pi\varepsilon_0}\frac{[(-\mathbf{v}/c^2)\times\mathbf{m}]\cdot\mathbf{r}}{r^3}.$$

Comparing this expression with that for an electric dipole (see the previous problem) in the same linear approximation in $v \ll c$:

$$\phi'_{\mathrm{e}} = \gamma\frac{1}{4\pi\varepsilon_0}\frac{\mathbf{p}\cdot\mathbf{r}}{r^3} \approx \frac{1}{4\pi\varepsilon_0}\frac{\mathbf{p}\cdot\mathbf{r}}{r^3},$$

We see that they coincide if we take

$$\mathbf{p} = -\frac{\mathbf{v}\times\mathbf{m}}{c^2}. \qquad (***)$$

Hence, in the low-velocity approximation, the field of a static magnetic dipole **m** looks, for a moving observer, as a system consisting of this dipole plus a co-located electric dipole with the moment (***). (As was discussed in the model solution of the previous problem, the reciprocal statement for an electric dipole is incorrect.)

***Problem* 9.22.** Assuming that the magnetic monopole does exist and has magnetic charge g, calculate the change $\Delta\Phi$ of the magnetic flux in a superconductor ring due to the passage of single monopole through it. Evaluate $\Delta\Phi$ for the monopole charge conjectured by P Dirac, $g = ng_0 \equiv n(2\pi\hbar/e)$, where n is an integer; compare the result with the magnetic flux quantum Φ_0 (6.62) with $|q| = e$, and discuss their relation.

Hint: For simplicity, you may consider the monopole's passage along the symmetry axis of a thin, round superconducting ring, in the otherwise free space.

Solution: The monopole charge is defined so that its density ρ_{m} would make the Maxwell equations (6.99) completely symmetric:

$$\nabla\times\mathbf{E} + \frac{\partial\mathbf{B}}{\partial t} = \mathbf{j}_{\mathrm{m}}, \qquad \nabla\times\mathbf{H} - \frac{\partial\mathbf{D}}{\partial t} = \mathbf{j},$$
$$\nabla\cdot\mathbf{D} = \rho, \qquad \nabla\cdot\mathbf{B} = \rho_{\mathrm{m}},$$

[24] See, e.g. Eq. (A.48).

where $\mathbf{j}_m = \mathbf{v}\rho_m$ is the magnetic current due to the monopole's motion with velocity $\mathbf{v}$, and $\mathbf{D} = \varepsilon_0\mathbf{E}$, $\mathbf{B} = \mu_0\mathbf{H}$.

If a single monopole of charge g is at rest at origin, $\rho_m = g\delta(\mathbf{r})$, the last equation may be readily integrated[25] to give

$$\mathbf{B} = \frac{g}{4\pi}\frac{\mathbf{r}}{r^3},$$

in an evident analogy with the Coulomb law, rewritten for the field $\mathbf{D}$. Now let us see how the monopole's field look when viewed from a reference frame moving, with a constant velocity $\mathbf{u}$, along a straight line passing at the closest distance b from the monopole. Selecting axis x along this motion, axis y in such a way that the field observation point at time t has coordinates $\{vt, b, 0\}$ (see, e.g. figure 9.11a of the lecture notes, or its reproduction in the model solution of problem 9.19), and using the Lorentz transform (9.135), we obtain the following analog of Eqs. (9.139):

$$B_x = -\frac{g}{4\pi}\frac{u\gamma t}{(u^2\gamma^2t^2 + b^2)^{3/2}}, \qquad B_y = \frac{g}{4\pi}\frac{\gamma b}{(u^2\gamma^2t^2 + b^2)^{3/2}}, \qquad B_z = 0.$$

Due to the relativity of motion, and the angular symmetry of the problem, we may use the first of these results for any point of the lab frame's plane $[y, z]$, at distance b from its center, to calculate the magnetic flux of the monopole (moving with velocity $-\mathbf{u}$) through the area of the circle with $x = 0$, $y^2 + z^2 \leqslant R^2$:

$$\Phi(t) = 2\pi\int_0^R B_x b db = -\frac{g}{2}\int_0^R \frac{u\gamma t}{(u^2\gamma^2t^2 + b^2)^{3/2}} b db = -\frac{g}{4}u\gamma t\int_{b=0}^{b=R}\frac{d(u^2\gamma^2t^2 + b^2)}{(u^2\gamma^2t^2 + b^2)^{3/2}}$$

$$= \frac{g}{2}u\gamma t\frac{1}{(u^2\gamma^2t^2 + b^2)^{1/2}}\Bigg|_{b=0}^{b=R} = \frac{g}{2}u\gamma t\left[\frac{1}{(u^2\gamma^2t^2 + R^2)^{1/2}} - \frac{1}{u\gamma t}\right], \quad \text{for } t \leqslant 0.$$

As time increases from $t = -\infty$ to $t = 0$, this function is always negative, with a gradually increasing magnitude; at $t \to 0$, it tends to a non-vanishing value, $-g/2$. However, because of the divergence of the second term in square brackets at $t = 0$, we cannot formally follow this result to $t > 0$, and should rather make such extension using the symmetry considerations. Since B_x, for any $b > 0$, is an odd function of time, so should be its spatial integral, i.e. the change of flux. Hence the function $\Phi(t)$, evidently continuous, has the shape shown (schematically) in the figure below, i.e. performs a jump,

$$\Delta\Phi = -g,$$

[25] This may be done either by plugging the symmetry-imposed solution $\mathbf{B}(\mathbf{r}) = \mathbf{n}_r B(r)$ into that equation rewritten in spherical coordinates, or by applying the magnetic-field analog of the Gauss law (3.34),

$$\oint_S B_n d^2r = \int_V \rho_m d^3r,$$

to a sphere of radius r, with the center at the origin.

as a result of the monopole's passage through the ring. (It is possible to prove that this result does not depend on the shape of the ring, and on details of the monopole's trajectory.)

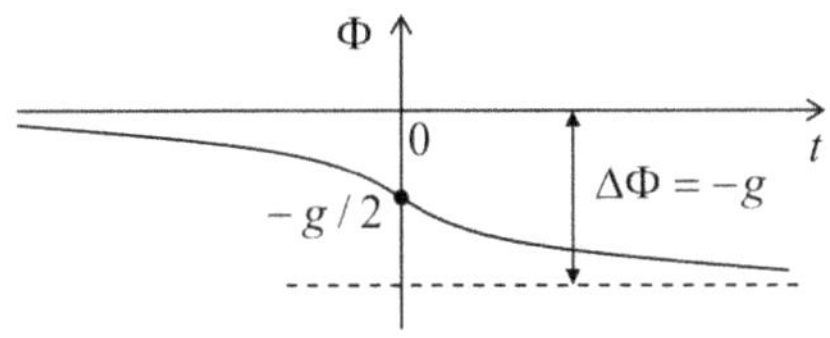

For the Dirac monopole, this jump equals

$$\Delta\Phi = ng_0 = n\frac{2\pi\hbar}{e},$$

i.e. would be the integer number of magnetic flux quanta Φ_0 for superconductors with the carrier charge $q = \pm e$ (and equals to $2n\Phi_0$ for known superconductors with $q = -2e$). This coincidence is by no means occasional. As was discussed in section 6.5 of the lecture notes, one flux quantum is equivalent to the shift of the phase of the wavefunction by 2π, indistinguishable in the ordinary matter. P Dirac has derived his value of g exactly from the requirement that usual particles, with their wavefunctions spread around the monopole's trajectory, would not be affected by its passage—otherwise the consequences of existence of even a single monopole for our Universe would be too drastic.

However, the change by $2\Phi_0$ (or a multiple of this amount) of the flux applied to a superconducting ring may be readily measured by a SQUID magnetometer—see section 6.5 again. Actually, a monopole search using SQUIDs was conducted at Stanford University in the 1980s, and one observed event was even claimed—but never confirmed either by this or other groups.

Problem* 9.23. Calculate the trajectory of a relativistic particle in a uniform electrostatic field **E**, for an arbitrary direction of its initial velocity **u**(0), using two different approaches—at least one of them different from the approach used in section 9.6 of the lecture notes for the case **u**(0) ⊥ **E**.

Solution: An elegant way to solve this problem is to integrate the four-vector equation (9.145),

$$\frac{dp^\alpha}{d\tau} = qF^{\alpha\beta}u_\beta,$$

directly, considering the proper time τ of the particle (rather than the lab-frame time t) as an argument. For the three non-zero components of the four-velocity u_β,[26] this gives us three equations which may be conveniently rewritten as follows:

[26] I am using the same coordinate system choice as in section 9.6 of the lecture notes, with the axis z directed along the electric field, and the axis x in the plane of the motion, so that $u_y = 0$ for any τ.

$$\frac{d(\gamma c)}{d(\Gamma\tau)} = \gamma u_z, \qquad \frac{d(\gamma u_x)}{d(\Gamma\tau)} = 0, \qquad \frac{d(\gamma u_z)}{d(\Gamma\tau)} = \gamma c,$$

where $\Gamma \equiv qE/cm$ is a constant parameter with the dimensionality of the reciprocal time (s^{-1}). The integration of the middle equation is elementary, and yields

$$\gamma u_x = \text{const} = \frac{cu_x(0)}{[c^2 - u_x^2(0) - u_z^2(0)]^{1/2}} \equiv C.$$

The remaining two equations may be combined (by the additional differentiation of any of them over τ and plugging the remaining equation into the result) to give two similar second-order differential equations

$$\frac{d^2}{d\tau^2}(\gamma c) = \Gamma^2(\gamma c), \qquad \frac{d^2}{d\tau^2}(\gamma u_z) = \Gamma^2(\gamma u_z),$$

with the solutions which differ only because of the different initial conditions:

$$\gamma c = A \cosh \Gamma(\tau - \tau_0), \qquad \gamma u_z = A \sinh \Gamma(\tau - \tau_0). \tag{*}$$

The constants A and τ_0, participating in these solutions, may be found from the initial conditions:

$$A = c\left[\frac{c^2 - u_z^2(0)}{c^2 - u_x^2(0) - u_z^2(0)}\right]^{1/2} = c\frac{\gamma(0)}{\gamma_z(0)}, \qquad \gamma_z(0) \equiv \frac{1}{[1 - u_z^2(0)/c^2]^{1/2}},$$

$$\tau_0 = -\frac{1}{\Gamma}\tanh^{-1}\frac{u_z(0)}{c}.$$

Physically, τ_0 is the proper time at which the particle reaches the lowest value of z.

Now we can find the tangent to the particle's trajectory:

$$\frac{dz}{dx} = \frac{u_z}{u_x} = \frac{A}{C}\sinh \Gamma(\tau - \tau_0),$$

but in order to integrate this equation we still need to express its right-hand side as a function of either x or z. For example, we can find $z(\tau)$, integrating the second of Eqs. (*):

$$\gamma u_z \equiv \gamma\frac{dz}{dt} \equiv \frac{dz}{d\tau} = A \sinh \Gamma(\tau - \tau_0),$$

which yields

$$z = A\int_0^\tau \sinh \Gamma(\tau - \tau_0)d\tau = \frac{A}{\Gamma}[\cosh \Gamma(\tau - \tau_0) - \cosh \Gamma\tau_0].$$

From here,

$$\cosh \Gamma(\tau - \tau_0) = \frac{\Gamma z}{A} + \cosh \Gamma \tau_0 = \frac{\Gamma z}{A} + \frac{\gamma(0)c}{A} = \frac{\Gamma z}{A} + \gamma_z(0),$$

$$\sinh \Gamma(\tau - \tau_0) = \left[\left(\frac{\Gamma z}{A} + \gamma_z(0)\right)^2 - 1\right]^{1/2},$$

and now we can finally find the trajectory's tangent

$$\frac{dz}{dx} = \frac{A}{C} \sinh \Gamma(\tau - \tau_0) = \frac{1}{C}\{[\Gamma z + A\gamma_z(0)]^2 + A^2\}^{1/2},$$

and the trajectory itself:

$$x = C\int_0^z \frac{dz}{\{[\Gamma z + A\gamma_z(0)]^2 - A^2\}^{1/2}} = \frac{C}{\Gamma}\left[\cosh^{-1}\left(\frac{\Gamma z}{A} + \gamma_z(0)\right) - \cosh^{-1}\gamma_z(0)\right]. \quad (**)$$

This is a natural generalization of the result derived in section 9.6 for the particular case $u_z(0) = 0$. Indeed, in that case $\gamma_z(0) = 1$, so that the above expressions yield $A = c\gamma(0)$, $A/\Gamma = \gamma(0)mc^2/qE \equiv \mathscr{E}(0)/qE$, $C/\Gamma = p(0)c/qE$, so that Eq. (**) is reduced to the form

$$x = \frac{p(0)c}{qE}\left[\cosh^{-1}\left(\frac{qEz}{\mathscr{E}_0} + 1\right)\right],$$

equivalent to Eq. (9.165) of the lecture notes.

An alternative way to solve this problem is to notice that Eq. (9.165) of the lecture notes,

$$z = \frac{\mathscr{E}(0)}{qE}\left(\cosh\frac{qEx}{cp(0)} - 1\right),$$

obtained for the case $\mathbf{u}(0) \perp \mathbf{E}$, may be used in the general case as well, if we shift the origins of x, z, and t (see the figure below),

$$z = z_0 + \frac{\mathscr{E}(0)}{qE}\left(\cosh\frac{qE(x - x_0)}{cp(0)} - 1\right),$$

and the 'only' remaining thing to do is to express the parameters $\mathscr{E}(0)$, $p(0)$, x_0, and z_0 via the components $u_x(0)$, $u_z(0)$ of the new initial velocity, which in this notation becomes $\mathbf{u}(t_0)$ However, the recalculation (leading of course to the same result as above) makes this method not much easier than the first, more elegant way. Still, it is highly recommended to the reader as a useful optional exercise.

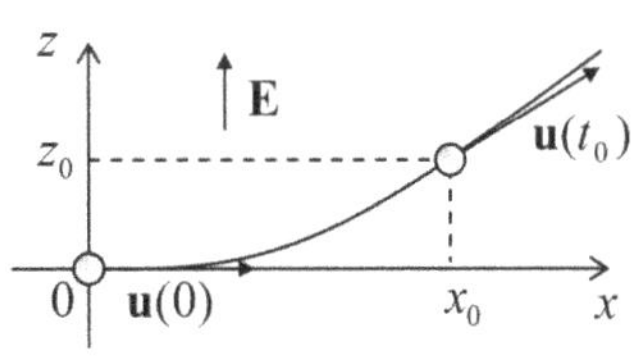

***Problem* 9.24.** A charged relativistic particle with velocity u performs a planar cyclotron rotation in a uniform external magnetic field B. How much would the velocity and the orbit radius change at a slow change of the field to a new magnitude B' ?

Solution: The initial orbit's radius may be found from Eq. (9.153) of the lecture notes:

$$R = \frac{p}{qB} \equiv \frac{\gamma m u}{qB} \equiv \frac{m}{qB} \frac{u}{(1 - u^2/c^2)^{1/2}},$$

where m is the particle's rest mass. Thus, the initial magnetic flux through the orbit circle's area $A = \pi R^2$ is

$$\Phi = \pi R^2 B = \pi \frac{m^2}{q^2 B} \frac{u^2}{1 - u^2/c^2}.$$

For the new field value B', the flux is described by a similar formula:

$$\Phi' = \pi R'^2 B' = \pi \frac{m^2}{q^2 B'} \frac{u'^2}{1 - u'^2/c^2},$$

where u' is the new velocity of the particle.

As was argued in section 9.7 (in the context of slow spatial variations of the magnetic field), the magnetic flux through the orbit is the adiabatic invariant of particle's motion. These arguments remain valid for slow temporal variations of the field as well, so that we may write $\Phi' = \Phi$, giving us the following equation for finding u':

$$\frac{1}{B'} \frac{u'^2}{1 - u'^2/c^2} = \frac{1}{B} \frac{u^2}{1 - u^2/c^2}.$$

This is a linear equation for u'^2, easy to solve. The result,

$$u' = \frac{u}{\left[\frac{B}{B'} + \frac{u^2}{c^2}\left(1 - \frac{B}{B'}\right)\right]^{1/2}},$$

becomes simpler in both the non-relativistic limit: $u' \approx u(B'/B)^{1/2}$, and the ultra-relativistic limit: $u' \approx u \approx c$. At arbitrary u, the field's increase results in an increase of particle's velocity. Physically, this change of the particle's velocity, and hence of its energy, is the result of the Faraday's induction e.m.f. (6.2) appearing during the field's increase.

For the orbit's radius, the same adiabatic invariance of flux, $\Phi' = \Phi$, provides the equation

$$B' R'^2 = B R^2,$$

which gives a simpler result, independent on particle's velocity:

$$R' = R\left(\frac{B}{B'}\right)^{1/2}.$$

It shows that a field's increase leads to a shrinkage of the orbit.

Problem* 9.25. Analyze the motion of a relativistic particle in uniform, mutually perpendicular fields **E** and **B**, for the particular case when E is *exactly* equal to cB.

Solution: Evidently, for this particular case, both velocities defined by Eqs. (9.168) and (9.174) yield $v = c$, so that the introduction of a reference frame moving with this velocity (the trick used in section 9.6(iii) of the lecture notes) is not too helpful, so that we better start from scratch.

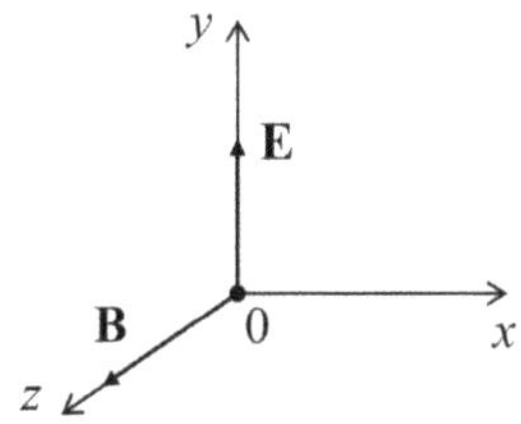

Directing the coordinate axes as in figure 9.12 of the lecture notes (partly reproduced above), so that $E_y = E$ and $B_z = B = E/c$ are the only non-zero field components, for an arbitrary velocity $\mathbf{u}$ of the particle we obtain $\mathbf{u} \times \mathbf{B} = \mathbf{n}_x u_y E/c - \mathbf{n}_y u_x E/c$, so that Eq. (9.144) of the lecture notes, decomposed into three scalar equations, yields

$$\frac{dp_x}{dt} = q\frac{u_y}{c}E, \qquad \frac{dp_y}{dt} = qE\left(1 - \frac{u_x}{c}\right), \qquad \frac{dp_z}{dt} = 0. \tag{*}$$

while Eq. (9.148) gives the following equation for the kinetic energy $\mathscr{E}$:

$$\frac{d\mathscr{E}}{dt} = qEu_y. \tag{**}$$

Combining the first of Eqs. (*) with Eq. (**), we obtain

$$\mathscr{E} - cp_x = \text{const} \equiv a, \tag{***}$$

while the last of Eqs. (*) yields $p_z = \text{const}$, so that we may form another (positive) constant of motion, $(mc^2)^2 + c^2p_z^2 \equiv b \geqslant 0$. Plugging these two constants (equal to a and b) into the basic equation (9.78) for the relativistic energy of the particle, rewritten as

$$\mathscr{E}^2 - c^2p_x^2 \equiv (\mathscr{E} - cp_x)(\mathscr{E} + cp_x) = (mc^2)^2 + c^2p_y^2 + c^2p_z^2,$$

we obtain

$$a(\mathscr{E} + cp_x) = c^2p_y^2 + b. \tag{****}$$

Now considering Eqs. (***) and (****) as a system of two linear equations for $\mathscr{E}$ and p_x, we may express these variables via p_y (and the constants of motion):

$$\mathscr{E} = \frac{c^2p_y^2 + b}{2a} + \frac{a}{2}, \qquad cp_x = \frac{c^2p_y^2 + b}{2a} - \frac{a}{2}.$$

The first of these relations may be plugged into the product of $\mathscr{E}$ by the derivative dp_y/dt, expressed from the differential equation (*) for p_y, and then transformed using the universal relativistic relation $\mathbf{p} = (\mathscr{E}/c^2)\mathbf{u}$ which was mentioned in section 9.3:

$$\mathscr{E}\frac{dp_y}{dt} = \mathscr{E}\, qE\left(1 - \frac{u_x}{c}\right) \equiv qE\left(\mathscr{E} - \mathscr{E}\frac{u_x}{c}\right) \equiv qE(\mathscr{E} - cp_x) \equiv qEa. \qquad (*****)$$

The result of this substitution,

$$\left(\frac{c^2p_y^2 + b}{2a} + \frac{a}{2}\right)\frac{dp_y}{dt} \equiv \frac{d}{dt}\left[\frac{c^2}{6a}p_y^3 + \left(\frac{b}{2a} + \frac{a}{2}\right)p_y\right] = qEa,$$

may be readily integrated over time to obtain the time dependence of p_y in an implicit form:

$$\frac{c^2}{6a}p_y^3 + \left(\frac{b}{2a} + \frac{a}{2}\right)p_y = qEa(t - t_0),$$

where t_0 is the time moment when $p_y = 0$.

Even without solving this cubic equation, we may use the relation $dt/\mathscr{E} = dp_y/qEa$ (which follows from Eq. (*****) above) to express particle's coordinates as functions of p_y:

$$x \equiv \int u_x dt = \int \frac{c^2p_x}{\mathcal{E}}dt = \frac{c}{qEa}\int cp_x dp_y = \frac{c}{qEa}\int\left(\frac{c^2p_y^2 + b}{2a} - \frac{a}{2}\right)dp_y$$

$$= \frac{c}{qEa}\left[\frac{c^2}{6a}p_y^3 + \left(\frac{b}{2a} - \frac{a}{2}\right)p_y\right] + \text{const},$$

$$y \equiv \int u_y dt = \int \frac{c^2p_y}{\mathscr{E}}dt = \frac{c}{qEa}\int cp_y dp_y = \frac{c^2}{2qEa}p_y^2 + \text{const},$$

$$z = \int u_z dt = \int \frac{c^2p_z}{\mathscr{E}}dt = \frac{c}{qEa}\int cp_z dp_y = \frac{c^2p_z}{qEa}p_y + \text{const}.$$

Since z is a linear function of p_y, the first two of these relations yield the particle's trajectory in the form $x(z)$, $y(z)$. Note that at large times, the magnitude of p_y (and hence of z) grows at $(t - t_0)^{1/3}$, the coordinate y grows faster, as $(t - t_0)^{2/3}$, while the motion along axis x (i.e. in the direction perpendicular to both vectors **E** and **B**) becomes the fastest one:

$$x \to \frac{c^3}{6qEa^2}p_y^3 \to c(t - t_0).$$

This is essentially the drift which was discussed in section 9.6(iii), whose velocity formally obeys Eq. (9.168) and (9.174), but since in the reference frame moving with this velocity ($v = c$) both the electric and magnetic fields disappear, the frequency (9.172) vanishes, so that the particle does not perform the simultaneous cyclotron motion.

Problem* 9.26. Find the law of motion of a relativistic particle in uniform, parallel, static fields **E** and **B**.

Solution: For an arbitrary direction of the initial momentum **p**(0) of the particle, we can always select such a reference frame that

$$\mathbf{E} = \{0,\ 0,\ E\}, \qquad \mathbf{B} = \{0,\ 0,\ B\}, \qquad \mathbf{p}(0) = \{p_x(0),\ 0,\ p_z(0)\}.$$

Then the four-vector equation (9.145) of motion of the particle,

$$\frac{du^\alpha}{d\tau} = \frac{q}{m} F^{\alpha\beta} u_\beta,$$

may be separated into four Cartesian components as

$$\frac{d(\gamma c)}{d\tau} = \Gamma(\gamma u_z), \qquad \frac{d(\gamma u_x)}{d\tau} = -\omega(\gamma u_y), \qquad \frac{d(\gamma u_y)}{d\tau} = \omega(\gamma_u u_x), \qquad \frac{d(\gamma u_z)}{d\tau} = \Gamma(\gamma c), \quad (*)$$

where $\Gamma \equiv qE/cm$, and $\omega \equiv qB/m$. (Note that ω so defined is *not* the usual relativistic cyclotron frequency (9.151), $\omega_c \equiv qB/M = qB/\gamma m$: the denominator in the definition of ω does not include the Lorentz factor $\gamma = (1 - \beta^2)^{-1/2}$, as it would if we dealt not with the proper time τ of the particle, but with the lab time t).

Now we see that the first and the last equations of Eqs. (*) form an independent system, which may be readily reduced to one second-order equation (by the usual trick of differentiation of one of the initial equations over τ, and then plugging the counterpart equation into the result)[27]. As a result we obtain two similar equations,

$$\frac{d^2}{d\tau^2}(\gamma c) = \Gamma^2(\gamma c), \qquad \frac{d^2}{d\tau^2}(\gamma u_z) = \Gamma^2(\gamma u_z),$$

which may be readily integrated, and integration constants found from the initial values of γ and u_z. The result is

$$\gamma(\tau) = \gamma(0)\left[\cosh \Gamma\tau + \beta_z(0)\sinh \Gamma\tau\right], \qquad u_z(\tau) = c\frac{\beta_z(0)\cosh \Gamma\tau + \sinh \Gamma\tau}{\cosh \Gamma\tau + \beta_z(0)\sinh \Gamma\tau}. \quad (**)$$

The second and third of Eqs. (*) form a similar system (in which the electric field does not participate), which is exactly the same as for the non-relativistic cyclotron

[27] Note that since this system of two equations does not depend on the magnetic field, it gives an alternative way of solving the problem discussed in section 9.6(ii) of the lecture notes.

motion. It may be readily integrated to give (at the appropriate choice of origins of the proper time τ, and of the axis y in the plane of the initial velocity):

$$\gamma u_x = -\frac{p_x(0)}{m}\sin\omega\tau, \qquad \gamma u_y = \frac{p_x(0)}{m}\cos\omega\tau.$$

Since $\gamma u_x \equiv \gamma dx/dt = dx/d\tau$ (and similarly for y), these equations may be readily integrated again, giving

$$x = R\cos\omega\tau + \text{const}, \qquad y = R\cos\omega\tau + \text{const},$$

where the cyclotron orbit radius $R \equiv p_x(0)/qB$ is not affected by the initial momentum of the particle in the z direction.

The relation between the proper time τ and the lab time t may be found by plugging the first of Eqs. (**) into the fundamental relation $dt = \gamma d\tau$, and integrating it. The result,

$$t = \frac{\gamma(0)}{\Gamma}\left[\sinh\Gamma\tau + \beta_z(0)\cosh\Gamma\tau\right] + \text{const},$$

shows that at large times ($\tau \gg 1/\Gamma$), the proper time scales as the logarithm of the lab time. So, as viewed from the laboratory frame, the cyclotron motion gradually slows down, due to the gradual increase of the relativistic mass $M = \gamma m$ of the particle—because of the increase of its energy by the electric field.

***Problem* 9.27.** Neglecting relativistic effects, calculate the smallest voltage V that has to be applied between the anode and cathode of a magnetron (see figure 9.13 of the lecture notes and its discussion) to enable electrons to reach the anode, at negligible electron–electron interactions (including the space-charge effects), and collisions with the residual gas molecules. You may:

(i) model the cathode and anode as two coaxial round cylinders, of radii R_1 and R_2, respectively;
(ii) assume that the magnetic field $\mathbf{B}$ is uniform and directed along their common axis; and
(iii) neglect the initial velocity of the electrons emitted by the cathode.

(After the solution, estimate the validity of the last assumption and of the non-relativistic approximation for reasonable values of parameters.)

Solution: In the non-relativistic limit, we may restrict the Taylor expansion

$$\frac{1}{\gamma} \equiv \left(1 - \frac{u^2}{c^2}\right)^{1/2} = 1 - \frac{u^2}{2c^2} + \cdots$$

to the two leading terms spelled up above, and hence approximate Eq. (9.183) of the lecture notes as

$$\mathscr{L} = mc^2 + \frac{m}{2}u^2 - q\phi + q\mathbf{u}\cdot\mathbf{A}. \qquad (*)$$

In our current problem, in the cylindrical coordinates $\mathbf{r} = \{\rho, \varphi, z\}$ (with the origin of ρ placed on the common axis of the two electrodes), the magnetic field has only a z-component, $\mathbf{B} = \mathbf{n}_z B$, while the electric field has only a radial component: $\mathbf{E} = -\mathbf{n}_\rho d\phi(\rho)/\partial\rho$. Since the emitted electrons have negligible initial velocities, neither component of the Lorentz force (5.10) drives them in the z-direction. As a result, each electron moves only in a plane normal to the common axis, and its velocity has only two components:

$$\mathbf{u} = \mathbf{n}_\rho \dot{\rho} + \mathbf{n}_\varphi \rho\dot{\varphi},$$

so that

$$u^2 = \dot{\rho}^2 + \rho^2\dot{\varphi}^2.$$

Selecting the vector-potential's gauge so that it has the natural axially symmetric form $\mathbf{A} = \mathbf{n}_\varphi A(\rho)$, and using the well-known expression for the curl operator of such a vector in the cylindrical coordinates[28], $\nabla \times \mathbf{A} = \mathbf{n}_z[\partial(\rho A)/\partial\rho]/\rho$, we see that the potential's definition, $\nabla \times \mathbf{A} = \mathbf{B}$, with $\mathbf{B} = \mathbf{n}_z B = \text{const}$, is satisfied if $A(\rho) = B\rho/2$. So the scalar product, participating in the Lagrangian function (*), is

$$\mathbf{u} \cdot \mathbf{A} = \rho\dot{\varphi}\frac{B\rho}{2},$$

and the whole function (excluding its inconsequential rest-energy part mc^2) becomes

$$\mathscr{L} = \frac{m}{2}(\dot{\rho}^2 + \rho^2\dot{\varphi}^2) - q\phi(\rho) + q\dot{\varphi}\frac{B\rho^2}{2} + \text{const},$$

with $m = m_e$ and $q = -e$. Now using this function, in a regular way[29], to derive the Lagrange equations of motion of two generalized coordinates, ρ and φ, we obtain

$$\frac{d}{dt}(m\dot{\rho}) - m\rho\dot{\varphi}^2 + q\frac{d\phi}{d\rho} = 0, \quad \frac{d}{dt}\left(m\rho^2\dot{\varphi} + q\frac{B\rho^2}{2}\right) = 0, \qquad (**)$$

with the function $\phi(\rho)$ following from the electrostatics[30].

(*Parenthetic remark*: In our simple case of unconstrained motion, Eqs. (**) may be also obtained directly from the non-relativistic version of Eq. (9.144):

$$\frac{d}{dt}(m\mathbf{u}) = q(\mathbf{E} + \mathbf{u} \times \mathbf{B}),$$

spelled out in the polar coordinates, but technically this way is somewhat more cumbersome.)

[28] See, e.g. Eq. (A.63) with $\mathbf{f} = \mathbf{A}$, $A_\rho = A_z = 0$, and $\partial/\partial\varphi = \partial/\partial z = 0$.

[29] If you need a reminder (I hope you do not), please see *Part CM*, Eq. (2.19).

[30] See, e.g. Eq. (2.45) of the lecture notes.

The second of Eqs. (**) is the simplest of the two, immediately giving the following integral of motion[31]:

$$m\rho^2\dot{\varphi} + q\frac{B\rho^2}{2} = \text{const}.$$

The constant on the right-hand side of this relation may be evaluated from the condition of negligible velocity (including its azimuthal component) on the cathode's surface, i.e. at $\rho = R_1$. The result may be represented as

$$\dot{\varphi} = -\frac{qB}{2m}\left(1 - \frac{R_1^2}{\rho^2}\right). \qquad (***)$$

This equation may be combined with the law of conservation of the energy (9.196)[32]:

$$\frac{m}{2}u^2 + q\phi(\rho) \equiv \frac{m}{2}(\dot{\rho}^2 + \rho^2\dot{\varphi}^2) + q\phi(\rho) = \text{const}.$$

Applying this law to the points $\rho = R_1$ (where both components of velocity are negligibly small) and $\rho = R_2$, and taking $\phi(R_1) = 0$ and $\phi(R_2) = V$, where V is the voltage applied between the anode and the cathode, we obtain

$$\frac{m}{2}(\dot{\rho}^2 + \rho^2\dot{\varphi}^2)_{\rho=R_2} + qV = 0.$$

Plugging Eq. (***), written for $\rho = R_2$, into this equality, we may solve it for the voltage V:

$$V = -\frac{m}{2q}\left[\dot{\rho}^2|_{\rho=R_2} + R_2^2\left(\frac{qB}{2m}\right)^2\left(1 - \frac{R_1^2}{R_2^2}\right)^2\right]. \qquad (****)$$

(For electrons, with their negative charge $q = -e$, this voltage is positive.)

If V is lower than the threshold value, the particles move along loop trajectories not reaching the anode (see the—very schematic—dashed line in the figure below), with the radial velocity $\dot{\rho}$ vanishing at some point, most distant from the axis, but still smaller than R_2. The threshold voltage V_t we are looking for may be defined as the value of V at which this most distant point of the trajectory exactly equals R_2—see the solid line in the same figure. For this value, Eq. (****), with $q = -e$, and $m = m_e$, yields

$$V_t = \frac{eB^2}{8m_e}R_2^2\left(1 - \frac{R_1^2}{R_2^2}\right)^2.$$

[31] This integral may also be derived by applying Eq. (9.201) to a circular contour of radius ρ, pierced by a constant, normal magnetic field **B**, so that $\Phi = \pi\rho^2 B$.

[32] As we know from section 9.7, this law may be derived from particle's equation of motion—in our particular case, from Eqs. (**), so it is no surprise we would not need to use explicitly the first of these equations.

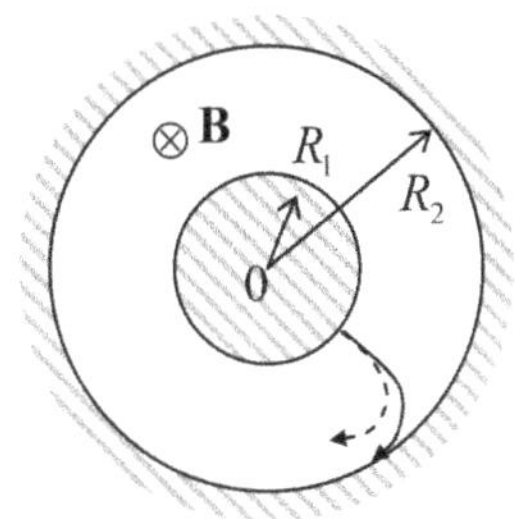

For typical magnetrons (say, those used in kitchen microwave ovens) with $R_2 \sim 1$ cm, $R_1 \sim R_2/2$ and $B \sim 0.1$ T,[33] this expression yields realistic values of V_t, of the order of 10 kV. Note that on the scale of the energy ~10 keV provided by this field, the random thermal energies $\sim k_B T$ of the electrons emitted by the cathode (with the typical temperature $T \sim 3{,}000$ K), of the order of 0.3 eV, are indeed negligible. On the other hand, such kinetic energy is much lower than the rest energy $m_e c^2 \approx 0.5$ MeV of an electron, so that the relativistic effects are negligible as well.

***Problem* 9.28.** A charged, relativistic particle has been injected into a uniform electric field whose magnitude oscillates in time with frequency ω. Calculate the time dependence of the particle's velocity, as observed from the lab reference frame.

Solution: Directing axis x along the electric field, and axis y perpendicular to it, but within the plane of the initial velocity of the particle (so that $\beta_z \equiv 0$), we can rewrite the relativistic equation (9.144) of its motion in the lab frame as

$$\frac{d(\gamma\beta_x)}{d(\omega t)} = \kappa \cos \omega t, \qquad \frac{d(\gamma\beta_y)}{d(\omega t)} = 0,$$

with the only dimensionless parameter

$$\kappa \equiv \frac{qE_\omega}{mc\omega},$$

E_ω being the electric field's amplitude. These equations may be readily integrated:

$$\gamma\beta_x = \gamma_0\beta_{x0} + \kappa \sin \omega t, \qquad \gamma\beta_y = \gamma_0\beta_{y0}. \qquad (*)$$

Since

$$\gamma^2 \equiv \frac{1}{1-\beta^2}, \quad \text{i.e.} \quad \gamma^2(1-\beta^2) = 1, \quad \text{i.e.} \quad \gamma^2 = 1 + (\gamma\beta)^2 = 1 + (\gamma\beta_x)^2 + (\gamma\beta_y)^2,$$

we may use Eq. (*) to calculate the Lorentz factor as an explicit function of time:

$$\gamma = \left[1 + (\gamma\beta_x)^2 + (\gamma\beta_y)^2\right]^{1/2} = \left[1 + (\gamma_0\beta_{x0} + \kappa \sin \omega t)^2 + (\gamma_0\beta_{y0})^2\right]^{1/2}. \qquad (**)$$

[33] This value is chosen so that the average frequency $\omega = \bar{\dot\varphi}$ of the electron motion, following from Eq. (***), would be close to the own frequency of the resonating cavities on the device (see figure 9.13 of the lecture notes), and hence to the frequency of the microwaves it generates. This frequency is mandated to be in a narrow range around 2.45 GHz, to avoid the microwave spectrum contamination.

Now we can combine Eqs. (*) and (**) to find both reduced velocities as functions of time:

$$\beta_x = \frac{\gamma_0\beta_{x0} + \kappa \sin \omega t}{\left[1 + (\gamma_0\beta_{x0} + \kappa \sin \omega t)^2 + (\gamma_0\beta_{y0})^2\right]^{1/2}},$$

$$\beta_y = \frac{\gamma_0\beta_{y0}}{\left[1 + (\gamma_0\beta_{x0} + \kappa \sin \omega t)^2 + (\gamma_0\beta_{y0})^2\right]^{1/2}},$$

$$\beta \equiv \left[\beta_x^2 + \beta_y^2\right]^{1/2} = \left[\frac{(\gamma_0\beta_{x0} + \kappa \sin \omega t)^2 + (\gamma_0\beta_{y0})^2}{1 + (\gamma_0\beta_{x0} + \kappa \sin \omega t)^2 + (\gamma_0\beta_{y0})^2}\right]^{1/2}.$$

For the simplest particular case of no initial velocity ($\beta_{x0} = \beta_{y0} = 0$), the last expression is reduced to

$$\beta = \frac{\kappa \sin \omega t}{(1 + \kappa^2 \sin^2 \omega t)^{1/2}}.$$

The results show that if the field is relatively weak,

$$\kappa \ll 1, \quad \text{i.e.} \quad qE_\omega a \ll mc^2,$$

where $a \equiv c/\omega$ is the natural scale of the particle's oscillation amplitude, the oscillations are virtually sinusoidal—natural for a free non-relativistic particle driven by a sinusoidal force. However, in the opposite limit $\kappa \gg 1$, $|\beta| \approx 1$, i.e. $|u| \approx c$, almost all the oscillation period, besides very narrow time intervals during which the velocity 'switches' from $+c$ to $-c$ and back.

***Problem* 9.29.** Analyze the motion of a non-relativistic particle in a region where the electric and magnetic fields are both uniform and constant in time, but not necessarily parallel or perpendicular to each other.

Solution: Let us select the direction of the vector **B** for the axis z, and direct the axis y so that the vector **E** resides in the [y, z] plane—see the figure below. Then the non-relativistic equation of motion (see Eq. (9.144) of the lecture notes with $\gamma_u = 1$, i.e. $\mathbf{p} = m\mathbf{u}$) has the following Cartesian components:

$$m\ddot{x} = q\dot{y}B, \qquad m\ddot{y} = q\left(E_y - \dot{x}B\right), \qquad m\ddot{z} = qE_z. \qquad (*)$$

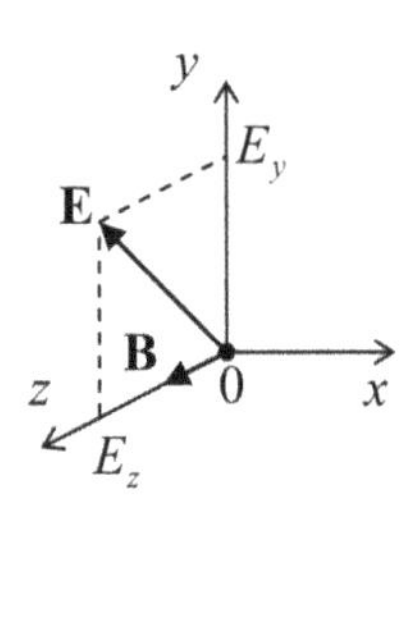

The last equation is independent of its counterparts, and is easily integrated to give the usual 'free fall' motion along the axis z, with the constant acceleration $a_z = qE_z/m$. The first two equations may be merged by the introduction of the rotational 2D velocity

$$\boldsymbol{u} \equiv \mathbf{n}_x(\dot{x} - u_d) + \mathbf{n}_y\dot{y}, \quad \text{with } u_d \equiv \frac{E_y}{B}.$$

Differentiating this vector, and using Eqs. (*),

$$\dot{\boldsymbol{u}} = \mathbf{n}_x\ddot{x} + \mathbf{n}_y\ddot{y} = \mathbf{n}_x\frac{q\dot{y}B}{m} + \mathbf{n}_y\frac{(E_y - \dot{x}B)}{m} = \frac{qB}{m}\left[\mathbf{n}_x\dot{y} - \mathbf{n}_y(\dot{x} - u_d)\right]$$
$$= \omega_c(\mathbf{n}_x\boldsymbol{u}_y - \mathbf{n}_y\boldsymbol{u}_x),$$

we get a differential equation similar to Eq. (9.150) of the lecture notes:

$$\dot{\boldsymbol{u}} = \omega_c\boldsymbol{u} \times \mathbf{n}_z,$$

with the cyclotron frequency (9.151): $\omega_c = qB/m$. This means that the particle's projection on the plane $[x, y]$ performs a circular motion around a point that moves along the axis x with a constant drift velocity $u_d = E_y/B$, i.e. essentially the same motion (along a trochoidal trajectory) as for the case of mutually perpendicular fields ($E_y = E$, $E_z = 0$), which was analyzed in the lecture notes—see figure 9.12 and its discussion.

Thus, the only essentially new feature resulting from the arbitrary angle between the fields is an accelerated motion of the particle along the direction of the vector **B**. This is qualitatively true for a relativistic particle as well, but quantitatively in this case the motion is more complex, because the velocity components become coupled via the Lorentz factor γ_u.

***Problem* 9.30.** A static distribution of electric charge in otherwise free space has created a time-independent distribution **E**(**r**) of the electric field. Use two different approaches to express the energy density u' and the Poynting vector **S**$'$, as observed from a reference frame moving with constant velocity **v**, via components of the vector **E**. In particular, is **S**$'$ equal to $(-\mathbf{v}u')$?

Solution:

Approach 1. In the lab frame, in which the charges rest, their magnetic field equals zero, so that the Lorentz transform formulas (9.135) are reduced to

$$\mathbf{E}'_{||} = \mathbf{E}_{||}, \qquad \mathbf{B}'_{||} = 0,$$
$$\mathbf{E}'_{\perp} = \gamma\mathbf{E}_{\perp}, \quad \mathbf{B}'_{\perp} = -\gamma\frac{\mathbf{v} \times \mathbf{E}_{\perp}}{c^2},$$

where the indices || and ⊥ mark the field components parallel and perpendicular to the velocity vector **v**. As a result, the field energy density (9.225), as observed in the moving frame, is

$$\begin{aligned} u' &= \frac{\varepsilon_0}{2}E'^2 + \frac{1}{2\mu_0}B'^2 \\ &= \frac{\varepsilon_0}{2}\left(E'^2_{\|} + E'^2_{\perp}\right) + \frac{1}{2\mu_0}\left(B'^2_{\|} + B'^2_{\perp}\right) = \frac{\varepsilon_0}{2}\left(E^2_{\|} + \gamma^2 E^2_{\perp}\right) + \frac{1}{2\mu_0}\left(\gamma\frac{vE_{\perp}}{c^2}\right)^2 \\ &= \frac{\varepsilon_0}{2}E^2_{\|} + \left(\frac{\varepsilon_0}{2} + \frac{1}{2\mu_0 c^2}\beta^2\right)\gamma^2 E^2_{\perp} = \frac{\varepsilon_0}{2}\left(E^2_{\|} + \frac{1+\beta^2}{1-\beta^2}E^2_{\perp}\right), \end{aligned} \qquad (*)$$

while the corresponding Poynting vector (6.114) is

$$\begin{aligned} \mathbf{S}' &= \frac{1}{\mu_0}\mathbf{E}' \times \mathbf{B}' = \frac{1}{\mu_0}\left(\mathbf{E}'^2_{|} + \mathbf{E}'_{\perp}\right) \times \left(\mathbf{B}'^2_{\|} + \mathbf{B}'_{\perp}\right) \\ &= \frac{1}{\mu_0}(\mathbf{E}_{\|} + \gamma\mathbf{E}_{\perp}) \times \left(0 - \gamma\frac{\mathbf{v} \times \mathbf{E}_{\perp}}{c^2}\right) \\ &= -\varepsilon_0\gamma\left[\mathbf{E}_{\|} \times (\mathbf{v} \times \mathbf{E}_{\perp}) + \gamma\mathbf{E}_{\perp} \times (\mathbf{v} \times \mathbf{E}_{\perp})\right]. \end{aligned}$$

Applying to both double vector products the *bac minus cab* rule[34], and then using the mutual parallelism of the vectors **v** and $\mathbf{E}_{\|}$, and their normality to vector $\mathbf{E}_{\perp}$, we may reduce the last expression to a simpler one:

$$\begin{aligned} \mathbf{S}' &= -\varepsilon_0\gamma\left[\mathbf{v}\,(\mathbf{E}_{\|} \cdot \mathbf{E}_{\perp}) - \mathbf{E}_{\perp}(\mathbf{v} \cdot \mathbf{E}_{\|}) + \gamma\mathbf{v}\,(\mathbf{E}_{\perp} \cdot \mathbf{E}_{\perp}) - \gamma\mathbf{E}_{\perp}(\mathbf{E}_{\perp} \cdot \mathbf{v})\right] \\ &= \varepsilon_0\gamma\left(\mathbf{E}_{\perp} vE_{\|} - \mathbf{v}\gamma E^2_{\perp}\right). \end{aligned} \qquad (**)$$

Approach 2. Let us use the standard choice of coordinates (see, e.g. figure 9.1 of the lecture notes), so that $\mathbf{E}_{\|} \equiv \mathbf{n}_x E_{\|} = \mathbf{n}_x E_x$, and $\mathbf{E}_{\perp} = \mathbf{n}_y E_y + \mathbf{n}_z E_z$, i.e. $E_{\perp}{}^2 = E_y{}^2 + E_z{}^2$. According to Eqs. (9.225)–(9.229) of the lecture notes, in our case the symmetric four-tensor (9.221) of the energy–momentum of the field in the lab reference frame (in which the charges rest and hence $\mathbf{B} = 0$) is

$$\theta = \frac{\varepsilon_0}{2}\begin{pmatrix} E^2_{\|} + E^2_{\perp} & 0 & 0 & 0 \\ 0 & -E^2_{\|} + E^2_{\perp} & -2E_{\|}E_y & -2E_{\|}E_z \\ 0 & -2E_{\|}E_y & -E^2_y + E^2_{\|} + E^2_z & -2E_yE_z \\ 0 & -2E_{\|}E_z & -2E_yE_z & -E^2_z + E^2_{\|} + E^2_y \end{pmatrix},$$

[34] See, e.g. Eq. (A.47).

Applying to this tensor the Lorentz transform (9.109), just as it was spelled out in the model solution of problem 9.15, we obtain

$$u' \equiv \theta'^{00} = \gamma^2[\theta^{00} - \beta(\theta^{01} + \theta^{10}) + \beta^2\theta^{11}] = \gamma^2\frac{\varepsilon_0}{2}\left[\left(E_{\|}^2 + E_{\perp}^2\right) - \beta\cdot 0 + \beta^2\left(-E_{\|}^2 + E_{\perp}^2\right)\right]$$

$$\equiv \frac{1}{1-\beta^2}\frac{\varepsilon_0}{2}\left[(1-\beta^2)E_{\|}^2 + (1+\beta^2)E_{\perp}^2\right],$$

i.e. an expression equivalent to Eq. (*) above, which was obtained using approach 1. Similarly, the same Lorentz transform yields

$$\frac{S'_x}{c} = \theta'^{01} = \gamma^2[\theta^{01} - \beta(\theta^{00} + \theta^{11}) + \beta^2\theta^{10}]$$

$$= \gamma^2\frac{\varepsilon_0}{2}\left\{0 - \beta\left[\left(E_{\|}^2 + E_{\perp}^2\right) + \left(-E_{\|}^2 + E_{\perp}^2\right)\right] + \beta^2 0\right\} \equiv -\gamma^2\varepsilon_0\beta E_{\perp}^2,$$

$$\frac{S'_y}{c} = \theta'^{02} = \gamma(\theta^{02} - \beta\theta^{12}) = \gamma\left[0 - \beta\frac{\varepsilon_0}{2}\left(-2E_{\|}E_y\right)\right] \equiv \gamma\beta\varepsilon_0 E_{\|}E_y,$$

$$\frac{S'_z}{c} = \theta'^{03} = \gamma(\theta^{03} - \beta\theta^{13}) = \gamma\left[0 - \beta\frac{\varepsilon_0}{2}\left(-2E_{\|}E_z\right)\right] \equiv \gamma\beta\varepsilon_0 E_{\|}E_z.$$

The set of these three scalar relations is obviously equivalent to the vector equality (**), obtained using approach 1.

The obtained expression for the Poynting vector $\mathbf{S}'$ shows that in the general case, when $\mathbf{E}_{\|} \neq 0$, it has not only a component directed along the velocity vector $\mathbf{v}$, as the apparent result $(-\mathbf{v}u')$ does, but also a component directed along the vector $\mathbf{E}_{\perp}$, i.e. normal to the vector $\mathbf{v}$.

***Problem* 9.31.** A plane wave, of frequency ω and intensity S, is normally incident on a perfect mirror, moving with velocity v in the same direction as the wave.

(i) Calculate the reflected wave's frequency, and
(ii) use the Lorentz transform of the fields to calculate the reflected wave's intensity,

both as observed from the lab reference frame.

Solutions:

(i) According to Eq. (9.44) of the lecture notes, the incident radiation, as perceived in the reference frame moving with the mirror, is

$$\omega' = \omega\left(\frac{1-\beta}{1+\beta}\right)^{1/2}, \quad \text{with } \beta \equiv \frac{v}{c}.$$

From the point of view of the mirror-bound observer, the frequency does not change upon reflection, so that the mirror may be treated as a source of radiation of frequency ω'. From the point of view of the lab-frame observer, this source moves from the lab-frame observer with the same velocity v. Hence, we may apply Eq. (9.44) again to calculate the reflected wave's frequency as perceived by the observer:

$$\omega_\mathrm{r} = \omega'\left(\frac{1-\beta}{1+\beta}\right)^{1/2} = \omega\frac{1-\beta}{1+\beta} \equiv \omega\frac{c-v}{c+v}. \qquad (*)$$

(ii) According to Eqs. (7.6) and (7.8) of the lecture notes, in the incident plane wave, propagating in free space along direction $\mathbf{n}$,

$$\mathbf{B} = \mu_0\mathbf{H} = \mu_0\frac{\mathbf{n}\times\mathbf{E}}{Z_0} \equiv \frac{\mathbf{n}\times\mathbf{E}}{c}, \quad \text{because } \mu_0\frac{1}{Z_0} \equiv \mu_0\left(\frac{\varepsilon_0}{\mu_0}\right)^{1/2} \equiv (\varepsilon_0\mu_0)^{1/2} = \frac{1}{c}. \qquad (**)$$

Now applying Eqs. (9.135) to this transverse wave (with $E_{||} = 0$, $B_{||} = 0$, $\mathbf{E}_\perp = \mathbf{E}$, and $\mathbf{B}_\perp = \mathbf{B}$), we may calculate its fields as perceived in the reference frame moving with the mirror:

$$\mathbf{E}' = \gamma(\mathbf{E} + \mathbf{v}\times\mathbf{B}) = \gamma\left[\mathbf{E} + v\frac{\mathbf{n}\times(\mathbf{n}\times\mathbf{E})}{c}\right]$$
$$= \gamma\mathbf{E}\left(1 - \frac{v}{c}\right) \equiv \mathbf{E}\frac{1-\beta}{(1-\beta^2)^{1/2}} = \mathbf{E}\left(\frac{1-\beta}{1+\beta}\right)^{1/2},$$
$$\mathbf{B}' = \gamma\left(\mathbf{B} - \frac{\mathbf{v}\times\mathbf{E}}{c^2}\right) = \gamma\left[\frac{\mathbf{n}\times\mathbf{E}}{c} - \frac{\mathbf{v}\times\mathbf{E}}{c^2}\right]$$
$$= \gamma\frac{\mathbf{n}\times\mathbf{E}}{c}\left(1 - \frac{v}{c}\right) \equiv \frac{\mathbf{n}\times\mathbf{E}}{c}\left(\frac{1-\beta}{1+\beta}\right)^{1/2},$$

so that the relation between vectors $\mathbf{E}'$ and $\mathbf{B}'$ is also given by Eq. (**)—as could be expected. Calculating the Poynting vector (6.114), we obtain

$$\mathbf{S}' \equiv \mathbf{E}'\times\mathbf{H}' = \frac{\mathbf{E}'\times\mathbf{B}'}{\mu_0} = \frac{\mathbf{E}\times\mathbf{B}}{\mu_0}\left(\frac{1-\beta}{1+\beta}\right)$$
$$= \mathbf{E}\times\mathbf{H}\left(\frac{1-\beta}{1+\beta}\right) = \mathbf{S}\left(\frac{1-\beta}{1+\beta}\right) \equiv \mathbf{S}\left(\frac{c-v}{c+v}\right).$$

the last factor giving the relation between the wave intensities in the reference frames.

As was discussed in section 7.4, at wave's reflection from a perfect mirror, the wave intensity does not change, so that the last relation (with the opposite sign because of the change of wave's direction) is also valid for the reflected wave—as observed from the moving reference frame. Now repeating the above calculation, we can now find the intensity of the reflected wave as observed from the lab frame[35]:

$$S_r = S'\left(\frac{c-v}{c+v}\right) = S\left(\frac{c-v}{c+v}\right)^2.$$

Note that all relations in this solution, including the final results (*) and (***), are valid for any sign of the mirror's velocity v. In particular, Eq. (***) shows that the wave's intensity is decreased if the mirror moves in the same direction as the incident wave ('running away' from it) and is increased in the opposite case. Since the intensity increase in the latter system (the mirror moving toward wave's source) is frequency-insensitive, it may be in some sense considered as an amplifier of the power of an arbitrary waveform. However, this system is rather different from the usual amplifiers, because the power gain in it is accompanied by an upward shift of all frequency components of the signal—see Eq. (*).

***Problem* 9.32.** Carry out the second task of the previous problem by using the relations between the wave's energy, power, and momentum.

Hint: As a by-product, this approach should also give you the pressure exerted by the wave on the moving mirror.

Solution: Let us direct axis x along the incident wave's propagation, and plot the position of several representative fronts (i.e. the planes of equal total phase) of this wave and the reflected wave, as well as that of the mirror, as functions of time—see the figure below. As the plots show, during an arbitrary time interval Δt, the mirror is reached by the incident wave that was initially spread over volume

$$\Delta V \equiv A\Delta x = A(c\Delta t - v\Delta t) = A(c - v)\Delta t,$$

where A is the wave front's area, while by the end of the period, the reflected wave is spread over volume

$$\Delta V_r \equiv A\Delta x_r = A(c\Delta t + v\Delta t) = A(c + v)\Delta t.$$

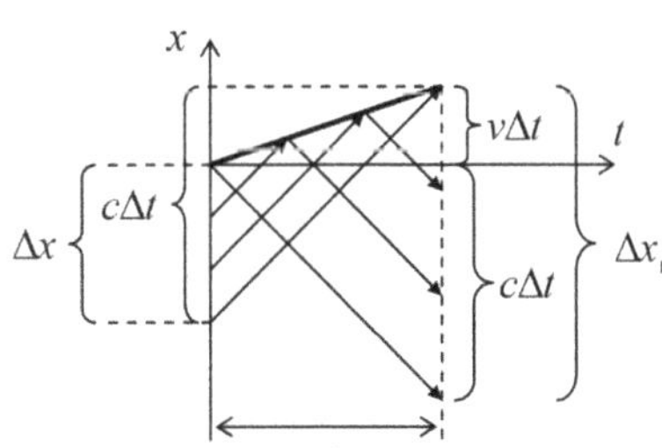

(Although the figure is drawn for $v > 0$, it is straightforward to check that the above formulas are valid for any mirror's velocity with $|v| \leqslant c$.)

[35] This result may be also obtained by a completely different approach, using Eq. (9.237) of the lecture notes—see the next problem.

Let u be the energy density of the incident wave, and u_r—that of the reflected wave (both as measured in the lab frame). Then the initial wave's energy within the volume ΔV is $u\Delta V = uA(c - v)\Delta t$, and that of the reflected wave, within volume ΔV_r, is $u_r\Delta V_r = u_rA(c + v)\Delta t$. Due to the energy conservation, the difference of these two energies has to be equal to the work, $\tau A\Delta x_m = \tau A v\Delta t$, performed by the wave's pressure force τA at mirror's displacement $\Delta x_m = v\Delta t$—see the figure above. This energy balance gives us, after the cancellation of A and Δt, the following equation for u_r, and τ:

$$u(c - v) - u_m(c + v) = \tau v. \qquad (*)$$

Another equation for the same quantities may be obtained from the conservation of total momentum of the system: the force impulse $\mathbf{F}\Delta t = \tau A\Delta t\mathbf{n}_x$ given to the mirror by the wave, i.e. the change of mirror's momentum during time interval Δt, should be equal and opposite to the change of wave's momentum, resulting from its reflection, during the same time interval. But according to Eqs. (9.237) and (7.9) of the lecture notes (the latter one with $v = c$), for a plane wave propagating in free space along an axis $\mathbf{n}$, the momentum per unit volume is just

$$\mathbf{g} = \frac{\mathbf{S}}{c^2} = \frac{u}{c}\mathbf{n}.$$

Hence, for our system's geometry (see the figure below), the momentum balance is

$$\tau A\Delta t\mathbf{n}_x = -\left[\frac{u_r}{c}(-\mathbf{n}_x)\Delta V_r - \frac{u}{c}\mathbf{n}_z\Delta V\right].$$

Now plugging in the above expressions for ΔV and ΔV_r, and canceling $\mathbf{n}_x$, A, and Δt, we obtain

$$\tau = \frac{u(c - v) + u_r(c + v)}{c}. \qquad (**)$$

Solving the simple system of two equations (*) and (**), we obtain

$$u_r = \left(\frac{c - v}{c + v}\right)^2 u, \quad \text{so that } S_r = \left(\frac{c - v}{c + v}\right)^2 S,$$

i.e. the same result as by the Lorentz-transform approach used in the model solution of the previous problem. In addition to that, the same system of equations yields the following expression for wave's pressure:

$$\tau = 2u\frac{c - v}{c + v} \equiv \frac{2S}{c}\frac{c - v}{c + v}.$$

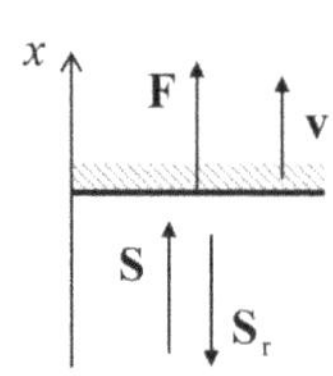

As a sanity check, for an immobile mirror ($v = 0$), this formula, with the account of Eq. (7.9b), gives the result (9.245)–(9.246).

***Problem* 9.33.** Consider the simple model of capacitor charging by a lumped current source, shown in the figure below, and prove that the momentum given by a uniform, stationary, external magnetic field **B** to the current-carrying conductor is equal and opposite to the momentum of the electromagnetic field that the current $I(t)$ builds up in the capacitor. (You may assume that the capacitor is planar and very broad, and neglect the fringe field effects.)

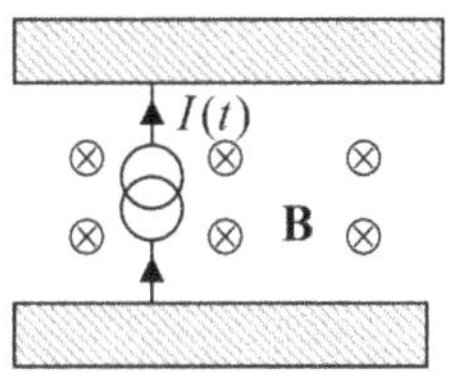

Solution: In the geometry shown in the figure, with positive $I(t)$, the total magnetic force (5.8) exerted on the wire is directed to the left and has magnitude $F(t) = I(t)Bd$, where d is the wire's length, i.e. the distance between the capacitor's plates. Integrating this expression over time, for the momentum transferred from the field to the wire (and hence to the capacitor) we obtain

$$p_{\text{capacitor}} = \int F(t)dt = Bd\int I(t)dt \equiv BdQ, \qquad (*)$$

where $Q \equiv \int I(t)dt$ is capacitor's charge built up by the current.

If $Q > 0$, the electric field this charge creates is directed, in our figure, from the top plate down, so that the electromagnetic field's momentum density vector (9.237) is directed to the right. The total field momentum has the magnitude

$$p_{\text{field}} = \int g d^3r = gAd = \frac{S}{c^2}Ad = \frac{EH}{c^2}Ad = \frac{(V/d)(B/\mu_0)}{c^2}Ad \equiv \frac{\varepsilon_0 A}{d}VBd, \qquad (**)$$

where A is the capacitor's area, and $V = Ed$ is the final voltage between its plates. But the fraction in the last form of Eq. (**) is just the capacitance C, so that its product by V equals the same Q which participates in Eq. (*). Thus the two momenta, $p_{\text{capacitor}}$ and p_{field} are indeed equal and opposite, so that the total momentum of the whole system (capacitor + field) is conserved.

***Problem* 9.34.** Consider an electromagnetic plane wave packet propagating in free space, with the electric field represented as the Fourier integral

$$\mathbf{E}(\mathbf{r}, t) = \text{Re}\int_{-\infty}^{+\infty} \mathbf{E}_k e^{i\psi_k} dk, \quad \text{with } \psi_k \equiv kz - \omega_k t, \quad \text{and } \omega_k \equiv c\,|\,k\,|.$$

Express the full linear momentum (per unit area of wave's front) of the packet via the complex amplitudes $\mathbf{E}_k$ of its Fourier components. Does the momentum depend on time? (In contrast with problem 7.7, in this case the wave packet is not necessarily narrow.)

Solution: As Eq. (9.237) of the lecture notes shows, we need to calculate the time dependence of

$$\frac{G}{A} \equiv \int_{-\infty}^{+\infty} g_z(z, t)dz = \frac{1}{c^2}\int_{-\infty}^{+\infty} S_z(z, t)dz = \frac{1}{c^2}\int_{-\infty}^{+\infty} [\mathbf{E}(z, t) \times \mathbf{H}(z, t)]_z dz. \quad (*)$$

According to Eqs. (7.6) and (7.7), the magnetic field of the packet may be represented as

$$\mathbf{H}(z, t) = \text{Re}\int_{-\infty}^{+\infty} \mathbf{H}_{k'} e^{i\psi_{k'}} dk' \equiv \frac{1}{2}\left[\int_{-\infty}^{+\infty} \mathbf{H}_{k'} e^{i\psi_{k'}} dk' + \text{c.c.}\right],$$

$$\text{with } \mathbf{H}_k = \frac{\mathbf{n}_z \times \mathbf{E}_k}{Z_0}\,\text{sgn}(k), \quad \text{and} \quad Z_0 \equiv \left(\frac{\mu_0}{\varepsilon_0}\right)^{1/2},$$

where $\psi_{k'} \equiv k'z - \omega_{k'}t$. Plugging the expressions for **E** and **H** into Eq. (*), taking into account that $1/cZ_0 \equiv \varepsilon_0$, and changing the order of integration, we obtain[36]

$$\frac{G}{A} = \frac{\varepsilon_0}{4c}\int_{-\infty}^{+\infty} dk\int_{-\infty}^{+\infty} dk' \,\text{sgn}\,(k)\Big[\mathbf{E}_k \times (\mathbf{n}_z \times \mathbf{E}_{k'})\int_{-\infty}^{+\infty} e^{i(\psi+\psi')}dz$$
$$+ \mathbf{E}_k \times (\mathbf{n}_z \times \mathbf{E}^*_{k'})\int_{-\infty}^{+\infty} e^{i(\psi-\psi')}dz + \text{c.c.}\Big]. \quad (**)$$

Using Eq. (A.88), we may readily work out both internal integrals, which do not depend on $\mathbf{E}_k$:

$$\int_{-\infty}^{+\infty} e^{i(\psi_k \pm \psi_{k'})}dz \equiv \exp\{-i(\omega_k \pm \omega_{k'})t\}\int_{-\infty}^{+\infty} e^{i(k\pm k')z}dz$$
$$= 2\pi \exp\{-i(\omega_k \pm \omega_{k'})t\}\delta(k \pm k')$$
$$= 2\pi \times \begin{cases} \exp\{-2i\omega_k t\}\;\delta(k + k'), & \text{for the upper sign,} \\ \delta(k - k'), & \text{for the lower sign,} \end{cases}$$

so that the integral over k' in Eq. (*) is also elementary, and that equality is reduced to

$$\frac{G}{A} = \frac{\pi\varepsilon_0}{2c}\int_{-\infty}^{+\infty} [-\mathbf{E}_k \times (\mathbf{n}_z \times \mathbf{E}_{-k})\exp\{-2i\omega_k t\} + \mathbf{E}_k \times (\mathbf{n}_z \times \mathbf{E}^*_k) + \text{c.c.}]\text{sgn}\,(k)dk.$$

Taking also into account that in a plane wave, propagating along axis z, both vectors $\mathbf{E}_k$ and $\mathbf{E}_{-k}$ are perpendicular to the vector $\mathbf{n}_z$,[37] we may rewrite this expression in a simpler form:

---

[36] Cf. the derivation of Eq. (7.42) in section 7.2.

[37] A word of caution: the directions of vectors $\mathbf{E}_k$ and $\mathbf{E}_{-k}$ (in the plane normal to vector $\mathbf{n}_z$) do not necessarily coincide (see, e.g. the last part of section 7.1 of the lecture notes), so that using their scalar product is important.

$$\frac{G}{A} = -\frac{\pi\varepsilon_0}{2c}\int_{-\infty}^{+\infty}[\mathbf{E}_k \cdot \mathbf{E}_{-k}\exp\{-2i\omega_k t\} + \text{c.c.}]\text{sgn}\,(k)dk$$
$$+\frac{\pi\varepsilon_0}{2c}\int_{-\infty}^{+\infty}[\mathbf{E}_k \cdot \mathbf{E}_k^* + \text{c. c.}]\text{sgn}\,(k)dk. \qquad (***)$$

Since with our definition of ω_k it is an even function of k,[38] the first term on the right-hand side is an integral, in symmetric limits, of an odd function of k, and hence equals zero. For the second term, this is not necessarily true; for example, if the wave packet is narrow in the momentum space, and propagates in the direction of larger z, the only significant amplitudes $|\mathbf{E}_k|$ correspond to values of k close to a certain positive value k_0. In that second term, the scalar product $\mathbf{E}_k \cdot \mathbf{E}_k^*$ equals just $E_k E_k^* \equiv |E_k|^2$,[39] so that taking into account the (equal) complex conjugate term, we finally obtain

$$\frac{G}{A} = \frac{\pi\varepsilon_0}{c}\int_{-\infty}^{+\infty} E_k E_k^* \text{sgn}\,(k)dk \equiv \frac{2\pi\varepsilon_0}{c}\int_0^{+\infty} E_k E_k^* dk.$$

Note that in contrast to the wave packet's energy (whose calculation was the subject of problem 7.7), the cancellation of the rapidly oscillating terms in Eq. (***) is exact for the packet of any form.

***Problem* 9.35.** Calculate the pressure exerted on well-conducting walls of a waveguide with a rectangular ($a \times b$) cross-section, by a wave propagating along it in the fundamental (H_{10}) mode. Give an interpretation of the result.

Solution: According to Eq. (7.131) of the lecture notes (with $n = 1$ and $m = 0$) and Eqs. (7.137)–(7.138), the broader walls of the waveguide (in figure 7.22, with $y = 0, b$) are exposed to both the electric and magnetic fields of the wave, with complex amplitudes

$$E_x = 0, \quad E_y = i\frac{ka}{\pi}ZH_l \sin\frac{\pi x}{a}, \quad E_z = 0;$$
$$H_x = -i\frac{k_z a}{\pi}H_l \sin\frac{\pi x}{a}, \quad H_y = 0, \quad H_z = H_l \cos\frac{\pi x}{a};$$

while at the narrower walls ($x = 0, a$), only the longitudinal magnetic field is non-vanishing, with the complex amplitude equal to $\pm H_l$. The wave vector components in these formulas are related by Eqs. (7.122) and (7.132), which in our case ($k_x = \pi/a$, $k_y = 0$) may be rewritten as

$$\left(\frac{ka}{\pi}\right)^2 - \left(\frac{k_z a}{\pi}\right)^2 = 1.$$

[38] An additional question for the reader: what would the initial expression for $\mathbf{E}(\mathbf{r}, t)$ describe if we took $\omega_k = ck$, without the modulus sign?

[39] If there any doubt in this statement: since the complex-number vector $\mathbf{E}_k$ is normal to the propagation axis z, we may represent it as $\mathbf{n}_x E_x + \mathbf{n}_y E_y = \mathbf{n}_x A_x \exp\{i\varphi_x\} + \mathbf{n}_y A_y \exp\{i\varphi_y\}$, where $E_{x,y}$ may be still complex, but $A_{x,y}$ and $\varphi_{x,y}$ are real. Now scalar-multiplying this vector by its complex-conjugate, we obtain $A_x^2 + A_y^2$, i.e. the same result if we calculated $|E_k|^2 \equiv |E_x|^2 + |E_y|^2$.

Plugging these relations into Eqs. (9.239) and (9.241), for the average pressure on the walls in terms of the longitudinal field amplitude H_l, we get:

$$\overline{\frac{dF}{dA}}\bigg|_{y=0,b} = \frac{\mu_0}{4}|H_l|^2\left(\cos^2\frac{\pi x}{a} - \sin^2\frac{\pi x}{a}\right), \qquad \overline{\frac{dF}{dA}}\bigg|_{x=0,a} = \frac{\mu_0}{4}|H_l|^2,$$

where the averaging over time gave an extra factor ½.

The pressure on the narrower walls is positive, due to the magnetic field effect, while that on the broader walls is negative in the middle (at $x = a/2$, $\sin^2(\pi x/a) = 1$, $\cos^2(\pi x/a) = 0$), due to the electric field effect (attraction). On the average over x, these two pressures on the broader wall compensate each other, so that the net forces exerted on them are equal zero.

These results may be readily explained in terms of the fundamental relation between the flux $c\mathbf{g} = \mathbf{S}/c$ of the wave's momentum and its pressure on the mirrors that reflects them. Indeed, as was discussed in the model solution of problem 7.19, the H_{10} wave may be represented as a sum of two plane waves, propagating with wave vectors $\mathbf{k}_\pm = \{\pm\pi/a, 0, k_z\}$, which are reflected from the narrower walls of the waveguide. These reflections create an outward pressure on the walls, similar to that at the normal incidence—see Eq. (9.245) of the lecture notes and its discussion. On the other hand, since the H_{10} waves have $k_y = 0$, they are not reflected from the broader walls of the waveguide, and as a result do not exert a net force on these walls.

Reference

[1] Fraunfelder H 1963 *The Mössbauer Effect* (Benjamin)

Chapter 10

Radiation by relativistic charges

***Problem* 10.1.** Derive Eqs. (10.10) of the lecture notes from Eqs. (10.1) by a direct (but careful!) integration.

Solution: Let us first recognize the error that led to the wrong results (10.4). A direct substitution of (10.1b) into Eq. (10.1a) yields

$$\phi(\mathbf{r}, t) = \frac{q}{4\pi\varepsilon_0}\int d^3r'' \frac{1}{|\mathbf{r} - \mathbf{r}''|}\delta(\mathbf{r}'' - \mathbf{r}'). \qquad (*)$$

The naïve integration over $\mathbf{r}''$, giving the first of Eqs. (10.4), ignores the fact that according to Eq. (10.1), the particle's position $\mathbf{r}'$ has to be taken at the time $t - R/c \equiv t - |\mathbf{r} - \mathbf{r}''|/c$, i.e. at the point which depends on the integration variable as well.

In order to overcome this difficulty, let us rewrite the first of Eqs. (10.1a) in a different, but equivalent form by introducing an additional integration over some auxiliary time argument t'', and compensating it with a delta-function at the appropriate point $t'' = t - R/c \equiv t - |\mathbf{r} - \mathbf{r}''|$

$$\phi(\mathbf{r}, t) = \frac{1}{4\pi\varepsilon_0}\int d^3r'' \int dt'' \frac{\rho(\mathbf{r}'', t'')}{|\mathbf{r} - \mathbf{r}''|}\delta\left(t'' - t + \frac{|\mathbf{r} - \mathbf{r}''|}{c}\right).$$

Now changing the integration order, and plugging in the first of Eqs. (10.1b), we obtain

$$\phi(\mathbf{r}, t) = \frac{q}{4\pi\varepsilon_0}\int dt'' \int d^3r'' \frac{\delta[\mathbf{r}'' - \mathbf{r}'(t'')]}{|\mathbf{r} - \mathbf{r}''|}\delta\left(t'' - t + \frac{|\mathbf{r} - \mathbf{r}''|}{c}\right).$$

The internal integral may be limited by the vicinity of the point $\mathbf{r}'(t'')$, were it includes only a standard 3D delta-function (multiplied by a smooth[1] function of $\mathbf{r}''$), and may be taken as usual, leaving us with

$$\phi(\mathbf{r}, t) = \frac{q}{4\pi\varepsilon_0}\int dt'' \frac{1}{|\mathbf{r} - \mathbf{r}'(t'')|}\,\delta\left(t'' - t + \frac{|\mathbf{r} - \mathbf{r}'(t'')|}{c}\right). \qquad (**)$$

[1] The function is singular only if $\mathbf{r} = \mathbf{r}'$, i.e. if the field observation point exactly coincides with the charge's position—the case we are not interested in.

This is also not a standard integral of a delta-function, but at least it is now one-dimensional. In order to work it out, we may use the following property of the 1D delta-function: if a monotonic scalar function $f(\xi)$ at some point ξ_0 has a non-zero derivative $df/d\xi$, then

$$\delta(f - f_0) = \frac{\delta(\xi - \xi_0)}{|df/d\xi|_{\xi=\xi_0}}, \quad \text{where } f_0 \equiv f(\xi_0). \qquad (***)$$

The proof of this relation is easy: it is sufficient to integrate the right-hand side of Eq. (***) over any interval of f, using the simple chain rule:

$$\int \frac{\delta(\xi - \xi_0)}{|df/d\xi|_{\xi=\xi_0}} d\xi = \int \frac{\delta(\xi - \xi_0)}{|df/d\xi|_{\xi=\xi_0}} \frac{df}{d\xi} d\xi$$

The last form of this expression shows that if the integration interval contains the point ξ_0, then the integral equals 1,[2] while if the interval does not include this point, then the integral equals zero. Since this property is essentially the definition of the delta-function $\delta(f - f_0)$,[3] it proves Eq. (***).

Let us apply this identity to our case, taking $\xi \equiv t''$, $f(\xi) \equiv f(t'') \equiv t'' - t + |\mathbf{r} - \mathbf{r}'(t'')|/c$, and the point $\xi_0 \equiv (t'')_0$ to correspond to the retarded time t_{ret}, i.e. to the physically meaningful root (with $t_{\text{ret}} < t$) of equation (10.5):

$$t_{\text{ret}} - t + \frac{|\mathbf{r} - \mathbf{r}'(t_{\text{ret}})|}{c} = 0.$$

By our definition of the function $f(t'')$, at this point it turns to zero, i.e. $f_0 = 0$. The derivative df/dt'' at this point may be calculated exactly as was done in Eqs. (10.11)–(10.16) of the lecture notes (just replacing the notation t_{ret} with t'', and returning to t_{ret} after the calculation), giving

$$\left|\frac{df}{dt''}\right|_{t''=t_{\text{ret}}} = \frac{1}{\partial t_{\text{ret}}/\partial t} = (1 - \boldsymbol{\beta} \cdot \mathbf{n})_{\text{ret}} > 0.$$

Plugging Eq. (***), with these specifications, into Eq. (**), we obtain

$$\phi(\mathbf{r}, t) = \frac{q}{4\pi\varepsilon_0} \int dt'' \frac{\delta[t'' - t + |\mathbf{r} - \mathbf{r}'(t'')|/c]}{|\mathbf{r} - \mathbf{r}'(t'')|} \equiv \frac{q}{4\pi\varepsilon_0} \int dt'' \frac{\delta(f - f_0)}{|\mathbf{r} - \mathbf{r}'(t'')|}$$
$$= \frac{q}{4\pi\varepsilon_0 \, |df/dt''|_{t''=t_{\text{ret}}}} \int dt'' \frac{\delta(t'' - t_{\text{ret}})}{|\mathbf{r} - \mathbf{r}'(t'')|} = \frac{q}{4\pi\varepsilon_0 (1 - \boldsymbol{\beta} \cdot \mathbf{n})_{\text{ret}}} \int dt'' \frac{\delta(t'' - t_{\text{ret}})}{|\mathbf{r} - \mathbf{r}'(t'')|}.$$

Since t_{ret} does not depend on t'', this is a standard integral of a delta-function multiplied by a smooth function[4], so it may be taken in the regular way, giving:

[2] The sign of the derivative $df/d\xi$ at the point ξ_0 does not affect the result, because the delta-function behaves as a symmetric one (i.e. may be considered as an approximation of a very short symmetric pulse).

[3] See, e.g. Eq. (A.85).

[4] Actually, the same approach might be applied directly to Eq. (*), but due to the 3D character of the involved delta-function, the necessary calculations are somewhat longer.

$$\phi(\mathbf{r}, t) = \frac{q}{4\pi\varepsilon_0(1 - \boldsymbol{\beta} \cdot \mathbf{n})_{\text{ret}}} \frac{1}{|\mathbf{r} - \mathbf{r}'(t_{\text{ret}})|} \equiv \frac{q}{4\pi\varepsilon_0(1 - \boldsymbol{\beta} \cdot \mathbf{n})_{\text{ret}}} \frac{1}{R_{\text{ret}}}.$$

This is Eq. (10.10a) of the lecture notes; the derivation of Eq. (10.10b) may be carried out absolutely similarly.

***Problem* 10.2.** Derive the radiation-related parts of Eqs. (10.19) and (10.20) of the lecture notes from the Liénard–Wiechert potentials (10.10) by direct differentiation.

Solution: We need to plug Eqs. (10.10),

$$\phi(\mathbf{r}, t) = \frac{q}{4\pi\varepsilon_0}\left[\frac{1}{R(1 - \boldsymbol{\beta} \cdot \mathbf{n})}\right]_{\text{ret}}, \qquad \mathbf{A}(\mathbf{r}, t) = \frac{\mu_0 qc}{4\pi}\left[\frac{\boldsymbol{\beta}}{R(1 - \boldsymbol{\beta} \cdot \mathbf{n})}\right]_{\text{ret}},$$

into the general equations (6.7)

$$\mathbf{E} = -\frac{\partial \mathbf{A}}{\partial t} - \nabla\phi, \qquad \mathbf{B} = \nabla \times \mathbf{A}, \qquad (*)$$

and perform the prescribed temporal and spatial differentiation. However, since we need to calculate only the radiative parts of the fields, which dominate at $R \to \infty$, at these operations we may skip the derivatives of the $(1/R_{\text{ret}})$ factor.

Indeed, let us consider, for example, the derivative[5]

$$\frac{\partial \phi(\mathbf{r}, t)}{\partial t} = \frac{q}{4\pi\varepsilon_0} \frac{\partial}{\partial t}\left[\frac{1}{R(1 - \boldsymbol{\beta} \cdot \mathbf{n})}\right]_{\text{ret}}.$$

As was discussed in section 10.1 of the lecture notes, at the fixed radius-vector **r** of the observation point, the retarded time t_{ret} is a unique function of t, so that for this partial differentiation we may use the simple chain rule:

$$\frac{\partial \phi}{\partial t} = \frac{q}{4\pi\varepsilon_0} \frac{\partial}{\partial t_{\text{ret}}}\left[\frac{1}{R(1 - \boldsymbol{\beta} \cdot \mathbf{n})}\right]_{\text{ret}} \frac{\partial t_{\text{ret}}}{\partial t}, \qquad (**)$$

where the last derivative is given by Eq. (10.16) of the lecture notes:

$$\frac{\partial t_{\text{ret}}}{\partial t} = \left(\frac{1}{1 - \boldsymbol{\beta} \cdot \mathbf{n}}\right)_{\text{ret}}.$$

Now, taking the first derivative in Eq. (**) exactly, at the second step using Eq. (10.13), we obtain

$$\begin{aligned}\frac{\partial}{\partial t_{\text{ret}}}\left[\frac{1}{R(1 - \boldsymbol{\beta} \cdot \mathbf{n})}\right]_{\text{ret}} &= -\frac{\partial(1 - \boldsymbol{\beta} \cdot \mathbf{n})_{\text{ret}}/\partial t_{\text{ret}}}{[R(1 - \boldsymbol{\beta} \cdot \mathbf{n})^2]_{\text{ret}}} - \frac{\partial R_{\text{ret}}/\partial t_{\text{ret}}}{[R^2(1 - \boldsymbol{\beta} \cdot \mathbf{n})]_{\text{ret}}} \\ &= \left[\frac{\dot{\boldsymbol{\beta}} \cdot \mathbf{n}}{R(1 - \boldsymbol{\beta} \cdot \mathbf{n})^2} + \frac{c\boldsymbol{\beta} \cdot \mathbf{n}}{R^2(1 - \boldsymbol{\beta} \cdot \mathbf{n})}\right]_{\text{ret}}.\end{aligned} \qquad (***)$$

[5] Actually, we do not need this particular derivative for the field calculation, but it is the simplest of all the derivatives of this type, while being conceptually similar to them.

This expression shows that at $\mathbf{r} \to \infty$, i.e. at $R_{\text{ret}} \equiv |\mathbf{r} - \mathbf{r}'(t_{\text{ret}})| \to \infty$, namely, at

$$R_{\text{ret}} \gg c\left[\frac{\boldsymbol{\beta} \cdot \mathbf{n}}{\dot{\boldsymbol{\beta}} \cdot \mathbf{n}}(1 - \boldsymbol{\beta} \cdot \mathbf{n})\right]_{\text{ret}} \sim c\tau,$$

where τ is the time scale of the particle's acceleration in the observer's direction[6], the second term in Eq. (**), which was the result of differentiation of $(1/R_{\text{ret}})$, may be neglected. Similar estimates are valid for all partial derivatives contributing to the fields (*).

Upon this condition, the required differentiation of the Cartesian components of the Liénard–Wiechert potentials,

$$\phi = \frac{q}{4\pi\varepsilon_0}\left[\frac{1}{R(1 - \boldsymbol{\beta} \cdot \mathbf{n})}\right]_{\text{ret}},$$
$$A_j = \frac{\mu_0 qc}{4\pi}\left[\frac{\beta_j}{R(1 - \boldsymbol{\beta} \cdot \mathbf{n})}\right]_{\text{ret}} \equiv \frac{q}{4\pi\varepsilon_0 c}\left[\frac{\beta_j}{R(1 - \boldsymbol{\beta} \cdot \mathbf{n})}\right]_{\text{ret}}, \quad \text{with } j = 1, 2, 3,$$

becomes easy. For example,

$$\frac{\partial A_j}{\partial t} = \frac{q}{4\pi\varepsilon_0 c}\left\{\frac{\partial\left(\beta_j\right)_{\text{ret}}/\partial t_{\text{ret}}}{[R(1 - \boldsymbol{\beta} \cdot \mathbf{n})]_{\text{ret}}} - \frac{\left(\beta_j\right)_{\text{ret}}\partial(1 - \boldsymbol{\beta} \cdot \mathbf{n})_{\text{ret}}/\partial t_{\text{ret}}}{[R(1 - \boldsymbol{\beta} \cdot \mathbf{n})^2]_{\text{ret}}}\right\}\frac{\partial t_{\text{ret}}}{\partial t}$$
$$= \frac{q}{4\pi\varepsilon_0 c}\left[\frac{\dot{\beta}_j}{R(1 - \boldsymbol{\beta} \cdot \mathbf{n})^2} + \frac{\beta_j\ (\dot{\boldsymbol{\beta}} \cdot \mathbf{n})}{R(1 - \boldsymbol{\beta} \cdot \mathbf{n})^3}\right]_{\text{ret}}.$$

The similar spatial differentiations are enabled by the relation similar to Eq. (10.16), cited in a footnote in section 10.1, whose jth Cartesian component is

$$\frac{\partial t_{\text{ret}}}{\partial r_j} = -\left[\frac{n_j}{c(1 - \boldsymbol{\beta} \cdot \mathbf{n})}\right]_{\text{ret}}.$$

For example,

$$\frac{\partial\phi}{\partial r_j} = \frac{q}{4\pi\varepsilon_0}\frac{\partial}{\partial t_{\text{ret}}}\left[\frac{1}{R(1 - \boldsymbol{\beta} \cdot \mathbf{n})}\right]_{\text{ret}}\frac{\partial t_{\text{ret}}}{\partial r_j} = -\frac{q}{4\pi\varepsilon_0 c}\left[\frac{n_j\ (\dot{\boldsymbol{\beta}} \cdot \mathbf{n})}{R(1 - \boldsymbol{\beta} \cdot \mathbf{n})^3}\right]_{\text{ret}}.$$

[6] Note that this condition is never fulfilled at zero acceleration. This conclusion is in accordance with the result of analysis, in section 9.5 of the lecture notes, of the fields induced by a uniformly moving particle, which are essentially its Lorentz-transformed Coulomb field—see Eqs. (9.139) and (9.140).

From these results, for the jth component of the electric field we obtain

$$E_j = -\frac{\partial A_j}{\partial t} - \frac{\partial \phi}{\partial r_j} = \frac{q}{4\pi\varepsilon_0 c}\left[-\frac{\dot{\beta}_j}{R(1-\boldsymbol{\beta}\cdot\mathbf{n})^2} + \frac{\left(n_j - \beta_j\right)\dot{\boldsymbol{\beta}}\cdot\mathbf{n}}{R(1-\boldsymbol{\beta}\cdot\mathbf{n})^3}\right]_{\text{ret}}$$

$$\equiv \frac{q}{4\pi\varepsilon_0 c}\left[\frac{(\mathbf{n}-\boldsymbol{\beta})_j(\dot{\boldsymbol{\beta}}\cdot\mathbf{n}) - \dot{\beta}_j(1-\boldsymbol{\beta}\cdot\mathbf{n})}{R(1-\boldsymbol{\beta}\cdot\mathbf{n})^3}\right]_{\text{ret}}.$$

But this is exactly the result given by the radiative part of Eq. (10.19):

$$\mathbf{E} \equiv \frac{q}{4\pi\varepsilon_0 c}\left[\frac{\mathbf{n}\times\{(\mathbf{n}-\boldsymbol{\beta})\times\dot{\boldsymbol{\beta}}\}}{R(1-\boldsymbol{\beta}\cdot\mathbf{n})^3}\right]_{\text{ret}}.$$

This fact may be verified either by the Cartesian components, or (even simpler) by using, for the nominator of the last fraction, the *bac minus cab* rule[7], and then taking into account that $\mathbf{n}$ is a unit vector, so that $\mathbf{n}\cdot\mathbf{n} = 1$:

$$\mathbf{n}\times\{(\mathbf{n}-\boldsymbol{\beta})\times\dot{\boldsymbol{\beta}}\} = (\mathbf{n}-\boldsymbol{\beta})(\mathbf{n}\cdot\dot{\boldsymbol{\beta}}) - \dot{\boldsymbol{\beta}}\,[\mathbf{n}\cdot(\mathbf{n}-\boldsymbol{\beta})]$$
$$\equiv (\mathbf{n}-\boldsymbol{\beta})\,(\dot{\boldsymbol{\beta}}\cdot\mathbf{n}) - \dot{\boldsymbol{\beta}}\,(1-\boldsymbol{\beta}\cdot\mathbf{n}).$$

Evidently, the jth component of this vector is exactly the nominator in our result for E_j.

An absolutely similar calculation of the derivatives $\partial A_j/\partial r_{j'}$, and then the Cartesian components of the vector $\mathbf{B} = \nabla\times\mathbf{A}$,

$$B_{j''} = \left(\frac{\partial A_j}{\partial r_{j'}} - \frac{\partial A_{j'}}{\partial r_j}\right)\varepsilon_{jj'j''},$$

shows that the radiative part of the magnetic field is indeed related to the similar part of the electric field by Eq. (10.20): $\mathbf{B} = \mathbf{n}_{\text{ret}}\times\mathbf{E}/c$.

***Problem* 10.3.** A point charge q that had been in a stationary position on a circle of radius R, is carried over, along the circle, to the opposite position on the same diameter (see the figure below) as fast as only physically possible, and then is kept steady at this new position. Calculate and sketch the time dependence of the electric field $\mathbf{E}$ at the center of the circle.

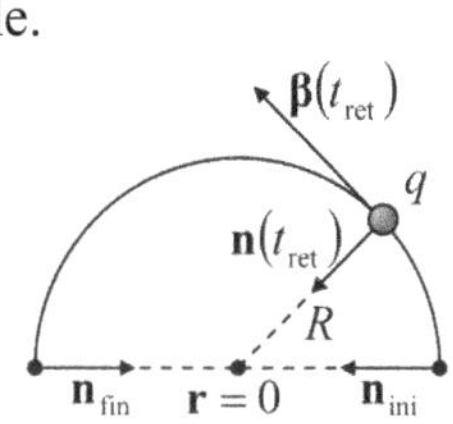

[7] If you still do not remember it, please see Eq. (A.47) again.

Solution: In this simple system, the distance R between the field observation point (which we may take for $\mathbf{r} = 0$) and the charge's position $\mathbf{r}'$ remains constant (equal to the circle's radius), so that the delay between the retarded time t_{ret} (defined by Eq. (10.5) of the lecture notes) and the field observation time t remains constant:

$$t - t_{\text{ret}} = \frac{R}{c}.$$

Let us take the retarded time of the beginning of charge's motion for zero. The length of semi-circle's arc is πR, and the highest possible speed of charge motion is $u \to c$; hence the time of its arrival to the final point is $\pi R/u \to \pi R/c$—with all times as measured in the lab reference frame. The corresponding interval of the field observation times is

$$t_{\text{ini}} < t < t_{\text{fin}}, \qquad \text{with } t_{\text{ini}} = \frac{R}{c}, \qquad t_{\text{fin}} = \frac{\pi R}{c} + \frac{R}{c} \equiv (\pi + 1)\frac{R}{c}. \qquad (*)$$

Since the charge does not move at $t_{\text{ret}} < 0$ and at $t_{\text{ret}} > \pi R/c$, the field observed outside of the interval (*) may be calculated just from the Coulomb law:

$$\mathbf{E}(0, t) = \frac{q}{4\pi\varepsilon_0 R^2} \times \begin{cases} \mathbf{n}_{\text{ini}}, & \text{for } t < t_{\text{ini}}, \\ \mathbf{n}_{\text{fin}}, & \text{for } t > t_{\text{fin}}, \end{cases} \qquad (**)$$

where $\mathbf{n} \equiv \mathbf{R}/R$ is the unit vector pointing from the charge's position to the observation point—see the figure above. Within the interval (*), $\mathbf{E}(0, t)$ has to be found from the retarded-field formula (10.19)[8]:

$$\mathbf{E}(\mathbf{r}, t) = \frac{q}{4\pi\varepsilon_0}\left[\frac{\mathbf{n} - \boldsymbol{\beta}}{\gamma^2(1 - \boldsymbol{\beta}\cdot\mathbf{n})^3 R^2} + \frac{\mathbf{n} \times \{(\mathbf{n} - \boldsymbol{\beta}) \times \dot{\boldsymbol{\beta}}\}}{(1 - \boldsymbol{\beta}\cdot\mathbf{n})^3 cR}\right]_{\text{ret}}, \qquad (***)$$

where the index 'ret' marks the variable values at the retarded time $t_{\text{ret}} = t - R/c$. Since the parameter $\gamma \equiv (1 - \beta^2)^{-1/2}$ diverges at $\beta \to 1$, in our case the first term in square brackets is negligible. In the second term, the scalar product $\boldsymbol{\beta} \cdot \mathbf{n}$ in the denominator also vanishes, because at any instant t_{ret} of the circular motion of the charge, the vectors $\boldsymbol{\beta}(t_{\text{ret}})$ and $\mathbf{n}(t_{\text{ret}})$ are mutually perpendicular (see the figure above). In order to calculate the nominator of that term, we may use the well-known formula for the acceleration at a uniform circular motion:

$$\dot{\boldsymbol{\beta}} \equiv \frac{\dot{\mathbf{u}}}{c} = \frac{\mathbf{n}u^2/R}{c} \to \frac{\mathbf{n}c}{R}.$$

With this, and taking into account (twice) that $\mathbf{n} \times \mathbf{n} = 0$ and also that $\mathbf{n} \cdot \mathbf{n} = 1$, the double vector product in the second term may be simplified (for example, using the *bac minus cab* rule[9]) as follows:

$$\mathbf{n} \times \{(\mathbf{n} - \boldsymbol{\beta}) \times \dot{\boldsymbol{\beta}}\} = \frac{c}{R}\mathbf{n} \times \{(\mathbf{n} - \boldsymbol{\beta}) \times \mathbf{n}\} = \frac{c}{R}\mathbf{n} \times \{(-\boldsymbol{\beta}) \times \mathbf{n}\} = -\frac{c}{R}\boldsymbol{\beta},$$

[8] This formula, with $\boldsymbol{\beta} \equiv 0$, of course yields Eq. (**) as well.

[9] If you need a reminder, see Eq. (A.47).

so that Eq. (***) is reduced to

$$\mathbf{E}(0, t) = -\frac{q}{4\pi\varepsilon_0 R^2}\boldsymbol{\beta}(t_{\text{ret}}) \equiv -\frac{q}{4\pi\varepsilon_0 R^2}\boldsymbol{\beta}\left(t - \frac{R}{c}\right), \quad \text{for } t_{\text{ini}} < t < t_{\text{fin}}. \qquad (****)$$

Since in our approximation $u \to c$, the length of vector $\boldsymbol{\beta}$ equals 1, the magnitude of this field stays constant, equal to that of the static Coulomb values (**), but its direction rotates in time as the figure below shows. The instant leaps of the field vector at moments t_{ini} and t_{fin}, shown by dashed arrows, are not quite surprising, taking into account that the problem assignment implies an instant (and hence not quite physical[10]) acceleration/deceleration of the charge at these moments.

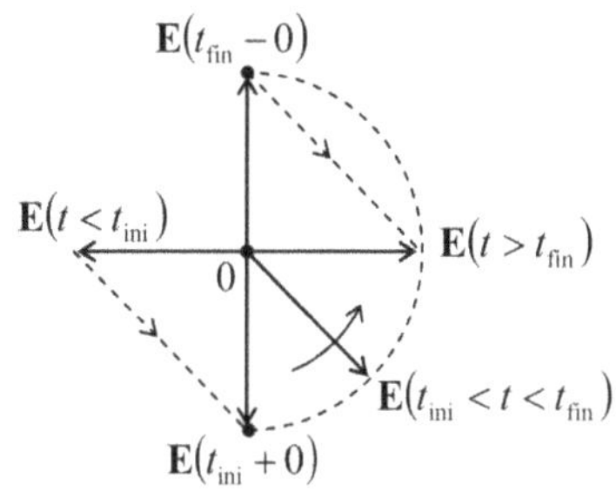

Note also that while Eq. (****) results from the second, 'radiative' term of Eq. (***), the field it gives is proportional to $1/R^2$—in contrast to the real radiation fields, which are proportional to $1/R$. This is also not quite surprising, because according to the discussion in section 10.3 of the lecture notes (see, e.g. figure 10.7a), the main radiation of an ultra-relativistic particle at its circular motion is localized within a very narrow cone, centered to vector $\boldsymbol{\beta}(t_{\text{ret}})$, and is always directed out of the circle—never toward its center.

***Problem* 10.4.** Express the total power of electromagnetic radiation by a relativistic particle with electric charge q and rest mass m, moving with velocity $\mathbf{u}$, via the external Lorentz force $\mathbf{F}$ exerted on the particle.

Solution: According to Eqs. (9.70) and (9.144) of the lecture notes, we may write

$$m\frac{d}{dt}(\gamma\mathbf{u}) = \mathbf{F}, \quad \text{where } \mathbf{F} = q(\mathbf{E} + \mathbf{u} \times \mathbf{B}). \qquad (*)$$

By an explicit differentiation, we obtain

$$\frac{d}{dt}(\gamma\mathbf{u}) \equiv c\frac{d}{dt}(\gamma\boldsymbol{\beta}) \equiv c\frac{d}{dt}\frac{\boldsymbol{\beta}}{(1-\beta^2)^{1/2}} = c\left[\frac{\dot{\boldsymbol{\beta}}}{(1-\beta^2)^{1/2}} + \frac{\boldsymbol{\beta}}{2(1-\beta^2)^{3/2}}\frac{d}{dt}(\beta^2)\right].$$

[10] Note that in this assignment, at least the continuity equation (4.5) is obeyed, i.e. the electric charge is conserved. Unfortunately, some textbooks pose various 'blinking charge' problems, in which this condition (required by the Maxwell equations) is violated. While such problems may be useful as mathematical exercises, the reader should recognize very clearly that their solutions may not represent physical reality.

For what follows, it is convenient to differentiate β^2 in the last term as the scalar product $\boldsymbol{\beta} \cdot \boldsymbol{\beta}$, obtaining

$$\frac{d}{dt}(\gamma \mathbf{u}) = c\left[\frac{\dot{\boldsymbol{\beta}}}{(1-\beta^2)^{1/2}} + \frac{\boldsymbol{\beta}}{2(1-\beta^2)^{3/2}}(2\boldsymbol{\beta}\cdot\dot{\boldsymbol{\beta}})\right] \equiv c\gamma^3[(1-\beta^2)\,\dot{\boldsymbol{\beta}} + \boldsymbol{\beta}(\boldsymbol{\beta}\cdot\dot{\boldsymbol{\beta}})],$$

so that the first of Eqs. (*) becomes

$$mc\gamma^3[(1-\beta^2)\,\dot{\boldsymbol{\beta}} + \boldsymbol{\beta}(\boldsymbol{\beta}\cdot\dot{\boldsymbol{\beta}})] = \mathbf{F}. \tag{**}$$

Now scalar-multiplying both sides of this equation by the vector $\boldsymbol{\beta}$, we obtain

$$mc\gamma^3[(1-\beta^2)(\boldsymbol{\beta}\cdot\dot{\boldsymbol{\beta}}) + \beta^2(\boldsymbol{\beta}\cdot\dot{\boldsymbol{\beta}})] \equiv mc\gamma^3\,\boldsymbol{\beta}\cdot\dot{\boldsymbol{\beta}} = \boldsymbol{\beta}\cdot\mathbf{F},$$

$$\text{so that}\quad \boldsymbol{\beta}\cdot\dot{\boldsymbol{\beta}} = \frac{\boldsymbol{\beta}\cdot\mathbf{F}}{mc\gamma^3}. \tag{***}$$

Plugging the last expression back into Eq. (**), we obtain a simple relation,

$$mc\gamma^3\left[(1-\beta^2)\,\dot{\boldsymbol{\beta}} + \boldsymbol{\beta}\frac{\boldsymbol{\beta}\cdot\mathbf{F}}{mc\gamma^3}\right] \equiv mc\gamma\,\dot{\boldsymbol{\beta}} + \boldsymbol{\beta}(\boldsymbol{\beta}\cdot\mathbf{F}) = \mathbf{F},$$

which yields an explicit expression for the reduced acceleration via the Lorentz force:

$$\dot{\boldsymbol{\beta}} = \frac{\mathbf{F} - \boldsymbol{\beta}(\boldsymbol{\beta}\cdot\mathbf{F})}{mc\gamma}. \tag{****}$$

(In the non-relativistic limit, $\boldsymbol{\beta} \to 0$, the second term in the denominator is negligible, and $\gamma \approx 1$, so that this expression is reduced to the usual Newton law $\dot{\mathbf{u}} = \mathbf{F}/m$.) Now plugging Eqs. (***) and (****) into the last form of the Liénard extension (10.37), we obtain the final answer[11]:

$$\begin{aligned}
\mathscr{P} &= \frac{Z_0 q^2}{6\pi}\frac{\gamma^4}{(mc\gamma)^2}\left\{[\mathbf{F} - \boldsymbol{\beta}(\boldsymbol{\beta}\cdot\mathbf{F})]^2 + \frac{1}{\gamma^2}(\boldsymbol{\beta}\cdot\mathbf{F})^2\right\} \\
&\equiv \frac{Z_0 q^2}{6\pi}\left(\frac{\gamma}{mc}\right)^2\{[F^2 + \beta^2(\boldsymbol{\beta}\cdot\mathbf{F})^2 - 2\mathbf{F}\cdot\boldsymbol{\beta}(\boldsymbol{\beta}\cdot\mathbf{F})] + (1-\beta^2)(\boldsymbol{\beta}\cdot\mathbf{F})^2\} \\
&\equiv \frac{Z_0 q^2}{6\pi}\left(\frac{\gamma}{mc}\right)^2[F^2 - (\boldsymbol{\beta}\cdot\mathbf{F})^2]\,.
\end{aligned}$$

Note also that the magnetic part of the Lorentz force is always normal to vector $\boldsymbol{\beta} \equiv \mathbf{u}/c$, so that it drops out of the second term of this formula, and we may express this result directly via the fields driving the particle's motion:

$$\mathscr{P} = \frac{Z_0 q^4}{6\pi}\left(\frac{\gamma}{mc}\right)^2[(\mathbf{E} + \mathbf{u}\times\mathbf{B})^2 - (\mathbf{u}\cdot\mathbf{E}/c)^2]\,.$$

Alternatively, this expression may be also obtained by plugging Eqs. (9.145) and (9.125) into the first (four-vector) form of Eq. (10.36).

[11] An additional exercise: obtain this result by transforming Eq. (10.39) of the lecture notes.

***Problem* 10.5.** A relativistic particle with the rest mass m and electric charge q, initially at rest, is accelerated by a constant force **F** until it reaches a certain velocity u, and then moves by inertia. Calculate the total energy radiated during the acceleration.

Solution: Evidently, the particle so accelerated moves along a straight line. The electromagnetic wave power radiated at such linear acceleration is given by Eq. (10.40) of the lecture notes:

$$\mathscr{P} = \frac{Z_0 q^2}{6\pi m^2 c^2}\left(\frac{dp}{dt_{\text{ret}}}\right)^2,$$

where the particle's momentum p and the radiation emission time t_{ret} are both measured in the lab reference frame. Because of that, according to Eq. (9.144), $dp/dt_{\text{ret}} = F$, and the power may be expressed as

$$\mathscr{P} = \frac{Z_0 q^2 F^2}{6\pi m^2 c^2},$$

so it is time-independent through the acceleration process, if F retains constant. Hence the total radiated energy is

$$\mathscr{E}_{\text{rad}} = \frac{Z_0 q^2 F^2}{6\pi m^2 c^2}\Delta t, \qquad (*)$$

where $\Delta t = p_{\max}/F$ is the acceleration time. According to Eq. (9.70), as the particle reaches its final, maximum value u, its momentum becomes

$$p_{\max} = m\gamma u \equiv m\frac{u}{(1 - u^2/c^2)^{1/2}},$$

so that Eq. (*) yields

$$\mathscr{E}_{\text{rad}} = \frac{Z_0 q^2 F}{6\pi m c^2}\frac{u}{(1 - u^2/c^2)^{1/2}}.$$

This result shows that at fixed finite velocity, the radiated energy increases proportionally to the applied force F.

***Problem* 10.6.** Calculate

(i) the instantaneous power, and
(ii) the power spectrum

of the radiation emitted into a unit solid angle by a relativistic particle with charge q, performing 1D harmonic oscillations with frequency ω_0 and displacement amplitude a.

Solutions:

(i) For the first task, we may directly use Eq. (10.30) of the lecture notes:

$$\frac{d\mathscr{P}}{d\Omega} = \frac{Z_0 q^2}{(4\pi)^2}\frac{|\mathbf{n} \times [(\mathbf{n} - \boldsymbol{\beta}) \times \dot{\boldsymbol{\beta}}]|^2}{(1 - \mathbf{n} \cdot \boldsymbol{\beta})^5},$$

where all the variables have to be evaluated at the retarded time t_{ret}. Selecting the reference frame as shown in figure 10.4a of the lecture notes, reproduced below, we may write

$$\mathbf{n} = \{\sin\theta,\ 0,\ \cos\theta\}, \qquad \mathbf{r}'(t_{\text{ret}}) = \{0,\ 0,\ a\cos\omega_0 t_{\text{ret}}\},$$
$$\boldsymbol{\beta}(t_{\text{ret}}) = \{0,\ 0,\ -\beta_0 \sin\omega_0 t_{\text{ret}}\}, \qquad \dot{\boldsymbol{\beta}}(t_{\text{ret}}) = \{0,\ 0,\ -\beta_0\omega_0 \cos\omega_0 t_{\text{ret}}\},$$

where β_0 is the reduced amplitude of the particle's velocity,

$$\beta_0 \equiv \frac{a\omega_0}{c},$$

and a is particle's displacement amplitude. (Note that a and ω_0 have to satisfy the evident condition $\beta_0 \equiv a\omega_0/c \leqslant 1$.) Calculating the vector products we need,

$$\mathbf{n}\cdot\boldsymbol{\beta} = -\beta_0 \cos\theta \sin\omega_0 t_{\text{ret}},$$
$$(\mathbf{n} - \boldsymbol{\beta}) \times \dot{\boldsymbol{\beta}} = \mathbf{n} \times \dot{\boldsymbol{\beta}} = \{0,\ \beta_0\omega_0 \sin\theta \cos\omega_0 t_{\text{ret}},\ 0\},$$
$$\mathbf{n} \times [(\mathbf{n} - \boldsymbol{\beta}) \times \dot{\boldsymbol{\beta}}] = -\beta_0\omega_0 \sin\theta \cos\omega_0 t_{\text{ret}}\{\cos\theta,\ 0,\ \sin\theta\},$$

so that $|\mathbf{n} \times [(\mathbf{n} - \boldsymbol{\beta}) \times \dot{\boldsymbol{\beta}}]\,|^2 = (\beta_0\omega_0 \sin\theta \cos\omega_0 t_{\text{ret}})^2,$

we get

$$\frac{d\mathscr{P}}{d\Omega} = \frac{Z_0 q^2}{(4\pi)^2} \frac{(\beta_0\omega_0 \sin\theta \cos\omega_0 t_{\text{ret}})^2}{(1 + \beta_0 \cos\theta \sin\omega_0 t_{\text{ret}})^5}.$$

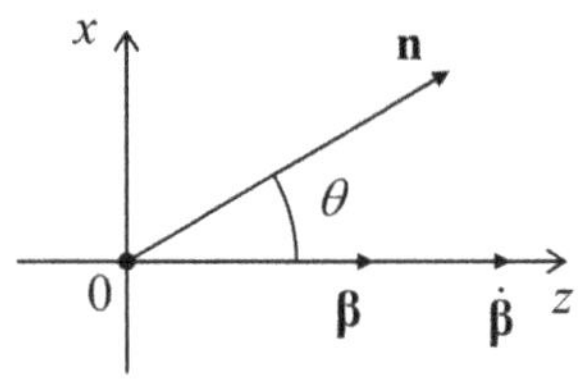

For a non-relativistic oscillator ($\beta_0 \ll 1$), the denominator of the last fraction approaches 1, and our result, after integration over the solid angle, is reduced to the Larmor law—see Eq. (8.27) of the lecture notes, with $Z = Z_0$, $v = c$, and

$$\ddot{p} = q\ddot{r}' = q\dot{u} = qc\dot{\beta} = qc\omega_0\beta_0 \cos\omega_0 t.$$

In particular, the radiation intensity is proportional to $\sin^2\theta$, and its time dependence ($\mathscr{P} \propto \cos^2\omega_0 t_{\text{ret}}$) shows that the fields of the radiated wave are sinusoidal functions of the observation time t, with the frequency ω_0. (Indeed, in the non-relativistic limit, the difference (10.5) between the times t and t_{ret} is virtually a constant.)

However, in the case $\beta_0 \to 1$, our result describes periodic radiation peaks (two per the oscillation period, at $\sin\omega_0 t_{\text{ret}} = \pm 1$), very narrow both in time and space (angle θ), emitted at the moments when the particle's speed approaches the speed of light, in the direction of motion. The angular structure of such a peak is the same as at the linear acceleration (see section 10.2 of the lecture notes, in particular figure 10.4b), while its exact temporal shape depends on whether we speak of the intensity as a function of t_{ret} or t, because exactly at their emission moments, the difference between these times is particularly significant—see Eq. (10.16). Hence, this difference needs to be taken into account at the analysis of the radiation spectrum—see below.

(ii) Since the radiation is periodic, so that its spectrum may only consist of harmonics of the motion frequency ω_0. Their amplitudes may be recovered from Eq. (10.61) of the lecture notes (which has been derived with the continuous spectrum in mind) by a careful juggling delta-functions, but it is more prudent (and instructive) to recast all basic equations of the spectral theory, outlined in section 10.3, for the periodic time dependence of the radiated field. In particular, Eq. (10.51) needs to be replaced with a sum over integer harmonic numbers (say, k—not to be confused with the wave number, which does not explicitly participate in this solution):

$$\mathbf{E}(t) = \sum_{k=-\infty}^{+\infty} \mathbf{E}_k \exp\{-ik\omega_0 t\}.$$

The magnetic field of the radiated wave (which, at a sufficiently large distance from the source, is locally plane) may be expanded into a similar series,

$$\mathbf{H}(t) = \sum_{k'=-\infty}^{+\infty} \mathbf{H}_{k'} \exp\{-ik'\omega_0 t\},$$

with complex amplitudes $\mathbf{H}_{k'} = \mathbf{n} \times \mathbf{E}_{k'}/Z_0$, so at time averaging over the period $\mathcal{T}$, the Poynting vector falls into a sum of independent harmonic components:

$$\begin{aligned}
\overline{S_n} = \overline{E(t)H(t)} &= \frac{1}{Z_0}\sum_{k,k'=-\infty}^{+\infty} E_k E_{k'} \overline{\exp\{-i\,(k+k')k\omega_0 t\}} \\
&\equiv \frac{1}{Z_0}\sum_{k,k'=-\infty}^{+\infty} E_k E_{k'} \frac{1}{\mathcal{T}}\int_0^{+\mathcal{T}} \exp\{-i\,(k+k')k\omega_0 t\}dt \\
&= \frac{1}{Z_0}\sum_{k,k'=-\infty}^{+\infty} E_k E_{k'}\delta_{k+k',0} = \frac{1}{Z_0}\sum_{k=-\infty}^{+\infty} E_k E_{-k} = \frac{1}{Z_0}\sum_{k=-\infty}^{+\infty} E_k E_k^* = \frac{2}{Z_0}\sum_{k=1}^{+\infty} |\mathbf{E}_k|^2.
\end{aligned}$$

Thus instead of Eq. (10.54) of the lecture notes, we can write the following expression for the average radiation power per unit solid angle:

$$\frac{d\overline{\mathcal{P}}}{d\Omega} = \sum_{k=1}^{\infty} \frac{d\overline{\mathcal{P}_k}}{d\Omega}, \qquad \text{with } \frac{d\overline{\mathcal{P}_k}}{d\Omega} = R^2\overline{S_n} = \frac{2R^2}{Z_0}|\mathbf{E}_k|^2. \qquad (*)$$

Into the last expression we have to plug in the Fourier amplitudes $\mathbf{E}_k$ from the reciprocal Fourier transform, for example over a symmetric time interval:

$$\mathbf{E}_k = \frac{1}{\mathcal{T}}\int_{-\mathcal{T}/2}^{+\mathcal{T}/2} \mathbf{E}(t)\exp\{ik\omega_0 t\}dt = \frac{\omega_0}{2\pi}\int_{-\pi/\omega_0}^{+\pi/\omega_0} \mathbf{E}(t)\exp\{ik\omega_0 t\}dt,$$

so that

$$\frac{d\overline{\mathscr{P}_k}}{d\Omega} = \frac{R^2\omega_0^2}{2\pi^2 Z_0}\left|\int_{-\pi/\omega_0}^{+\pi/\omega_0} \mathbf{E}(t)\exp\{ik\omega_0 t\}dt\right|^2.$$

Plugging in the $\mathbf{E}(t)$ given by the second (radiative) term of Eq. (10.19), and repeating the transformation from integration over t to that over t_{ret}, discussed in section 10.3 of the lecture notes, instead of Eq. (10.61) we obtain

$$\frac{d\overline{\mathscr{P}_k}}{d\Omega} = \frac{Z_0 q^2 k^2 \omega_0^4}{32\pi^4}\left|\int_{-\pi/\omega_0}^{+\pi/\omega_0}\left[\mathbf{n}\times(\mathbf{n}\times\boldsymbol{\beta})\exp\left\{ik\omega_0\left(t - \frac{\mathbf{n}\cdot\mathbf{r}'}{c}\right)\right\}\right]_{\text{ret}} dt_{\text{ret}}\right|^2. \quad (**)$$

As a sanity check, let us calculate the particle's energy loss per one period of motion, due to the radiation into a unit solid angle:

$$-\frac{d\mathscr{E}}{d\Omega} = \mathcal{T}\frac{d\overline{\mathscr{P}}}{d\Omega} = \frac{2\pi}{\omega_0}\frac{d\overline{\mathscr{P}}}{d\Omega} = \frac{2\pi}{\omega_0}\sum_{k=1}^{\infty}\frac{d\overline{\mathscr{P}_k}}{d\Omega}.$$

We may use this formula for non-periodic motion as well, by following the limit $\mathcal{T}\to\infty$, i.e. $\omega_0\to 0$, while keeping the observation frequency $\omega \equiv k\omega_0$ finite. In this limit, the sum over $k = \omega/\omega_0$ may be replaced by an integral, and we obtain

$$-\frac{d\mathscr{E}}{d\Omega} \to \frac{2\pi}{\omega_0}\int_0^\infty \frac{d\overline{\mathscr{P}_k}}{d\Omega}dk = \frac{2\pi}{\omega_0}\int_0^\infty \frac{d\overline{\mathscr{P}_{\omega/\omega_0}}}{d\Omega}d\left(\frac{\omega}{\omega_0}\right)$$

$$\equiv \int_0^\infty I(\omega)d\omega, \quad \text{where } I(\omega) \equiv \frac{2\pi}{\omega_0^2}\frac{d\overline{\mathscr{P}_{\omega/\omega_0}}}{d\Omega}.$$

Now using Eq. (**), for the function $I(\omega)$, defined just like in the first of Eqs. (10.54), we obtain the expression,

$$I(\omega) \equiv \frac{2\pi}{\omega_0^2}\frac{d\overline{\mathscr{P}_{\omega/\omega_0}}}{d\Omega} = \frac{Z_0 q^2\omega^2}{16\pi^3}\left|\int_{-\pi/\omega_0}^{+\pi/\omega_0}\left[\mathbf{n}\times(\mathbf{n}\times\boldsymbol{\beta})\exp\left\{i\omega\left(t - \frac{\mathbf{n}\cdot\mathbf{r}'}{c}\right)\right\}\right]_{\text{ret}} dt_{\text{ret}}\right|^2,$$

which at $\omega_0 \to 0$ duly tends to Eq. (10.61).

Plugging the expressions for $\mathbf{n}$ and $\boldsymbol{\beta}$, discussed in part (i) of the solution, into Eq. (**), performing the vector multiplications, and introducing the dimensionless integration variable $\xi \equiv \omega_0 t_{\text{ret}}$, we obtain

$$\frac{d\overline{\mathscr{P}_k}}{d\Omega} = \frac{Z_0 q^2\beta_0^2 k^2\omega_0^2\sin^2\theta}{32\pi^4}|I_k|^2,$$

where I_k is a dimensionless integral:

$$I_k \equiv \int_{-\pi}^{+\pi} \sin\xi \exp\{ik(\xi - \alpha\cos\xi)\}d\xi, \quad \text{with } \alpha \equiv \beta_0 \cos\theta < 1.$$

Transforming the integral as

$$I_k = \frac{1}{2i}\int_{-\pi}^{+\pi} (e^{i\xi} - e^{-i\xi})\, e^{ik\xi}e^{-ik\alpha\cos\xi}\, d\xi \equiv \frac{1}{2i}\int_{-\pi}^{+\pi} [e^{i(k+1)\xi} - e^{i(k-1)\xi}]\, e^{-ik\alpha\cos\xi}d\xi,$$

we may notice that the integrals of the imaginary parts of $\exp\{i(k \pm 1)\xi\} \equiv \cos(k \pm 1)\xi + i\sin(k \pm 1)\xi$, in our symmetric limits, vanish, so that

$$\begin{aligned} I_k &= \frac{1}{2i}\int_{-\pi}^{+\pi} [\cos(k+1)\xi - \cos(k-1)\xi]e^{-ik\alpha\cos\xi}d\xi \\ &\equiv \frac{1}{i}\int_{0}^{+\pi} \cos(k+1)\xi \; e^{-ik\alpha\cos\xi}d\xi - \frac{1}{i}\int_{0}^{+\pi} \cos(k-1)\xi e^{-ik\alpha\cos\xi}d\xi, \end{aligned}$$

Apart from constant coefficients, these integrals are the standard integral representations of the Bessel functions of the first kind[12], so that we obtain

$$I_k = \frac{\pi}{i}[i^{k+1}J_{k+1}(k\alpha) - i^{m-1}J_{m-1}(k\alpha)] = i^m\pi[J_{k+1}(k\alpha) + J_{k-1}(k\alpha)] = i^m 2\pi\frac{k}{\alpha}J_k(k\alpha),$$

where for the last step, the first of the recurrence relations (2.142) has been used. So, we have

$$\frac{d\overline{\mathscr{P}_k}}{d\Omega} = \frac{Z_0 q^2\omega_0^2}{8\pi^2}\tan^2\theta \;\; k^4 J_k^2\left(k\beta_0\cos\theta\right), \quad \text{for } k \geqslant 1. \tag{***}$$

The figure below shows, on the semi-log scale, the dependence of the power of a few harmonics of radiation on the effective reduced amplitude $\zeta \equiv \beta_0\cos\theta = (a\omega_0/c)\cos\theta$ of the particle's motion, as given by Eq. (***).

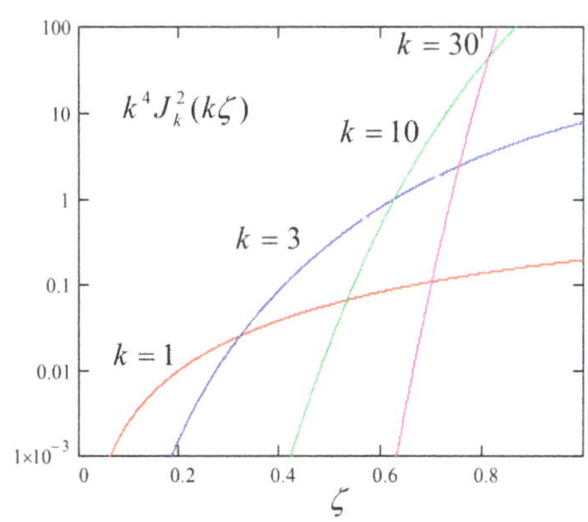

[12] See, e.g. Eq. (A.42a).

In the non-relativistic limit ($\beta_0 \to 0$), only the first harmonic ($k = 1$) is non-vanishing, i.e. the radiation takes place only at the frequency of particle motion, and our result for it,

$$\frac{d\overline{\mathscr{P}_1}}{d\Omega} = \frac{Z_0 q^2 \omega_0^2}{8\pi^2} \tan^2\theta \; J_1^2(\beta_0 \cos\theta)$$

$$\to \frac{Z_0 q^2 \omega_0^2}{8\pi^2} \tan^2\theta \left(\frac{\beta_0 \cos\theta}{2}\right)^2 \equiv \frac{Z_0 q^2 \omega_0^2}{32\pi^2} \beta_0^2 \sin^2\theta,$$

coincides with the time-averaged result of the first part of the problem, and hence complies with the non-relativistic Larmor formula (8.28).

On the other hand, when the maximum value β_0 of the normalized velocity of the particle approaches 1, i.e. the particle becomes ultra-relativistic during two parts of each period of its motion, the intensity of high harmonics going in the direction of motion ($\cos\theta \approx 1$) grows rapidly, reflecting the narrow radiation cones generated during these parts—see figure 10.4 and its discussion. (The additional factor $\tan^2\theta$ in Eq. (***) makes these cones hollow, just as in the constant-acceleration case.)

A good additional exercise: review these results for the case when the force driving the particle oscillation (rather than the particle displacement as above) is sinusoidal. The results would be different only in the relativistic case, due to the increase of the effective mass (9.71) of the particle at the periods of its fastest motion, and the corresponding reduction of its acceleration.

***Problem* 10.7.** Analyze the polarization and the spectral contents of the synchrotron radiation propagating in the direction perpendicular to the particle's rotation plane. How do the results change if not one, but $N > 1$ similar particles move around the circle, at equal angular distances?

Solution: It is instrumental here to represent the double product in the basic Eq. (10.19),

$$\mathbf{n} \times [(\mathbf{n} - \boldsymbol{\beta}) \times \dot{\boldsymbol{\beta}}],$$

describing the wave's electric field, as the sum of two terms:

$$\mathbf{n} \times (\mathbf{n} \times \dot{\boldsymbol{\beta}}) - \mathbf{n} \times (\boldsymbol{\beta} \times \dot{\boldsymbol{\beta}}), \qquad (*)$$

because in our case the second term vanishes. (Indeed, both vectors $\boldsymbol{\beta}$ and $\dot{\boldsymbol{\beta}}$ move within the particle's rotation plane, so that the inner vector product $\boldsymbol{\beta} \times \dot{\boldsymbol{\beta}}$ is perpendicular to the plane, as well as vector $\mathbf{n}$, i.e. the last two vectors are parallel to each other, and their vector product equals zero.)

The double vector product in the remaining, first term of Eq. (*) is opposite in direction to the instant direction of the vector $\dot{\boldsymbol{\beta}}$, i.e. it rotates, within the plane, with the angular velocity ω_c of the particle's motion. So, the vector $\mathbf{E}$ rotates (about wave's propagation direction $\mathbf{n}$) with the same angular frequency, i.e. the synchrotron radiation in the direction out of the rotation plane is circularly polarized, in the angular direction of the particle's rotation.

Now note that in our case a different double vector product, $\mathbf{n} \times (\mathbf{n} \times \boldsymbol{\beta})$, also rotates similarly, with the same angular velocity ω_c, while the scalar product $\mathbf{n} \cdot \mathbf{r}'$ vanishes. According to Eq. (10.61) of the lecture notes, this means that the out-of-plane radiation has only one frequency component, ω_c—even in the ultra-relativistic case, when the in-plane radiation has $\sim\gamma^3 \gg 1$ harmonics of this basic frequency!

Now if we have $N > 1$ similar particles moving around the same circle at equal angular distances $\Delta\varphi = 2\pi/N$, the net electric field is the vector sum of N vectors, with equal magnitudes, and the same angular distances $\Delta\varphi$. As was already discussed in the model solution of problem 8.18[13], such a sum equals zero, so that in this direction the system's radiation intensity vanishes.

***Problem* 10.8.** Calculate and analyze the time dependence of the energy of a charged relativistic particle performing synchrotron motion in a constant and uniform magnetic field **B**, and hence emitting the synchrotron radiation. Qualitatively, what is the particle's trajectory?

Hint: You may assume that the energy loss is relatively slow ($-d\mathscr{E}/dt \ll \omega_c\mathscr{E}$), but should spell out the condition of validity of this assumption.

Solution: With Eq. (9.153) of the lecture notes for the cyclotron orbit's radius R,

$$R = \frac{m\gamma u}{qB} \equiv \frac{mc}{qB}\beta\gamma, \qquad (*)$$

and Eq. (9.73) for the particle's energy, Eq. (10.45) for the radiative energy loss yields

$$-\frac{d\mathscr{E}}{dt} = \frac{Z_0 q^4 B^2}{6\pi m^2}\beta^2\gamma^2 \equiv \frac{Z_0 q^4 B^2}{6\pi m^2}(\gamma^2 - 1) = \frac{\mathscr{E}^2 - (mc^2)^2}{2mc^2\tau}, \quad \text{where } \tau \equiv \frac{3\pi m^3 c^2}{Z_0 q^4 B^2}. \qquad (**)$$

Separating the variables of this differential equation,

$$\frac{d\varepsilon}{\varepsilon^2 - 1} = -\frac{dt}{2\tau}, \qquad \text{where } \varepsilon \equiv \frac{\mathscr{E}}{mc^2} \geq 1,$$

we can readily integrate it[14]:

$$\frac{1}{2}\ln\frac{\varepsilon - 1}{\varepsilon + 1} = -\frac{t}{2\tau} + \text{const.}$$

[13] And as follows, for example, from Eq. (A.21).

[14] As a math reminder, the integral on the left-hand side may be reduced to two integrals of the type $\int d\xi/\xi = \ln\xi$, by representing the involved fraction as the following sum:

$$\frac{1}{\varepsilon^2 - 1} = \frac{1}{2(\varepsilon - 1)} - \frac{1}{2(\varepsilon + 1)}.$$

Solving this simple equation for ε, we finally obtain

$$\mathscr{E} = mc^2 \coth\left(\frac{t}{2\tau} + \text{const}\right). \qquad (***)$$

This expression shows that the kinetic energy $T \equiv \mathscr{E} - mc^2$ of the particle is gradually lost on the time scale given by the constant τ. (As was already discussed in the model solution of problem 8.2, for an electron in a typical magnetic field of 1 T, this constant is close to 2.6 s.) At the finite stage of this process, at $t \gg \tau$, the argument of coth is large, and this function may be approximated as

$$\coth \xi \equiv \frac{e^{\xi} + e^{-\xi}}{e^{\xi} - e^{-\xi}} \equiv \frac{1 + e^{-2\xi}}{1 - e^{-2\xi}} \approx 1 + 2e^{-2\xi}, \quad \text{for } \xi \gg 1,$$

so that Eq. (***) yields the exponential decay law, already obtained in the solution of problem 8.2:

$$T = mc^2\left[\coth\left(\frac{t}{2\tau} + \text{const}\right) - 1\right] \approx \text{const} \times \exp\left\{-\frac{t}{\tau}\right\}.$$

If the initial value of T is much smaller than mc^2, i.e. the particle is non-relativistic, such exponential law is obeyed at all times.

In the opposite limit when the particle is initially ultra-relativistic, the initial decay law is different. Indeed, in this case the function coth in Eq. (***) has to be large, so it may be approximated as

$$\coth \xi \equiv \frac{\cosh \xi}{\sinh \xi} \approx \frac{1}{\xi}, \quad \text{for } \xi \gg 1,$$

and the solution is reduced to

$$\mathscr{E}(t) \approx mc^2 \frac{2\tau}{t + \text{const}}.$$

The constant participating in this expression may be readily related to the initial energy $\mathscr{E}(0)$ of the particle, finally giving

$$\mathscr{E}(t) = \frac{\mathscr{E}(0)}{1 + \mathscr{E}(0)t/2\tau mc^2} \approx \mathscr{E}(0)\left[1 - \frac{\mathscr{E}(0)}{2mc^2}\frac{t}{\tau}\right], \qquad \text{for } \mathscr{E}(t) \gg mc^2.$$

This result (which may be also readily derived directly from Eq. (**)), describes an even faster initial decrease of energy, with an effective time constant $\tau' = [2mc^2/\mathscr{E}(0)]\tau \ll \tau$.

The above theory, which assumes that each orbit is approximately circular, is only valid if $\omega_c\tau \gg 1$. Rewriting Eq. (9.151) of the lecture notes as

$$\omega_c = \frac{qB}{\gamma m},$$

we may represent the validity condition as

$$\gamma \ll \frac{m^2c^2}{Z_0 q^3 B}.$$

For electrons in the field of 1 T, the right-hand side of this relation is of the order of 10^{12}, so that is all practical systems it is well satisfied.

As the kinetic energy goes down due to the radiative losses, so does the parameter $\beta\gamma$, so that, as Eq. (*) shows, the cyclotron radius R decreases as well. Hence particle's trajectory is a converging spiral.

***Problem* 10.9.** Analyze the polarization of the synchrotron radiation propagating within the particle's rotation plane.

Solution: The second term of Eq. (10.19) of the lecture notes shows that the instant electric field **E** of the radiated wave is oriented as the vector

$$\mathbf{n} \times \left[(\mathbf{n} - \boldsymbol{\beta}) \times \frac{d\boldsymbol{\beta}}{dt}\right]$$

at the retarded point, i.e. at the charge position. At the synchrotron radiation, the vectors $\boldsymbol{\beta}$ and $d\boldsymbol{\beta}/dt$ are always in the plane of particle's rotation, and perpendicular to each other. This is why if the observer (vector **n**) is also in that plane, the inner product's vector $(\mathbf{n} - \boldsymbol{\beta}) \times d\boldsymbol{\beta}/dt$ is perpendicular to that plane, so that the outer product vector, and hence the vector **E**, are within the rotation plane again. (This conclusion could be also made from the evident mirror symmetry of the problem relative to the rotation plane.)

Problem* 10.10. The basic quantum theory of radiation shows that the electric dipole radiation by a particle is allowed only if the change of its angular momentum's magnitude L at the transition is of the order of the Planck's constant $\hbar$.

(i) Estimate the change of L of an ultra-relativistic particle due to its emission of a typical single photon of the synchrotron radiation.
(ii) Does the quantum mechanics forbid such a radiation? If not, why?

Solutions:

(i) According to Eq. (9.70) of the lecture notes, the magnitude of the linear momentum of a particle is $p = \gamma mu$, so that its orbital angular momentum at the circular motion with radius R is $L = pR = \gamma muR$. From the basic geometry, the cyclotron orbit radius is $R = u/\omega_c$, so that

$$L_z = \frac{\gamma mu^2}{\omega_c},$$

where axis z is normal to the rotation plane. On the other hand, according to the basic Eq. (9.73), the particle's energy is $\mathscr{E} = \gamma mc^2$, so that in the ultra-relativistic limit ($u \to c$) we may write

$$\mathscr{E} = L_z \omega_c. \qquad (*)$$

This means that at the angular momentum change allowed by the usual quantum-mechanical selection rule, $|\Delta L_z| = \hbar$, the change of the particle's energy is

$$|\Delta \mathscr{E}| = \hbar\omega_c.$$

On the other hand, Eq. (*) means that the most typical frequency (10.72) of the synchrotron radiation, giving photons with energy

$$|\Delta \mathscr{E}| = \hbar\omega_{\max} \sim \frac{\gamma^3}{2}\hbar\omega_c,$$

corresponds to a much larger angular momentum change:

$$|\Delta L_z| = \frac{|\Delta \mathscr{E}|}{\omega_c} \sim \frac{\gamma^3}{2}\hbar \gg \hbar. \qquad (**)$$

(ii) The above result means that the selection rule $|\Delta L_z| = \hbar$, which is sometime perceived as a universal law, is invalid. This apparent violation of quantum mechanics is only superficial, because the standard selection rule is valid only in the non-relativistic electric-dipole approximation[15]. As we know from section 8.2, this approximation (and, as the result, the Larmor formula (8.27) as well) is only valid if the size scale of the radiating system (e.g. a in figure 8.1) is much smaller than the radiation wavelength $\lambda = 2\pi/k = 2\pi c/\omega = 2\pi c\hbar/|\Delta \mathscr{E}|$. However, for the synchrotron radiation by an ultra-relativistic particle, the system's size scale is $R \approx c/\omega_c$, and the relation is opposite:

$$\lambda \sim \frac{2\pi c}{\omega_{\max}} = \frac{2\pi c\hbar}{|\Delta \mathscr{E}|} \sim \frac{2}{\gamma^3}\frac{2\pi c}{\omega_c} \approx \frac{4\pi}{\gamma^3}R \ll R,$$

so that the standard selection rule is invalid.

The intrinsic reason of this difference is as follows. For a small radiating system (with $a \ll \lambda$), its orbital momentum and that of the radiated photon may be referred to the same origin—the system's center of mass. Hence the angular momentum $\pm\hbar$ carried away by a circularly polarized photon has to be compensated by the equal and opposite change of particle's L_z, giving the selection rule $|\Delta L_z| = \hbar$. On the other hand, as was discussed above, the most typical synchrotron radiation photon has the effective size $\sim \lambda \ll R$, and (as was discussed in detail in section 10.3) is emitted tangentially to particle's orbit, so that its angular momentum about the orbit center is not $\hbar$ (as it is about photon's propagation axis), but rather

$$L_{\text{photon}} \sim p_{\text{photon}} R = \left(\hbar\omega_{\text{photon}}/c\right)R \sim \hbar\frac{\gamma^3}{2}\omega_c\frac{R}{c} \approx \hbar\frac{\gamma^3}{2} \gg \hbar,$$

thus allowing a similarly large change (**) of the particle's angular momentum.

[15] See, e.g. its derivation in *Part QM* section 9.3.

***Problem* 10.11.** A relativistic particle moves along axis z, with velocity u_z, through an undulator—a system of permanent magnets providing (in the simplest model) a perpendicular magnetic field, whose distribution near the axis z is sinusoidal[16]:

$$\mathbf{B} = \mathbf{n}_y B_0 \cos k_0 z.$$

Assuming that the field is so weak that it causes only very small deviations of the particle's trajectory from the straight line, calculate the angular distribution of the resulting radiation. What condition does this assumption impose on the system's parameters?

Solution: Using the assumption of a nearly linear trajectory of the particle, $\mathbf{u} \approx \mathbf{n}_z u_z$, the Lorentz force acting on the particle has only one Cartesian component:

$$\mathbf{F} = q\mathbf{u} \times \mathbf{B} = (\mathbf{n}_z \times \mathbf{n}_y)\, qu_z\, B_0 \cos k_0\, z = -\mathbf{n}_x\, qu_z\, B_0 \cos k_0\, z,$$

so that the equations of its motion (with all variables measured in the lab frame) are

$$\dot{p}_x = -qu_z B_0 \cos k_0\, z, \qquad \text{with } p_x = m\gamma u_x = \frac{mu_x}{\left[1 - \left(u_x^2 + u_z^2\right)/c^2\right]^{1/2}},$$

$$\dot{p}_z = 0, \qquad \text{with } p_z = m\gamma u_z = \frac{mu_z}{\left[1 - \left(u_x^2 + u_z^2\right)/c^2\right]^{1/2}}.$$

In the first (linear) approximation in small u_x, the equations are simplified:

$$m\gamma\dot{u}_x \approx -qu_z B_0 \cos k_0\, z, \qquad m\gamma\dot{u}_z \mathbf{n}_z \approx 0, \qquad \text{with } \gamma = \frac{1}{\left(1 - u_z^2/c^2\right)^{1/2}} = \text{const},$$

so that, with an appropriate choice of the time origin, $z \approx u_z t$, and

$$\dot{u}_x \approx -\frac{qu_z B_0}{m\gamma} \cos(k_0 u_z t), \quad \text{i.e. } \dot{\boldsymbol{\beta}} \approx -\mathbf{n}_x \frac{q\beta_z B_0}{m\gamma} \cos(k_0 u_z t), \quad \text{and } \boldsymbol{\beta} \approx \mathbf{n}_z \beta_z. \qquad (*)$$

Hence the mutual orientation of the vectors $\boldsymbol{\beta}$ and $\dot{\boldsymbol{\beta}}$ is the same as at the synchrotron radiation—see figure 10.5 and Eq. (10.46) of the lecture notes. This means that for the instantaneous power we may reuse Eq. (10.47):

$$\frac{d\mathscr{P}}{d\Omega} = \frac{2Z_0 q^2}{\pi^2} |\dot{\boldsymbol{\beta}}|^2 \gamma^6 f(\theta, \varphi), \qquad (**)$$

where the angular distribution function,

$$f(\theta, \varphi) \equiv \frac{1}{8\gamma^6(1 - \beta \cos\theta)^3}\left[1 - \frac{\sin^2\theta \cos^2\varphi}{\gamma^2(1 - \beta\cos\theta)^2}\right], \qquad (***)$$

has been discussed in section 10.3 of the lecture notes and plotted (for the ultra-relativistic case) in figure 10.6. The only (but very substantial) difference of our

[16] As the Maxwell equation for $\nabla \times \mathbf{H}$ shows, such stationary field distribution cannot be created in any non-vanishing volume of free space. However, it may be created on a line—e.g. on the particle's trajectory.

current problem is that particle's acceleration's magnitude is not constant in time (as it is at the synchrotron radiation) but oscillates sinusoidally with frequency $\omega = k_0 u_z$. Time-averaging Eq. (**) and using the second of Eqs. (*), we obtain

$$\frac{d\overline{\mathscr{P}}}{d\Omega} = \frac{Z_0 q^4 B_0^2}{\pi^2 m^2} \beta_z^2 \gamma^4 f(\theta, \varphi).$$

It may look that this approximate analysis is valid if the amplitude

$$u_0 = \frac{q u_z B_0}{m\gamma\omega} \equiv \frac{q B_0}{m\gamma k_0}$$

of the transverse velocity $u_x(t)$, following from the above calculations, is much less than u_z. Using Eq. (9.151) for the cyclotron frequency, this condition may be recast a very simple and natural form $\omega_c \ll \omega \equiv k_0 u_z$. However, since the angular distribution (***) has the width scale $\sim 1/\gamma$, the actual condition of theory's validity is $u_0 \ll u_z/\gamma$, i.e.

$$\omega_c \ll \frac{\omega}{\gamma}.$$

In modern undulators, this condition limits the magnet period $\lambda = 2\pi/k_0$ to a few centimeters—for example numbers, see section 10.3 of the lecture notes, in particular figure 10.10.

Note, however, that the above solution, based on Eq. (10.30) of the lecture notes, neglects the possible coherent addition of radiation induced by different periods of undulator's magnetic field—see the next problem. (If such coherence is important, we need to use Eq. (10.19) for the electric field instead, integrating it for all particle's trajectory, and only then calculating its total power.)

***Problem* 10.12.** Discuss possible effects of the interference of the undulator radiation from different periods of its static field distribution. In particular, calculate the angular positions of the power density maxima.

Solution: As was discussed in detail in section 8.4 of the lecture notes, the constructive interference of the radiation from two coherent, in-phase sources takes place when the difference Δt of their times of wave propagation toward the observer (in the figure below $\Delta t = (d\cos\theta)/c$) equals the integer number of wave periods $\mathscr{T} = \lambda/c$. However, in the case of the undulator with a period d (equal to $2\pi/k_0$ of the previous problem) with the particles propagating along it with velocity u, the radiation induced by adjacent periods of the system has an additional time shift d/u, due to the difference of the particle arrival times, so that the constructive interference condition becomes

$$\frac{d}{u} - \frac{d\cos\theta}{c} = n\frac{\lambda}{c}, \qquad (*)$$

with any integer n. On the other hand, the wavelength λ of the kth harmonic of the undulator's radiation (where $k = 1, 2, \ldots$) equals $c\mathscr{T}_k = c\mathscr{T}_1/k = c(d/u)/k$.

Combining these expressions, we see that the condition for the angles of the radiation maxima does not depend on d:

$$\cos\theta = \frac{1}{\beta}\left(1 - \frac{n}{k}\right), \quad \text{where } \beta \equiv \frac{u}{c}. \tag{**}$$

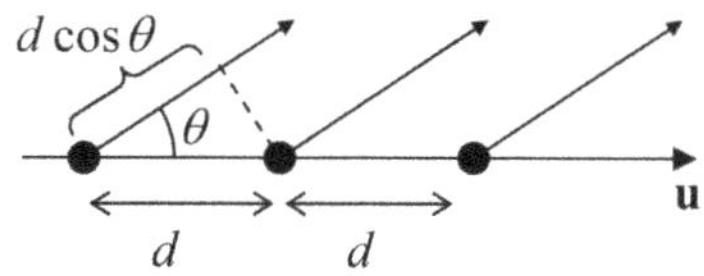

Evidently, since $1/\beta$ cannot be smaller than 1, the real angles θ, with $-1 \leqslant \cos\theta \leqslant +1$, may satisfy this condition only for certain positive values of the integer n.

In particular, for ultra-relativistic particles with $\gamma \gg 1$, used in standard undulators, the analogy of the undulator radiation and the synchrotron radiation (see the model solution of previous problem) allows us to use Eq. (10.72) of the lecture notes to conclude that the most intense radiation harmonics have numbers $k \sim \gamma^3/2 \gg 1$. Since for such particles $1/\beta \equiv 1/(1 - 1/\gamma^2)^{1/2} \approx 1 + 1/2\gamma^2$, the condition (**) becomes

$$\cos\theta_n \approx \left(1 + \frac{1}{2\gamma^2}\right)\left(1 - \frac{2n}{\gamma^3}\right) \approx 1 - \frac{1}{2\gamma^2}\left(\frac{4n}{\gamma} - 1\right).$$

Using the Taylor expansion $\cos\theta \approx 1 - \theta^2/2$, valid for small angles θ, we see that the lowest-order maxima correspond to angles

$$\theta_n \approx \frac{1}{\gamma}\left(\frac{4n}{\gamma} - 1\right)^{1/2} = \frac{2}{\gamma^{3/2}}(n - n_{\min})^{1/2} \ll 1, \quad \text{where } n_{\min} \equiv \frac{\gamma}{4} \gg 1,$$

with a very small angular distance,

$$\Delta\theta_n \equiv (\theta_{n+1} - \theta_n)_{n\sim n_{\min}} \approx \left.\frac{d\theta_n}{dn}\right|_{n\sim n_{\min}} \approx \frac{1}{\gamma^{3/2} n_{\min}^{1/2}} = \frac{2}{\gamma^2} \ll \frac{1}{\gamma},$$

between the adjacent maxima, completely masked by the relativistic width $\Delta\theta \sim 1/\gamma$ of the radiation—again, see the model solution of the previous problem. In this case, all results of that solution, in which the interference has been neglected, are qualitatively valid.

On the other hand, for non- (or nearly) relativistic particles, particularly in 'soft' (sinusoidal or nearly sinusoidal) undulator field profiles, such as the sinusoidal profile of the previous problem, the radiation is nearly sinusoidal, i.e. the largest power corresponds to $k = 1$. In this case the distance between the adjacent interference maxima may be quite substantial:

$$\Delta(\cos\theta_n)_{k=1} = \frac{1}{\beta} \sim 1,$$

and they be readily resolved. As was briefly discussed in the model solution of the previous problem, the quantitative calculation of the radiation power distribution in this case (a good optional exercise for the reader) should start from Eq. (10.19), rather than Eq. (10.30) of the lecture notes.

If the radiation is produced by a short bunch of particles with a length ~d, rather than by a single particle, the result of such a power calculation depends on the bunch's shape. This fact enables using this effect for the experimental non-destructive measurements of the shape of very short (in time domain, femtosecond) electron bunches. It is interesting that for such experiments the undulator is not really necessary: the needed space-periodic field may be produced just by the flying charge's image in a periodic metallic structure—see the figure below. The standard diffraction gratings, with $d \sim 1$ μm, are quite convenient for this purpose, because with $\beta \sim 1$ (the condition which is easy to reach with electrons), the resulting radiation wavelengths $\lambda \sim d/\beta$ are in the visible light range, for which sensitive radiation detectors are readily available[17].

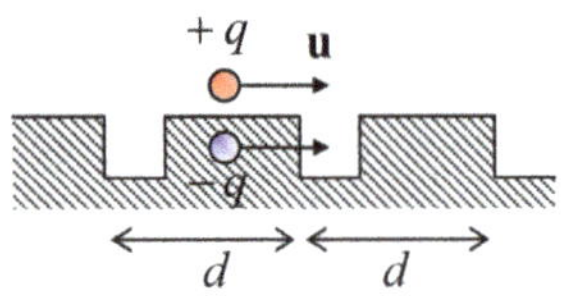

***Problem* 10.13.** An electron, launched directly toward a plane surface of a perfect conductor, is instantly absorbed by it at the collision. Calculate the angular distribution and the frequency spectrum of the electromagnetic waves radiated at this collision, if the initial kinetic energy T of the particle is much larger than the conductor's workfunction ψ.[18] Is your result valid near the conductor's surface?

Solution: As has been discussed in section 10.5 of the lecture notes in the context of transition radiation, at not very high frequencies ($\omega \ll \omega_p$),[19] the field outside of the conductor may be viewed as a result of a head-on collision of the particle with charge q with its mirror image, in this case with charge $(-q)$, moving toward each other—see the figure below.

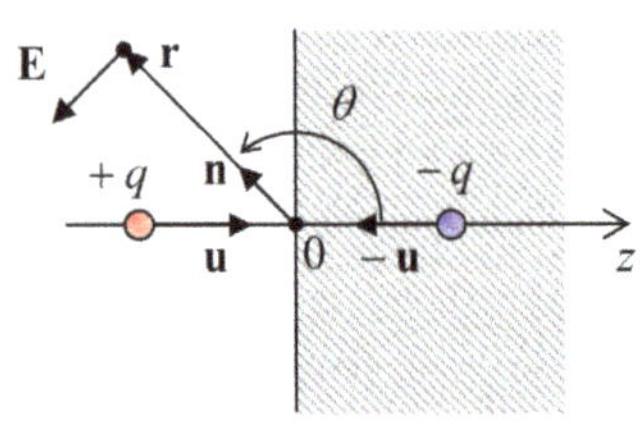

[17] This is exactly the arrangement which was used by S Smith and E Purcell in 1953 for the first observation of this *Smith–Purcell effect*. Historically this was the first step toward modern undulators and free-electron lasers, which use much stronger periodic fields and hence produce much more intense radiation.

[18] See section 2.9, in particular figure 2.27a.

[19] Taking the (admittedly, too vague) characterization of the conductor as 'perfect' in the assignment literally, we may assume that its free carrier density n is infinite, and so is the plasma frequency ω_p—see Eq. (7.37).

Due to the condition $\psi \ll T$, the change of the particle velocity, due to its Coulomb interaction with its image, may be neglected. Hence for the initial velocities of the particle and its image we may write

$$\boldsymbol{\beta}_{\text{charge}} = \frac{u}{c}\mathbf{n}_z, \qquad \boldsymbol{\beta}_{\text{image}} = -\frac{u}{c}\mathbf{n}_z.$$

At collision both the particle's charge, and that of its image, disappear—in our approximation, instantly. Hence, for the calculation of the emitted radiation we may use the low-frequency approximation described by Eq. (10.76), in which the term marked 'fin' disappears. Since the motion of these two charges is correlated, we have to sum up their electric fields rather than radiated powers—in Eq. (10.76) inside the modulus sign, taking into account the difference of their charge signs:

$$\begin{aligned} I(\omega) &= \frac{1}{4\pi^2 c}\frac{1}{4\pi\varepsilon_0}\left|\sum_j q_j \frac{\mathbf{n}\times(\mathbf{n}\times\boldsymbol{\beta}_j)}{1-\boldsymbol{\beta}_j\cdot\mathbf{n}}\right|^2 \\ &= \frac{1}{4\pi^2 c}\frac{1}{4\pi\varepsilon_0}\left| q_{\text{charge}}\frac{\mathbf{n}\times(\mathbf{n}\times\boldsymbol{\beta}_{\text{charge}})}{1-\boldsymbol{\beta}_{\text{charge}}\cdot\mathbf{n}} + q_{\text{image}}\frac{\mathbf{n}\times(\mathbf{n}\times\boldsymbol{\beta}_{\text{image}})}{1-\boldsymbol{\beta}_{\text{image}}\cdot\mathbf{n}}\right|^2 \\ &= \frac{1}{4\pi^2 c}\frac{q^2\beta^2}{4\pi\varepsilon_0}\left|\frac{\mathbf{n}\times(\mathbf{n}\times\mathbf{n}_z)}{1-\beta\ \mathbf{n}_z\cdot\mathbf{n}} + \frac{\mathbf{n}\times(\mathbf{n}\times\mathbf{n}_z)}{1+\beta\ \mathbf{n}_z\cdot\mathbf{n}}\right|^2 = \frac{1}{4\pi^2 c}\frac{q^2}{4\pi\varepsilon_0}\frac{\beta^2\sin^2\theta}{1-\beta^2\cos^2\theta}. \end{aligned}$$

So, the spectral density of the radiation is frequency-independent, corresponding to a short, delta-function-like pulse radiated at the collision, though in practice its time duration is spread by $\sim 1/\omega_{\text{p}}$, i.e. the frequency spectrum is limited to frequencies below $\sim\omega_{\text{p}}$. Concerning the angular distribution, for non-relativistic particles it follows the usual $\sin^2\theta$ pattern, but in the ultra-relativistic case is squeezed to the usual hollow cone with $\cos^2\theta \to 1$. Note that our result is only valid outside of the perfect conductor (into which the waves do not penetrate), so that the cone corresponds to angles $\theta \approx \pi$.

Since the function $I(\omega)$ does not vanish (and in the non-relativistic case even reaches its maximum) at $\theta = \pi/2$, i.e. in directions along conductor's surface, one may wonder whether this is also an artifact of some approximation. Actually (again, at frequencies $\omega \ll \omega_{\text{p}}$) it is not. Indeed, as we already know, the vector **E** of the electric field of the wave generated at any linear motion of the charge lies within the common plane of the observer and the line of motion—see, e.g. Eq. (10.19). On the other hand, it is perpendicular to the vector **n** of wave propagation—see the figure above. Hence at $\theta = \pi/2$, the wave's electric field is normal to the conductor's surface (while its magnetic field is parallel to the surface). As we know from the discussion in section 7.6, this is fully consistent with the boundary conditions (7.104) on a perfect conductor's surface. Note that this means that the propagation of the electromagnetic pulse from the collision point is accompanied by a ring-shaped pulses of the surface electric charge and vortex-like current, propagating from the collision point on the conductor's surface.

***Problem* 10.14.** A relativistic particle, with the rest mass m and electric charge q, flies ballistically, with velocity u, by an immobile point charge q', with an impact parameter b so large that the deviations of its trajectory from the straight line are negligible. Calculate the total energy loss due to the electromagnetic radiation during the passage. Formulate the conditions of validity of your result.

Solution: The easiest way to solve the problem is to use the formula obtained in the model solution of problem 10.4:

$$\mathscr{P} = \frac{Z_0 q^4}{6\pi}\left(\frac{\gamma}{mc}\right)^2 [(\mathbf{E} + \mathbf{u}\times\mathbf{B})^2 - (\mathbf{u}\cdot\mathbf{E}/c)^2]\,. \qquad (*)$$

In the lab frame, where the charge q' rests, it produces only the Coulomb electrostatic field,

$$\mathbf{E} = \frac{qq'}{4\pi\varepsilon_0 r^2}\frac{\mathbf{r}}{r},$$

directed along the vector $\mathbf{r}$ connecting the charges, and hence is a function of time—see the figure below. According to the conditions of our current problem, the velocity $\mathbf{u}$ of the charge q may be taken for a constant, so that from this sketch,

$$r \equiv |\mathbf{r}| = (b^2 + u^2t^2)^{1/2}, \qquad \cos\alpha = \frac{ut}{r} = \frac{ut}{(b^2 + u^2t^2)^{1/2}}, \qquad \frac{\mathbf{u}\cdot\mathbf{r}}{cr} = \beta\cos\alpha,$$

where t is the time referred to the instant of the closest approach of the particles.

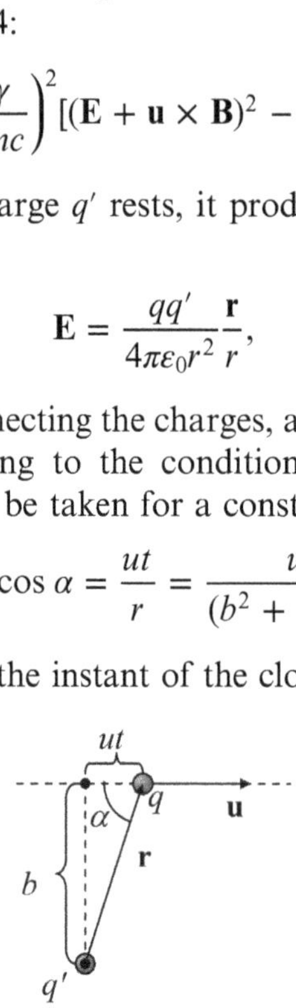

As a result, the expression in square brackets of Eq. (*) is reduced to

$$E^2 - (\mathbf{u}\cdot\mathbf{E}/c)^2 = \left(\frac{qq'}{4\pi\varepsilon_0 r^2}\right)^2 (1 - \beta^2\cos^2\alpha)$$
$$= \left(\frac{qq'}{4\pi\varepsilon_0}\right)^2 \frac{1}{(b^2 + u^2t^2)^2}\left(1 - \beta^2\frac{u^2t^2}{b^2 + u^2t^2}\right).$$

Integrating the power (*) over the whole duration of the particle passage, we obtain the full energy loss:

$$-\Delta\mathscr{E} = \int_{-\infty}^{+\infty} \mathscr{P}dt = \frac{Z_0 q^2}{6\pi}\left(\frac{\gamma}{mc}\right)^2\left(\frac{qq'}{4\pi\varepsilon_0}\right)^2 I,$$

where

$$I \equiv \int_{-\infty}^{+\infty} \frac{1}{(b^2 + u^2t^2)^2}\left(1 - \frac{\beta^2 u^2 t^2}{b^2 + u^2t^2}\right) dt$$

$$\equiv \frac{2}{ub^3}\left[\int_0^{+\infty} \frac{d\xi}{(1 + \xi^2)^2} - \beta^2 \int_0^{+\infty} \frac{\xi^2 d\xi}{(1 + \xi^2)^3}\right], \quad \text{with } \xi \equiv \frac{ut}{b}.$$

The first of these dimensionless integrals is well known[20], and equals $\pi/4$, while the second one may be readily calculated from it and a similar integral with $n = 3$, equal to $3\pi/16$:

$$\int_0^{+\infty} \frac{\xi^2 d\xi}{(1 + \xi^2)^3} = \int_0^{+\infty} \frac{[(1 + \xi^2) - 1]d\xi}{(1 + \xi^2)^3}$$

$$= \int_0^{+\infty} \frac{d\xi}{(1 + \xi^2)^2} - \int_0^{+\infty} \frac{d\xi}{(1 + \xi^2)^3} = \frac{\pi}{4} - \frac{3\pi}{16} = \frac{\pi}{16},$$

so that

$$I = \frac{2}{ub^3}\left(\frac{\pi}{4} - \beta^2 \frac{\pi}{16}\right) \equiv \frac{\pi}{2ub^3}\left(1 - \frac{\beta^2}{4}\right),$$

and, finally, the radiative energy loss is

$$-\Delta\mathscr{E} = \frac{Z_0 q^2}{6\pi}\left(\frac{\gamma}{mc}\right)^2 \left(\frac{qq'}{4\pi\varepsilon_0}\right)^2 \frac{\pi}{2ub^3}\left(1 - \frac{\beta^2}{4}\right) \equiv \frac{1}{12}\frac{Z_0 q^2}{m^2 b^3 c^3}\left(\frac{qq'}{4\pi\varepsilon_0}\right)^2 \frac{1 - \beta^2/4}{\beta\,(1 - \beta^2)}.$$

(Additional task for the reader: compare this result semi-qualitatively with the one that would follow from a combination of Eq. (10.79) with an interpretation of Eq. (10.94), suitable for our current problem.)

This result has been obtained by neglecting the change of particle's trajectory and velocity (and hence its energy) due to its interaction with the Coulomb center. This assumption is valid if the scale of the potential energy of interaction and the calculated energy loss are much smaller than the particle's kinetic energy:

$$-\Delta\mathscr{E},\ \frac{qq'}{4\pi\varepsilon_0 b} \ll mc^2\left[\frac{1}{(1 - \beta^2)^{1/2}} - 1\right].$$

[20] See, e.g. Eq. (A.32c) with $n = 2$.

Appendix A

Selected mathematical formulas

This appendix lists selected mathematical formulas that are used in this lecture course series, but not always remembered by students (and some instructors :-).

A.1 Constants

- Euclidean circle's *length-to-diameter ratio*:

$$\pi = 3.141\ 592\ 653\ldots; \qquad \pi^{1/2} \approx 1.77. \tag{A.1}$$

- *Natural logarithm base*:

$$e \equiv \lim_{n\to\infty}\left(1 + \frac{1}{n}\right)^n = 2.718\ 281\ 828\ldots; \tag{A.2a}$$

from that value, the logarithm base conversion factors are as follows ($\xi > 0$):

$$\frac{\ln\xi}{\log_{10}\xi} = \ln 10 \approx 2.303, \qquad \frac{\log_{10}\xi}{\ln\xi} = \frac{1}{\ln 10} \approx 0.434. \tag{A.2b}$$

- The *Euler* (or 'Euler–Mascheroni') *constant*:

$$\gamma \equiv \lim_{n\to\infty}\left(1 + \frac{1}{2} + \frac{1}{3} + \ldots\frac{1}{n} - \ln n\right) = 0.577\ 156\ 649\ 0\ldots; \quad e^{\gamma} \approx 1.781. \tag{A.3}$$

A.2 Combinatorics, sums, and series

(i) *Combinatorics*

- The number of different *permutations*, i.e. *ordered* sequences of k elements selected from a set of n distinct elements ($n \geqslant k$), is

$$^{n}P_k \equiv n\cdot(n-1)\cdots(n-k+1) = \frac{n!}{(n-k)!}; \tag{A.4a}$$

in particular, the number of different permutations of *all* elements of the set ($n = k$) is

$$^{k}P_k = k \cdot (k-1)\cdots 2 \cdot 1 = k!. \tag{A.4b}$$

- The number of different *combinations*, i.e. *unordered* sequences of k elements from a set of $n \geqslant k$ distinct elements, is equal to the binomial coefficient

$$^{n}C_k \equiv \binom{n}{k} \equiv \frac{^{n}P_k}{^{k}P_k} = \frac{n!}{k!(n-k)!}. \tag{A.5}$$

In an alternative, very popular 'ball/box language', $^{n}C_k$ is the number of different ways to put in a box, in an arbitrary order, k balls selected from n distinct balls.

- A generalization of the binomial coefficient notion is the multinomial coefficient,

$$^{n}C_{k_1,k_2,\ldots k_l} \equiv \frac{n!}{k_1!k_2!\ldots k_l!}, \quad \text{with } n = \sum_{j=1}^{l} k_j, \tag{A.6}$$

which, in the standard mathematical language, is a number of different permutations in a multiset of l distinct element types from an n-element set which contains k_j ($j = 1, 2,\ldots l$) elements of each type. In the 'ball/box language', the coefficient (A.6) is the number of different ways to distribute n distinct balls between l distinct boxes, each time keeping the number (k_j) of balls in the jth box fixed, but ignoring their order inside the box. The binomial coefficient $^{n}C_k$ (A.5) is a particular case of the multinomial coefficient (A.6) for $l = 2$ - counting the explicit box for the first one, and the remaining space for the second box, so that if $k_1 \equiv k$, then $k_2 = n - k$.

- One more important combinatorial quantity is the number $M_n^{(k)}$ of ways to place n *indistinguishable* balls into k distinct boxes. It may be readily calculated from Eq. (A.5) as the number of different ways to select ($k - 1$) partitions between the boxes in an imagined linear row of ($k - 1 + n$) 'objects' (balls in the boxes *and* partitions between them):

$$M_n^{(k)} = {}^{n-1+k}C_{k-1} \equiv \frac{(k-1+n)!}{(k-1)!n!}. \tag{A.7}$$

(ii) *Sums and series*

- *Arithmetic progression*:

$$r + 2r + \cdots + nr \equiv \sum_{k=1}^{n} kr = \frac{n(r+nr)}{2}; \tag{A.8a}$$

in particular, at $r = 1$ it is reduced to the sum of n first natural numbers:

$$1 + 2 + \cdots + n \equiv \sum_{k=1}^{n} k = \frac{n(n+1)}{2}. \tag{A.8b}$$

- Sums of squares and cubes of n first natural numbers:

$$1^2 + 2^2 + \cdots + n^2 \equiv \sum_{k=1}^{n} k^2 = \frac{n(n+1)(2n+1)}{6}; \tag{A.9a}$$

$$1^3 + 2^3 + \cdots + n^3 \equiv \sum_{k=1}^{n} k^3 = \frac{n^2(n+1)^2}{4}. \tag{A.9b}$$

- The *Riemann zeta function*:

$$\zeta(s) \equiv 1 + \frac{1}{2^s} + \frac{1}{3^s} + \cdots \equiv \sum_{k=1}^{\infty} \frac{1}{k^s}; \tag{A.10a}$$

the particular values frequently met in applications are

$$\begin{aligned} &\zeta\left(\frac{3}{2}\right) \approx 2.612, \qquad \zeta(2) = \frac{\pi^2}{6}, \qquad \zeta\left(\frac{5}{2}\right) \approx 1.341, \\ &\zeta(3) \approx 1.202, \qquad \zeta(4) = \frac{\pi^4}{90}, \qquad \zeta(5) \approx 1.037. \end{aligned} \tag{A.10b}$$

- Finite geometric progression (for real $\lambda \neq 1$):

$$1 + \lambda + \lambda^2 + \cdots + \lambda^{n-1} \equiv \sum_{k=0}^{n-1} \lambda^k = \frac{1 - \lambda^n}{1 - \lambda}; \tag{A.11a}$$

in particular, if $\lambda^2 < 1$, the progression has a finite limit at $n \to \infty$ (called the *geometric series*):

$$\lim_{n\to\infty} \sum_{k=0}^{n-1} \lambda^k = \sum_{k=0}^{\infty} \lambda^k = \frac{1}{1 - \lambda}. \tag{A.11b}$$

- *Binomial sum* (the 'binomial theorem'):

$$(1 + a)^n = \sum_{k=0}^{n} {}^nC_k a^k, \tag{A.12}$$

where ${}^nC_k$ are the binomial coefficients defined by Eq. (A.5).

- The *Stirling formula*:

$$\lim_{n\to\infty} \ln(n!) = n(\ln n - 1) + \frac{1}{2}\ln(2\pi n) + \frac{1}{12n} - \frac{1}{360n^3} + \ldots; \qquad \text{(A.13)}$$

for most applications in physics, the first term[1] is sufficient.

- The *Taylor* (or 'Taylor–Maclaurin') *series*: for any infinitely differentiable function $f(\xi)$:

$$\begin{aligned}\lim_{\tilde{\xi}\to 0} f(\xi + \tilde{\xi}) &= f(\xi) + \frac{df}{d\xi}(\xi)\,\tilde{\xi} + \frac{1}{2!}\frac{d^2f}{d\xi^2}(\xi)\,\tilde{\xi}^2 + \cdots \\ &= \sum_{k=0}^{\infty}\frac{1}{k!}\frac{d^kf}{d\xi^k}(\xi)\,\tilde{\xi}^k;\end{aligned} \qquad \text{(A.14}a\text{)}$$

note that for many functions this series converges only within a limited, sometimes small range of deviations $\tilde{\xi}$. For a function of several arguments, $f(\xi_1, \xi_2, \ldots, \xi_N)$, the first terms of the Taylor series are

$$\begin{aligned}\lim_{\tilde{\xi}_k\to 0} f(\xi_1 + \tilde{\xi}_1, \xi_2 + \tilde{\xi}_2, \cdots) = f(\xi_1, \xi_2, \cdots) \\ + \sum_{k=1}^{N}\frac{\partial f}{\partial \xi_k}(\xi_1, \xi_2, \cdots)\,\tilde{\xi}_k \\ + \frac{1}{2!}\sum_{k,k'=1}^{N}\frac{\partial^2 f}{\partial_k\xi\,\partial\xi_{k'}}\tilde{\xi}_k\tilde{\xi}_{k'} + \cdots\end{aligned} \qquad \text{(A.14}b\text{)}$$

- The *Euler–Maclaurin formula*, valid for any infinitely differentiable function $f(\xi)$:

$$\begin{aligned}\sum_{n=0}^{n-1} f(k) = \int_0^n f(\xi)dx - \frac{1}{2}[f(n) - f(0)] + \frac{1}{6}\cdot\frac{1}{2!}\left[\frac{df}{d\xi}(n) - \frac{df}{d\xi}(0)\right] \\ - \frac{1}{30}\cdot\frac{1}{4!}\left[\frac{d^3f}{d\xi^3}(n) - \frac{d^3f}{d\xi^3}(0)\right] \\ + \frac{1}{42}\cdot\frac{1}{6!}\left[\frac{d^5f}{d\xi^5}(n) - \frac{d^5f}{d\xi^5}(0)\right] + \cdots;\end{aligned} \qquad \text{(A.15}a\text{)}$$

the coefficients participating in this formula are the so-called *Bernoulli numbers*[2]:

$$\begin{aligned}&B_1 = \frac{1}{2}, \quad B_2 = \frac{1}{6}, \quad B_3 = 0, \quad B_4 = \frac{1}{30}, \quad B_5 = 0, \\ &B_6 = \frac{1}{42}, \quad B_7 = 0, \quad B_8 = \frac{1}{30}, \cdots\end{aligned} \qquad \text{(A.15}b\text{)}$$

---

[1] Actually, this leading term was derived by A de Moivre in 1733, before J Stirling's work.

[2] Note that definitions of B_k (or rather their signs and indices) vary even in the most popular handbooks.

A.3 Basic trigonometric functions

- Trigonometric functions of the sum and the difference of two arguments[3]:

$$\cos(a \pm b) = \cos a \cos b \mp \sin a \sin b, \qquad (A.16a)$$

$$\sin(a \pm b) = \sin a \cos b \pm \cos a \sin b. \qquad (A.16b)$$

- Sums of two functions of arbitrary arguments:

$$\cos a + \cos b = 2\cos\frac{a+b}{2}\cos\frac{b-a}{2}, \qquad (A.17a)$$

$$\cos a - \cos b = 2\sin\frac{a+b}{2}\sin\frac{b-a}{2}, \qquad (A.17b)$$

$$\sin a \pm \sin b = 2\sin\frac{a \pm b}{2}\cos\frac{\pm b - a}{2}. \qquad (A.17c)$$

- Trigonometric function products:

$$2\cos a\ \cos b = \cos(a+b) + \cos(a-b), \qquad (A.18a)$$

$$2\sin a\ \cos b = \sin(a+b) + \sin(a-b), \qquad (A.18b)$$

$$2\sin a\ \sin b = \cos(a-b) - \cos(a+b); \qquad (A.18c)$$

For the particular case of equal arguments, $b = a$, these three formulas yield the following expressions for the squares of trigonometric functions, and their product:

$$\cos^2 a = \frac{1}{2}(1 + \cos 2a), \qquad \sin a \cos a = \frac{1}{2}\sin 2a,$$
$$\sin^2 a = \frac{1}{2}(1 - \cos 2a). \qquad (A.18d)$$

- Cubes of trigonometric functions:

$$\cos^3 a = \frac{3}{4}\cos a + \frac{1}{4}\cos 3a, \qquad \sin^3 a = \frac{3}{4}\sin a - \frac{1}{4}\sin 3a. \qquad (A.19)$$

[3] I am confident that the reader is quite capable of deriving the relations (A.16) by representing the exponent in the elementary relation $e^{i(a \pm b)} = e^{ia}e^{\pm ib}$ as a sum of its real and imaginary parts, Eqs. (A.18) directly from Eqs. (A.16), and Eqs. (A.17) from Eqs. (A.18) by variable replacement; however, I am still providing these formulas to save his or her time. (Quite a few formulas below are included because of the same reason.)

- Trigonometric functions of a complex argument:

$$\begin{aligned}\sin(a+ib) &= \sin a \cosh b + i \cos a \sinh b,\\ \cos(a+ib) &= \cos a \cosh b - i \sin a \sinh b.\end{aligned} \tag{A.20}$$

- Sums of trigonometric functions of n equidistant arguments:

$$\sum_{k=1}^{n} \begin{Bmatrix} \sin \\ \cos \end{Bmatrix} k\xi = \begin{Bmatrix} \sin \\ \cos \end{Bmatrix} \left(\frac{n+1}{2}\xi\right) \sin\left(\frac{n}{2}\xi\right) \Big/ \sin\left(\frac{\xi}{2}\right). \tag{A.21}$$

A.4 General differentiation

- Full differential of a product of two functions:

$$d(fg) = (df)g + f(dg). \tag{A.22}$$

- Full differential of a function of several independent arguments, $f(\xi_1, \xi_2, \ldots, \xi_n)$:

$$df = \sum_{k=1}^{n} \frac{\partial f}{\partial \xi_k} d\xi_k. \tag{A.23}$$

- Curvature of the Cartesian plot of a function $f(\xi)$:

$$\kappa \equiv \frac{1}{R} = \frac{|\, d^2 f / d\xi^2 \,|}{[1 + (df/d\xi)^2]^{3/2}}. \tag{A.24}$$

A.5 General integration

- Integration *by parts* - immediately follows from Eq. (A.22):

$$\int_{g(A)}^{g(B)} f \; dg = fg \Big|_A^B - \int_{f(A)}^{f(B)} g \; df. \tag{A.25}$$

- Numerical (approximate) integration of 1D functions: the simplest *trapezoidal rule*,

$$\begin{aligned}\int_a^b f(\xi) d\xi &\approx h\left[f\left(a+\frac{h}{2}\right) + f\left(a+\frac{3h}{2}\right) + \cdots + f\left(b-\frac{h}{2}\right)\right]\\ &= h\sum_{n=1}^{N} f\left(a - \frac{h}{2} + nh\right), \qquad h \equiv \frac{b-a}{N}.\end{aligned} \tag{A.26}$$

has relatively low accuracy (error of the order of $(h^3/12)d^2f/d\xi^2$ per step), so that the following *Simpson formula*,

$$\int_a^b f(\xi)d\xi \approx \frac{h}{3}[f(a) + 4f(a+h) + 2f(a+2h) + \cdots + 4f(b-h) + f(b)], \quad h \equiv \frac{b-a}{2N}, \tag{A.27}$$

whose error per step scales as $(h^5/180)d^4f/d\xi^4$, is used much more frequently[4].

A.6 A few 1D integrals[5]

(i) *Indefinite integrals*:

- Integrals with $(1+\xi^2)^{1/2}$:

$$\int (1+\xi^2)^{1/2}d\xi = \frac{\xi}{2}(1+\xi^2)^{1/2} + \frac{1}{2}\ln|\xi + (1+\xi^2)^{1/2}|, \tag{A.28}$$

$$\int \frac{d\xi}{(1+\xi^2)^{1/2}} = \ln|\xi + (1+\xi^2)^{1/2}|, \tag{A.29a}$$

$$\int \frac{d\xi}{(1+\xi^2)^{3/2}} = \frac{\xi}{(1+\xi^2)^{1/2}}. \tag{A.29b}$$

- Miscellaneous indefinite integrals:

$$\int \frac{d\xi}{\xi(\xi^2 + 2a\xi - 1)^{1/2}} = \cos^{-1}\frac{a\xi - 1}{|\xi|(a^2+1)^{1/2}}, \tag{A.30a}$$

$$\int \frac{(\sin\xi - \xi\cos\xi)^2}{\xi^5}d\xi = \frac{2\xi\sin 2\xi + \cos 2\xi - 2\xi^2 - 1}{8\xi^4}, \tag{A.30b}$$

$$\int \frac{d\xi}{a + b\cos\xi} = \frac{2}{(a^2-b^2)^{1/2}}\tan^{-1}\left[\frac{(a-b)}{(a^2-b^2)^{1/2}}\tan\frac{\xi}{2}\right], \quad \text{for } a^2 > b^2. \tag{A.30c}$$

$$\int \frac{d\xi}{1+\xi^2} = \tan^{-1}\xi. \tag{A.30d}$$

[4] Higher-order formulas (e.g. the *Bode rule*), and other guidance including ready-for-use codes for computer calculations may be found, for example, in the popular reference texts by W H Press *et al* [1]. In addition, some advanced codes are used as subroutines in the software packages listed in the same section. In some cases, the Euler–Maclaurin formula (A.15) may also be useful for numerical integration.

[5] A powerful (and free) interactive online tool for working out indefinite 1D integrals is available at http://integrals.wolfram.com/index.jsp.

(ii) *Semi-definite integrals*:

- Integrals with $1/(e^{\xi} \pm 1)$:

$$\int_a^{\infty} \frac{d\xi}{e^{\xi} + 1} = \ln(1 + e^{-a}), \tag{A.31a}$$

$$\int_{a>0}^{\infty} \frac{d\xi}{e^{\xi} - 1} = \ln \frac{1}{1 - e^{-a}}. \tag{A.31b}$$

(iii) *Definite integrals*:

- Integrals with $1/(1 + \xi^2)$:[6]

$$\int_0^{\infty} \frac{d\xi}{1 + \xi^2} = \frac{\pi}{2}, \tag{A.32a}$$

$$\int_0^{\infty} \frac{d\xi}{(1 + \xi^2)^{3/2}} = 1; \tag{A.32b}$$

more generally,

$$\int_0^{\infty} \frac{d\xi}{(1 + \xi^2)^n} = \frac{\pi}{2} \frac{(2n - 3)!!}{(2n - 2)!!} \equiv \frac{\pi}{2} \frac{1 \cdot 3 \cdot 5 \ldots (2n - 3)}{2 \cdot 4 \cdot 6 \ldots (2n - 2)}, \quad \text{for } n = 2,\ 3,\ \ldots \tag{A.32c}$$

- Integrals with $(1 - \xi^{2n})^{1/2}$:

$$\int_0^1 \frac{d\xi}{(1 - \xi^{2n})^{1/2}} = \frac{\pi^{1/2}}{2n} \Gamma\left(\frac{1}{2n}\right) \Big/ \Gamma\left(\frac{n + 1}{2n}\right), \tag{A.33a}$$

$$\int_0^1 (1 - \xi^{2n})^{1/2} d\xi = \frac{\pi^{1/2}}{4n} \Gamma\left(\frac{1}{2n}\right) \Big/ \Gamma\left(\frac{3n + 1}{2n}\right), \tag{A.33b}$$

where $\Gamma(s)$ is the *gamma-function*, which is most often defined (for $\text{Re}\, s > 0$) by the following integral:

$$\int_0^{\infty} \xi^{s-1} e^{-\xi} d\xi = \Gamma(s). \tag{A.34a}$$

[6] Eq. (A.32a) follows immediately from Eq. (A.30d), and Eq. (A.32b) from Eq. (A.29b)—a couple more examples of the (intentional) redundancy in this list.

The key property of this function is the recurrence relation, valid for any $s \neq 0, -1, -2, \ldots$:

$$\Gamma(s+1) = s\Gamma(s). \qquad \text{(A.34}b\text{)}$$

Since, according to Eq. (A.34a), $\Gamma(1) = 1$, Eq. (A.34b) for non-negative integers takes the form

$$\Gamma(n+1) = n!, \qquad \text{for} \quad n = 0, 1, 2, \cdots \qquad \text{(A.34}c\text{)}$$

(where $0! \equiv 1$). Because of this, for integer $s = n + 1 \geqslant 1$, Eq. (A.34a) is reduced to

$$\int_0^\infty \xi^n e^{-\xi} d\xi = n!. \qquad \text{(A.34}d\text{)}$$

Other frequently met values of the gamma-function are those for positive semi-integer arguments:

$$\Gamma\left(\frac{1}{2}\right) = \pi^{1/2}, \qquad \Gamma\left(\frac{3}{2}\right) = \frac{1}{2}\pi^{1/2}, \qquad \Gamma\left(\frac{5}{2}\right) = \frac{1}{2}\cdot\frac{3}{2}\pi^{1/2},$$
$$\Gamma\left(\frac{7}{2}\right) = \frac{1}{2}\cdot\frac{3}{2}\cdot\frac{5}{2}\pi^{1/2}, \qquad \ldots. \qquad \text{(A.34}e\text{)}$$

- Integrals with $1/(e^\xi \pm 1)$:

$$\int_0^\infty \frac{\xi^{s-1}d\xi}{e^\xi + 1} = (1 - 2^{1-s})\,\Gamma(s)\zeta(s), \qquad \text{for } s > 0, \qquad \text{(A.35}a\text{)}$$

$$\int_0^\infty \frac{\xi^{s-1}d\xi}{e^\xi - 1} = \Gamma(s)\zeta(s), \qquad \text{for } s > 1, \qquad \text{(A.35}b\text{)}$$

where $\zeta(s)$ is the Riemann zeta-function—see Eq. (A.10). Particular cases: for $s = 2n$,

$$\int_0^\infty \frac{\xi^{2n-1}d\xi}{e^\xi + 1} = \frac{2^{2n-1} - 1}{2n}\pi^{2n}B_{2n}, \qquad \text{(A.35}c\text{)}$$

$$\int_0^\infty \frac{\xi^{2n-1}d\xi}{e^\xi - 1} = \frac{(2\pi)^{2n}}{4n}B_{2n}. \qquad \text{(A.35}d\text{)}$$

where B_n are the Bernoulli numbers—see Eq. (A.15). For the particular case $s = 1$ (when Eq. (A.35a) yields uncertainty),

$$\int_0^\infty \frac{d\xi}{e^\xi + 1} = \ln 2. \qquad \text{(A.35}e\text{)}$$

- Integrals with $\exp\{-\xi^2\}$:

$$\int_0^\infty \xi^s e^{-\xi^2} d\xi = \frac{1}{2}\Gamma\left(\frac{s+1}{2}\right), \quad \text{for } s > -1; \tag{A.36a}$$

for applications the most important particular values of s are 0 and 2:

$$\int_0^\infty e^{-\xi^2} d\xi = \frac{1}{2}\Gamma\left(\frac{1}{2}\right) = \frac{\pi^{1/2}}{2}, \tag{A.36b}$$

$$\int_0^\infty \xi^2 e^{-\xi^2} d\xi = \frac{1}{2}\Gamma\left(\frac{3}{2}\right) = \frac{\pi^{1/2}}{4}, \tag{A.36c}$$

though we will also run into the cases $s = 4$ and $s = 6$:

$$\begin{aligned} \int_0^\infty \xi^4 e^{-\xi^2} d\xi &= \frac{1}{2}\Gamma\left(\frac{5}{2}\right) = \frac{3\pi^{1/2}}{8}, \\ \int_0^\infty \xi^6 e^{-\xi^2} d\xi &= \frac{1}{2}\Gamma\left(\frac{7}{2}\right) = \frac{15\pi^{1/2}}{16}; \end{aligned} \tag{A.36d}$$

for odd integer values $s = 2n + 1$ (with $n = 0, 1, 2, \ldots$), Eq. (A.36a) takes a simpler form:

$$\int_0^\infty \xi^{2n+1} e^{-\xi^2} d\xi = \frac{1}{2}\Gamma(n+1) = \frac{n!}{2}. \tag{A.36e}$$

- Integrals with cosine and sine functions:

$$\int_0^\infty \cos(\xi^2)\, d\xi = \int_0^\infty \sin(\xi^2)\, d\xi = \left(\frac{\pi}{8}\right)^{1/2}. \tag{A.37}$$

$$\int_0^\infty \frac{\cos\xi}{a^2 + \xi^2} d\xi = \frac{\pi}{2a} e^{-a}. \tag{A.38}$$

$$\int_0^\infty \left(\frac{\sin\xi}{\xi}\right)^2 d\xi = \frac{\pi}{2}. \tag{A.39}$$

- Integrals with logarithms:

$$\int_0^1 \ln \frac{a + (1-\xi^2)^{1/2}}{a - (1-\xi^2)^{1/2}} d\xi = \pi[a - (a^2 - 1)^{1/2}], \quad \text{for } a \geqslant 1. \tag{A.40}$$

$$\int_0^1 \ln \frac{1 + (1-\xi)^{1/2}}{\xi^{1/2}} d\xi = 1. \tag{A.41}$$

- Integral representations of the Bessel functions of integer order:

$$J_n(\alpha) = \frac{1}{2\pi}\int_{-\pi}^{+\pi} e^{i(\alpha \sin \xi - n\xi)} d\xi, \quad \text{so that } e^{i\alpha \sin \xi} = \sum_{k=-\infty}^{\infty} J_k(\alpha) e^{ik\xi}; \qquad \text{(A.42}a\text{)}$$

$$I_n(\alpha) = \frac{1}{\pi}\int_0^{\pi} e^{\alpha \cos \xi} \cos n\xi \; d\xi. \qquad \text{(A.42}b\text{)}$$

A.7 3D vector products

(i) *Definitions*:

- *Scalar* ('dot-') *product*:

$$\mathbf{a} \cdot \mathbf{b} = \sum_{j=1}^{3} a_j b_j, \qquad \text{(A.43)}$$

where a_j and b_j are vector components in any orthogonal coordinate system. In particular, the vector squared (the same as its norm squared):

$$a^2 \equiv \mathbf{a} \cdot \mathbf{a} = \sum_{j=1}^{3} a_j^2 \equiv \| \mathbf{a} \|^2. \qquad \text{(A.44)}$$

- *Vector* ('cross-') *product*:

$$\mathbf{a} \times \mathbf{b} \equiv \mathbf{n}_1(a_2 b_3 - a_3 b_2) + \mathbf{n}_2(a_3 b_1 - a_1 b_3) + \mathbf{n}_3(a_1 b_2 - a_2 b_1) = \begin{vmatrix} \mathbf{n}_1 & \mathbf{n}_2 & \mathbf{n}_3 \\ a_1 & a_2 & a_3 \\ b_1 & b_2 & b_3 \end{vmatrix}, \qquad \text{(A.45)}$$

where $\{\mathbf{n}_j\}$ is the set of mutually perpendicular unit vectors[7] along the corresponding coordinate system axes[8]. In particular, Eq. (A.45) yields

$$\mathbf{a} \times \mathbf{a} = 0. \qquad \text{(A.46)}$$

(ii) *Corollaries* (readily verified by Cartesian components):

- Double vector product (the so-called *bac minus cab* rule):

$$\mathbf{a} \times (\mathbf{b} \times \mathbf{c}) = \mathbf{b}(\mathbf{a} \cdot \mathbf{c}) - \mathbf{c}(\mathbf{a} \cdot \mathbf{b}). \qquad \text{(A.47)}$$

[7] Other popular notations for this vector set are $\{\mathbf{e}_j\}$ and $\{\hat{\mathbf{r}}_j\}$.

[8] It is easy to use Eq. (A.45) to check that the direction of the product vector corresponds to the well-known 'right-hand rule' and to the even more convenient *corkscrew rule*: if we rotate corkscrew's handle from the first operand toward the second one, its axis moves in the direction of the product.

- Mixed scalar–vector product (the *operand rotation rule*):

$$\mathbf{a} \cdot (\mathbf{b} \times \mathbf{c}) = \mathbf{b} \cdot (\mathbf{c} \times \mathbf{a}) = \mathbf{c} \cdot (\mathbf{a} \times \mathbf{b}). \qquad \text{(A.48)}$$

- Scalar product of vector products:

$$(\mathbf{a} \times \mathbf{b}) \cdot (\mathbf{c} \times \mathbf{d}) = (\mathbf{a} \cdot \mathbf{c})(\mathbf{b} \cdot \mathbf{d}) - (\mathbf{a} \cdot \mathbf{d})(\mathbf{b} \cdot \mathbf{c}); \qquad \text{(A.49}a\text{)}$$

in the particular case of two similar operands (say, $\mathbf{a} = \mathbf{c}$ and $\mathbf{b} = \mathbf{d}$), the last formula is reduced to

$$(\mathbf{a} \times \mathbf{b})^2 = (ab)^2 - (\mathbf{a} \cdot \mathbf{b})^2. \qquad \text{(A.49}b\text{)}$$

A.8 Differentiation in 3D Cartesian coordinates

- Definition of the *del* (or 'nabla') vector-operator ∇:[9]

$$\nabla \equiv \sum_{j=1}^{3} \mathbf{n}_j \frac{\partial}{\partial r_j}, \qquad \text{(A.50)}$$

where r_j is a set of linear and orthogonal (*Cartesian*) coordinates along directions $\mathbf{n}_j$. In accordance with this definition, the operator ∇ acting on a *scalar* function of coordinates, $f(\mathbf{r})$,[10] gives its gradient, i.e. a new *vector*:

$$\nabla f \equiv \sum_{j=1}^{3} \mathbf{n}_j \frac{\partial f}{\partial r_j} \equiv \mathbf{grad}\, f. \qquad \text{(A.51)}$$

- The *scalar product* of del by a *vector* function of coordinates (a *vector field*),

$$\mathbf{f}(\mathbf{r}) \equiv \sum_{j=1}^{3} \mathbf{n}_j f_j(\mathbf{r}), \qquad \text{(A.52)}$$

compiled formally following Eq. (A.43), is a *scalar* function—the *divergence* of the initial function:

$$\nabla \cdot \mathbf{f} \equiv \sum_{j=1}^{3} \frac{\partial f_j}{\partial r_j} \equiv \mathbf{div}\, \mathbf{f}, \qquad \text{(A.53)}$$

[9] One can run into the following notation: $\nabla \equiv \partial/\partial\mathbf{r}$, which is convenient is some cases, but may be misleading in quite a few others, so it will be not used in these notes.

[10] In this, and four next sections, all scalar and vector functions are assumed to be differentiable.

while the *vector product* of ∇ and $\mathbf{f}$, formed in a formal accordance with Eq. (A.45), is a new vector - the *curl* (in European tradition, called rotor and denoted **rot**) of $\mathbf{f}$:

$$\nabla \times \mathbf{f} \equiv \begin{vmatrix} \mathbf{n}_1 & \mathbf{n}_2 & \mathbf{n}_3 \\ \dfrac{\partial}{\partial r_1} & \dfrac{\partial}{\partial r_2} & \dfrac{\partial}{\partial r_3} \\ f_1 & f_2 & f_3 \end{vmatrix} = \mathbf{n}_1\left(\frac{\partial f_3}{\partial r_2} - \frac{\partial f_2}{\partial r_3}\right) + \mathbf{n}_2\left(\frac{\partial f_1}{\partial r_3} - \frac{\partial f_3}{\partial r_1}\right) + \mathbf{n}_3\left(\frac{\partial f_2}{\partial r_1} - \frac{\partial f_1}{\partial r_2}\right) \equiv \mathbf{curl}\,\mathbf{f}. \qquad \text{(A.54)}$$

- One more frequently met 'product' is $(\mathbf{f}\cdot\nabla)\mathbf{g}$, where $\mathbf{f}$ and $\mathbf{g}$ are two arbitrary vector functions of $\mathbf{r}$. This product should be also understood in the sense implied by Eq. (A.43), i.e. as a vector whose jth Cartesian component is

$$[(\mathbf{f}\cdot\nabla)\,\mathbf{g}]_j = \sum_{j'=1}^{3} f_{j'} \frac{\partial g_j}{\partial r_{j'}}. \qquad \text{(A.55)}$$

A.9 The Laplace operator $\nabla^2 \equiv \nabla \cdot \nabla$

- Expression in Cartesian coordinates—in the formal accordance with Eq. (A.44):

$$\nabla^2 = \sum_{j=1}^{3} \frac{\partial^2}{\partial r_j^2}. \qquad \text{(A.56)}$$

- According to its definition, the Laplace operator acting on a *scalar* function of coordinates gives a new scalar function:

$$\nabla^2 f \equiv \nabla \cdot (\nabla f) = \mathbf{div}(\mathbf{grad}\, f) = \sum_{j=1}^{3} \frac{\partial^2 f}{\partial r_j^2}. \qquad \text{(A.57)}$$

- On the other hand, acting on a *vector* function (A.52), the operator ∇^2 returns another *vector*:

$$\nabla^2 \mathbf{f} = \sum_{j=1}^{3} \mathbf{n}_j \nabla^2 f_j. \qquad \text{(A.58)}$$

Note that Eqs. (A.56)–(A.58) are only valid in Cartesian (i.e. orthogonal and linear) coordinates, but generally not in other (even orthogonal) coordinates—see, e.g. Eqs. (A.61), (A.64), (A.67), and (A.70) below.

A.10 Operators ∇ and ∇^2 in the most important systems of orthogonal coordinates[11]

(i) *Cylindrical*[12] *coordinates* $\{\rho, \varphi, z\}$ (see figure below) may be defined by their relations with the Cartesian coordinates:

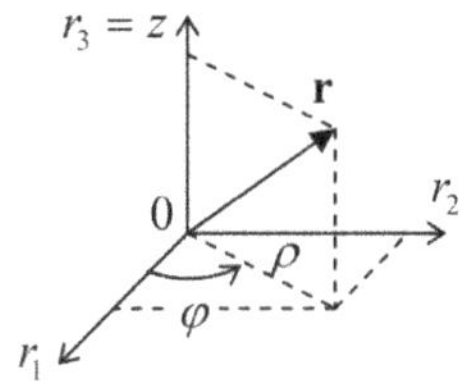

$$r_1 = \rho\cos\varphi,\quad r_2 = \rho\sin\varphi,\quad r_3 = z. \tag{A.59}$$

- Gradient of a scalar function:

$$\nabla f = \mathbf{n}_\rho\frac{\partial f}{\partial\rho} + \mathbf{n}_\varphi\frac{1}{\rho}\frac{\partial f}{\partial\varphi} + \mathbf{n}_z\frac{\partial f}{\partial z}. \tag{A.60}$$

- The Laplace operator of a scalar function:

$$\nabla^2 f = \frac{1}{\rho}\frac{\partial}{\partial\rho}\left(\rho\frac{\partial f}{\partial\rho}\right) + \frac{1}{\rho^2}\frac{\partial^2 f}{\partial\varphi^2} + \frac{\partial^2 f}{\partial z^2}, \tag{A.61}$$

- Divergence of a vector function of coordinates ($\mathbf{f} = \mathbf{n}_\rho f_\rho + \mathbf{n}_\varphi f_\varphi + \mathbf{n}_z f_z$):

$$\nabla\cdot\mathbf{f} = \frac{1}{\rho}\frac{\partial(\rho f_\rho)}{\partial\rho} + \frac{1}{\rho}\frac{\partial f_\varphi}{\partial\varphi} + \frac{\partial f_z}{\partial z}. \tag{A.62}$$

- Curl of a vector function:

$$\nabla\times\mathbf{f} = \mathbf{n}_\rho\left(\frac{1}{\rho}\frac{\partial f_z}{\partial\varphi} - \frac{\partial f_\varphi}{\partial z}\right) + \mathbf{n}_\varphi\left(\frac{\partial f_\rho}{\partial z} - \frac{\partial f_z}{\partial\rho}\right) + \mathbf{n}_z\frac{1}{\rho}\left(\frac{\partial(\rho f_\varphi)}{\partial\rho} - \frac{\partial f_\rho}{\partial\varphi}\right). \tag{A.63}$$

- The Laplace operator of a vector function:

$$\nabla^2\mathbf{f} = \mathbf{n}_\rho\left(\nabla^2 f_\rho - \frac{1}{\rho^2}f_\rho - \frac{2}{\rho^2}\frac{\partial f_\varphi}{\partial\varphi}\right) + \mathbf{n}_\varphi\left(\nabla^2 f_\varphi - \frac{1}{\rho^2}f_\varphi + \frac{2}{\rho^2}\frac{\partial f_\rho}{\partial\varphi}\right) + \mathbf{n}_z\,\nabla^2 f_z. \tag{A.64}$$

---

[11] Some other orthogonal curvilinear coordinate systems are discussed in *Part EM* section 2.3.

[12] In the 2D geometry with fixed coordinate z, these coordinates are called *polar*.

(ii) *Spherical coordinates* $\{r, \theta, \varphi\}$ (see figure below) may be defined as:

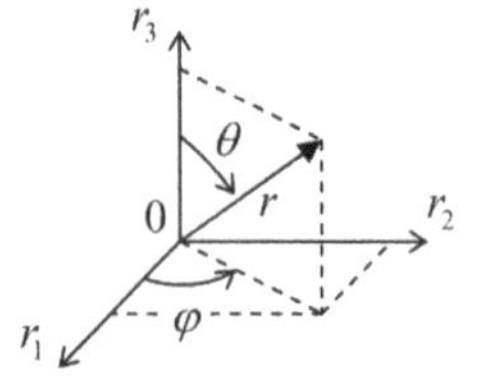

$$r_1 = r\sin\theta\cos\varphi,\quad r_2 = r\sin\theta\sin\varphi,\quad r_3 = r\cos\theta. \tag{A.65}$$

- Gradient of a scalar function:

$$\nabla f = \mathbf{n}_r\frac{\partial f}{\partial r} + \mathbf{n}_\theta\frac{1}{r}\frac{\partial f}{\partial\theta} + \mathbf{n}_\varphi\frac{1}{r\sin\theta}\frac{\partial f}{\partial\varphi}. \tag{A.66}$$

- The Laplace operator of a scalar function:

$$\nabla^2 f = \frac{1}{r^2}\frac{\partial}{\partial r}\left(r^2\frac{\partial f}{\partial r}\right) + \frac{1}{r^2\sin\theta}\frac{\partial}{\partial\theta}\left(\sin\theta\frac{\partial f}{\partial\theta}\right) + \frac{1}{(r\sin\theta)^2}\frac{\partial^2 f}{\partial\varphi^2}. \tag{A.67}$$

- Divergence of a vector function $\mathbf{f} = \mathbf{n}_r f_r + \mathbf{n}_\theta f_\theta + \mathbf{n}_\varphi f_\varphi$:

$$\nabla\cdot\mathbf{f} = \frac{1}{r^2}\frac{\partial(r^2 f_r)}{\partial r} + \frac{1}{r\sin\theta}\frac{\partial(f_\theta\sin\theta)}{\partial\theta} + \frac{1}{r\sin\theta}\frac{\partial f_\varphi}{\partial\varphi}. \tag{A.68}$$

- Curl of a similar vector function:

$$\nabla\times\mathbf{f} = \mathbf{n}_r\frac{1}{r\sin\theta}\left(\frac{\partial(f_\varphi\sin\theta)}{\partial\theta} - \frac{\partial f_\theta}{\partial\varphi}\right) + \mathbf{n}_\theta\frac{1}{r}\left(\frac{1}{\sin\theta}\frac{\partial f_r}{\partial\varphi} - \frac{\partial(rf_\varphi)}{\partial r}\right) + \mathbf{n}_\varphi\frac{1}{r}\left(\frac{\partial(rf_\theta)}{\partial r} - \frac{\partial f_r}{\partial\theta}\right). \tag{A.69}$$

- The Laplace operator of a vector function:

$$\begin{aligned}\nabla^2\mathbf{f} = {}&\mathbf{n}_r\left(\nabla^2 f_r - \frac{2}{r^2}f_r - \frac{2}{r^2\sin\theta}\frac{\partial}{\partial\theta}(f_\theta\sin\theta) - \frac{2}{r^2\sin\theta}\frac{\partial f_\varphi}{\partial\varphi}\right)\\ &+ \mathbf{n}_\theta\left(\nabla^2 f_\theta - \frac{1}{r^2\sin^2\theta}f_\theta + \frac{2}{r^2}\frac{\partial f_r}{\partial\theta} - \frac{2\cos\theta}{r^2\sin^2\theta}\frac{\partial f_\varphi}{\partial\varphi}\right)\\ &+ \mathbf{n}_\varphi\left(\nabla^2 f_\varphi - \frac{1}{r^2\sin^2\theta}f_\varphi + \frac{2}{r^2\sin\theta}\frac{\partial f_r}{\partial\varphi} + \frac{2\cos\theta}{r^2\sin^2\theta}\frac{\partial f_\theta}{\partial\varphi}\right).\end{aligned} \tag{A.70}$$

A.11 Products involving ∇

(i) *Useful zeros:*

- For any scalar function $f(\mathbf{r})$,

$$\nabla \times (\nabla f) \equiv \mathbf{curl}(\mathbf{grad}\, f) = 0. \tag{A.71}$$

- For any vector function $\mathbf{f}(\mathbf{r})$,

$$\nabla \cdot (\nabla \times \mathbf{f}) \equiv \mathbf{div}(\mathbf{curl}\, f) = 0. \tag{A.72}$$

(ii) The *Laplace operator* expressed via the curl of a curl:

$$\nabla^2\mathbf{f} = \nabla(\nabla \cdot \mathbf{f}) - \nabla \times (\nabla \times \mathbf{f}); \tag{A.73}$$

(iii) Spatial differentiation of a product of a *scalar* function by a *vector* function:

- The scalar 3D generalization of Eq. (A.22) is

$$\nabla \cdot (f\, \mathbf{g}) = (\nabla f) \cdot \mathbf{g} + f(\nabla \cdot \mathbf{g}). \tag{A.74a}$$

- Its vector generalization is similar:

$$\nabla \times (f\, \mathbf{g}) = (\nabla f) \times \mathbf{g} + f(\nabla \times \mathbf{g}). \tag{A.74b}$$

(iv) Spatial differentiation of products of *two vector* functions:

$$\nabla \times (\mathbf{f} \times \mathbf{g}) = \mathbf{f}(\nabla \cdot \mathbf{g}) - (\mathbf{f} \cdot \nabla)\mathbf{g} - (\nabla \cdot \mathbf{f})\mathbf{g} + (\mathbf{g} \cdot \nabla)\, \mathbf{f}, \tag{A.75}$$

$$\nabla(\mathbf{f} \cdot \mathbf{g}) = (\mathbf{f} \cdot \nabla)\mathbf{g} + (\mathbf{g} \cdot \nabla)\, \mathbf{f} + \mathbf{f} \times (\nabla \times \mathbf{g}) + \mathbf{g} \times (\nabla \times \mathbf{f}), \tag{A.76}$$

$$\nabla \cdot (\mathbf{f} \times \mathbf{g}) = \mathbf{g} \cdot (\nabla \times \mathbf{f}) - \mathbf{f} \cdot (\nabla \times \mathbf{g}). \tag{A.77}$$

A.12 Integro-differential relations

(i) For an *arbitrary surface* S limited by closed contour C:

- The *Stokes theorem*, valid for any differentiable vector field $\mathbf{f}(\mathbf{r})$:

$$\int_S (\nabla \times \mathbf{f}) \cdot d^2\mathbf{r} \equiv \int_S (\nabla \times \mathbf{f})_n d^2r = \oint_C \mathbf{f} \cdot d\mathbf{r} \equiv \oint_C f_\tau dr, \tag{A.78}$$

where $d^2\mathbf{r} \equiv \mathbf{n}d^2r$ is the elementary area vector (normal to the surface), and $d\mathbf{r}$ is the elementary contour length vector (tangential to the contour line).

(ii) For an *arbitrary volume* V limited by closed surface S:

- *Divergence* (or 'Gauss') *theorem*, valid for any differentiable vector field $\mathbf{f}(\mathbf{r})$:

$$\int_V (\nabla \cdot \mathbf{f})\, d^3r = \oint_S \mathbf{f} \cdot d^2\mathbf{r} \equiv \oint_S f_n d^2r. \qquad \text{(A.79)}$$

- *Green's theorem*, valid for two differentiable scalar functions $f(\mathbf{r})$ and $g(\mathbf{r})$:

$$\int_V (f\, \nabla^2 g - g\nabla^2 f)\, d^3r = \oint_S (f\, \nabla g - g\nabla f)_n d^2r. \qquad \text{(A.80)}$$

- An identity valid for any two scalar functions f and g, and a vector field $\mathbf{j}$ with $\nabla\cdot\mathbf{j} = 0$ (all differentiable):

$$\int_V [f(\mathbf{j} \cdot \nabla g) + g(\mathbf{j} \cdot \nabla f)]\; d^3r = \oint_S fgj_n d^2r. \qquad \text{(A.81)}$$

A.13 The Kronecker delta and Levi-Civita permutation symbols

- The *Kronecker delta symbol* (defined for integer indices):

$$\delta_{jj'} \equiv \begin{cases} 1, \text{ if } j' = j, \\ 0, \text{ otherwise.} \end{cases} \qquad \text{(A.82)}$$

- The *Levi-Civita permutation symbol* (most frequently used for 3 integer indices, each taking one of values 1, 2, or 3):

$$\varepsilon_{jj'j''} \equiv \begin{cases} +1, & \text{if the indices follow in the 'correct' ('even')} \\ & \text{order: } 1 \to 2 \to 3 \to 1 \to 2 \ldots, \\ -1, & \text{if the indices follow in the 'incorrect' ('odd')} \\ & \text{order: } 1 \to 3 \to 2 \to 1 \to 3 \ldots, \\ 0, & \text{if any two indices coincide.} \end{cases} \qquad \text{(A.83)}$$

- Relation between the Levi-Civita and the Kronecker delta products:

$$\varepsilon_{jj'j''}\varepsilon_{kk'k''} = \sum_{l,l',l''=1}^{3} \begin{vmatrix} \delta_{jl} & \delta_{jl'} & \delta_{jl''} \\ \delta_{j'l} & \delta_{j'l'} & \delta_{j'l''} \\ \delta_{j''l} & \delta_{j''l'} & \delta_{j''l''} \end{vmatrix}; \qquad \text{(A.84}a\text{)}$$

summation of this relation, written for 3 different values of $j = k$, over these values yields the so-called *contracted epsilon identity*:

$$\sum_{j=1}^{3} \varepsilon_{jj'j''}\varepsilon_{jk'k''} = \delta_{j'k'}\delta_{j''k''} - \delta_{j'k''}\delta_{j''k'}. \qquad \text{(A.84}b\text{)}$$

A.14 Dirac's delta-function, sign function, and theta-function

- Definition of 1D *delta-function* (for real $a < b$):

$$\int_a^b f(\xi)\delta(\xi)d\xi = \begin{cases} f(0), & \text{if } a < 0 < b, \\ 0, & \text{otherwise,} \end{cases} \tag{A.85}$$

where $f(\xi)$ is any function continuous near $\xi = 0$. In particular (if $f(\xi) = 1$ near $\xi = 0$), the definition yields

$$\int_a^b \delta(\xi)d\xi = \begin{cases} 1, & \text{if } a < 0 < b, \\ 0, & \text{otherwise.} \end{cases} \tag{A.86}$$

- Relation to the *theta-function* $\theta(\xi)$ and *sign function* $\text{sgn}(\xi)$

$$\delta(\xi) = \frac{d}{d\xi}\theta(\zeta) = \frac{1}{2}\frac{d}{d\xi}\text{sgn}(\xi), \tag{A.87a}$$

where

$$\theta(\xi) \equiv \frac{\text{sgn}(\xi) + 1}{2} = \begin{cases} 0, & \text{if } \xi < 0, \\ 1, & \text{if } \xi > 1, \end{cases}$$
$$\text{sgn}(\xi) \equiv \frac{\xi}{|\xi|} = \begin{cases} -1, & \text{if } \xi < 0, \\ +1, & \text{if } \xi > 1. \end{cases} \tag{A.87b}$$

- An important integral[13]:

$$\int_{-\infty}^{+\infty} e^{is\xi}ds = 2\pi\delta(\xi). \tag{A.88}$$

- 3D generalization of the delta-function of the radius-vector (the 2D generalization is similar):

$$\int_V f(\mathbf{r})\delta(\mathbf{r})d^3r = \begin{cases} f(0), \text{ if } 0 \in V, \\ 0, \quad \text{otherwise;} \end{cases} \tag{A.89}$$

it may be represented as a product of 1D delta-functions of Cartesian coordinates:

$$\delta(\mathbf{r}) = \delta(r_1)\delta(r_2)\delta(r_3). \tag{A.90}$$

---

[13] The coefficient in this equation may be readily recalled by considering its left-hand part as the Fourier-integral presentation of function $f(s) \equiv 1$, and applying Eq. (A.85) to the reciprocal Fourier transform

$$f(s) \equiv 1 = \frac{1}{2\pi}\int_{-\infty}^{+\infty} e^{-is\xi}[2\pi\delta(\xi)]d\xi.$$

A.15 The Cauchy theorem and integral

Let a complex function $f(z)$ be analytic within a part of the complex plane z, that is limited by a closed contour C and includes point z'. Then

$$\oint_C f(z)dz = 0, \tag{A.91}$$

$$\oint_C f(z)\frac{dz}{z - z'} = 2\pi i f(z'). \tag{A.92}$$

The first of these relations is usually called the *Cauchy integral theorem* (or the 'Cauchy–Goursat theorem'), and the second one—the *Cauchy integral* (or the 'Cauchy integral formula').

A.16 Literature

(i) Properties of some *special functions* are briefly discussed at the relevant points of the lecture notes; in the alphabetical order:

- Airy functions: *Part QM* section 2.4;
- Bessel functions: *Part EM* section 2.7;
- Fresnel integrals: *Part EM* section 8.6;
- Hermite polynomials: *Part QM* section 2.9;
- Laguerre polynomials (both simple and associated): *Part QM* section 3.7;
- Legendre polynomials, associated Legendre functions: *Part EM* section 2.8, and *Part QM* section 3.6;
- Spherical harmonics: *Part QM* section 3.6;
- Spherical Bessel functions: *Part QM* sections 3.6 and 3.8.

(ii) For *more formulas*, and their discussions, I can recommend the following handbooks (in the alphabetical order of the authors)[14]:

- *Handbook of Mathematical Formulas* [2][15];
- *Tables of Integrals, Series, and Products* [3];
- *Mathematical Handbook for Scientists and Engineers* [4];
- *Integrals and Series* volumes 1 and 2 [5].

A popular textbook *Mathematical Methods for Physicists* [6] may be also used as a formula manual. Many formulas are also available from the symbolic calculation parts of commercially available software packages listed in section (iv) below.

(iii) Probably the most popular collection of *numerical calculation codes* are the twin manuals by W Press *et al* [1]:

- *Numerical Recipes in Fortran* 77;

[14] On a personal note, perhaps 90% of all formula needs throughout my research career were satisfied by a tiny, wonderfully compiled old book: H. Dwight, *Tables of Integrals and Other Mathematical Data*, 4th ed., Macmillan, 1961, whose copies, rather amazingly, are still available on the Web.

[15] An updated version of this collection is now available online at http://dlmf.nist.gov/.

- *Numerical Recipes* [in C++ - KKL].

My lecture notes include very brief introductions into numerical methods of differential equation solution:

- ordinary differential equations: *Part CM* section 3.9, and
- partial differential equations: *Part CM* section 8.5, and *Part EM* section 2.8, which include references to literature for further reading.

(iv) The following are the most popular *software packages* for numerical and symbolic calculations, all with plotting capabilities (in the alphabetical order):

- *Maple* (http://www.maplesoft.com/);
- *MathCAD* (http://www.ptc.com/products/mathcad/);
- *Mathematica* (http://www.wolfram.com/products/mathematica/index.html);
- *MATLAB* (http://www.mathworks.com/products/matlab/).

References

[1] Press W *et al* 1992 *Numerical Recipes in Fortran 77* 2nd edn (Cambridge: Cambridge University Press); Press W *et al* 2007 *Numerical Recipes* 3rd edn (Cambridge: Cambridge University Press)

[2] Abramowitz M and Stegun I (eds) 1965 *Handbook of Mathematical Formulas* (New York: Dover)

[3] Gradshteyn I and Ryzhik I 1980 *Tables of Integrals, Series, and Products* 5th edn (New York: Academic)

[4] Korn G and Korn T 2000 *Mathematical Handbook for Scientists and Engineers* 2nd edn (New York: Academic)

[5] Prudnikov A *et al* 1986 *Integrals and Series* vol 1 (Boca Raton, FL: CRC Press); Prudnikov A *et al* 1986 *Integrals and Series* vol 2 (Boca Raton, FL: CRC Press)

[6] Arfken G *et al* 2012 *Mathematical Methods for Physicists* 7th edn (New York: Academic)

Appendix B

Selected physical constants[1]

Table B.1.

Symbol	Quantity	SI value and unit	Gaussian value and unit	Relative rms uncertainty
c	speed of light in free space	$2.99\ 792\ 458 \times 10^{8}$ m s^{-1}	$2.99\ 792\ 458 \times 10^{10}$ cm s^{-1}	0 (defined value)
G	gravitation constant	6.6741×10^{-11} m^3 kg^{-1} s^{-2}	6.6741×10^{-8} cm ^{3} g^{-1} s^{-2}	$\sim 5 \times 10^{-5}$
$\hbar$	Planck constant	$1.05\ 457\ 180 \times 10^{-34}$ J s	$1.05\ 457\ 180 \times 10^{-27}$ erg s	$\sim 1 \times 10^{-8}$
e	elementary electric charge	$1.6\ 021\ 762 \times 10^{-19}$ C	$4.803\ 203 \times 10^{-10}$ statcoulomb	$\sim 6 \times 10^{-9}$
m_e	electron's rest mass	$0.91\ 093\ 835 \times 10^{-30}$ kg	$0.91\ 093\ 835 \times 10^{-27}$ G	$\sim 1 \times 10^{-8}$
m_p	proton's rest mass	$1.67\ 262\ 190 \times 10^{-27}$ kg	$1.67\ 262\ 190 \times 10^{-24}$ G	$\sim 1 \times 10^{-8}$
μ_0	magnetic constant	$4\pi \times 10^{-7}$ N A^{-2}	–	0 (defined value)
ε_0	electric constant	$8.854\ 187\ 817 \times 10^{-12}$ F m^{-1}	–	0 (defined value)
k_B	Boltzmann constant	$1.380\ 648 \times 10^{-23}$ J K^{-1}	$1.3\ 806\ 509 \times 10^{-16}$ erg K^{-1}	$\sim 2 \times 10^{-6}$

Comments:

1. The fixed value of c was defined by an international convention in 1983, in order to extend the official definition of a second (as 'the duration of 9 192 631 770 periods of the radiation corresponding to the transition

[1] The listed numerical values of the constants are from the most recent (2014) International CODATA recommendation (see, e.g. http://physics.nist.gov/cuu/Constants/index.html), besides a newer result for k_B—see [1].

between the two hyperfine levels of the ground state of the cesium-133 atom') to that of a meter. The values are back-compatible with the legacy definitions of the meter (initially, as 1/40 000 000th of the Earth's meridian length) and the second (for a long time, as $1/(24 \times 60 \times 60) = 1/86\,400$th of the Earth's rotation period), within the experimental errors of those measures.

2. ε_0 and μ_0 are not really the fundamental constants; in the SI system of units one of them (say, μ_0) is selected arbitrarily[2], while the other one is defined via the relation $\varepsilon_0\mu_0 = 1/c^2$.
3. The Boltzmann constant k_B is also not quite fundamental, because its only role is to comply with the independent definition of the kelvin (K), as the temperature unit in which the triple point of water is exactly 273.16 K. If temperature is expressed in energy units k_BT (as is done, for example, in *Part SM* of this series), this constant disappears altogether.
4. The dimensionless *fine structure* ('Sommerfeld's) *constant* α is numerically the same in any system of units:

$$\alpha \equiv \begin{Bmatrix} e^2/4\pi\varepsilon_0\hbar c & \text{in SI units} \\ e^2/\hbar c & \text{in Gaussian units} \end{Bmatrix} \approx 7.297\,352\,566 \times 10^{-3}$$
$$\approx \frac{1}{137.035\,999\,14},$$

and is known with a much smaller rms uncertainty (currently, $\sim 3 \times 10^{-10}$) than those of the component constants.

References

[1] Gaiser C *et al* 2017 *Metrologia* **54** 280
[2] Newell D 2014 *Phys. Today* **67** 35–41

[2] Note that the selected value of μ_0 may be changed (a bit) in a few years—see, e.g. [2].

Bibliography

This section presents a partial list of the textbooks and monographs used in the work on the EAP series[1,2].

Part CM: Classical Mechanics

Fetter A L and Walecka J D 2003 *Theoretical Mechanics of Particles and Continua* (New York: Dover)
Goldstein H, Poole C and Safko J 2002 *Classical Mechanics* 3rd edn (Reading, MA: Addison Wesley)
Granger R A 1995 *Fluid Mechanics* (New York: Dover)
José J V and Saletan E J 1998 *Classical Dynamics* (Cambridge: Cambridge University Press)
Landau L D and Lifshitz E M 1976 *Mechanics* 3rd edn (Oxford: Butterworth-Heinemann)
Landau L D and Lifshitz E M 1986 *Theory of Elasticity* (Oxford: Butterworth-Heinemann)
Landau L D and Lifshitz E M 1987 *Fluid Mechanics* 2nd edn (Oxford: Butterworth-Heinemann)
Schuster H G 1995 *Deterministic Chaos* 3rd edn (New York: Wiley)
Sommerfeld A 1964 *Mechanics* (New York: Academic)
Sommerfeld A 1964 *Mechanics of Deformable Bodies* (New York: Academic)

Part EM: Classical Electrodynamics

Batygin V V and Toptygin I N 1978 *Problems in Electrodynamics* 2nd edn (New York: Academic)
Griffiths D J 2007 *Introduction to Electrodynamics* 3rd edn (Englewood Cliffs, NJ: Prentice-Hall)
Jackson J D 1999 *Classical Electrodynamics* 3rd edn (New York: Wiley)
Landau L D and Lifshitz E M 1984 *Electrodynamics of Continuous Media* 2nd edn (Auckland: Reed)
Landau L D and Lifshitz E M 1975 *The Classical Theory of Fields* 4th edn (Oxford: Pergamon)
Panofsky W K H and Phillips M 1990 *Classical Electricity and Magnetism* 2nd edn (New York: Dover)

---

[1] The list does not include the sources (mostly, recent original publications) cited in the ends of the chapters, and the mathematics textbooks and handbooks listed in appendix A, section A.16.

[2] Recently some high-quality teaching materials on advanced physics have become available online, including: R Fitzpatrick's text on classical electromagnetism (farside.ph.utexas.edu/teaching/jk1/Electromagnetism.pdf); B Simons' 'lecture shrunks' on advanced quantum mechanics (www.tcm.phy.cam.ac.uk/~bds10/aqp.html); and D Tong's lecture notes on several advanced topics (www.damtp.cam.ac.uk/user/tong/teaching.html).

doi:10.1088/978-0-7503-1407-7ch13

Stratton J A 2007 *Electromagnetic Theory* (New York: Wiley)
Tamm I E 1979 *Fundamentals of the Theory of Electricity* (Paris: Mir)
Zangwill A 2013 *Modern Electrodynamics* (Cambridge: Cambridge University Press)

Part QM: Quantum Mechanics

Abers E S 2004 *Quantum Mechanics* (London: Pearson)
Auletta G, Fortunato M and Parisi G 2009 *Quantum Mechanics* (Cambridge: Cambridge University Press)
Capri A Z 2002 *Nonrelativistic Quantum Mechanics* 3rd edn (Singapore: World Scientific)
Cohen-Tannoudji C, Diu B and Laloë F 2005 *Quantum Mechanics* (New York: Wiley)
Constantinescu F, Magyari E and Spiers J A 1971 *Problems in Quantum Mechanics* (Amsterdam: Elsevier)
Galitski V *et al* 2013 *Exploring Quantum Mechanics* (Oxford: Oxford University Press)
Gottfried K and Yan T-M 2004 *Quantum Mechanics: Fundamentals* 2nd edn (Berlin: Springer)
Griffith D 2005 *Quantum Mechanics* 2nd edn (Englewood Cliffs, NJ: Prentice Hall)
Landau L D and Lifshitz E M 1977 *Quantum Mechanics (Nonrelativistic Theory)* 3rd edn (Oxford: Pergamon)
Messiah A 1999 *Quantum Mechanics* (New York: Dover)
Merzbacher E 1998 *Quantum Mechanics* 3rd edn (New York: Wiley)
Miller D A B 2008 *Quantum Mechanics for Scientists and Engineers* (Cambridge: Cambridge University Press)
Sakurai J J 1994 *Modern Quantum Mechanics* (Reading, MA: Addison-Wesley)
Schiff L I 1968 *Quantum Mechanics* 3rd edn (New York: McGraw-Hill)
Shankar R 1980 *Principles of Quantum Mechanics* 2nd edn (Berlin: Springer)
Schwabl F 2002 *Quantum Mechanics* 3rd edn (Berlin: Springer)

Part SM: Statistical Mechanics

Feynman R P 1998 *Statistical Mechanics* 2nd edn (Boulder, CO: Westview)
Huang K 1987 *Statistical Mechanics* 2nd edn (New York: Wiley)
Kubo R 1965 *Statistical Mechanics* (Amsterdam: Elsevier)
Landau L D and Lifshitz E M 1980 *Statistical Physics, Part 1* 3rd edn (Oxford: Pergamon)
Lifshitz E M and Pitaevskii L P 1981 *Physical Kinetics* (Oxford: Pergamon)
Pathria R K and Beale P D 2011 *Statistical Mechanics* 3rd edn (Amsterdam: Elsevier)
Pierce J R 1980 *An Introduction to Information Theory* 2nd edn (New York: Dover)
Plishke M and Bergersen B 2006 *Equilibrium Statistical Physics* 3rd edn (Singapore: World Scientific)
Schwabl F 2000 *Statistical Mechanics* (Berlin: Springer)
Yeomans J M 1992 *Statistical Mechanics of Phase Transitions* (Oxford: Oxford University Press)

Multidisciplinary/specialty

Ashcroft W N and Mermin N D 1976 *Solid State Physics* (Philadelphia, PA: Saunders)
Blum K 1981 *Density Matrix and Applications* (New York: Plenum)
Breuer H-P and Petruccione E 2002 *The Theory of Open Quantum Systems* (Oxford: Oxford University Press)
Cahn S B and Nadgorny B E 1994 *A Guide to Physics Problems, Part 1* (New York: Plenum)

Cahn S B, Mahan G D and Nadgorny B E 1997 *A Guide to Physics Problems, Part 2* (New York: Plenum)

Cronin J A, Greenberg D F and Telegdi V L 1967 *University of Chicago Graduate Problems in Physics* (Reading, MA: Addison Wesley)

Hook J R and Hall H E 1991 *Solid State Physics* 2nd edn (New York: Wiley)

Joos G 1986 *Theoretical Physics* (New York: Dover)

Kompaneyets A S 2012 *Theoretical Physics* 2nd edn (New York: Dover)

Lax M 1968 *Fluctuations and Coherent Phenomena* (London: Gordon and Breach)

Lifshitz E M and Pitaevskii L P 1980 *Statistical Physics, Part 2* (Oxford: Pergamon)

Newbury N *et al* 1991 *Princeton Problems in Physics with Solutions* (Princeton, NJ: Princeton University Press)

Pauling L 1988 *General Chemistry* 3rd edn (New York: Dover)

Tinkham M 1996 *Introduction to Superconductivity* 2nd edn (New York: McGraw-Hill)

Walecka J D 2008 *Introduction to Modern Physics* (Singapore: World Scientific)

Ziman J M 1979 *Principles of the Theory of Solids* 2nd edn (Cambridge: Cambridge University Press)

www.ingramcontent.com/pod-product-compliance
Ingram Content Group UK Ltd.
Pitfield, Milton Keynes, MK11 3LW, UK
UKHW051043120926
8556IPUK00031B/871

9 780750 314084